Theo. J. Kneip

TECHNIQUES AND INSTRUMENTATION IN ANALYTICAL CHEMISTRY — VOLUME 5

ATOMIC ABSORPTION SPECTROMETRY

TECHNIQUES AND INSTRUMENTATION IN ANALYTICAL CHEMISTRY

Volume 1　Evaluation and Optimization of Laboratory Methods and Analytical Procedures. A Survey of Statistical and Mathematical Techniques
　　　　　　by D.L. Massart, A. Dijkstra and L. Kaufman
Volume 2　Handbook of Laboratory Distillation
　　　　　　by E. Krell
Volume 3　Pyrolysis Mass Spectrometry of Recent and Fossil Biomaterials Compendium and Atlas
　　　　　　by H.L.C. Meuzelaar, J. Haverkamp and F.D. Hileman
Volume 4　Evaluation of Analytical Methods in Biological Systems
　　　　　　Part A. Analysis of Biogenic Amines
　　　　　　edited by G.B. Baker and R.T. Coutts
Volume 5　Atomic Absorption Spectrometry
　　　　　　edited by J.E. Cantle

TECHNIQUES AND INSTRUMENTATION IN ANALYTICAL CHEMISTRY — VOLUME 5

ATOMIC ABSORPTION SPECTROMETRY

Edited by

John Edward Cantle

VG Isotopes Ltd., Winsford, Cheshire, U.K.

ELSEVIER SCIENTIFIC PUBLISHING COMPANY
Amsterdam — Oxford — New York   1982

ELSEVIER SCIENTIFIC PUBLISHING COMPANY
Molenwerf 1
P.O. Box 211, 1000 AE Amsterdam, The Netherlands

Distributors for the United States and Canada:

ELSEVIER SCIENCE PUBLISHING COMPANY INC.
52, Vanderbilt Avenue
New York, N.Y. 10017

Library of Congress Cataloging in Publication Data
Main entry under title:

Atomic absorption spectrometry.

(Techniques and instrumentation in analytical chemistry ; v. 5)
Includes bibliographies and index.
1. Atomic absorption spectrometry. I. Cantle, John Edward, 1949-   . II. Series.
QD96.A8A86   1982   543'.0858   82-13747
ISBN 0-444-42015-0

ISBN 0-444-42015-0 (Vol. 5)
ISBN 0-444-41744-3 (Series)

© Elsevier Scientific Publishing Company, 1982
All rights reserved. No part of this publication may be reproduced, stored in a retrieval system or transmitted in any form or by any means, electronic, mechanical, photocopying, recording or otherwise without the prior written permission of the publisher, Elsevier Scientific Publishing Company, P.O. Box 330, 1000 AH Amsterdam, The Netherlands.

Printed in The Netherlands

PREFACE

Experience gained during seven years with a leading instrumentation supplier proved to me that manufacturers' 'cook-books' were too restrictive — limited to their producers hardware — and that the majority of other published analytical methodologies were either not specific enough or by their disparate availability required the routine analyst to do a lot of library work.

This multi-author text provides the reader with a methods compendium that should find a home next to every atomic absorption spectrometer.

My own contributions are based on and developed from the training seminars in AAS that I was personally responsible for and derived so much pleasure from whilst with Instrumentation Laboratory.

I believe these teaching programmes were well received and raised the standard of instrument use in a wide area of application. I gratefully acknowledge my employer at that time and their permission for me to produce this text. In addition I acknowledge with thanks the section on the 'Use of Perchloric Acid' which was prepared by Ron Rooney of Rooney Laboratories and which was received with such enthusiasm when presented at an AAS meeting many years ago.

This book is fundamentally and deliberately a methods manual and it is for this reason that modern electronic wizardry is treated briefly. I do not however underestimate the contribution that microprocessor technology will make to routine analysis; versatility, convenience and ease of use are but three characteristics that have already been demonstrated. For a similar reason the theoretical aspects are presented in sufficient detail for a basic understanding of the principles involved, rather than a rigorous treatment.

I believe that the aims and intentions I had in preparing this book have been realised. I hope you agree!

April 1982
J. E. CANTLE
Winsford, Chesire, Gt. Britain

CONTENTS

Preface vii

Chapter 1. Basic principles, by J. W. Robinson 1
I. Historical. 1
II. The ability of free atoms to absorb 1
    A. Absorption by free atoms 2
        1. Absorption laws, 4-2. Oscillator strength, 6-
    B. Atomic absorption line widths 7
    C. Hollow cathodes 8
III. Quantitative measurements 9
IV. Background correction 10
    A. Hydrogen (deuterium) background corrector 11
    B. Double wavelength background corrector 11
    C. Zeeman background corrector. 12
V. Available lines 12
References 14

Chapter 2. Instrumental requirements and optimisation, by J. E. Cantle . . 15
I. Introduction 15
II. The components of the spectrometer 16
    A. The atomiser 16
        1. Atomisation in flames, 16-2. Microsampling cup apparatus, 21-3. Atomisation without flames — electrothermal atomisation, 22-4. Vapour generation techniques, 24-
    B. Light sources 25
        1. Operating conditions, 28-2. Multi-element lamps, 28-3. Electrodeless discharge lamps, 29-
    C. Monochromator and optics 30
        1. The optical layout, 31-2. Dual channel optics, 33-
    D. Electronics and readout 34
        1. Photomultiplier, 34-2. Source modulation, 34-3. The readout, 35-4. The use of a microprocessor, 35-

Chapter 3. Practical techniques, by J. E. Cantle 37
I. Introduction 37
II. Sample requirements and general preparation techniques 37
    A. Aqueous solutions 38
    B. Organic liquids 39
    C. Inorganic solids 39
    D. Organic solids 39
    E. Gases 39

III.	The use of perchloric acid for sample digestion.	40
	A. Precautions in use	40
	B. Wet oxidation procedures	42
	1. Wet oxidation of dried herbage, blood, meat, custard powder or other cellulose/protein low fat materials, 43-	
IV	Calibration	44
	A. Beer's Law.	44
	B. Reasons for non-linearity of calibration graphs.	45
	C. Terminology	47
	1. Sensitivity, 47-2. Detection limit, 47-3. Precision, 49-	
	D. Working with calibration curves	50
	E. De-optimisation.	50
V.	Methods of concentration.	51
	A. Concentration by evaporation.	51
	B. Solvent extraction of trace metals.	51
	C. Ion exchange.	52
VI.	Interferences in atomic absorption analysis.	52
	A. Spectral interference.	53
	B. Emission interference	53
	C. Chemical interference	53
	D. Matrix interference	54
	E. Non-specific scatter interference.	54
	F. Ionisation interferences.	55
VII.	Electrothermal atomisation, ETA	55
	A. Selection and optimisation of furnace operating conditions	57
	B. Calibration	58
	C. Sample handling and preparation	60
	1. Avoidance of contamination, 60-2. Essential precautions, 60-3. Desirable precautions, 61-4. Use of micropipettes, 62-5. Injection of organic solvents, 63-6. Sample preparation, 64-7. Chemical separations, 65-	
VIII.	Atomic emission spectrometry	66

Applications of atomic absorption spectrometry

Chapter 4a. Water and effluents, by B. J. Farey and L. A. Nelson.		67
I.	Introduction	67
	A. Characteristics of water and effluents	67
	B. Speciation of metals in water	67
	C. Public health aspects.	68
	D. Atomic absorption spectrometry as applied to the analysis of waters and effluents	68
II.	Sampling and storage.	69
	A. Sample collection	69
	B. Filtration.	69
	C. Sample preservation	70
III.	Analysis.	70

	A. Sample pretreatment	71
	1. Nitric acid method, 71-2. Sulphuric acid—nitric acid—perchloric acid method, 71-	
	B. Preconcentration	72
	1. Evaporation, 72-2. Chelation—solvent extraction, 72-3. Ion exchange 75-4. Co-precipitation, 76-	
	C. Flame techniques	77
	1. Preparation of standards, 77-2. Direct aspiration of sample, 78-3. Use of nitrous oxide—acetylene flame, 78-4. Interferences, 78-	
	D. Specialised techniques	81
	1. Mercury by cold-vapour atomic absorption, 81-2. Arsenic and selenium by hydride generation, 84-	
IV.	Flameless atomic absorption	86
	A. Application to waters and effluents	86
	B. Methodology	87
	C. Interferences	88
References		93

Chapter 4b. Marine analysis by atomic absorption spectrometry, by H. Haraguchi and K. Fuwa 95

I.	Introduction	95
II.	Seawater	96
	A. Storage of seawater samples	101
	B. Preconcentration	103
	1. Solvent extraction, 103-2. Coprecipitation, 109-3. Ion exchange, 110-4. Other preconcentration techniques, 112-	
III.	Marine organisms	114
IV.	Sediment	118
V.	Conclusion	119
References		119

Chapter 4c. Analysis of airborne particles in the workplace and ambient atmospheres, by T. J. Kneip and M. T. Kleinman 123

I.	Introduction	123
	A. Comparison of multielement methods	124
II.	Applications	124
	A. Flame AAS	125
	B. Electrothermal AAS	125
III.	General methods	126
	A. Sampling	127
	B. Sample preparation	129
	1. Dissolution method, 129-2. Atomization conditions, 131-	
IV.	Ambient air	134
V.	Workplace atmospheres	134
VI.	Conclusions	137
References		137

Chapter 4d. Application of atomic absorption spectrometry to the analysis of foodstuffs, by M. Ihnat 139
I. Introduction 139
   A. Literature on applications of atomic absorption spectrometry to food analysis................................... 139
   B. Scope of chapter..................................... 141
   C. Elements of interest 143
II. Sampling and preparation of analytical samples 143
   A. Sampling... 143
   B. Preparation of analytical samples 146
      1. Recommended procedures for preparation of analytical samples, 148-
III. Sample treatment 152
   A. Recommended procedures of sample treatment 153
      1. Digestion with nitric and perchloric acids, 153-2. Digestion with nitric, perchloric and sulfuric acids, 153-3. Dry ashing, 154-
IV. Recommended analytical procedures 155
   A. General analytical protocol.......................... 155
      1. Apparatus and reagents, 155-2. Standard solutions, 159-3. Reference materials, 160-4. Spectrometer operating parameters, 161-5. Atomic absorption spectrometry, 167-6. Reagent blanks, 168-7. Interferences, 169-8. Calibration and calculations, 169-
   B. Determination of specific elements................... 170
      1. Sodium and potassium, 170-2. Magnesium and calcium, 172-3. Boron, 174-4. Aluminium, 175-5. Vanadium, 176-6. Chromium, 177-7. Manganese, 178-8. Iron, 179-9. Cobalt, 182-10. Nickel, 184-11. Copper, 185-12. Zinc, 187-13. Arsenic and selenium, 189-14. Molybdenum, 191-15. Cadmium, 192-16. Tin, 195-17. Mercury, 196-18. Lead, 199-
References... 201

Chapter 4e. Applications of atomic absorption spectrometry in ferrous metallurgy, by K. Ohls and D. Sommer.................. 211
I. Introduction 211
II. Analysis of iron, steel and alloys....................... 214
   A. General solution methods 215
   B. Trace-element analysis............................... 217
      1. Solid sample technique, 219-2. Preconcentration, 222-
   C. Interference of sample matrix 224
   D. Precision and accuracy 227
III. Analysis of ores, slags and other oxides................ 229
   A. General solution methods 230
      1. Direct dissolution, 230-2. Universal fusion technique, 230-
   B. Trace-element analysis............................... 232
      1. Graphite tube methods (low salt content), 233-2. Injection technique (high salt content), 234-
   C. Interference of sample matrix 235
   D. Precision and accuracy 236

IV.	Special AAS methods	236
	A. Development of methods for gases	236
	1. Determination of mercury in iron ores, dust and slurry, 237-	
	B. Metal determination in organic solvents	239
V.	Calibration techniques	242
	A. Calibration using standard solutions	242
	B. Recalibration procedures	243
	C. Background correction	244
	D. Calculations and data handling	245
References		246

Chapter 4f. The analysis of non-ferrous metals by atomic absorption spectrometry, by F. J. Bano ... 251

I.	Introduction	251
II.	Determination of impurities in alloys based on various non-ferrous metals	251
	A. Aluminium	251
	B. Cobalt	252
	C. Copper	253
	D. Lead	254
	E. Nickel	256
	F. Zinc	257
III.	Electrothermal techniques	257
References		259

Chapter 4g. Atomic absorption methods in applied geochemistry, by M. Thompson and S. J. Wood ... 261

I.	Introduction	261
	A. Analytical requirements in applied geochemistry	261
	B. Atomic absorption spectrometry in applied geochemistry	261
	C. Instrumental requirements	263
	D. Interference effects	264
	E. Sample attack methods	265
II.	General aspects of sample preparation methods	265
	A. Digestion vessels	265
	B. Heating equipment	266
	C. Dispensers and diluters	268
	D. Reagents and calibrators	269
	E. Safety	269
	F. Mechanical sample preparation	269
	G. Analytical quality control	270
III.	Sample attack methods	270
	A. Nitric acid attack for soil or stream sediment	270

      B. Nitric acid—perchloric acid attack for rock, soil or stream sediment 271
      C. Hydrofluoric, nitric and perchloric acid attack for rock, soil or sediment 272
      D. Hydrofluoric acid—boric acid attack for silicon determination in rock, soil, or sediment 273
      E. Lithium metaborate fusion attack for silicon determination in rock, soil or sediment 275
      F. Chelation—solvent extraction method for determining trace metals in water samples 276
      G. Nitric acid—perchloric acid attack for herbage samples 278
      H. Mercury in rocks, soils and sediments 279
      I. Tin in rocks, soils and sediments 281
References 284

Chapter 4h. Applications of atomic absorption spectrometry in the petroleum industry, by W. C. Campbell 285
I. Introduction 285
    A. Flame atomisation 285
    B. Electrothermal atomisation 285
II. Sampling 286
    A. Sample contamination 286
    B. Reagent impurities 287
    C. Sample storage 287
    D. Sample preparation 287
III. Standards for petroleum analysis 288
IV. Applications 290
    A. Crude and residual fuel oils 290
      1. The determination of metals in crude and residual fuel oil by dilution and flame analysis, 290-2. The determination of metals in crude and residual fuel oils by ashing and flame analysis, 292-
    B. Fuel and gas oils 294
      1. The determination of metals in fuel and gas oils by flame analysis, 294-2. The determination of low levels of nickel and vanadium in fuel oils using electrothermal atomisation, 295-
    C. Unused lubricating oils 297
      1. The determination of calcium, barium, magnesium and zinc in unused lubricating oil by dilution and flame analysis, 297-
    D. Used lubricating oils 299
      1. The determination of wear metals in used lubricating oils, 300-
    E. Gasoline 301
      1. The determination of lead in leaded gasoline, 302-2. The determination of lead in lead free gasoline using electrothermal atomisation, 303-3. The determination of manganese in gasoline, 305-
References 306

Chapter 4i. Methods for the analysis of glasses and ceramics by atomic spectrometry, by W. M. Wise, J. P. Williams and R. A. Burdo 307
I. Introduction 307
II. Apparatus 307
   A. Flame emission and absorption 307
   B. Plasma emission 308
III. Chemicals 308
IV. Reagent solutions 308
V. Standards 309
VI. Sample preparation 312
VII. Determination of Li, Na, K, Mg, Ca, Sr, Ba, Al, Fe, Zn, Cd and Pb by AES or AAS 313
   A. Procedure for acidic decomposition 313
   B. Procedure for fusion 314
   C. Measurement of analytes 315
   D. Discussion 315
      1. Alkalies, 315-2. Alkaline earths, 316-3. Aluminium, 317-4. Iron, 317-5. Zinc and cadmium, 317-6. Lead, 317-
VIII. Determination of silicon by AAS 317
   A. Procedure for sample decomposition and AAS measurement .. 317
   B. Discussion 318
IX. Determination of boron by AES 318
   A. Procedure for sample decomposition and AES measurement .. 318
   B. Discussion 319
References 319

Chapter 4j. Clinical applications of flame techniques, by B. E. Walker . 321
I. Introduction 321
II. Sample collection 322
   A. Contamination 322
   B. Effect of occlusion 323
   C. Diurnal variation 324
III. Analysis 324
   A. Effect of protein content 324
   B. Interference 326
IV. Individual elements 327
   A. Calcium 327
      1. The determination of calcium in plasma and urine, 328-
   B. Copper 328
      1. The determination of copper in plasma and urine, 330-
   C. Iron 330
      1. The determination of plasma iron and iron binding capacity, 331-
   D. Lithium 332
      1. The determination of lithium in plasma and whole-blood, 332-

E. Magnesium 333
    1. The determination of magnesium in plasma and urine, 333-
F. Potassium 334
    1. The determination of potassium in plasma, 334-
G. Sodium 335
    1. The determination of sodium in plasma, 335-
H. Zinc... 336
    1. The determination of zinc in plasma, whole-blood and urine, 337-
References.. 338

Chapter 4k. Elemental analysis of body fluids and tissues by electrothermal atomisation and atomic absorption spectrometry by, H. T. Delves... 341
I. Introduction 341
II. Electrothermal atomisation for atomic absorption 341
    A. Physical interferences 342
    B. Chemical interferences 343
    C. Spectral interferences 343
III. Sample preparation for electrothermal atomisation—AAS........ 346
    A. Diluent solutions for matrix modification................ 347
    B. Acid extraction of metals from proteins 347
    C. Dissolution of tissues................................... 348
    D. Complete oxidation of tissues and fluids................. 349
        1. Wet and dry ashing, 349-2. Vapour phase and pressure digestion, 350-
        3. Low-temperature ashing, 350-
IV. Essential trace elements..................................... 351
    A. Zinc... 352
    B. Copper 353
    C. Iron... 355
    D. Manganese.. 356
    E. Selenium 357
    F. Chromium... 358
    G. Molybdenum.. 359
    H. Cobalt... 360
    I. Vanadium 361
V. Trace metals used therapeutically 362
    A. Aluminium 362
    B. Gold 364
    C. Bismuth.. 365
    D. Gallium 366
    E. Platinum 366
VI. Non-essential, toxic trace elements 367
    A. Lead 367
    B. Cadmium.. 372

	C. Nickel	373
	D. Beryllium	375
	E. Arsenic	375
	F. Mercury, antimony, tellurium	376
VII.	Conclusion	376
References.		377

Chapter 4l. Forensic science, by I. M. Dale 381
I. Introduction 381
II. Biological material and poisoning 382
    A. Sampling 382
    B. Pretreatment 382
    C. Freeze drying 383
    D. Low-temperature ashing 383
    E. Sample digestion 384
III. Toxic elements 384
    A. Arsenic 384
        1. Introduction, 384-2. Symptoms, 385-3. Fatal dose, 385-4. Sampling, 386-5. Analytical methods, 386-6. Arsine-entrained air flame, 386-7. Arsine-heated quartz cell, 386-8. Arsenic-heated graphite atomiser, 387- 9. Arsenic in human tissue — normal levels, 388-10. Arsenic in human tissue — exposed subjects, 388-
IV. Gunshot residue analysis 390
    A. Introduction 390
    B. Cautionary note 392
    C. Sampling 392
    D. Analysis 394
References. 394

Chapter 4m. Fine, industrial and other chemicals, by L. Ebdon 395
I. Introduction 395
II. Chemicals 395
    A. Inorganic 395
        1. Fine chemicals and analytical reagents, 395-2. Industrial chemicals, 406-
    B. Organic 407
        1. Organometallic, 407-2. Impurities in organic compounds, 409-
    C. Phosphors 410
III. Photographic film and chemicals 411
IV. Catalysts 412
V. Semi-conductors 413
VI. Electroplating solutions 414
VII. Cosmetics, detergents and household products 416
VIII. Pharmaceutical products and drugs 417
    A. Determination of metallic concentrations of significance 418
    B. Determination of impurities 421
    C. Indirect methods 423

IX.	Fungicides	424
X.	Paints and pigments	424
XI.	Paper and pulp	427
XII.	Textiles, fibres and leather	428
XIII.	Polymers and rubber	430
XIV.	Nuclear industry	432
XV.	Waste materials	433
XVI.	Archaeology	434
XVII.	Gases	437
XVIII.	Isotopes	438
References		440
Index		447

Chapter 1

Basic principles

J. W. ROBINSON
Department of Chemistry, Louisiana State University, Baton Rouge, Louisiana 70803 (U.S.A.)

I. HISTORICAL

Historically, absorption by atoms has been documented for a long period of time. For example, Fraunhofer [1] observed dark lines in the radiation from the sun. These were later identified as the absorption lines of metal atoms and found to coincide with the emission lines of those same atoms [2]. For many years the observation was of little more than academic interest and that interest resided with the atomic physicist. It was not until 1955 that Alan Walsh, working in Australia, published the first paper demonstrating the use of atomic absorption spectrometry (AAS) as an analytical tool [3]. He foresaw the widespread analytical utility of the phenomenon and in his early work made major contributions which solved problems which could have prevented its use in quantitative work [4—6].

As the name implies, AAS is the measurement of absorption of radiation by free atoms. The total amount of absorption depends on the number of free atoms present and the degree to which the free atoms absorb the radiation. The key to understanding the application of atomic asborption spectrometry to analytical chemistry lies in understanding the factors which affect (1) the ability of the atoms to absorb and (2) the factors which affect the generation and loss of free atoms from a particular atom population. These two factors will be considered separately.

II. THE ABILITY OF FREE ATOMS TO ABSORB

A given population of free atoms exists at various electronic energy levels. The distribution of atoms in the energy levels is given by the Boltzmann distribution equation, i.e.

$$\frac{N_2 E}{\tau} = \frac{N_1 E g_1}{\tau g_2} e^{-E/kT} \tag{1}$$

where N_1 = number of atoms in the ground state; N_2 = number of atoms in the excited state; E = energy difference between the ground state and the excited state; τ = lifetime in the excited state; g_1 and g_2 = statistical weights of atoms in the ground state and the excited state; k = Boltzmann distribution constant; and T = temperature of the system.

References p. 14

Further,

$$E = h\nu \tag{2}$$

where h = Planck's constant; ν = the frequency of the absorbed radiation
and

$$\nu = c/\lambda \tag{3}$$

where c = the speed of light; and λ = the wavelength of the absorbed radiation.

The equilibrium between atoms in the high energy state and atoms in the low energy state depends on the energy required to excite the atoms and the temperature of the system. If the energy is great, the number of excited atoms is small and vice versa. If the temperature is high, the number of excited atoms is increased.

At room temperature and at the temperatures obtained by most flame atomisers the number of atoms in the excited state is a very small fraction of those in the ground state and for absorption purposes can be ignored. We will discuss this relationship later.

A. Absorption by free atoms

If radiation from a continuous light source, such as a hydrogen lamp, is passed through a population of free atoms and then through a monochromator to a detection system, the absorption spectrum of the atoms in the light path could then be measured. For all elements the spectrum consists of a few very narrow absorption lines. The great majority of light passing through the atom population is unabsorbed and passes through to the monochromator. The energy of this unabsorbed light is at the wrong wavelength (see eqn. 2) for absorption and thus does not register in the absorption measurement.

In order to measure absorption by free atoms it is first necessary to choose the correct wavelength at which this absorption can be measured. That wavelength is a critical property of the atoms and is given by the equation

$$E = E_1 - E_2 = h\nu = hc/\lambda \tag{4}$$

where E_1 = the energy of the higher energy state, or the energy after absorption; E_2 = the energy of the lower energy state, or the energy before absorption.

This relationship follows directly from quantum theory. It can be stated simply: the energy of the radiation which the atom can absorb must equal the energy difference between the energy state the atom exists in before absorption and the energy state the atom exists in after absorption.

The electronic states of atoms are well defined by atomic theory. As we proceed up the periodic table by increasing atomic numbers, electrons progressively fill the inner and outer shells. For the element sodium, two inner shells are completely filled and there is one electron in the outer third shell. This electron is said to be in an *s* orbital. However, the remaining orbitals of the third shell and all the orbitals of the fourth, fifth and sixth shells, etc., are empty. When the outer electron of sodium is in the *s* orbital the atom is said to be in the "ground state" or the "unexcited state". If the atom absorbs radiation the electron undergoes a transition to one of the empty orbitals at the higher energy levels. After many years of study the energies of the atomic orbitals are now well known.

Since the energies of the ground state orbital and any one of the upper orbitals are well defined then the energy difference can be calculated. This energy difference is given by $E_1 - E_2$ and from eqn. (4) the wavelength of light absorbed by the atom when it undergoes a transition from E_2 (ground state) to E_1 (excited state) can be calculated.

These energy levels can be illustrated by Grotrian diagrams [7]. These diagrams may be quite complicated and for simplicity a partial Grotrian diagram is illustrated in Figure 1. From this diagram it can be seen that the *s* electron can undergo transition to various *p* orbitals. These in turn exhibit fine structure as a result of the electron in a *p* orbital spinning in either of two possible directions within the orbital. There is a slight difference in the energy of such an electron depending on its direction of spin, i.e. the spin quantum number.

In the case of sodium atoms two slightly different energy levels exist in the excited state and two slightly different absorption wavelengths are observed, at 589 nm and 589.5 nm. These wavelengths coincide exactly with the yellow sodium doublet emission lines observed when sodium atoms are heated.

From the Grotrian diagram of an element, the absorption wavelengths of the atoms of that element can be deduced. Theoretically, any transitions between two states permitted by quantum theory can be used in AAS. In practice, however, it is found that this is not the case because the number of atoms in the upper excited state is extremely small and the degree of absorption by these few excited atoms is scarcely measurable. Consequently, absorption lines relating to transitions between an excited state and a higher excited state are insensitive and not analytically useful.

The population of the upper states can be readily calculated from eqn. (1), the Boltzmann distribution. Some calculated data for different temperatures are shown in Table 1. It can be seen that the number of atoms in the excited state is vanishingly small. If absorption by these few atoms was measured, many orders of magnitude in sensitivity would be lost. Thus, AAS invariably utilizes transitions between the ground state (where the vast majority of atoms reside) and an upper excited state. For the highest sensitivity the transition between ground state and the first excited state is used and if lower sensitivity is required, e.g. to analyze samples of increased concentration, then a transition between the ground state and an upper excited state

References p. 14

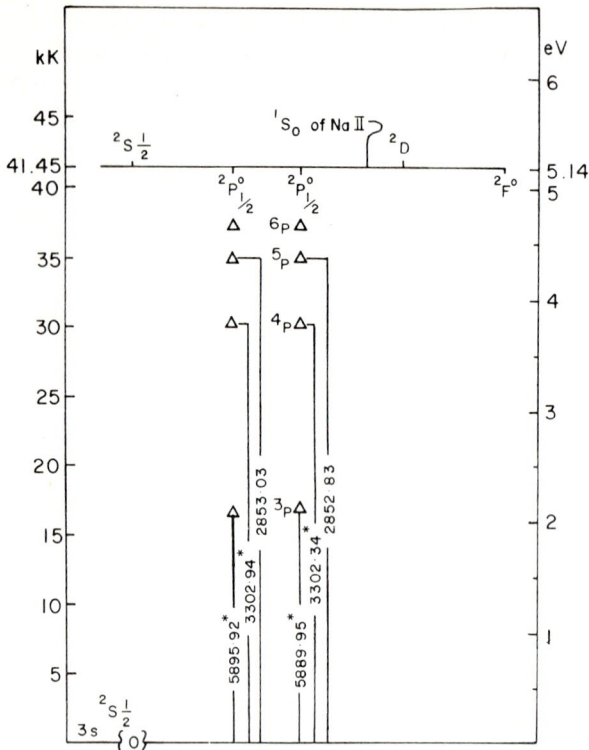

Fig. 1. Partial Grotrian diagram for sodium.

TABLE 1

THE RELATIONSHIP BETWEEN TEMPERATURE, EXCITATION WAVELENGTH, AND THE NUMBER OF EXCITED ATOMS

Excitation wavelength (Å)	Number of excited atoms per unit population at	
	3000 K	3500 K
2000	10^{-10}	$10^{-8.6}$
3000	$10^{-6.7}$	$10^{-5.7}$
6000	$10^{-3.3}$	$10^{-3.0}$

is used. These energy levels and the associated radiation wavelength can be obtained from the Grotrian diagram of the element in question [7].

1. *Absorption laws*

To a first approximation, absorption by free atoms is similar to absorption by molecules and there is a linear relationship between absorbance and the "concentration" of the sample. This relationship is given by the Beer–Lambert Law which can be written in several ways such as follows: $T = I_1/I_0$ where T = transmittance, I_0 is the initial radiation intensity and I_1 is the intensity often passing through the sample. Note that the ratio I_1/I_0 is independent of the light intensity. A sample of concentration c and path length b may absorb 30% of the light in which case $I_0 = 100$, $I_1 = 70$. If we put a second sample in line, the light intensity entering it will be 70. This will be I_0 for the second sample. But since T is constant for that sample the light intensity leaving the second sample will be $70 \times 70/100 = 49$. There is a logarithmic relationship between I_1 and path length. Similarly, there is a logarithmic relationship with concentration. This can be summarized as

$$I_1 = I_0 10^{-abc} \quad \text{or} \quad I_1/I_0 = 10^{-abc}$$

where a = absorptivity, b = sample pathlength, c = concentration
From this

$$-\log I_1/I_0 = abc$$

We define $-\log I_1/I_0$ as A, where A = absorbance
Hence

$$A = -\log T = abc \tag{5}$$

This can be summarized as $A = abc$, which is the Beer–Lambert Law stating that the relationship between concentration or path length and absorbance is linear.

In atomic absorption this relationship holds true in the sense that if we have two samples, one with a concentration twice the other, then the absorbance of the former will be approximately twice the latter within the limits of Beer's Law. However, in practice it is found that this relationship cannot be sustained because flame atomizers are generally used as the "sample cell" and the population of free atoms in a flame is far from homogeneous. Homogeneity of the sample is a basic requirement for the application of Beer's Law.

A much more accurate relationship is given in eqn. (6) below.

$$\int_0^\infty K_\nu d\nu = \frac{\pi e^2}{mc} Nf \tag{6}$$

where e = charge of the electron, m = mass of the electron and c = the speed of light, N is the number of free absorbing atoms in the light path and f is the oscillator strength of the absorption line. It can be seen that there are a number of constants in this equation. The only variables are N, the total number of atoms in light path, and f the oscillator strength. The relationship between them and the total amount of light absorbed is a basis for quantitative analysis. The degree of absorption for each element and each absorption line depends on the oscillator strength which is a direct measure of how strongly each atom will absorb at that wavelength.

References p. 14

2. Oscillator strength

The oscillator strength in emission spectroscopy is a measure of how closely an atom resembles a classical oscillator in its ability to emit radiation. The greater the oscillator strength the greater the emission intensity for a given set of conditions. This is given by the equation

$$A_{ij} = 8\pi^2 e^2 f_{ij}/\lambda^2 mc \tag{7}$$

where A_{ij} = the transition probability between energy levels i and j, f_{ij} = the oscillator strength of the associated emission line; λ = wavelength of the emission line.

The relationship between the absorption oscillator strength and the emission oscillator strength is given by the equation

$$g_j f_{ji} = g_i f_{ij} \tag{8}$$

where g_j and g_i are the statistical weights of the atoms in states i and j.

Oscillator strengths recorded in tables are usually given in 'gf' values. Typically, gf values are between 0.1 and 1.0 but values greater and less than these are not uncommon.

Short wavelength lines require relatively high energy to excite; therefore, in emission the oscillator strength of short wavelength lines is low.

In contrast, the absorption oscillator strengths of atoms are fairly uniform in all regions of the ultraviolet spectrum. The oscillator strengths of elements with absorption lines at short wavelengths are just as high as elements with absorption lines at long wavelengths. In consequence, elements such as zinc (213.8 nm) absorb just as strongly as elements such as sodium, which absorbs at 589 nm.

It is important to remember that the oscillator strength is a physical property of the atom and does not vary under normal experimental conditions. Examination of eqn. (6) shows that the total absorption is equal to a number of constants times the oscillator strength times the number of atoms in the light path.

The phenomenon of absorption is to a first approximation independent of temperature and wavelength. This is in contrast to emission spectroscopy where there is frequently a severe loss of emission intensity at shorter wavelengths of the ultraviolet part of the spectrum.

It should be noted that temperature plays an important part in the actual production of free atoms so that although the physical process of absorption is independent of temperature, the efficiency of production of atoms is not independent of temperature and is a major variable in experimental work.

In summary, atomic absorption is a physical property of any given element. The total degree of absorption depends on a number of constants and the number of atoms in the light path. The one constant which varies between elements is the oscillator strength. The values for oscillator strengths are available and can be found in the literature [13]. The most commonly used absorption lines are those originating from the ground state because the vast majority of free atoms reside in the ground state under normal experimental conditions. Upper excited states are very sparsely populated and the total degree of absorption is therefore very limited and not analytically useful. For a given element the oscillator strength is highest for the transition between the ground state and the first excited state and thus this transition is the most sensitive analytically.

For all metals and some metalloids the amount of energy required to excite an atom from the ground state to the first excited state is in the ultraviolet region of the spectrum. However, for the non-metals the amount of energy required is in the vacuum ultraviolet, so they do not absorb in the ultraviolet and cannot be observed with commercially available equipment. They could be observed if the equipment was redesigned and air eliminated. This is feasible, of course, but has never been undertaken by any instrument manufacturer on a commercial basis. Non-metals are therefore not at present directly determinable by AAS. However, some procedures have been developed whereby the non-metals can be determined indirectly. For example, the halides can be determined by reacting them with silver and measuring the silver in a precipitate or measuring the silver lost from a solution after precipitation with the halide.

B. Atomic absorption line widths

Earlier an experiment was described in which atomic absorption lines were observed using a hydrogen lamp radiation source. Experimentally it can be shown that these absorption line widths are extremely narrow and can only be isolated under conditions of very high spectral resolution.

The natural line width is the width of the absorption line not exposed to any broadening effects. It is a hypothetical case because the lines are essentially always broadened by experimental conditions. However the "natural line width" indicates the lower limit of the absorption lines width. It can be calculated from the uncertainty principle which states that

$$\Delta E \, \Delta \tau = h/2\pi \tag{9}$$

where $\Delta \tau$ is the lifetime of the excited state and ΔE the range of energy over which the line emits, i.e. the line width in terms of energy.
From eqn. (2)

$$\Delta E = h \Delta \nu \tag{10}$$

where $\Delta \nu$ is the frequency range over which the light is emitted.

Since $\Delta \tau$ is of the order of 10^{-8} s, it can be calculated that the natural line width is about (10^{-5} nm). In practice this natural line width is broadened by several effects. These include the Doppler effect caused by the motion of absorbing atoms in their environment (usually a flame). At any given instance

References p. 14

some atoms are traveling away from the light source and some atoms will be traveling toward the light source. They appear to absorb at slightly different wavelengths giving a broadening of the absorption line. Under normal circumstances this broadening is gaussian in distribution because the distribution of relative motion of the atoms is random, but their velocity relative to the dilator is gaussian.

Other forms of broadening include Lorentz broadening, which is a result of collision between the absorbing atoms and molecules or atoms of other species. Collision with foreign molecules may cause some broadening brought about by the fact that the foreign molecules may relax the atoms thus shortening their lifetime, $\Delta\tau$, and therefore increasing ΔE, the line width (see eqn. 9).

Stark broadening results from splitting of the electronic levels of the atoms caused by the presence of strongly varying electric fields. They may be caused by fast moving electrons, close to the atoms. Zeeman broadening is caused by varying magnetic fields in the vicinity of the atoms. In practice the Stark and Zeeman effects are minor.

These and less important effects which cause line broadening may increase the absorption line to values up to (10^{-3} nm). Although this is a considerable increase over the natural line width it is still very narrow in practice and very difficult to observe with conventional equipment. Consequently it was not practical to use a continuous radiation source such as a hydrogen lamp. Not only would it be very difficult to isolate the absorption line but the total amount of energy radiated by the light source over such a narrow absorption band would be very small and difficult to measure using conventional detectors.

This problem was overcome by Walsh, who successfully demonstrated the use of the hollow cathode as a light source.

C. Hollow cathodes

An illustration of the hollow cathode is given in Figure 2. In this system the metal of interest is used as the material from which the cathode is made. The light source is filled with an inert gas, such as neon, which is ionized by the anode. The positively charged neon ions are then attracted by the negative charged cathode and accelerated towards it. On arrival at the cathode the neon strikes the surface of the cathode. If it has sufficient energy it causes atoms of the cathode to be ejected. This process is called "sputtering". The sputtered atoms are invariably excited and emit radiation characteristic of the cathode metal(s). The emitted lines are generally very narrow in band width.

The sample atoms absorb only at their own characteristic wavelengths. It is therefore essential that the light source emit at exactly the same wavelength. This can be accomplished by using a hollow cathode made of the same element as the element being determined. In practice this has been

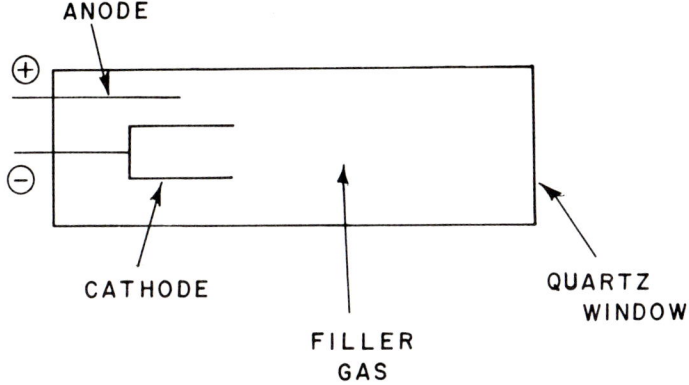

Fig. 2. Schematic diagram of a hollow cathode.

found very advantageous, since it has led to a high degree of freedom from interference from absorption by other elements because they absorb at other characteristic wavelengths. A disadvantage of the system, however, is that for each element to be determined a separate light source must be used. The instrument must be tuned to that wavelength and a separate detector operated for absorption measurements. The practical rule therefore, is that for each element to be determined, a separate hollow cathode must be used. This leads to a large inventory of hollow cathodes, particularly if a number of different samples and elements are to be analyzed. Some multi-element hollow cathodes are available and are widely used, but they have limitations, particularly in terms of lifetime and stability. Efforts to use continuous light sources have generally been unsuccessful, although they have found some application when high sensitivity is not required.

III. QUANTITATIVE MEASUREMENTS

The most accurate presentation of absorption by atoms is given by eqn. (6). However, this is difficult to apply in practice.

The process of absorption measurements in atomic absorption can be compared to absorption measurements in the standard colorimeter or ultraviolet/visible spectrophotometer. The equipment consists of a radiation source, sample cell and detector readout. The radiation from the source is measured without the sample in the sample cell and the intensity designated I_0. The sample is then placed in the sample cell and energy is absorbed. The new intensity of radiation is measured and designated I. $I_0 - I$ equals the amount of source radiation absorbed by the sample.

In atomic absorption a similar process takes place, except the radiation source is usually a specific line source and the sample cell is an atomizer such as a flame. As with the colorimeter, the intensity of radiation is measured

References p. 14

without the sample in the cell and designated I_0. Sample is then introduced into the flame, ground state atoms are produced and absorption takes place. The intensity of radiation is measured and designated I. The same relationship holds for this absorption: $I_0 - I =$ the amount of source radiation absorbed by the sample atoms. In order to understand quantitative analysis by atomic absorption we must know how this absorption increases or decreases as the concentration of atoms in the sample is varied.

The quantitative relationship is expressed by the Beer—Lambert Law, $A = abc$, where A is absorbance, a the absorptivity constant, b the cell path length and c the sample concentration. More simply, it states that the amount of light that is absorbed by a sample is a function of the number of absorbing atoms in the light path. Clearly, the number of atoms is a function of the sample cell path length and the concentration of the sample.

In atomic absorption analysis, the sample cell is an atomizer which in most cases has a very reproducible path length, b, and Beer's Law relationships are followed in most cases. The percentage absorption reading is converted to absorbance, $\log_{10}(I_0/I_1)$ and then related to the sample concentration. For example, a sample absorption of 12.9% = 0.06 absorbance, and 1% absorption = 0.0044 absorbance. The most important point to note about the Beer—Lambert Law is that the concentration of the sample is directly proportional to absorbance A and not to percentage absorption.

The atomic absorption process can be summarized as follows: radiant energy is emitted from a hollow cathode lamp and passed through a flame. The flame sampling system produces ground-state atoms from the sample. The intensity of radiation before and after sample atomization is measured and the result shown on a meter readout or digitally.

IV. BACKGROUND CORRECTION

The atomic absorption lines of each element are very narrow and are easily distinguished from the atomic absorption lines of other elements. They are so narrow in fact that isotopes of the same element absorb at slightly different wavelengths and can be distinguished from each other.

However, molecular absorption is broadband and may take place over a wide range of wavelengths (e.g. 100 nm), frequently including the wavelengths of atomic absorption lines. Broadband absorption may be caused by unburned fuel in flames or unburned solvent or fragments of water molecules introduced from the sample and consequently broadband (also termed background) absorption is common in flames. In order to collect accurate data it is frequently necessary to correct for broadband molecular absorption. There are several ways of doing this; the most common methods include: (1) Using a broadband radiation source, such as a hydrogen lamp or a deuterium lamp. (2) Measuring absorption of a non-resonance line very close to the wavelength of the absorption line. (3) Using a Zeeman background corrector. (4) Measuring the absorption of a blank sample.

Each of these methods is based on the fact that broadband absorption is virtually the same at the resonance line as at a wavelength very close to the resonance line. If the background absorption is measured close to the resonance line then a correction can be made for the background absorption at the resonance line.

In practice, first the absorption of the resonance line is measured. This is a measure of the absorption signal equivalent to the atomic absorption plus molecular background absorption. Secondly, absorption due to the molecular background is measured and the difference between the two is the true atomic absorption measurement. The bases of the techniques for background absorption are as follows.

A. Hydrogen (deuterium) background corrector

The radiation from a hydrogen lamp is measured at the same normal wavelength as the resonance line [8]. The radiation from this lamp fills the spectral slit and therefore a waveband of approximately 0.1 nm or greater will reach the detector. Absorption by molecules in the flame is across the entire spectral slit width, providing a measurement of the molecular background absorption. Atomic absorption also takes place at this wavelength. However, the lines are very narrow (10^{-3} nm) and the total amount of energy absorbed is very small compared with the molecular energy absorbed. If an atomic line is completely absorbed from a waveband of 0.1 nm the total absorption is only 1%. This is a negligible amount and may be ignored. Consequently, the absorption of the hydrogen lamp is a measure of the background absorption.

B. Double wavelength background corrector

In the double wavelength background corrector two lines from the hollow cathode are examined [9, 10]. First, the resonance line is used to measure the sum of the absorption by the free atoms and the molecular background. The absorption of a non-resonance line, close in wavelength to the resonance line is then measured. This gives a measure of the molecular absorption. The difference between these measurements is the atomic absorption signal. Some error may be involved in this procedure if the wavelength of the resonance and non-resonance lines are significantly different, because the molecular background may vary between the two wavelengths. The lines used may originate from the metal or the inert gas in the hollow cathode. Any non-resonance line is perfectly satisfactory to use since it will not be absorbed by the free atoms but will undergo molecular absorption.

References p. 14

C. Zeeman background corrector

In the Zeeman effect [11] the energy levels of a molecule are split under the influence of a varying strong magnetic field. If the atoms are not in a changing magnetic field they will absorb at a single wavelength (e.g., the resonance line). A strong magnetic field may then be turned on and the absorption line is split into fine structure with wavelengths greater than and less than the resonance line. If the magnetic field is strong enough the resonance line is eliminated entirely.

This process may be applied in practice by one of two methods. First, the hollow cathode may be exposed to an alternating magnetic field, causing the hollow cathode to emit alternately the resonance line and then hyperfine structure lines on each side of the resonance line. The resonance line is absorbed by the atoms and the molecular background, but the hyperfine lines are only absorbed by the molecular bands. The difference in absorption between the resonance line and the hyperfine lines is equal to the net absorption by the free atoms. One distinct advantage of this procedure is that the intensity of the lamp does not change and no adjustment has to be made to the instrument. The same detector operating under the same conditions can be used to make this measurement. Another advantage with this procedure is that only one light source is used and therefore the light path is identical when the sample and background measurements are made. This can be a problem which must always be addressed in other procedures.

An alternative procedure for the Zeeman background corrector is to operate the hollow cathode continuously but to expose the sample to an alternating magnetic field. The sample atoms absorb at the resonance line when not exposed to the magnetic field, but develop hyperfine structure and do not absorb the resonance line when the magnet is turned on. With the magnetic field on, absorption of the resonance line is a measure of molecular background. From the combined data, the net atomic absorption can be measured.

In modern instruments background correction is carried out automatically. This is a dangerous simplification. Frequently the operator has no knowledge of the degree of background correction made which may be more than 90% leading to serious quantitative errors. Unsuspected problems are often concealed. The operator must be alert to detect these problems and correct for them by some other means.

V. AVAILABLE LINES

Theoretically, any absorption wavelength equivalent to an energy transition state in a Grotrian diagram should be usable for AAS. However, the only heavily populated state is the ground state of the atoms. In practice all useful transitions originate in the ground states and terminate in a higher excited state. Equation 6 states that the total absorption is equal to a con-

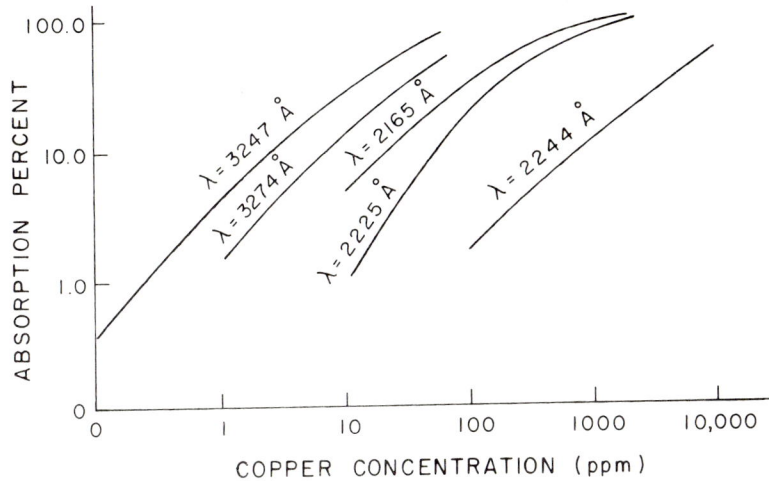

Fig. 3. Calibration curves for different copper absorption lines.

stant times the oscillator strength times N, the number of atoms which can absorb. This is the number of atoms in the lower state of the two energy states involved in the transition. (Examples of the relative population are given in Table 1.) The number of atoms in the upper energy states is very small and consequently any transition between two excited states will be virtually unobservable because the population of the lower of any two excited states is very small.

The analytical sensitivity is directly proportional to the oscillator strength, f. For a given atomic species the transition with the highest oscillator strength is the ground state to the first excited state. The oscillator strength steadily decreases with transition from the ground state to higher and higher excited states. If the highest sensitivity is necessary, the transition between the ground state to the first excited state must be chosen. This is the absorption line routinely used. If the expected sample concentration is high, absorption may be too strong for quantitative analysis. The degree of absorption will be at the high end of the absorption calibration curves and subject to serious error. Analytical determinations of these high concentrations can be made by using an absorption line with a lower oscillator strength. An example of several lines which can be used is shown in Fig. 3, which shows the absorption trace for copper. It can be seen that the useful analytical ranges for these lines are very wide.

Not all elements have numerous absorption lines available but many do. The wavelengths of these lines and the oscillator strengths and analytical sensitivities are given in a number of books and reviews [12—17]. A good rule of thumb is that a quantitative determination can be carried out with reasonable accuracy at concentrations five to ten times as high as the sensitivity limits (1% absorption).

References p. 14

TABLE 2

ABSORPTION LINES OF COBALT AND THEIR RELATIVE SENSITIVITIES IN AN AIR—ACETYLENE FLAME

Absorption wavelength (nm)	Sensitivity (ppm)	Analytical range (ppm)
240.7	0.15	0.6—5
242.5	0.23	
252.1	0.38	
241.1	0.55	
352.7	4.1	
345.3	4.2	

An illustration of the numerous lines available, using cobalt as an example, is given in Table 2. Similar tables are available in the literature for other elements in the periodic table [12—17].

REFERENCES

1. W. H. Wollaston, Phil. Trans. R. Soc., London, Ser. A, 92 (1802) 365.
2. D. Brewster, Report 2nd meeting, Brit. Assoc., (1832) 320.
3. A. Walsh, Spectrochim. Acta, 7 (1955) 108.
4. W. G. Jones and A. Walsh, Spectrochim. Acta, 16 (1960) 249.
5. J. V. Sullivan and A. Walsh, Spectrochim. Acta, 21 (1965) 721.
6. B. J. Russel and A. Walsh, Spectrochim. Acta, 15 (1959) 883.
7. C. Candler, Atomic Spectra, D. Van Nostrand, N.J., 2nd edn., 1964.
8. H. Kahn, At. Absorpt. Newsl., 7 (1968) 40.
9. A. C. Menzies, Anal. Chem., 32 (1960) 898.
10. J. W. Robinson, Anal. Chem., 33 (1961) 1226.
11. T. Hadeishi and R. D. McLaughlin, Science, 174 (1971) 404.
12. J. W. Robinson, Atomic Absorption Spectroscopy, M. Dekker, New York, 2nd edn., 1975.
13. G. F. Kirkbright, Atomic Absorption and Fluorescence Spectroscopy, M. Sargent, Academic Press, London, 1974.
14. Atomic Absorption Manual, Perkin-Elmer Corp., Norwalk, CT, U.S.A.
15. Atomic Absorption Manual, Instrumentation Labs, DE, U.S.A.
16. Atomic Absorption Manual, Varian Association, Palo Alto, CA, U.S.A.
17. Atomic Absorption Manual, Fisher Scientific, Waltham, MA, U.S.A.

Chapter 2

Instrumental requirements and optimisation

J. E. CANTLE

VG-Isotopes Ltd., Ion Path, Road 3, Winsford, Cheshire CW7 3BX (Gt. Britain)

I. INTRODUCTION

The general construction of an atomic absorption spectrometer, which need not be at all complicated, is shown schematically in Fig. 1. The most important components are the light source (A), which emits the characteristic narrow-line spectrum of the element of interest; an 'absorption cell' or 'atom reservoir' in which the atoms of the sample to be analysed are formed by thermal molecular dissociation, most commonly by a flame (B); a monochromator (C) for the spectral dispersion of the light into its component wavelengths with an exit slit of variable width to permit selection and isolation of the analytical wavelength; a photomultiplier detector (D) whose function it is to convert photons of light into an electrical signal which may be amplified (E) and eventually displayed to the operator on the instruments readout, (F).

When considered in this way the atomic absorption spectrometer is seen to be an emission flame photometer with a light source irradiating the flame. In atomic absorption, however, the flame is ideally an absorption cell in which the sample is atomised to produce atoms in their ground state only. In contrast, the flame used in emission photometry should produce as many thermally excited atoms as possible, thus ensuring maximum intensity of the emission spectrum produced from the elements of interest. High temperature (3000°C) vaporisation cells are therefore required for the successful atomic emission determination of all but the alkali and possibly the alkaline earth elements. The atomic ground state is highly populated even in relatively cool flames and this provides atomic absorption with high sensitivity and stability. Notwithstanding this, the technique of flame atomic emission has of course enjoyed the improvement in instrumentation that has occurred during the rapid growth of its absorption counterpart, and most instrument manufacturers now consider the requirements of both techniques in the design of a modern spectrometer. This will be discussed further in a later section discussing monochromation.

The following sections consider each of the major instrument components in turn, highlighting its function and discussing the procedures for optimisation in relation to the other components to produce the best possible instrument performance commensurate with the analytical need.

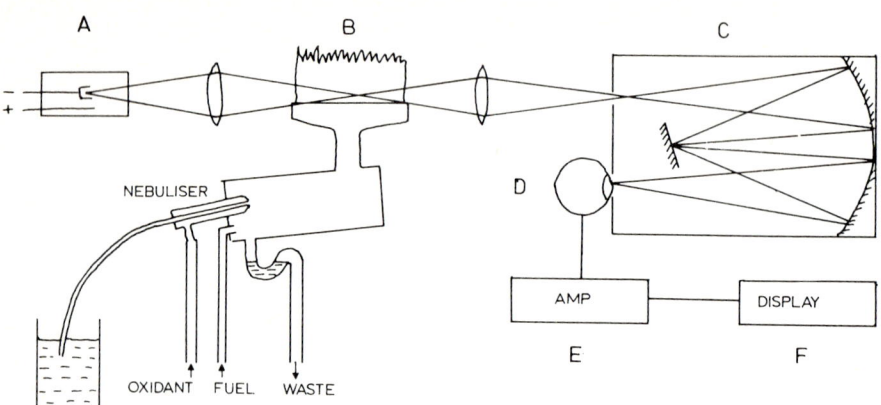

Fig. 1. General schematic of an atomic absorption spectrometer.

II. THE COMPONENTS OF THE SPECTROMETER

A. The atomiser

1. *Atomisation in flames*

The most commonly used atomiser is the chemical flame, based upon the combination of a fuel gas (e.g., acetylene) with an oxidant (e.g., air or nitrous oxide). The sample solution is introduced into the flame using a nebuliser in which the passage of the oxidant creates a partial vacuum by the venturi effect and thus the sample solution is drawn up through a capillary. Thus, an aerosol is produced having a wide variety of droplet sizes. This process is shown in Fig. 2.

Two basic types of flame atomising systems have been used for atomic absorption. Firstly, the total consumption or turbulent burner system in which the total sample aerosol in the oxidant stream and the fuel gas are fed separately through concentric tubes to the burner jet, where the flame is burned. Considerable turbulence, both optical and acoustic, takes place. On the positive side these burner systems are very simple in construction and thus were cheap to manufacture, did not flash-back and could handle virtually any mixture of gases. However, this system is now obsolete.

In the premix or laminar flow system the sample aerosol, oxidant and fuel are mixed in an inert chamber such that the larger droplets of sample are broken up or drained off before entering the flame. In this way a quieter and more stable flame is produced which is supported on a, typically, 10 cm path-length burner head. Because only the fine mist and evaporated sample reaches the flame an even burning takes place. This produces better atomisation and reduces interferences. If the nebuliser is adjustable, and most will

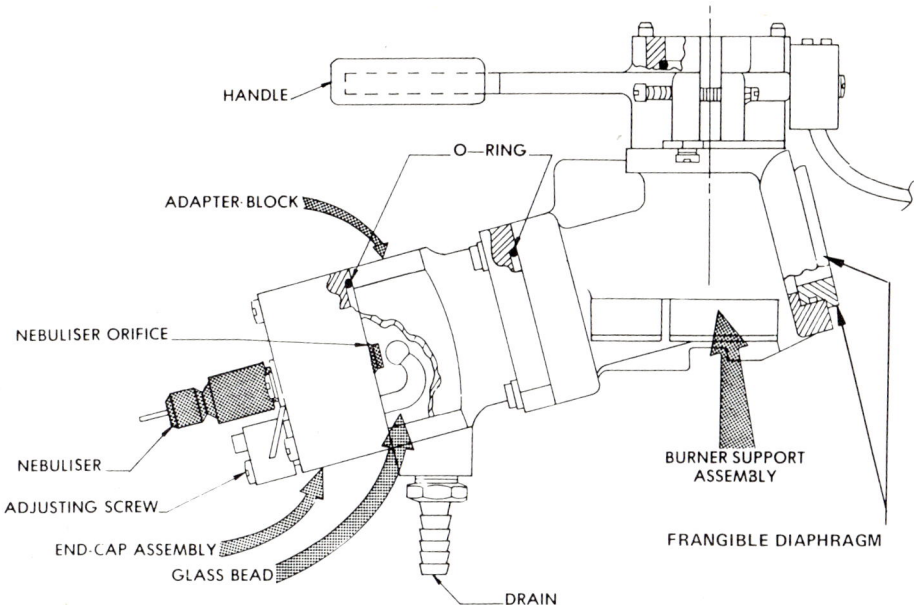

Fig. 2. Cross-section of premix chamber.

be, then the uptake rate maybe optimised for sample solutions of varying viscosity and surface tension. The procedure for optimising the nebuliser is quite straight forward. The operator should set up the spectrometer for the determination of an 'easy' element such as copper, which shows little dependence on flame stoichiometry. A sample solution of representative viscosity containing one or two parts per million of copper should then be aspirated. By carefully turning the adjusting screw on the nebuliser anti-clockwise, a point will soon be reached when no nebulisation occurs and bubbles of air are seen in the sample beaker. At this point the instrument's display will show zero. Slowly turn the nebuliser screw clockwise to increase the aspiration rate. The display should be set to be sufficiently responsive with a time constant of no longer than one second. As the uptake rate increases so will the absorbance displayed on the readout. When a maximum is reached the control should then be gently reversed, i.e. some slight anti-clockwise movement again, and by combining these adjusting motions the optimum signal-to-noise ratio will quickly be achieved. The highest absorbance obtained for a certain set of parameters should then be recorded and used for day-to-day performance comparisons. As a guide, a solution containing 1 ppm copper is quite capable of producing an absorbance signal of 0.150—0.180 absorbance on a modern instrument.

Proper alignment of the flame in the light path is obviously important. The instrument should provide a sturdy mount for the sampling system and should provide for vertical, horizontal and rotational control. Quite clearly the horizontal or lateral adjustment is the most critical assuming a reasonable burner height (4—10 mm below light beam) has been set. The rotational position can generally be set by eye. The recommended procedure for correct burner positioning is to insert a white card into the light path prior to flame ignition in order to identify the source light path. The burner position can then be quickly adjusted such that the slot is a few millimeters squarely below the hollow cathode lamp beam. Final burner position optimisation must be done with the flame lit and whilst aspirating an appropriate standard solution. Once again, a fairly responsive instrument display is maximised by moving the burner head in the available planes, paying particular attention to the lateral adjustment for obvious reasons.

The flame is a chemical reaction which takes place in the gas phase. The ideal flame for atomic absorption would generate the correct amount of thermal energy to dissociate the atoms from their chemical bonds. The most commonly used flames are air—acetylene and nitrous oxide—acetylene. The choice of oxidant depends upon the flame temperature and composition required for the production of free atoms. These temperatures vary the molecular or chemical form of the element. Air and acetylene produce flame temperatures of about 2300°C and permit the analysis by atomic absorption of some thirty or so elements. The nitrous oxide—acetylene flame is some 650°C hotter and extends the atomic absorption technique to around 66 elements. It also permits the successful analysis of most elements by flame atomic emission, in many cases at fractional parts per million levels, providing adequate spectral resolution is available.

The fuel and oxidant mixture must be controlled to provide the proper flame conditions for the element being analysed. A modern spectrometer should have a gas control system providing the precise and safe regulation which is important if reproducible results are to be obtained, particularly for those elements that show great dependence on flame stoichiometry.

When a fuel gas such as acetylene and an oxidant such as air are mixed and ignited a flame is produced, the structure of which is well defined. The mixture of gases enters at the base of the flame forming the inner cone, ignites and generates the very hot reaction zone. Most of the chemical reaction of combustion takes place in the reaction zone and is completed in the outer zone of the flame. Several steps occur as the sample leaves the container, enters the flame, is atomised and has its absorption measured. Firstly, the sample solution is nebulised to a fine cloud-like mist. In the premix system the mist mixes with fuel and oxidant before entering the base of the flame. Solvent evaporation occurs at this point. The efficiency of solvent evaporation depends upon four factors. (a) Drop size; small uniform droplets give rapid evaporation. (b) The chemical nature of the solvent affects the evaporation. Obviously, volatile solvents evaporate more quickly than less volatile solvents such as water. (c) A third factor affecting evaporation is flame temperature. High temperatures result in high evaporation rates. (d) Finally, solvent flow rate can be too slow and thus limit performance or too fast thereby flooding the flame and causing cooling.

Rapid and complete solvent evaporation is required for optimum performance. Atomisation occurs in the flame reaction zone, i.e. the conversion of sample molecules into atoms. Three factors affect the number of atoms formed. Firstly, the anion with which the metal atom is combined. Calcium chloride for instance is more easily dissociated than calcium phosphate. The second factor is flame temperature. Higher temperatures cause more rapid decomposition and, indeed, are often specifically required for elements which form refractory oxides. Finally, gas composition may affect the rate of atomisation if the constituents in the gas react with the sample or its derivatives. In the outer zone of the flame the atoms are burned to oxides. In this form they no longer absorb radiation at the wavelength of the uncombined ground state atoms.

Thus, it can be seen that there are several variables associated with the flame atomiser that must be optimised to achieve the best sensitivity and detection limit. The flame must be correctly positioned with respect to the light path. The fuel:oxidant ratio should be investigated to establish the optimum chemical environment for atomisation. The nebuliser and impact bead (where fitted) must be optimised to produce, overall, the best signal-to-noise ratio.

It is clear then that the chemical flame is an effective means by which a free, neutral atom population may be produced from a sample solution for analysis by atomic absorption spectrometry. The fact that flames were inherited from the older technique of flame emission spectrometry may account in part for their popularity, although they also have the following advantages for use in AAS:

(a) They are convenient to use, reliable and relatively free from a tendency to memory effects. Most flames in common use can be made noiseless and safe to operate.
(b) Burner systems are small, durable and inexpensive. Sample solutions are easily and rapidly handled by the use of relatively simple nebuliser assemblies.
(c) A wide variety of flames is available to allow the selection of optimum conditions for many different analytical purposes. Table 1 lists various fuel/oxidant combinations that have received attention.
(d) The signal-to-noise ratios obtainable are sufficiently high to allow adequate sensitivity and precision to be obtained in a wide range of analyses at different wavelengths in the range 190—900 nm.

Flame atomisation systems have some disadvantages, however, which limit their potential and convenience in use. These drawbacks have led many workers to devise techniques for the atomisation of samples for analysis that are not based entirely on nebuliser/flame systems. Some of the possible drawbacks of flames for analytical work are:

TABLE 1

CHARACTERISTICS OF VARIOUS FLAMES

Oxidant	Fuel	Approximate temperature (°C)	Typical burning velocity
Argon/diffused air	Hydrogen	400—1000	
Air	Coal gas	1840	55
Air	Propane	1930	45
Air	Hydrogen	2050	320
Air	Acetylene	2300	160
Nitrous oxide	Hydrogen	2650	390
Nitrous oxide	Acetylene	2950	285
Oxygen	Acetylene	3100	1130*

* Because of the high combustion velocity the flame cannot be burned on a premix burner system.

(a) The sample volume available may be less than that required for continuous use with an indirect nebuliser system. For low analyte concentration it may not be possible to dilute the solution to overcome this limitation. The discrete volume method of nebulisation overcomes this limitation to a certain extent whereby tiny volumes of 50—200 μl can be aspirated to produce a series of transient peaks. Where sample dilution is not possible because of sensitivity problems, this technique is certainly the best way to get the most analytical data from a limited volume of sample.

(b) Pneumatic nebulisers used in premix flame systems are only approximately 10% efficient in getting the sample into the flame.

(c) Flame cells are only rarely able to atomise solid samples and viscous liquid samples directly.

(d) Flame background absorption and emission at the wavelength of the analysis or thermal emission from the analyte or concomitant matrix at this wavelength may give rise to unacceptable signal noise with consequent loss of precision.

(e) In some locations it may be inconvenient to use high pressure cylinders of support and fuel gas, and in automated systems with no operator in attendance it may not be desirable to use a flame as the atom cell.

In addition to these practical disadvantages, other more fundamental factors act in flames to limit the sensitivity and selectivity that may be achieved. These are:

(a) The attainable atomic density in flames is limited by the dilution effect of the relatively high flow-rate of unburnt gas used to support the flame and to transport small volumes of sample solution to the flame. The optical density is also limited by the flame gas expansion that occurs on combustion. This factor has been estimated to be about 10^4 when unit volume of acetylene burns in air.

(b) Precise control over the chemical environment of the analyte and concomitant atoms in flame cells is not possible. The degree of control of chemical composition that can be achieved by variation of the fuel-to-oxidant ratio is accompanied by simultaneous changes in the flame temperature and its spectral absorption and emission characteristics. For many elements, particularly those that form thermally stable oxides, the efficiency of free-atom production from the sample introduced into the flame is low. Clearly, for the analysis of small liquid or solid samples and for the determination of trace amounts of many elements in larger samples it would be advantageous to achieve a higher concentration of atoms in a small cell volume than is possible with flames when solution nebulisation systems are used.

2. *Microsampling cup apparatus*

In this technique, illustrated in Fig. 3, the sample (typically $10\,\mu l$) is introduced into an air—acetylene flame in a nickel cup. The atoms pass through a hole into the alumina tube which is mounted above the burner directly in the light path of the instrument. The tube increases the residence time of the atoms in the lightpath. This apparatus is almost entirely reserved for the determination of volatile metals in biological materials such as whole blood and its use will be described in a later chapter. (See Chapter 4k.)

Fig. 3. Delves' cup atomiser.

3. *Atomisation without flames — electrothermal atomisation*

Various electrically heated furnaces have been described in recent years; a fixed sample volume is introduced into the furnace and after thermal pretreatment is rapidly atomised. This results in a transient signal whose height or area is proportional to the quantity of element under study. It has been demonstrated that in a graphite furnace atom cell a substantially higher peak concentration of atoms may be expected compared with a flame. This gain results directly from avoidance of the dilution and expansion effects that occur in flame cells. To assist the formation and maintenance of a dense free atom fraction of the element for atomic absorption analysis it is also an advantage that the chemical environment can be controlled by the use of an inert gas atmosphere, generally argon. It is apparent therefore that electrothermally heated atomisation devices such as graphite furnace atomisers will find widespread application in AAS provided their simplicity and reliability are comparable with those of flames.

Fig. 4. Atomic absorption spectrometer fitted with a furnace atomiser.

A furnace atomic absorption set-up is shown in Fig. 4. The graphite tube is mounted in the furnace workhead which replaces the burner in the spectrometer, and must be carefully aligned in the light path. It is after all, performing the same function as the flame, but 100—1000 times more efficiently in terms of sensitivity improvement. The power supply or programmer is a means of developing a series of carefully controlled voltages across the ends of the graphite tube, the purposes of which are as follows.

(i) *Purposes of the furnace temperature programme*
  (1) Drying liquid samples. (2) Melting solid samples. (3) Removing or partially removing the matrix. (4) Catalysing chemical changes to enhance

atomisation. (5) Catalysing chemical changes to ensure that only one form of the analyte is present at atomisation. (6) Catalysing reactions intentionally used for matrix modification. (7) Atomisation. (8) Cleaning the furnace.

The selection and optimisation of furnace operating conditions will be dealt with in more detail in section VIIA of the next chapter (page 57).

Electrothermal techniques are very sensitive, as already noted. For example, a 10 ppm lead standard produces an absorption signal of about 0.44 absorbance at 217 nm (flame AAS). This same instrument deflection can be achieved from a 20 μl injection of 10 ppb lead solution (furnace AAS), i.e. a thousand-fold lower in concentration.

However, furnace atomic absorption has certain disadvantages. The flame is more precise, faster and much less trouble and should always be used if sensitivity is adequate. The electrothermal method takes longer and requires more skill. Precision for comparable absorbances will generally be poorer. Moreover, contamination can be a real problem at these ultra-trace levels and all possible sources of analyte contamination have to be scrutinised.

Recent developments in the field of automatic sample injectors have done much to increase the ease of use, reliability, and performance levels of electrothermal atomisers. The sampling process may now be automated, thereby minimising the tedium associated with manual sample injection. The furnace autosamplers presently available fall into two categories; firstly, those that transfer a volume of solution into the furnace mechanically and then commence the normal temperature programme and secondly, an apparatus that produces an aerosol from the solution using a nebuliser and directs the spray into the tube furnace which is maintained at elevated temperature so that solvent evaporation occurs continuously with sample injection. These processes are illustrated in Fig. 5. Since the amount of sample deposited is controlled by time, (typically the rate is about $1 \mu l \, s^{-1}$) accurate depositions of very small volumes are possible. This allows samples of relatively high concentration to be analysed without dilution. At the same time, when very dilute solutions are being analysed, long deposit times allow practical injection of volumes much larger than tube furnaces can normally accommodate, thereby greatly improving sensitivity. The aerosol injection technique is ideally suited to the analysis of oils and other organic liquids. The fine spray is dried instantly on contact with the tube thereby overcoming the tendency of these materials to creep out of the system. Multipoint calibration curves may be generated from a single standard by dispensing different volumes into the furnace, by varying the aerosol deposition time.

Electrothermal atomic absorption analysis has developed significantly, complementary to flame atomic absorption, and present apparatus is easier to use and has greater performance than that available even three years ago. Reliable temperature-controlled heating of the furnace and furnace autosampling accessories will certainly assist the technique quickly to assume a similar degree of instrumental maturity and enable electrothermal methods to attain their deserved place in modern trace and ultra-trace metal analysis.

AUTOMATIC ANALYSIS WITH FURNACE AUTOSAMPLER
Diagram of Device for Aerosol Deposition of Sample

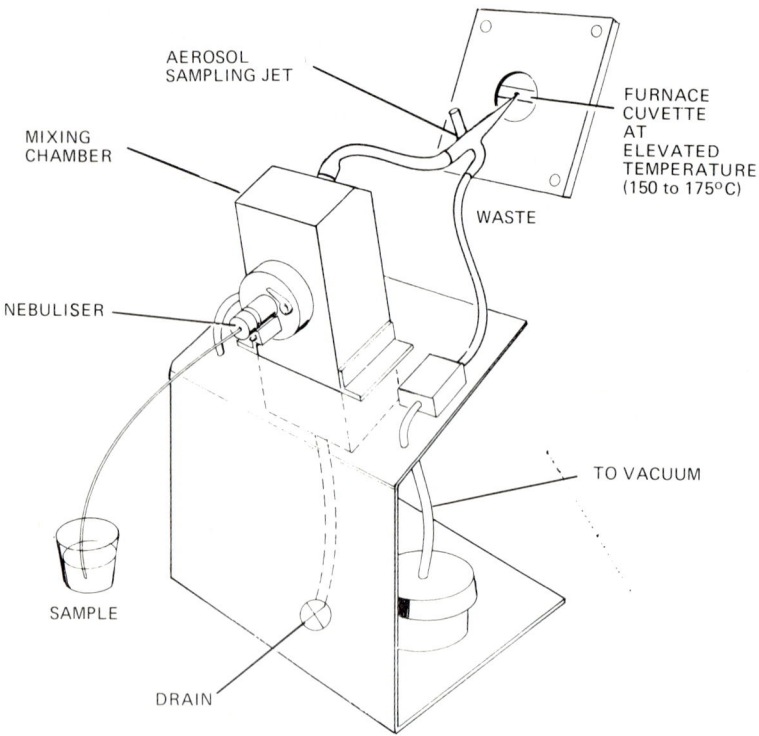

Fig. 5. Automatic analysis by furnace atomisation.

4. *Vapour generation techniques*

The chemical properties of some elements are such that special methods may be used both to separate them from the sample matrix before their introduction into the light path and to convert them into an atomic vapour once there. These elements include antimony, arsenic, bismuth, germanium, selenium and tellurium which readily form volatile hydrides upon reduction with acidified sodium borohydride, and mercury which is unique among the elements in possessing a high vapour pressure at room temperature and exists as a monatomic vapour. In this last example no atomiser is required. The sample under examination is reduced, generally by stannous chloride and hydrochloric acid, and the vapour swept into the light path of a mercury hollow cathode lamp. Most commonly a closed loop system is employed where an electric pump circulates the mercury vapour, and atomic absorption measurements are made using a quartz-ended gas cell in the light path.

In the former case, that of hydride generation, the evolved gas is swept to the atomiser by a stream of argon, nitrogen or air depending upon the type of atomiser employed. A number of flames can be used, the most successful being those burning hydrogen by entraining air from the atmosphere, thereby creating a diffusion flame. A support gas of argon or nitrogen is used. An air—acetylene flame may be used but causes reduction in analytical sensitivity. Alternatively the hydride(s) may be swept into a heated silica tube mounted above an air—acetylene burner supporting its usual flame. This method of atomisation has very high sensitivity, particularly for tin, and eliminates certain types of molecular absorption observed when the liberated hydrides are passed directly into the flame.

Where appropriate these methods of atomisation will be discussed again in the later chapters describing applications. These elements may be determined by conventional nebulisation systems with severely reduced sensitivity.

B. Light sources

The most familiar type of light source is the 'continuum' or white light source. The domestic filament bulb is an example, having an emission spectrum over a wide wavelength range starting at about 300 nm and extending into the infra-red. A continuum source which has a useful function in atomic absorption analysis is the hydrogen or deuterium-filled hollow cathode lamp. This source emits strongly in the UV part of the spectrum. The spectral characteristics of these continuum lamps are illustrated in Fig. 6.

However, since the absorption phenomenon being measured is occurring over an extremely narrow part of the spectrum (0.01 nm) it would require very high resolution to measure any significant absorption from a continuum lamp. This is why atomic absorption spectrometry as an analytical tool made

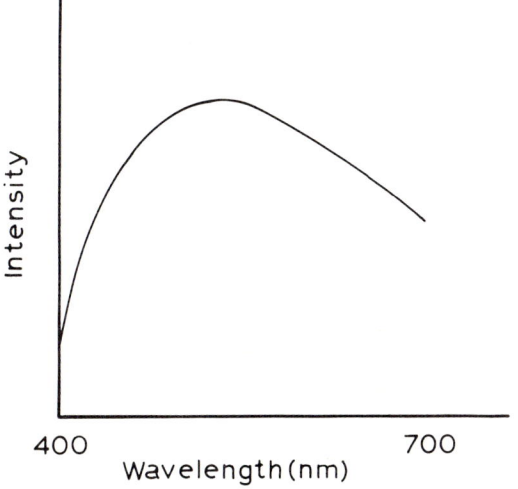

Fig. 6. Spectral characteristics of a continuum source.

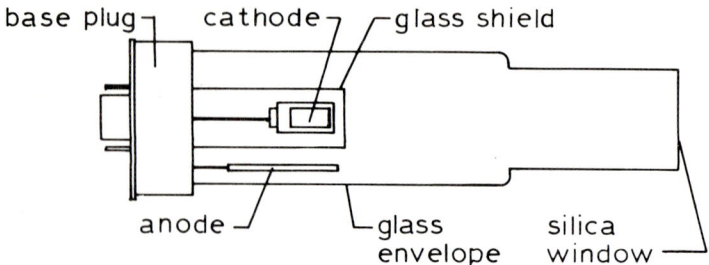

Fig. 7. Cross section through a hollow-cathode lamp.

very little progress until Walsh described the hollow-cathode lamp line source in 1955. A line source emits only at discrete wavelengths. The spectral lines are narrower than the absorption lines being measured, thus high resolution is not required. A typical hollow-cathode lamp is shown in cross section in Fig. 7 (see also Fig. 2 of Chapter 1, page 9). The lamp consists of a glass envelope containing a cathode and an anode. The cathode is a metal cup or cylinder which is made of the chemical element for whose analysis it will be used. Thus, for a copper analysis, a lamp will be used having a pure copper or brass cathode. In the second case the lamp spectrum will also contain zinc lines. The sealed glass envelope contains an inert gas, usually neon at low pressure. When a high voltage, 300—600 V is applied across the electrodes, positively charged neon ions bombard the cathode and dislodge (sputter) atoms of the cathode element. These atoms are excited by collisional processes and the emission spectrum of the element or elements is produced. Thus, by choosing the appropriate chemical element for the cathode, the atomic spectrum of that element may be readily generated. The cathode is usually surrounded by an insulating shield of mica, ceramic material or glass. This ensures that the discharge is confined to the interior of the cathode and results in an improvement in the intensity of the emitted lines.

The anode can be of almost any shape. It can be formed as an annular ring around the mouth of the cathode, or as a 'flag' near the mouth of the cathode, or as a wire or rod located in a convenient position. In some designs the anode serves the dual purpose in that it also acts as a mechanical support for the insulating shield.

The material used for the lamp window is important since it must transmit the spectral line(s) of the element being studied. Because quartz glass transmits over the full wavelength range it is suitable for all lamps. The other glasses are less expensive, and they can be used for elements whose resonance lines lie above 300 nm.

The fill gas is usually argon or neon at a pressure of 10—15 torr, i.e. about 1/50th of atmospheric pressure. Neon is preferred because it produces a higher signal intensity than argon, but where a neon line occurs in close proximity to the element's resonance line, argon is used instead.

Hollow-cathode lamps are not, of course, the only light source capable of producing the line spectra of chemical elements, but they are the most universally accepted source for atomic absorption instruments. The reasons for their popularity are to be found in their attributes:

(a) They generate a very narrow line. In order to measure the peak absorption, the width of the source must be very much narrower than the width of the absorption line. The width of the absorption line of the atoms in the flame is largely determined by temperature and pressure, and is normally about 0.01 nm. The atoms in the hollow-cathode lamp are in an environment having a considerably lower pressure and temperature. Consequently, the width of the emission line is about one tenth of the absorption line width.

(b) Their emission line is fixed at precisely the same wavelength as the absorption line because it is generated by the same energy transition. This means that the light source inherently generates an emission signal at precisely the correct wavelength for optimum absorption.

(c) Hollow-cathode lamps can be made for all chemical elements that can be determined by atomic absorption. All of these elements are suitable for inclusion in the cathode by one means or another (machining or powder metallurgy).

(d) They are simple to operate. All that is necessary is to connect the electrodes to a suitable power supply and adjust the current to the prescribed value. No complicated adjustments are involved, and control of environmental conditions is unnecessary.

(e) They are both stable and intense. Adequate intensity of signal is available for most elements such that intensity is not the most common limiting factor. Absorption is independent of source intensity so that increasing the intensity of the signal will not increase the degree of absorption. It must be pointed out, however, that intensity and noise level are inextricably linked in atomic absorption, as in most spectroscopic techniques. An increase in intensity will often yield a reduction in noise level, or provide an improvement in signal-to-noise ratio. This is usually because less electronic amplification of the signal is required and therefore less noise is generated in the detector/amplifier system. But, in flame absorption spectrometry, the flame itself can be a source of noise. Since the flame is a dynamic system, it presents an absorption which is not constant but fluctuating. These fluctuations in absorption are caused by flame movement due to drafts or imperfect gas mixing. Even very minor fluctuations will affect the intensity of the lamp emission reaching the detector and cause a varying or 'noisy' signal. This noise component is independent of lamp intensity. For many elements, the flame absorption fluctuation contributes most to the total noise. For these elements, therefore, any increase in source intensity would offer little advantage. Noise is only one aspect of stability. The other is drift, either during the warm-up period or during operation. Most hollow-cathode lamps are sufficiently stable for routine analysis after about five minutes warm-up, but for high precision analysis a longer warm-up period is desirable.

(f) Hollow-cathode lamps are economical. Most manufacturers guarantee their lamps for a life of 5000 milliampere hours or two years. This means that if a lamp is operated at 5 mA, it is guaranteed to last at least 1000 hours of operation. In routine use this would be adequate for many thousands of determinations.

1. *Operating conditions*

One particularly useful feature of the hollow-cathode lamp is that only one parameter needs attention, i.e. the operating current. Manufacturers generally recommend a suitable operating current. This is seldom highly critical and small departures from it will have relatively small effects on sensitivity. The only absolute limitation is that the maximum current specified must not be exceeded. The current actually recommended for a particular lamp is always a compromise since there is no clearly defined optimum and no specific lamp current corresponding to peak performance.

The most obvious effect of altering lamp current is the effect on intensity. Lamp intensity by itself is not important in absorption spectrometry (because the absorption signal does not depend on absolute intensity), but the secondary effect of improving the signal-to-noise ratio seems desirable. But this is not the only consideration. If it were, lamps would always be operated at their maximum rated current. There are two adverse effects of operating a lamp at high current and these require that a compromise be made in the selection of operating current. The first of these effects is that at higher currents the resonance line becomes broadened and distorted. This phenomenon is known as 'self-absorption broadening' and is caused by atoms in the discharge absorbing at the resonance wavelength emitted by similar atoms. The result is that the absorption sensitivity is degraded and the calibration curvature increases. Figure 8 illustrates the change in curvature for magnesium when the lamp current is increased. The second adverse effect is that lamp life is shortened as the operating current is increased.

With these factors in mind, it is good operating practice to use the manufacturer's recommendation as a starting point and then determine empirically the current which gives the optimum combination of signal-to-noise ratio and calibration linearity. The major consideration is not to exceed the maximum rated current.

2. *Multi-element lamps*

By combining two or more elements of interest into one cathode, it is possible to produce a hollow cathode lamp that can be used for the analysis of more than one element. Such a lamp is obviously convenient for the analyst, but there are some limitations to this approach. Some combinations of elements cannot be used because their resonance lines are so close that they interfere with each other. This makes it impossible to resolve the line

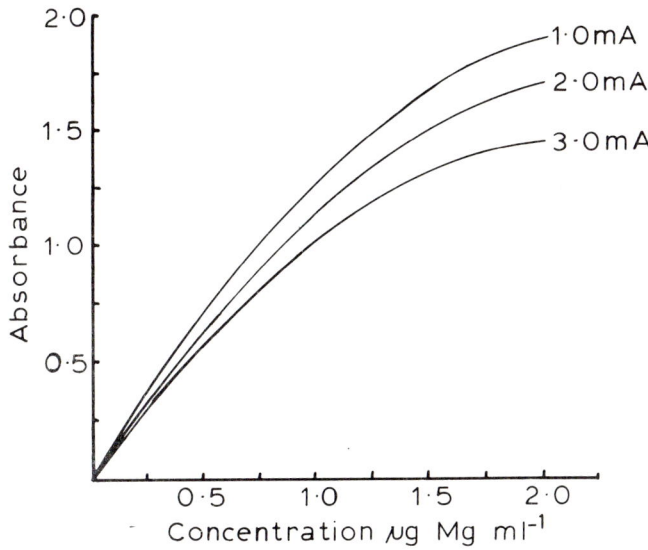

Fig. 8. Graph showing absorbance as a function of lamp current.

required and the combination is therefore unworkable. Other combinations cannot be used simply because of manufacturing difficulties in trying to incorporate elements of widely differing physical characteristics into a common cathode. However, for chemical elements to which such limitations do not apply, multi-element lamps do provide the analyst with a convenient source for the routine determination of several elements.

Continuum sources producing the hydrogen spectrum, as hinted earlier have an important role to play in the correction for non-atomic absorption. This is discussed fully in Chapter 3.

3. *Electrodeless discharge lamps*

For a number of years electrodeless discharge tubes excited by microwave frequencies have stimulated interest. The biggest claimed advantage of EDLs is the increased intensity of the line spectrum, by several orders of magnitude compared with the hollow-cathode lamp. They consist of a sealed quartz tube several centimeters in length and about 5—10 mm in diameter, filled with a few milligrams of the element of interest (as pure metal, halide or metal with added iodine) under an argon pressure of a few torr. The tube is mounted within the coil of a high-frequency generator at around 2400 MHz and excited by an output of a few Watts up to 200 Watts. Various views exist on the advantages or otherwise of electrodeless discharge lamps in AAS. Higher radiation intensity does not influence the sensitivity but the signal-to-noise ratio can occasionally be improved leading to better precision. Radio frequency EDLs operating at 27.12 MHz are now available which

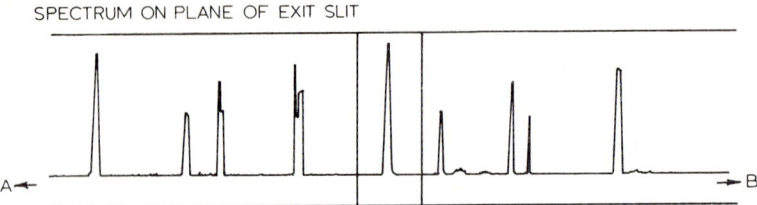

Fig. 9. Function of the monochromator in atomic absorption analysis. The resonance line is isolated from unwanted nearby radiation.

although having lower light output, have proved to give better long and short term stability than their microwave equivalents.

C. Monochromator and optics

The monochromator separates, isolates and controls the intensity of radiant energy reaching the detector. In effect it may be seen as an adjustable filter which selects a specific, narrow region of the spectrum for transmission to the detector and rejects all wavelengths outside this region. (See Figs. 9 and 10.) Ideally, the monochromator should be capable of isolating the resonance line only and excluding all other wavelengths. For some elements this is relatively easy, for others more difficult. The copper spectrum, for instance is relatively uncluttered, the nearest line being 2.7 nm from the 324.7 nm line. Nickel on the other hand has quite strong lines at 231.7 nm and 232.1 nm, one each side of the 232.0 nm primary line. The ability to discriminate between different wavelengths (i.e., resolution) is thus a very important characteristic of the monochromator.

Several methods may be used to isolate the required part of the spectrum, but the diffraction grating monochromator is now universally used in atomic absorption instruments. A schematic of a diffraction grating monochromator is shown in Fig. 10. The radiation from the hollow-cathode lamp enters the monochromator through the entrance slit and is focussed onto the grating. The grating disperses the radiation into individual wavelengths. By rotating the grating the analytical wavelength of interest will pass through the exit slit and be focused onto the detector.

The entrance slit, occurring as it does prior to dispersion, controls the amount of light entering the monochromator and should ideally be as large as possible. The exit slit determines the spectral bandwidth, i.e. the width of the tiny part of the spectrum transmitted to the detector. As we have seen already, this can be larger in some cases (Cu) than others (Ni). In practice the two slits are ganged together so that the choice of slit width is always a compromise between high light throughput and therefore excellent signal-to-noise characteristics and having the required degree of line separation to prevent the detector 'seeing' more than it should.

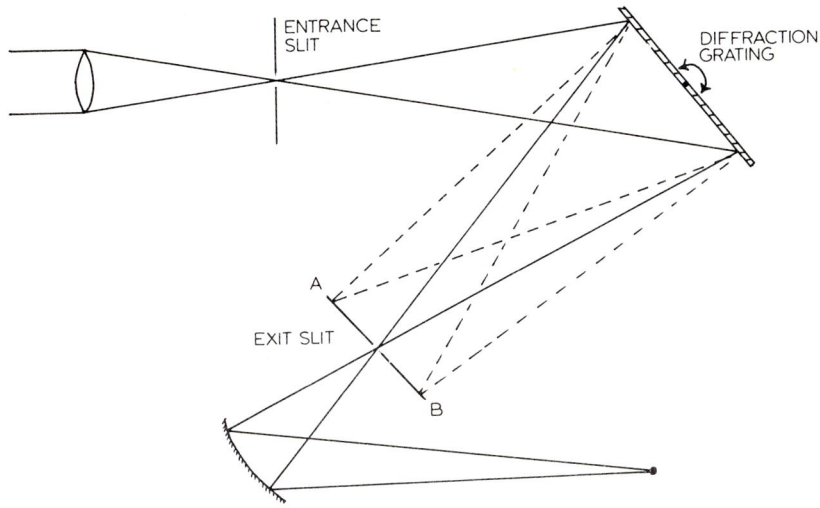

Fig. 10. Schematic of a monochromator as used in an atomic absorption spectrometer.

Since the hollow-cathode lamp spectra used in AAS are relatively simple, spectral bandwidths narrower than 0.1 nm are seldom if ever used. In atomic emission analysis, however, higher resolving power is often essential, particularly when the excitation source (e.g. the nitrous oxide—acetylene flame) is producing a complex spectrum. The instrument should, therefore, provide a wide range of slit settings and a convenient digital display of the wavelength in use for the operator.

1. *The optical layout*

It would be logical to arrange the components so far described in a straight line optically, and indeed this is done in the most successful instruments producing the so-called single beam system represented in Fig. 11.

The flame generates a ground state atomic vapour from the sample. This is irradiated by specific radiation from the hollow-cathode lamp. Some of this radiation is quantitatively absorbed at discrete wavelengths and the monochromator isolates the part of the spectrum over which that absorption is most precisely measured. The baseline, or background transmission of the system against which any absorption will be measured is set up by aspirating distilled water and adjusting the instruments readout to zero absorbance. This baseline will have various degrees of uncertainty associated with it. There are various contributions of noise in the system and certain variables can cause long term drift. For the most precise and accurate results the baseline zero should be as stable as possible, and on the whole modern spectrometers satisfy this condition. However, this was not always the case and in the past hollow-cathode lamps were prone to serious drift during use,

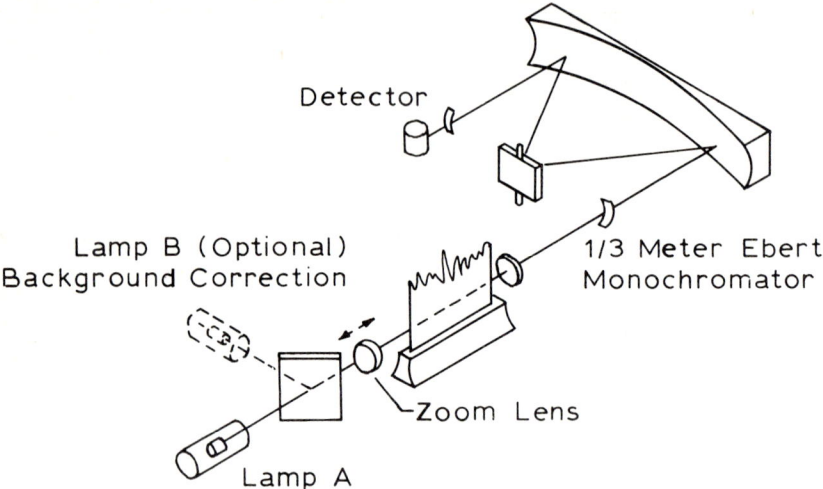

Fig. 11. Single beam optical arrangement.

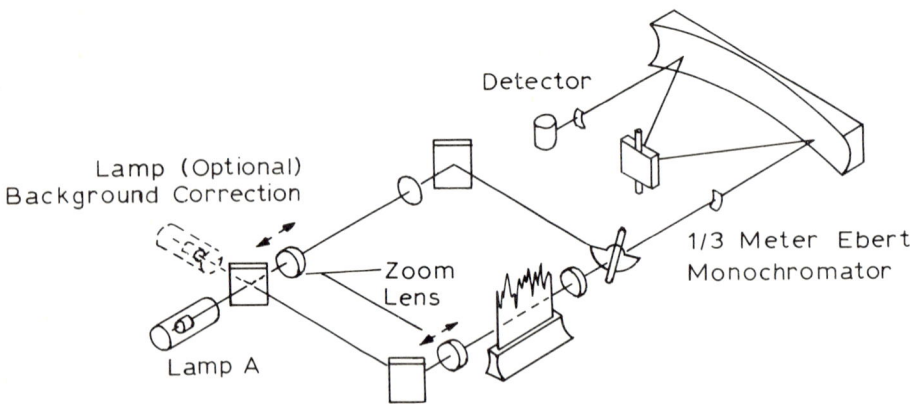

Fig. 12. Double beam optical arrangement.

thus requiring frequent adjustment of the instrument baseline. The double-beam optical configuration was introduced in an attempt to overcome the limitation imposed by contemporary lamps. In this arrangement, shown in Fig. 12, the radiation from the source is split into two beams, a sample beam going through the atomiser and a reference beam bypassing the atomiser and therefore of unvarying magnitude. The beams would be similar in intensity when the baseline zero was being established. Upon introduction of sample into the flame, the sample beam would be attenuated and the instrument would compute $\log I_0/I$ to measure the absorbance. Any variation in output from the hollow-cathode lamp would be equivalently translated in both beams and the end reading would be unaffected. It must be noted, however, that the double beam configuration will not compensate for fluctuation in atomiser efficiency, or in general for any variation in the flame/premix chamber/nebuliser areas.

Many double-beam instruments may be used in a single-beam mode, the reference beam not being sampled by the detector. Statistically, improved precision should result from the selection of single-beam operation if the source lamp is free from drift. In terms of light throughput, of course, the single-beam arrangement uses light most efficiently. However, it must be pointed out that ideally the light throughput for a double-beam system will be no worse than the equivalent single-beam instrument when equipped with a background corrector, which, as will be evident, requires the insertion of an extra source, one emitting a continuum. The incorporation of the background correction source into the instrument will always result in some loss of transmission. This loss would be 50% when static beam splitters are employed.

2. *Dual channel optics*

Figure 13 represents an instrument having dual-channel optics. This means that there are facilities for two line sources and one continuum source, each line source having its own monochromator and detector. Many workers have advocated the usefulness of this type of instrument. Certainly the recently introduced microcomputerised dual channel models are easier to use than their earlier counterparts, so it is worthwhile to summarise the possible attractions of simultaneous dual element analysis. One obvious possibility is the analysis of two elements in a sample at the same time, thus halving analysis time. A less obvious attraction is the possibility of analysing via an internal standard. This is where a second element, either already present or added to the sample, is measured and ratioed to the analytical element. The

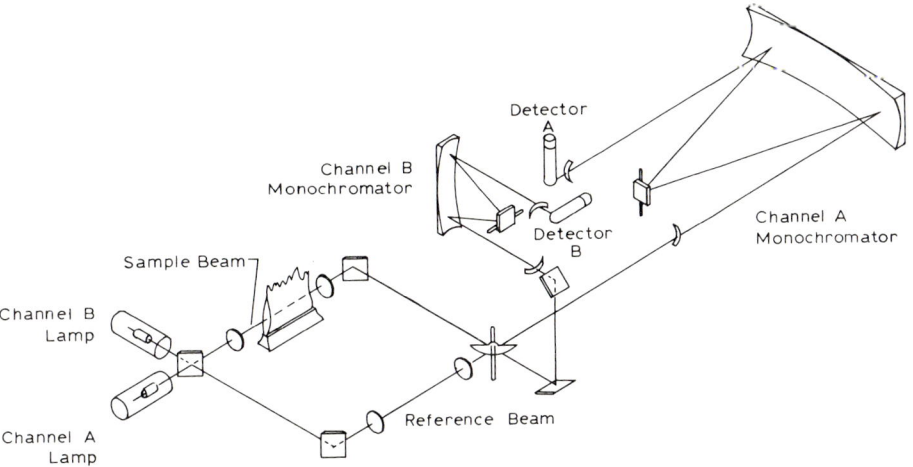

Fig. 13. Dual channel optical arrangement.

analyte display is now independent of physical variations in atomiser characteristics. Additionally, if the internal standard is already present within the sample homogeneously, it means that the sample preparation need not be quantitative. Weighings and dilution ratios may be quite approximate since the two elements' concentrations will always be fixed with respect to each other. Some examples of internal standard analysis will be presented in the application chapters.

D. Electronics and readout

The radiant energy from the source must be converted into an electrical signal for amplification and measurement by the readout system. This conversion is undertaken almost universally by a photomultiplier tube.

1. *Photomultiplier*

Photons from the radiation source bombard a cathode containing a photoemissive substance. This causes electrons to be dislodged from the cathode which then travel to the anode. The photomultiplier consists of a photoemissive cathode and an anode to collect the displaced electrons. Between the cathode and anode are additional photoemissive plates called dinodes. Each dinode collects electrons from the cathode or previous dinode. The bombarding electrons dislodge several electrons from the next dinode producing an extremely high flow of electrons at the anode. The operator controls the voltage between anode and cathode and thus sets the 'gain' of the detector. This voltage will vary from about 200—1000 volts to produce a wide range of gain settings. Since random processes within the tube will be expanded also, the lowest voltage that is practical should always be used to avoid excessive noise.

Generally speaking a manufacturer will fit a photomultiplier tube which has adequate response across the whole spectrum. This means that at the extremes of the wavelength range the performance will be tailing off and it may be worth considering more powerful photomultiplier tubes for dedicated work at the red end, for cesium and rubidium for instance. Photomultiplier response curves are available from manufacturers such as Hammamatzu and Philips.

2. *Source modulation*

It is necessary that the hollow-cathode lamp source be pulsed or modulated at a certain frequency and for the amplifier to be locked in to this frequency to permit discrimination against the continuous emission signal coming from the atomiser. Only the resonance radiation from the lamp must be seen. In modern digital electronic instruments the lamp cycle is controlled by a sophisticated electronic clock which is sampled to provide the short pulses of power to switch the lamps on and off and the amplifier in phase. The flame background emission is measured during periods of time when no lamp is on. It should be remembered, however, that the discrimination occurs at a point after all the light has fallen upon the photomultiplier tubes. This can result in saturation or breakthrough, and conditions must be modified in this case to reduce the level of flame emission. This problem will be considered in some detail in the section on interferences in Chapter 3.

3. *The readout*

The modern practice in scientific instruments is to present the results digitally. This avoids errors in scale readings on meters through parallax, misinterpolation between scale divisions etc. However, before the amplified output from the photomultiplier is displayed it must be converted to read in absorbance units, the logarithmic function of percent absorption. A logarithmic amplifier is used to make this conversion. Most current instruments will integrate the signal over a selected period, typically 0.1 s up to 99.9 s so that an unvarying result appears on the display after the selected time. Clearly, the precision of successive integrated results will improve as the integration period lengthens, so the operators choice will be a compromise. With an update time of 0.1 s the output fed to a chart recorder will appear equivalent to the analogue output which was a characteristic of older equipment. It is important that these very fast times are available for use with electrothermal atomisers which produce very short-lived absorption signals.

Very small signals may be expanded continuously using the instrument's scale expansion control. This facilitates the reading of small absorbances but it should be remembered that any fluctuations in the signal will be scale-expanded also. It is not always immediately grasped that readout directly in concentration units is achieved via the scale expand function. After all, if 10 ppm lead is aspirated versus a distilled water blank then an absorbance of about 0.500 would result. The readout could then be made to read directly in concentration units (ppm) by employing a scale expand factor of 2 and shifting the decimal point one place to the right. This is what concentration readout consists of. However to use this facility usefully, deviation from linearity must be compensated for. The practicality of this will be discussed in Chapter 3; suffice it to say at this stage that non-linearity in the calibration graph can be corrected for either using some manual device, or automatically using the microprocessor controlled instruments now on the market.

4. *The use of a microprocessor*

The introduction of a simple microprocessor into an atomic absorption spectrometer has increased flexibility and ease of operation. The following benefits have been introduced: (a) Easier operation via a keyboard and push

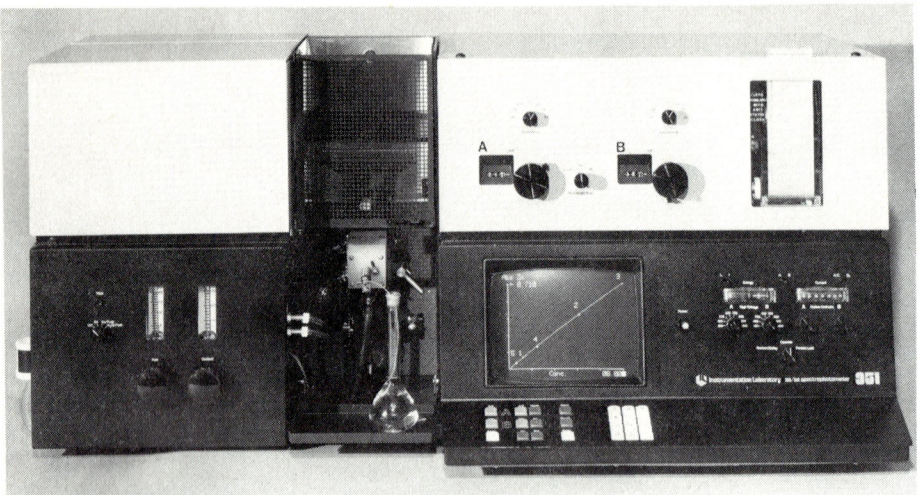

Fig. 14. Atomic absorption spectrometer equipped with a VDU.

buttons. (b) A greater range of parameters, especially integration periods. (c) Automatic linearisation of calibration graphs using a blank and up to five standards, using various well proven equations. (d) Statistical interpretation of data, including percent relative standard deviation on a series of measurements, thus providing important information about the analytical precision. (e) New media for presentation, e.g. video display (see Fig. 14). (f) Curve storage and memory. Calibration graphs once set up may be stored in the memory and recalled for use when required. One standard is then used to set the slope for that particular occasion.

The technology is now available for many more instrument functions to be selected both on the main instrument and peripheral devices such as an electrothermal atomiser or autosampler. Programmes for instrument setting and data processing can be stored, for example, on magnetic cards. Although, as already indicated, the actual speed of analysis may not be vastly improved, the advantages lie in the better reliability and accuracy obtainable and in the possibility of more efficient use of the time of a skilled analyst.

Chapter 3

Practical techniques

J. E. CANTLE

VG-Isotopes Ltd., Ion Path, Road 3, Winsford, Cheshire CW7 3BX (Gt. Britain)

I. INTRODUCTION

Atomic absorption spectrometry (AAS) is a virtually universal method for the determination of the majority of metallic elements and metalloids in both trace and major concentrations. The form of the original samples is not important provided that it can be brought into either an aqueous or a non-aqueous solution. This situation has been brought about by considerable improvements in instrumentation and also, perhaps partly as a result of this, a better understanding among analysts of the types of interference effect that may modify the expected response of a given element.

Atomic absorption methods combine the specificity of other atomic spectral methods with the adaptability of wet chemical methods. High specificity means that elements can be determined in the presence of each other. Separations, which are necessary with almost all other forms of wet analysis, are reduced to a minimum and often avoided altogether, making a typical atomic absorption analytical procedure attractively simple. This fact, combined with the ease of handling a modern atomic absorption spectrometer, makes it possible for routine analyses to be carried out quickly and economically by relatively junior laboratory staff.

Usually, separations are required for only one of two reasons, to remove a major cause of interference or to concentrate the elements to be determined should they be present in amounts less than their detection limit. While separation procedures must therefore be quantitative for the elements concerned, they do not necessarily have to be specific as it is possible to determine a number of elements together in one solution. This concept leads to the separation of groups of elements rather than individuals, and indeed to a general philosophy of chemical preparation and samples for atomic absorption in which as many elements as possible are brought together for determination in the final analysis solution. This should always be the aim in method development.

II. SAMPLE REQUIREMENTS AND GENERAL PREPARATION TECHNIQUES

Consideration will be given to the general principles involved in the preparation of various types of sample for atomic absorption analysis. A sample received in the laboratory can be placed into one of the following categories: (A) Aqueous solutions. (B) Organic solutions. (C) Inorganic solids. (D) Organic solids. (E) Gases. If these materials are to be aspirated then solids must be solubilised, gas streams filtered and liquid samples must satisfy certain criteria:

(1) The viscosity and solids content of the solution must be such as permits nebulisation without giving rise to problems associated with burner blocking or nebuliser 'salting-out'. The long path air—acetylene burner head will accommodate 2—3% solids at 40 psi oxidant pressure, and up to 10% solids by progressively reducing the oxidant pressure. The nitrous oxide burner head has a lower tolerance and 1% solids should be regarded as maximum for continuous aspiration. A viscous solution will be nebulised less efficiently than a less viscous solution and should be analysed by an addition procedure.
(2) Solid particles should be removed, preferably by centrifugation, particularly in trace level determinations where contamination will be more significant.
(3) The acid concentration should be as low as possible. The nebuliser fitted as standard may have a stainless steel capillary tube which will be attacked by acid over a period of time. Solutions containing more than 5% mineral acid should be nebulised using a nebuliser having a platinum/iridium capillary. A corrosion-resistant nebuliser having a plastic throat is required for solutions containing hydrofluoric acid. Additionally, in this instance, a teflon impact bead must replace the standard glass one where these are employed.
(4) Organic solvents must be chosen carefully.
(5) Interferences should be removed or compensated for (a general description of interferences is given in section V, page 52).
(6) The metal concentration must not be so high as to fall in a grossly curved part of the calibration graph. Beer's law is obeyed for most elements up to an absorbance of about 0.4; calibration graph curvature is discussed later.

The five sample types are now discussed in turn.

A. Aqueous solutions

Typical of these samples are raw and treated waters, seawater, biological fluids, beer, wines, plating solutions, effluents, etc. With this type of sample very little preparation is usually required. If the solution is suitable for aspiration then its approximate concentration can be determined, to check whether dilution with water is necessary. Degassing may be required, and/or the addition of releasing agents, ionisation suppressants, complexing agents, etc., as required for interference compensation. Concentration methods will be described later.

B. Organic liquids

These will mainly be petroleum products, many of which can be aspirated directly, or following viscosity adjustment with suitable organic solvents, which should be chosen according to certain criteria, i.e. the solvent should: (i) Dissolve or mix with the sample; (ii) Burn well, but in a controlled manner; (iii) Be available in a pure state, and not contain species having molecular absorption bands in the ultra-violet; (iv) Be innocuous and produce no harmful by-products upon combustion; (v) Be inexpensive.

Some examples of often-used solvents are *p*-xylene, n-heptane, cyclohexane, 10% isopropanol—white spirit mixture, methyl isobutyl ketone, methyl ethyl ketone and cyclohexanone.

Standardisation should be via organometallic standards, which are now available for a range of metals from B.D.H. and Hopkin & Williams, as well as from specialised oil-standard organisations such as Conostan.

C. Inorganic solids

Typically, these will be alloys, rocks, fertilisers, ceramics, etc. These materials are taken into solution using suitable aqueous/acid media, according to solubility: hot water, dilute acid, acid mixtures, concentrated acids, prolonged acid digestion using hydrofluoric acid if necessary, alkali fusion (e.g. using lithium metaborate), Teflon bomb dissolution. Fusion and 'bomb' methods are usually reserved for complex siliceous materials, traditionally reluctant to yield to solubilisation.

D. Organic solids

Typically foods, feedstuffs, leaves, plants, biological solids, tissue, polymers, etc. Prior to solubilisation these types of sample generally require destruction via wet digestion or ashing in a muffle furnace. A typical procedure featuring a nitric/perchloric acid mixture is reproduced below.

E. Gases

Atomic absorption techniques can be used to analyse gases indirectly, as liquid samples. To prepare the liquid sample the metals are removed from the gas stream or atmospheric sample using a filter medium such as a millipore filter disc. This is then either dissolved or washed in nitric acid and the solution analysed by standard additions. These procedures are now extensively used by Health & Safety Executive Inspectors to monitor (particularly) heavy metals in working environments. The reader is referred to Chapter 4C.

III. THE USE OF PERCHLORIC ACID FOR SAMPLE DIGESTION

Perchloric acid has achieved a largely undeserved reputation for being a highly dangerous and unpredictable substance; it is in fact a very useful reagent, but in common with many others needs to be handled with due respect. The properties which make it so useful are in general those which call for precautions in handling; a reasonable knowledge of the substance will thus allow the analyst to derive maximum benefit from its use, and ensure safe handling.

The properties of interest to the analyst are: (i) It is a strong acid, indeed probably the strongest. (ii) When dilute it is a normal reducing acid, but this changes with simple evaporation. (iii) It is one of the most powerful oxidising agents known, when hot and concentrated or when anhydrous. When hot, it is also a dehydrating acid. (iv) It is difficult to obtain the anhydrous acid, but by no means impossible, and it is this fact that has caused some of the problems encountered with the acid in the past.

A. Precautions in use

The precautions needed flow naturally from the properties; in addition to the precautions usual when handling any strong acid, it is necessary to ensure that perchloric acid does not come into contact with easily oxidisable organic matter or strongly reducing substances under circumstances where it can become hot or anhydrous. It is failure to observe these precautions that leads to problems, and a few examples of what to do and what not to do are given below:—

(1) One of the most common causes of perchloric acid explosions is the practice which judging by the number of literature reports appears to be widespread in academic laboratories; it is the making of perchlorate esters or perchlorate salts of organic bases. To mix organic matter and perchloric acid is dangerous, and no more intimate mixture can be obtained than to have the two incompatible entities in the same molecule. There are numerous reports in the literature from workers who have further compounded their error by attempting to dry the reaction products, usually leading at least to the destruction of the oven used.

(2) A rather less obvious precaution is to prevent perchloric acid becoming anhydrous. If an aqueous solution is evaporated, it finally boils at 203° C, as an azeotrope containing about 72% of $HClO_4$; this solution is a powerful oxidising agent, especially when hot, but is not in itself inherently hazardous. Anhydrous perchloric acid, obtainable by dehydration procedures is a quite different proposition. It is said to be spontaneously explosive, and although it is the belief of the author that "spontaneous" explosions are probably due to contamination of vessels or ingress of foreign matter such as dust, the potential for hazard is obviously too great for any but the most closely

controlled use. Any operation which can lead to dehydration should therefore be avoided; mixing of perchloric acid with acetic anhydride, for example, will give a mixture prone to spontaneous explosion on standing, even at room temperature.

(3) If a fume cupboard is used for perchloric acid evaporations, it may be assumed that there will be condensed acid in the upper parts of the cupboard and its ducts; if the same area is used for ammonia, the whole cupboard will also become coated with ammonium perchlorate, which is an equally good oxidising agent. Cupboards used for such purposes therefore must be constructed of inert materials, and especially not of wood. Wooden fume-cupboard frames exposed over long periods to perchloric acid fumes will eventually become impregnated with perchloric acid and perchlorates, and liable to spontaneous ignition from stray heat sources, such as a burner or an electric hot plate placed close enough to raise the surface temperature to a little over 100° C. Obviously the better the draught through the cupboard the longer may it be used before Nemesis overtakes it, but the day of reckoning will eventually come. Many fume cupboard manufacturers offer stainless steel lined cupboards for perchloric acid work, and although they are effective, it is a matter of opinion as to whether they are the best solution to the problem. One of the better solvent acids for stainless steel is hot perchloric acid, and such a cupboard could obviously introduce grave blank problems into low iron, nickel and chromium determinations. In the author's laboratory, asbestos lined cupboards coated with chlorinated rubber paint have given many years of trouble-free service; both ducts and walls are washed down at intervals. A further point to note here is that many laboratory fume exhaust systems use common ducts to a number of cupboards, and the mixing of exhausts from fuming perchloric acid and, say, the evaporation of alcohols in a separate cupboard could lead to an unexpected reaction in the ducts.

Having stressed some of the precautions to be taken, it is perhaps necessary to list the advantages of using perchloric acid, with particular reference to its use in wet oxidation procedure.

(1) All common perchlorates except those of potassium and ammonium are freely soluble in aqueous media. This is of great advantage when samples are rich in calcium, where if H_2SO_4 based reaction mixtures are used $CaSO_4$ can be deposited and cause co-precipitation, especially of lead and bismuth.

(2) The oxidation temperature is lower than with H_2SO_4, and losses of volatile elements are less likely. The only elements normally lost during $HClO_4$ oxidations are Hg, Re as Re_2O_7, Os as OsO_4 and Cr as $CrOCl_2$ if the mixture is allowed to fume too long at high temperature (see below); As will be lost if the mixture is allowed to fume to dryness.

(3) Once an oxidation procedure has been worked out and proven for a particular sample type, it can usually be run on a routine basis in quite large batches. The 2 g foodstuff samples described below can be handled in batches

of 20 or 50 with no difficulty, and this is a routine practice in the author's laboratory.

B. Wet oxidation procedures

The essence of a successful perchloric acid based wet oxidation is to arrange for the redox potential of the solution to rise from a level where only the most labile materials are oxidised to the final fuming perchloric acid state. This is easily arranged by diluting the perchloric acid with a less powerful oxidiser, usually nitric acid, and then raising the temperature. Provided that the sample dissolves in the hot acid mixture, and that the amount of nitric acid and the rate of temperature rise are such that the more easily oxidised materials are destroyed during the evaporation, the reaction will proceed in a smooth and controlled manner to completion.

Some materials cause difficulty; for example, the lower alcohols or terpene based oils will react violently with nitric acid alone, either in the cold or as soon as heat is applied and local hot spots develop. In these circumstances, the sample can often be treated with hydrochloric acid first, then nitric acid added, so that the strong oxidising action of the nitric acid is ameliorated. Free fats or oils do not dissolve in the acid mixture and float on the surface. If they form an unbroken film, this can lead to superheating and "bumping", splashing hot $HNO_3/HClO_4$ mixture; even if this does not happen, it is not uncommon for virtually no oxidation to take place until the nitric acid has evaporated and the beaker then contains a fat layer on top of hot concentrated perchloric acid. This represents a considerable hazard; the most likely effect is not an explosion, but a violent reaction at the interface discharging hot and often burning oil accompanied by perchloric acid over a large area. The problem is best avoided by continuously swirling the beaker during the evaporation of the nitric acid. The oil layer is thus dispersed as droplets with a reasonably large surface area to volume ratio, and the initial oxidation is therefore encouraged. This procedure can be used to oxidise even such unlikely materials as hydrocarbon oils; no more than 0.5 g of sample should be used at a time, and it is wise to carry out an initial experiment with 0.05—0.1 g, until experience and confidence have been gained.

A common mistake is to use too little perchloric acid; even with acid soluble materials posing no particular difficulty, there is considerable energy given off during the actual perchloric oxidation, and the acid is both consumed as oxidant and vigorously boiled off. If the total amount of acid available is insufficient, the mixture will go to dryness before the oxidation is complete and the dry residues will catch fire. True explosions on the 1—2 g sample scale are rare at this stage, but the beaker may shatter because of the thermal shock, and in any case the analysis will be ruined.

1. *Wet oxidation of dried herbage, blood, meat, custard powder or other cellulose/protein low fat materials*

Weigh 2 g of sample into a suitable beaker (a 350 or 400 ml conical beaker is very convenient), and add 15 ml of HNO_3 and 10 ml $HClO_4$. (If the material is known to be vigorously attacked by HNO_3, add 5 ml of HCl before the other acids.) If Sn is to be determined, add 0.5—1.0 ml of H_2SO_4 at this stage to avoid loss of the metal as $SnCl_4$.

Cover with a pyrex glass cover (to keep Pb, Al and Zn blanks low), and heat to boiling. There may be some effervescence but it will usually rapidly dissipate as the oxidation proceeds. Copious brown fumes are often evolved at this stage, and are an indication that oxidation of labile materials is proceeding satisfactorily. As the nitric acid evaporates, the brown fumes often clear, indicating the end of the destruction of the easily oxidised entities, followed by a further evolution of brown fumes as the solution temperature rises to about 150° C and the more resistant substances begin to be attacked. Observe the solution at this stage, and if an oily film separates, swirl the beaker to disperse the oily layer into the acid.

As the last nitric acid evaporates, the nuclear boiling of the liquid may change to effervescence again, which may become vigorous. If the reaction rate appears to be increasing too quickly for comfort, remove the beaker from the hotplate. If the reaction does not abate in 30 to 60 s, add 1—2 ml of nitric acid to lower the temperature and hence the redox potential, and then allow the reaction to proceed again; it may be necessary to heat the solution if the reaction has subsided too far.

As the vigorous oxidation proceeds, the solution may darken or become black in colour. If this happens, add nitric acid dropwise to the fuming solution until the dark colour clears; repeat the process as necessary.

The end of the oxidation is indicated by the final solution becoming clear, pale yellow to water-white, and boiling instead of effervescing. If chromium is to be determined, the beaker should be removed from the hotplate at this stage, since the next stage in the evaporation is for the remaining water to boil out and give the azeotropic mixture. At its boiling point of about 203° C, the redox potential is high enough to oxidise Cr to Cr(VI), which will then react with the chloride formed as a reduction product and be lost as chromyl chloride. If a solution containing 50 μg of Cr and 5 ml of perchloric acid is fumed to dryness, over 90% of the Cr is lost. For most other elements, the solution can now be evaporated to low bulk to reduce the acid concentration for subsequent analysis.

Finally, allow to cool, and to the cold residue add 1—2 ml of HCl. Warm gently to decolorise any dark brown colour (due to manganese) or any orange (due to Cr(VI)), cool again and make up to volume. In the author's laboratory the routine method uses a 2 g sample and a final dilution to 20 ml. The solution thus obtained can be used for atomic absorption or flame emission measurements of most of the metals, with techniques such as hydride generation for As, Se, Sn, Sb, Bi where necessary, or for polarographic analysis after suitable adjustment of solution conditions.

Digestions using hydrogen peroxide and sulphuric acid are also popular (refs.; The Analyst, 92 (1967) 403; and 101 (1976) 62). A 5—10 g sample is weighed into a 250 ml conical flask (dry samples are wetted with water), 5 ml of concentrated sulphuric acid is added followed by dropwise addition of 50% hydrogen peroxide solution, initially at room temperature and finally with use of a hot plate. The clear digest is usually diluted to 100 ml with water and an aliquot taken for analysis.

IV. CALIBRATION

A. Beer's Law

Beer's Law states that the absorbance of an absorbing species is proportional to its concentration and is represented in fairly simple terms in the idealised calibration graph in Fig. 1 which shows a plot of absorbance versus concentration. The straight line joining the origin with the single calibration point is referred to as the calibration graph. The concentration of the unknown solution which has absorbance A is read off the graph and produces an answer corresponding to concentration C. This assumes that the calibration is linear between the origin and the absorbance/concentration of the

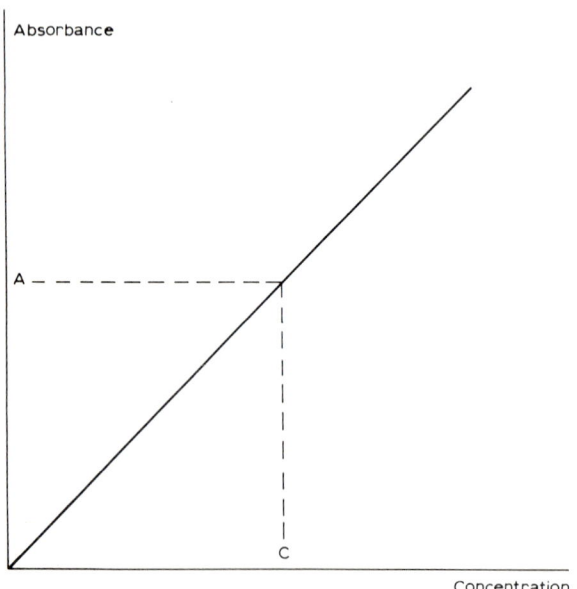

Fig. 1. Idealised calibration graph.

standard producing the single point on the graph. As stated previously, this is only strictly true at low absorbance values. Deviations from linearity are usually apparent as the absorbance increases. The calibration curve bending towards the concentration axis. The reasons for this are discussed in the following section.

B. Reasons for non-linearity of calibration graphs

(1) Unabsorbed radiation, stray light. All light must be absorbable to the same extent.
(2) Hollow-cathode lamp linewidth broadening due either to the age of the source or the use of high lamp currents.
(3) If the monochromator slit is too wide, more than one line may be transmitted to the detector. In this case the calibration graph will show greater curvature than would be the result if only the desired line were transmitted to the detector.

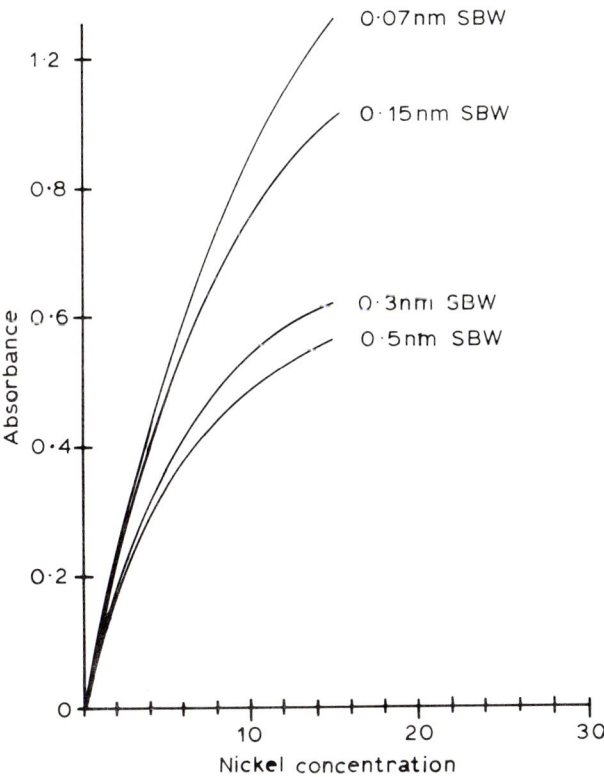

Fig. 2. Improvement in linearity and sensitivity of nickel determination at higher resolution settings.

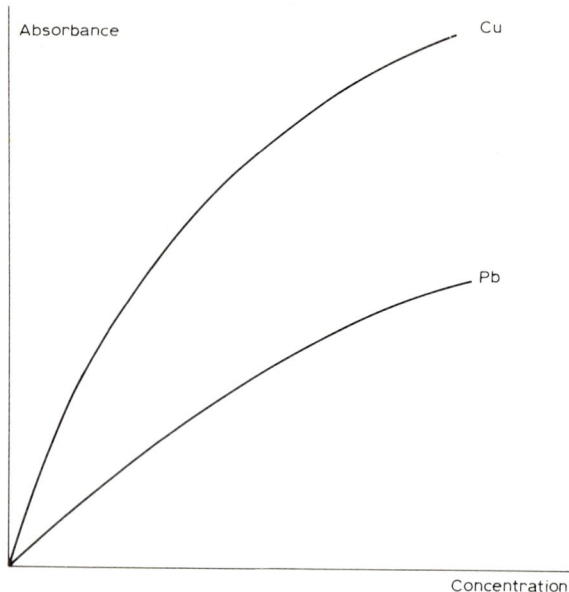
Fig. 3. Calibration graphs for lead and copper.

(4) Disproportionate decomposition of molecular species at high concentration. This results in a lower proportion of free atoms being available at higher concentrations for a constant atomisation temperature.

Clearly, these factors must be considered if the calibration is to be as linear as possible in the desired concentration range. An example of point 3 is represented in Fig. 2 which shows a series of calibration graphs for nickel. The main resonance line for nickel occurs at 232.0 nm in a densely populated spectral region; thus, for improved linearity narrow slits must be used. It is apparent that graphical curvature is inevitable and must be accommodated either by reducing the concentration range being studied or by resorting to electronic means of compensation. Electronic curvature correction, whilst apparently producing linear concentration read-out could tempt the analyst into working in a region of the absorbance—concentration relationship best avoided because of the severe curvature in this area. Calibration graphs for lead and copper are shown in Fig. 3. It is apparent that one element produces greater absorbance for a given concentration than the other. This leads us into the next section which will introduce certain general terminology used in atomic absorption analysis.

C. Terminology

1. *Sensitivity*

Sensitivity is defined as the concentration of a solution (typically in ppm) of an element needed to produce a signal of 0.0044 absorbance. This is equivalent to a 1% decrease in the transmitted radiation. The absorbance (not % absorption) is proportional to concentration. This is Beer's Law. The sensitivities of some commonly analysed elements are given in Table 1. The sensitivity figure is a useful performance index in that it gives the operator information about how well the machine has been optimised. For example, the sensitivity for copper is 0.03 ppm. This means that a solution containing this concentration of copper would be expected to produce a reading equivalent to 0.0044 absorbance. Since this is a very small deflection it is convenient when optimising to use solutions of greater strength so that, for example, by extrapolation a solution of copper containing 3 ppm would be expected to produce a deflection of 0.440 absorbance (440 milliabsorbance units). Similarly, the sensitivity for lead is 0.1 ppm, a higher figure than for copper. This means that a higher concentration is required for lead than for copper to give an equivalent absorbance. Thus, when optimising lead the operator would expect to see a deflection of 440 milliabsorbance units or greater from 10 ppm of lead. It can be seen that the sensitivity figure is a measure of the slope of the calibration graph and gives the operator useful information about how well the atomic absorption instrument has been optimised. The sensitivity figure however gives the operator no clue as to the minimum detectable amount of the element. This requires a second performance index.

2. *Detection limit*

The detection limit (or limit of detection) is defined as the lowest concentration of an element in solution which can be detected with 95% certainty. It is therefore the concentration which will produce a deflection equal to twice the standard deviation of a series of readings (typically ten). Historically it has also been defined as the concentration which produces a deflection from the base line of a chart recorder which is equal to twice the peak to peak variability of that base line. The important thing to note with detection limit information is that the method of calculation should be defined when original work is being published.

The detection limit is a theoretical figure and one would never attempt to measure routinely concentrations at the detection limit for real samples. As a guide to whether an element can be routinely analysed the detection limit figure should be multiplied by a factor, which in many cases will be as great as ten, in order to show whether a successful analysis would result. For example, the detection limit for lead is 0.01 ppm by flame atomic

TABLE 1

SOME TYPICAL SENSITIVITY DATA FOR FLAME AAS

Element		Sensitivity ($\mu g\ ml^{-1}$)	Detection limit ($\mu g\ ml^{-1}$)
Ag	Silver	0.04	0.003
Al	Aluminum	0.43	0.04
As	Arsenic	0.63	0.21
			0.001*
Au	Gold	0.1	0.014
B	Boron	8.0	2.0
Ba	Barium	0.17	0.015
Be	Beryllium	0.014	0.0026
Bi	Bismuth	0.22	0.07
Ca	Calcium	0.02	0.001
Cd	Cadmium	0.01	0.002
Co	Cobalt	0.05	0.005
Cr	Chromium	0.04	0.003
Cs	Cesium	0.058	0.019
Cu	Copper	0.03	0.004
Er	Erbium	0.47	0.07
Fe	Iron	0.05	0.003
Hg	Mercury		0.0005*
K	Potassium	0.02	0.003
Li	Lithium	0.023	0.0016
Mg	Magnesium	0.004	0.0002
Mn	Manganese	0.03	0.003
Mo	Molybdenum	0.28	0.046
Na	Sodium	0.005	0.00023
Ni	Nickel	0.05	0.007
Pb	Lead	0.1	0.01
Pr	Praseodymium	28.1	7.3
Pt	Platinum	1.0	0.098
Rb	Rubidium	0.02	0.0034
Re	Rhenium	8.5	0.73
Rh	Rhodium	0.12	0.022
Sb	Antimony	0.28	0.06
			0.006*
Sc	Scandium	0.25	0.03
Se	Selenium	0.11	0.16
			0.001*
Si	Silicon	1.3	0.35
Sn	Tin	0.88	0.16
Sr	Strontium	0.06	0.001
Ta	Tantalum	12.0	0.94
Te	Tellurium	0.21	0.07
Ti	Titanium	1.5	0.12
Tl	Thallium	0.23	0.03
V	Vanadium	0.96	0.05
W	Tungsten	16.9	1.4
Y	Yttrium	1.69	0.3
Yb	Ytterbium	0.077	0.005
Zn	Zinc	0.007	0.0026

absorption. One could reasonably expect therefore to measure lead at ten times this value, i.e. 0.1 ppm, in real samples using a background correction technique if necessary.

3. Precision

The precision of an analysis is most conveniently defined in terms of percent relative standard deviation. The standard deviation is relatively easily calculated following a series of discrete measurements either of absorbance or of concentration. The relative standard deviation is then defined as the standard deviation expressed as a percentage of the mean of the data used to calculate the standard deviation.

One of the attractions of flame atomic absorption is that relative standard deviations are often better then 1% in ideal situations and only marginally poorer than this when lower levels are being analysed. By definition the precision of measurement is 50% relative standard deviation at the detection limit, because it is at this point that the signal to noise ratio equals 2.

The standard deviation (sigma) is calculated according to the formula

$$\sigma = [(\bar{x} - x)^2/(n - 1)]^{1/2} \qquad (1)$$

where σ = standard deviation, x = analytical value, $\bar{x}$ = arithmetic mean, n = number of values taken. The standard deviation concept is a very useful one for expressing analytical confidence in the answer obtained from the instrument. Most analysts would be content working with an analytical confidence of 95%. Thus if one's analytical result were 5 ppm, for example, and the calculated sigma were 0.1 ppm then with a 95% confidence the answer would be 5 ppm plus or minus 2 × 0.1 i.e. an analytical range of 4.8—5.2. In order to provide greater confidence of an answer around 5 ppm the analyst would have to improve the precision, that is, make sigma smaller than 0.1 ppm. This would involve a re-examination of the instrumental parameters to check the possibilities for obtaining more stable analytical results.

Note to Table 1: All sensitivities and detection limits were obtained using an air—acetylene or nitrous oxide—acetylene flame unless otherwise noted. Potassium was added to the aqueous standards to suppress ionisation for easily ionised elements.

*Hydride Generator

D. Working with calibration curves

Calibration graph curvature cannot be avoided completely though it can be minimised by paying attention to the points made in section IIIB. The analyst should prepare a range of standards covering the concentration range of interest, optimise the instrument, spray the standards and note the absorbances and produce a calibration plot. Depending upon the shape of the graph he might either decide to reduce the number of standards or to increase the number of standards in order to define the curve as accurately as possible. If curvature is severe it may at this stage be worth examining some of the variables that might control the magnitude of curvature, such as lamp current, spectral bandwidth and of course the wavelength of analysis.

As a general guide, in the author's opinion, if there is more than 10% curvature then two standards and a blank are not sufficient to define the graph accurately for electronic curve fitting purposes. Three, four or even five standards and a blank should be prepared in order to give the instrument's microprocessor sufficient information to fit the curve appropriately. If the instrument has a manual curvature correction function then the manufacturer should be consulted for his recommended method of use, but generally speaking the range of standards would be aspirated and their absorbances noted and the amount of curvature from origin to top-standard calculated. The amount of curvature correction introduced by the curve-correct control would then be progressively increased and at each stage a blank and all standards would be aspirated again and the amount of curvature recalculated. This procedure would be repeated until the amount of curvature were acceptable and the level of acceptability of course would depend upon the degree of accuracy required by the analyst. The author considers this a better approach to working with curved calibration graphs than attempting to linearise the graph by a trial and error procedure.

E. De-optimisation

If the absorbance produced from the sample is too high to permit accurate analysis in the working range of the standards normally employed, then the analyst has to decide which of three options to take. Firstly, the sample can be diluted to bring its absorbance into the optimum working range for that element's wavelength, or secondly, an alternative wavelength having a lower absorptivity may be used. The majority of elements analysed by atomic absorption offer the analyst a range of wavelengths. Of course in most cases, for accurate analyses, the most sensitive wavelengths would be chosen, but for analysis at higher concentrations the use of a less sensitive line offers an alternative to diluting the samples. The third alternative is for the analyst to rotate the burner head by the required amount to desensitise the analysis, thus reducing the absorbance to whatever extent is necessary to bring the sample into an optimum absorbance range. On most instruments the burner can be rotated through any number of degrees from 0 to 90.

The three methods given above are the best ways to de-optimise the analytical system and thus reduce curvature. As a general guide, if the top standard is arranged to provide an absorbance of around 0.4 or 0.5 absorbance units then a successful analysis will result with more than adequate precision. There is seldom need to work with higher absorbances. The absorbance from the sample could of course also be reduced by generally de-tuning the instrument by moving the burner head or reducing the nebuliser uptake rate for instance. These methods are not recommended, however; the analyst should adhere to the three general solutions given above.

V. METHODS OF CONCENTRATION

Where the concentration of an element in a sample falls below the detection limit for that element or is low enough to make a precise direct measurement impossible, other techniques must be used to pre-concentrate the element or remove the matrix. The possibilities given below are an alternative to using an electrothermal atomiser where the sensitivity is of the order of 100 to 1000 times greater; see section VI.

A. Concentration by evaporation

This is the simplest technique but is prone to contamination or element loss by evaporation. Also the sample matrix may become too concentrated to pass through the nebuliser and burner without deposition. Since the matrix is also concentrated the final sample aspirated may suffer from matrix interference. This should be investigated and if necessary a method of standard additions used; see section V.

B. Solvent extraction of trace metals

This is probably the most widely used separation technique as it can be reduced to its simplest form. It is possible and often desirable in atomic absorption to extract and therefore concentrate more than one element at one operation. Specificity resides in the measuring technique, as has been stated previously. The choice of collating or complexing reagent is therefore not limited, as in colorimetry, to one which gives a strong colour for the metal being determined and complicated methods involving extractions and back extractions in order to improve specificity are avoided. A number of advantages result from the extraction of APDC—metal complexes into a suitable organic phase. The metal may be concentrated by as much as a hundredfold if desired. Wanted metals can be separated from high concentrations of other solutes which may cause difficulties in nebulisation and atomisation. The atomic absorption signal for nearly all metals is further enhanced by a factor of maybe threefold when aspirated in an organic solvent instead of an aqueous solution. APDC complexes are soluble in a number of ketones. Methyl isobutyl ketone, which is a recommended solvent for atomic absorption, allows a concentration factor of ten times.

Further details of specific extraction systems will be given in the later application chapters.

C. Ion exchange

Although invariably slower than solvent extraction, ion exchange techniques have been used to separate certain groups of metals from an undesirable matrix. Perhaps the most useful separation of this type is of trace heavy metals from higher concentrations of alkali metals. There is a very large number of examples in the literature in which ion exchange has been used to separate an interfering matrix either by retaining the analyte elements or by retaining the matrix element. In the latter case relatively large amounts of the ion exchanger may be required and in this case no actual concentration of the analyte is achieved in the process. The reader is again directed to the specialist literature for more information on the kind of separation that may be applicable to atomic absorption methods.

VI. INTERFERENCES IN ATOMIC ABSORPTION ANALYSIS

This topic will be treated fairly generally because the authors of the subsequent application chapters will be describing more specific examples.

The newcomer to AAS could easily be led into believing that he has been misled when informed that this analytical technique is free from interferences. This impression unfortunately arises from early work in the technique when, of course, only a few applications had been studied. With the increase of interest, a wider range of applications was studied and consequently more problems were encountered. However, the interferences encountered in atomic absorption spectrometry are now extremely well documented and many which were reported early in the literature were found to be due to instrumental imperfections and have now virtually disappeared. All interferences can be overcome by the use of simple techniques.

Interferences encountered in AAS can be separated into the following categories: (A) Spectral. (B) Flame emission. (C) Chemical. (D) Matrix. (E) Non-specific scatter. (F) Ionisation. The majority of difficulties that the analyst can expect to encounter arise from chemical, matrix, light-scattering and ionisation interferences.

A. Spectral interference

Spectral interference is rarely encountered in atomic absorption spectrometry. Spectral interferences in the past were experienced typically if, in a given solution, element A was being determined in the presence of element B. If the source contained both elements and the absorption lines of these could not be resolved by the monochromator, element B would cause an interference. In some early hollow-cathode lamps this was a well-known phenomenon. It could be overcome, however, by using an alternative absorption line, the probability of two lines coinciding again being extremely remote. Most of the interferences have now disappeared due to improvements in the purification techniques of the cathodes.

B. Emission interference

Emission interference was common in many early instruments which were accessories for UV/visible spectrophotometers, which operated in most instances on a d.c. system. The interference was caused by emission of the element at the same wavelength as that at which absorption was occurring. All modern instruments use a.c. systems which are of course 'blind' to the continuous emission from the flame. However, if the intensity of the emission is high, the 'noise' associated with the determination will increase, since the noise of a photomultiplier detector varies with the square root of the radiation falling upon it.

This effect can be reduced by either increasing the source current or by closing down the slit, both methods resulting in an increase in the signal-to-noise ratio.

C. Chemical interference

This is by far the most frequently encountered interference in AAS. Basically, a chemical interference can be defined as anything that prevents or suppresses the formation of ground state atoms in the flame. A common example is the interference produced by aluminium, silicon and phosphorus in the determination of magnesium, calcium, strontium, barium and many other metals. This is due to the formation of aluminates, silicates and phosphates which, in many instances, are refractory in the analytical flame being used.

In order to overcome this type of interference, two techniques may be emphasised, both of which release the element under investigation. The first relies upon the application of chemistry, in the knowledge that, in many instances, a compound may be added which will lead to the release of the element that we are interested in by the formation of a preferential complex. Thus, a chelate such as EDTA can be added to complex the cation thus preventing its association with an anion that could lead to the formation of a refractory compound. Alternatively, a reagent can be added that will preferentially form a compound with the interfering anion, again leading to the 'release' of the cation; for example, the addition of lanthanum chloride to solutions of calcium containing the phosphate anion. The calcium is 'released' due to the preferential formation of lanthanum phosphate.

Secondly, virtually all chemical interferences may be overcome by using the high-temperature nitrous oxide—acetylene flame.

D. Matrix interference

This is a general term covering: (i) Enhancement of sensitivity due to the presence of an organic solvent in the aqueous solution. (ii) Depression of sensitivity due to the sample having a greater viscosity than the standard solutions. (iii) Depression of the result due to a high salt content.

These interferences can be readily overcome by using one of the following techniques: (i) The method of additions. (ii) Matching the matrix of the standards with that of the sample. (iii) Solvent extraction to remove the cation to be determined from the interfering matrix. (iv) Relating the erroneous value obtained to an accurate value by using a factor determined by other means.

E. Non-specific scatter interference

This causes the enhancement of an analytical result at the ppm or sub-ppm levels due to the solution containing a high concentration of dissolved salts. The effect is due to the presence of dried and semi-dried salt particles in the flame which scatter and absorb the incident radiation from the source. Since the intensity of the transmitted radiation will be decreased, there will be an increase in the absorption signal. This non-specific scatter effect is wavelength dependent and is more pronounced at shorter wavelengths. It is most significant below 250 nm.

The effect can be overcome by one of the following techniques: (i) Solvent extraction to remove the element from the interfering matrix. (ii) Repeating the determination at a nearby non-absorbing line and subtracting the scatter reading from the signal obtained at the absorbing line. This technique has several limitations; one being that the precision at the non-absorbing wavelength will probably be worse than at the absorbing line. Also, some elements do not have close suitable non-absorbing lines. (iii) By using a deuterium background corrector. This makes use of the fact that the sample beam is ratioed to the reference beam and that a UV continuum source will behave in the same way as a non-absorbing line, in that it will enable scatter-only measurements to be made. By arranging for it to be out of phase with the sample beam, scatter readings will automatically be subtracted from the erroneous absorption measurement at the detector stage.

F. Ionisation interferences

To understand ionisation interferences, it is necessary to appreciate what is occurring in the flame during the aspiration of a sample. The flame is being used as a source of energy to convert elements in the solution droplets created by the nebuliser into ground-state atoms.

$$MX \xrightarrow{E} MX \xrightarrow{E} M^0 + X^0$$
solution　　　　salt　　　　　ground-state atoms
　　　　　　particles

Many determinations require the use of the nitrous oxide—acetylene flame and it is usually under these conditions that ionisation interferences occur. They arise from the energetic nature of the flame which gives ground-state atoms but also excites some atoms to such an extent that one or more electrons are lost and ionisation occurs.

$$M^0 \xrightarrow{E} M^+ + e^-$$

This effect will obviously be greatest with elements having low ionisation potentials such as the alkali and alkaline earth metals, e.g. barium is approximately 80% ionised in the nitrous oxide flame. Since the ground state therefore becomes depopulated, the sensitivity will decrease.

An equally important effect arises when an easily ionised element is being determined in the presence of another. There will be an enhancement of sensitivity compared with pure aqueous standards. This arises from the presence of excess free electrons which suppress further ionisation.

$$M^+ + e^- \rightarrow M^0$$
　　　excess

This effectively increases the population of ground state atoms. In practical applications, some use may be made of this phenomenon. By adding an excess of a readily ionised salt to samples and standards, an increase in sensitivity may be achieved. Potassium chloride is usually chosen for this purpose owing to its high purity, low ionisation potential and lack of visible emission in the flame.

VII. ELECTROTHERMAL ATOMISATION, ETA

In addition to the normal requirements of a good atomic absorption spectrometer two parameters are of paramount importance when employing a furnace as the atomising source. These are: (a) the response time of the signal handling circuitry must be sufficiently rapid to capture the transient absorption signals which are characteristic of furnace atomisers, and (b) the background corrector must be extremely effective, since non-atomic absorption is more of a problem in ETA than in flame cells.

Certainly all of the development work using these devices and indeed much of today's routine analyses were performed using a chart recorder to display the analytical peaks. The value of such a dynamic visual display cannot be overemphasised, for the following reasons.

(a) The drying process can be monitored in that very often the droplet will disturb the light path and alternate it. This will show as a deflection on the chart when background correction is not being employed. As the solution is gradually dried the baseline will be recovered. Erratic pen movements at this stage would be evidence of a hasty drying stage. As a guide a 20 μl aliquot manually injected will require a drying time of about 45 s, depending upon its salt content.

(b) At the intermediate stage(s) the pyrolysis temperature will ideally be high enough to decompose the sample matrix without volatilising the analyte. Again the effectiveness of this stage can be monitored using a chart recorder. Figure 4 shows an ideal trace. The drying stage has been accomplished steadily and the non-atomic peak corresponding to the pyrolysis products of the matrix is clearly resolved from the final atomic signal from the analyte upon atomisation.

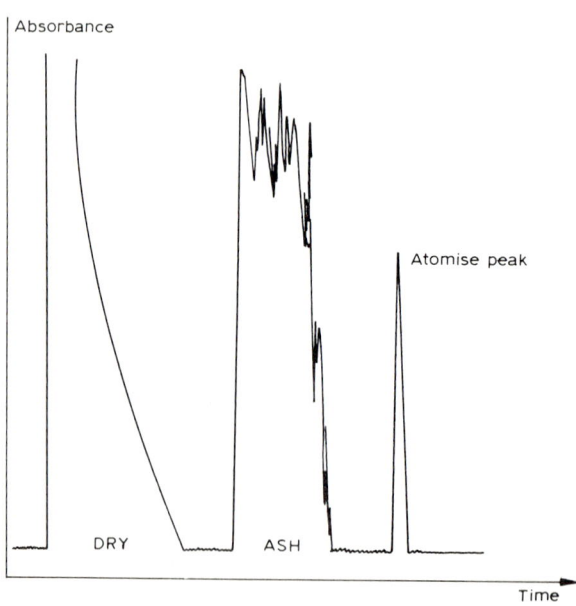

Fig. 4. Chart trace illustrating dry, ash and atomise events with background corrector off.

(c) Changes in peak shape of the atom signal can be spotted with a visual display provided the time constant is adequately small. For a chart recorder the time for full scale deflection should be 0.2—0.3 s. An oscilloscope is, of course, the ideal, with an effective time constant of zero.

As instruments have developed, the ability to capture transient signals digitally has been made possible in two main ways, either as a peak height absorbance or an integrated peak area absorbance. Since the exact manner in which these modes of operation function on individual instruments varies, the manufacturer's recommendations should be followed, but it should not be forgotten that the visual display, either on a chart or cathode ray tube is extremely valuable in setting up a new method or monitoring the success of an existing one from time to time.

A. Selection and optimisation of furnace operating conditions

The furnace power supply and controller enables the basic steps normally considered to occur in the flame method of atomisation to be carried out in a sequential form, each stage governed by the particular phase of the controller programme. In its simplest form the programme will consist of three readily identifiable stages, namely dry—ash—atomise. Each of these stages has to be carefully optimised to obtain the best results for any particular analysis. The importance of gradual drying has already been discussed. For the intermediate and final temperatures the construction of an ash/atomise curve may be beneficial, and of particular value when dealing with an analytical situation where no comparable studies can be found. The first stage is to prepare an aqueous solution containing only the analyte element at a concentration that should produce an optimised signal of approximately 0.1—0.2 absorbance. The metal should ideally be present as nitrate, sulphate or perchlorate since chloride salts are often volatile and can be lost prior to atomisation in molecular form. Having prepared the standard the drying time can be established quickly and the pyrolysis temperature ignored since no matrix is present. The atomisation temperature is then varied and a graph obtained plotting atomic absorption signal versus atomisation temperature. Many elements will produce a graph which increases with temperature and then reaches a plateau which indicates that there is no advantage in going beyond a certain atomisation temperature. Indeed, unnecessarily high temperatures are to be avoided in any case to maximise tube life and to minimise blackbody emission breakthrough to the photomultiplier.

A temperature which is on the plateau part of the graph is selected and fixed and the intermediate temperature gradually raised starting from 150° C with an arbitrary time of 30 s. The atomisation time will normally be 5 s. A graph of atomise peak height versus ash temperature is now plotted, which should result in something approximating to Fig. 5. The ash curve too has a plateau region, which turns downwards as the temperature at which the

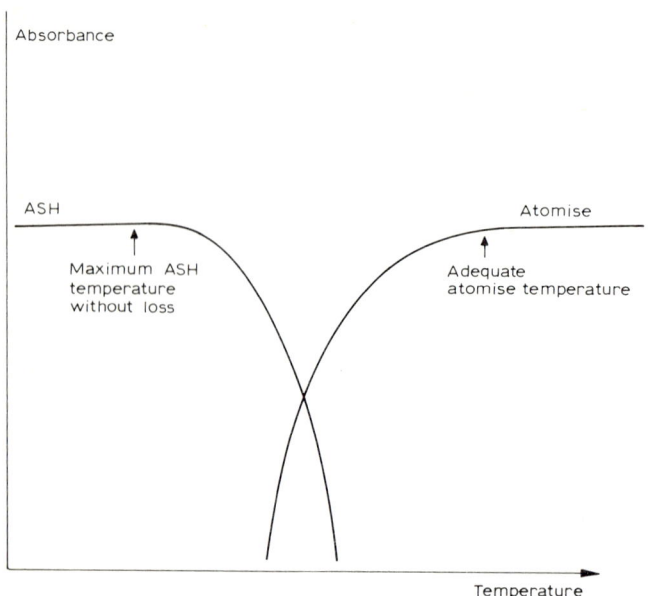

Fig. 5. Ash/atomise curves.

analyte is lost is exceeded. At this point the atomisation signal will rapidly fall, so the maximum ash temperature will be typically 100° C below this point.

One must bear in mind that should a longer ashing time be required (and this would depend on the sample matrix) then the ash curve might need modifying. Conditions must be rechecked at the longer ash time and the ash temperature changed accordingly. A successful ashing sequence will result in no smoke being evolved upon atomisation, but this is not always possible particularly when dealing with volatile metal analytes. The residual smoke formed during atomisation must be compensated for by using background correction if it is of manageable proportion (less than 0.5 absorbance) or by adjusting the sample amount to a smaller injection volume.

B. Calibration

In principle there are no differences between calibration procedures for flame and ETA methods although the latter case will take longer, as has already been pointed out. Calibration standards should match the samples as nearly as possible with respect to major components otherwise standard additions must be used. Indeed the method of standard additions will have to be used at some time to check accuracy, so this procedure will usually be tried first. If the standard additions graph is parallel to the direct calibration graph then freedom from matrix interferences would be indicated. The standard additions principle as applied to ETA is now described.

For more complex solutions and samples where it is not possible to remove the matrix during the ashing step it may be necessary to use the method of standard additions. Some workers advocate that this method should always be run initially to check for interference effects so that the best calibration procedure can be selected. If the standard additions graph and the direct calibration graph were parallel, freedom from interferences in the sample would be indicated, i.e. the element is in the same form in sample and standard immediately before atomisation, or the two forms give the same absorption response.

It is advisable that all readings be obtained in duplicate or triplicate. There are again variations of the procedure.

(1) Three or four aliquots of the sample solution are transferred to volumetric flasks and known, different amounts of a standard solution of the analyte element are added to each of these except one. The solutions in the flasks are then made up to volume. These solutions plus a reagent blank are then run in the electrothermal atomizer. Each reading should have the blank subtracted before plotting. The sample result is then itself blank-corrected.

(2) The standard additions can be made in the furnace tube itself thus avoiding dilution of the sample and possible contamination from volumetric ware. The selected volume of sample is injected into the tube, dried and ashed, the programme being stopped before atomisation. After cooling, a known amount of the analyte element is injected on top of the ashed sample and the complete atomisation programme run through. This is repeated for two more different additions and, of course, for the sample with no addition. The standard addition graph is plotted in the normal way, absorbance peak versus amount of added analyte element, and the amount of the element in the sample of different concentrations, different volumes of the same concentration, or multiple additions of the same concentration. The drying step should be run between multiple additions.

The validity of the method of standard additions depends on the forms of the analyte element in the sample and in the added standard responding in the same way during the atomisation step. This may not always be so in practice. For example, in determining lead in whole blood by this method it would have to be proved (or assumed) that lead added as a lead nitrate solution is atomised to the same extent as lead bound organically in the sample. This is most easily checked by running a certified standard material through the procedure. If no such reference standard exists, as with the above blood example, the sample should also be run after pre-treatment for removal of the matrix, e.g. wet or dry ashing, and the results compared.

Two further limitations apply to the use of the method of standard additions. Simultaneous background correction must be employed because of possibly varying amounts of matrix material present in the tube during the several firings needed to make one determination. Furthermore, all readings must be within the linear portion of the calibration graph in order that meaningful results may be obtained. It is, of course, perfectly permissible to use a linearisation function, either as provided in the atomic absorption spectrometer itself or as may be devised with a programmable desk-top calculator.

C. Sample handling and preparation

The increased sensitivity which is the main feature of electrothermal atomisation methods introduces a number of difficulties connected with the handling and preparation of samples. Some practical guidance on the avoidance of errors through contamination and on the choice and use of micropipettes is set forth in the following subsections. The analyst must appreciate, however, that we are dealing with a technique of ultramicroanalysis, and any advice or experience that he can make use of on that subject will be entirely relevant here.

1. *Avoidance of contamination*

The importance of clean-air rooms, or at least clean areas and benches, is a point often omitted from the earlier literature.

Atomic absorption with electrothermal atomisation typically involves measurement of less than 1 ng of an element during each analytical sequence. At such extremely low levels, contamination of the apparatus with detectable amounts of common elements is a severe problem. Contamination may occur at any stage in the procedure. It may arise from the reagents used in sample preparation, from the vessels used during preparation, or from the laboratory atmosphere at any stage in the procedure even when the sample is actually situated within the graphite tube. Sodium, magnesium and zinc are often detected as contaminants after conventional laboratory washing procedures, and zinc particularly is often found in cleaning agents. Other elements which frequently figure as contaminants are iron, copper and potassium. If the laboratory deals with a particular type of sample, then the matrix of the sample is a potential contaminant.

The precautions recommended to avoid contamination are detailed below and are divided into two sections, those considered essential to enable electrothermal atomisation to be carried out successfully, and those considered desirable.

2. *Essential precautions*

(1) All glass or plastic vessels to be used for electrothermal atomisation work should be washed, rinsed and then soaked in 2% v/v nitric acid for at least 24 h and then thoroughly rinsed in high purity deionised or double-distilled water.

(2) A 'clean' bench area should be reserved for solution preparation for use with the furnace.
(3) The volumetric and storage ware used for solutions for electrothermal atomisation should be kept separate from apparatus used for conventional laboratory work.
(4) Solutions of low concentration should be prepared immediately before use and after preparation should be transferred to a suitable plastic container for storage.
(5) When solutions of the same concentration of an element have to be prepared regularly on a routine basis it is advisable to keep the same apparatus for the same solutions.
(6) Efficient fume extraction.
(7) A high purity water supply, either deionised having a minimum resistivity of $10\,M\Omega\,cm^{-1}$, or double-distilled. It is preferable to produce water as it is required rather than to store it for later use. For this purpose a deionising system is usually more convenient.
(8) Micropipette tips may introduce contamination. If this is excessive it may be necessary to soak the tips in dilute nitric acid and then wash with high purity water before use. In any case, with each solution, the micropipette tip should be washed through twice with injections which are discarded into a beaker before it is used for the actual injection into the graphite tube.

3. Desirable precautions

(1) The complete electrothermal atomisation system should, preferably, be in a room separate from the general laboratory, and well away from sample preparation procedures.
(2) The room should be under a small positive pressure, supplied from a pump system with filtration for dust particles.
(3) The room should not be used by personnel as an access room for other parts of the laboratory.

Plastic ware

All solutions should be stored in plastic bottles, as glass vessels usually give greater contamination and also adsorption of the required element from the solution onto the glass surface. Polypropylene and polyethylene are the best general purpose storage vessels, but even these cause solution deterioration, noticeably with solutions having concentrations below the ppm level. This effect is activated by acidic solutions and soaking in water is preferable to dissolve any contaminating salt. If soaking in acid is necessary then acid concentration no greater than a few percent v/v should be used. Polystyrene vessels should not be used because they are particularly susceptible to adsorption effects. The best storage vessels are those made from PTFE or FEP materials, which give very few problems from contamination or formation of bonding sites. These are very expensive, however, compared with polyethylene and polypropylene.

Micropipette tips

Micropipette tips may give contamination from materials used in construction or from packing, but more usually contamination comes from handling in the laboratory. For this reason it is very good practice to put the tip on the micropipette from within the supplier's packet by handling the outside of the packet only. Also, when the pipette is put down between operations it is essential to rest it in such a way that the tip does not come into contact with any bench surfaces or other objects. Use of the same tip for many injections is acceptable providing one tip is used for only one solution as cross-contamination may otherwise result. It is good policy to dispose of the tip when changing solutions so that it is not left for later use, by which time it could be contaminated by contact or by atmospheric dust.

In some cases the dyes used for colouring tips have been found to contain contaminants. Pink tips have been found to contain cobalt, yellow tips may contain cadmium, and white tips may contain lead. This source of contamination is now less common and should not cause any significant problems if the tips are purchased from a reputable manufacturer.

As a check on contamination from this source it is essential, of course, to make at least two injections from each solution in order to check reproducibility.

4. *Use of micropipettes*

Unless a special type of autosampler can be used, micropipettes are an essential part of electrothermal atomisation techniques, as they give one of the most reliable methods of introducing small volumes of liquid samples into the graphite atomizer.

The pipettes are all based on air displacement with a simple plunger and are provided with non-wettable plastic (usually polypropylene) disposable tips to contain the solution, preventing any contamination of the pipette itself. Most micropipettes have a double action plunger system, i.e. calibration and overshoot positions, which ensures that the sample is completely dispensed. There are many manufacturers offering a complete range of volumes; in addition some have available a selection of pipettes with adjustable volumes.

Choice of micropipette tip

The tips used with the micropipette contribute greatly to the accuracy of the analysis and different types are better suited to certain uses and volumes. In general, the 'dead' air space between the sample and the plunger seal should be kept to a minimum since this can seriously affect both accuracy and precision, due to the air expanding in response to even very small changes in temperature, such as could be produced by a loose or a firm grip.

The dart-like tip is ideal for injections of smaller volumes, e.g. of 50 µl and less. This allows only a small 'dead' air space, and the tip reaches almost to the surface of the graphite tube to deposit the sample. A more cylindrical tip is best suited for higher volumes.

Operating technique

Correct operation of a micropipette is essential to enable volumes to be dispensed precisely. The plunger of the micropipette should be depressed slowly but firmly to the first stop position, and the tip just inserted in the solution to be tested. The plunger should be allowed to rise carefully over a period of 3—4 s. To inject the sample into the atomiser, the plunger is slowly (3—4 s) depressed fully to expel the sample completely. The pipette and tip should be completely removed from the tube before the plunger is gently released to avoid taking the sample back into the tip.

Accuracy and precision

Accuracy is usually within 1% of the stated volume, and most manufacturers will specify each pipette to be accurate within this limit. It is, however, important to maintain the instrument in good order according to instructions.

Since atomic absorption is a relative technique, relying on standard solutions for calibration, the absolute accuracy of the micropipette is not critical providing the same one is used to dispense both known standards and unknown samples; the prime requirement is therefore the ability to reproduce the volume, i.e. good precision. Precision should be within 1% RSD with volumes down to 20 µl, but will only be within 2% RSD at 10 µl. Volumes below 10 µl should not be used if at all possible because of the loss of precision. These figures represent results of sampling very dilute aqueous solutions. As the total dissolved solid content of the solution rises, so the viscosity will increase, and as this becomes significant the precision will deteriorate. The technique of reverse pipetting gives better precision but is not suited to injection in a graphite tube.

Pipetting precision of blood or serum is improved by taking up a volatile liquid, e.g. n-heptane, between the bottom and first stops of a double action micropipette. The sample is taken up in the calibrated part and both are then injected into the furnace. The heptane washes out the sample completely and is evaporated off at the drying stage.

5. *Injection of organic solvents*

Standard tips are made from polypropylene which has been treated to be non-wetting for aqueous solutions. When non-aqueous solvents such as ketones, alcohols and chlorinated hydrocarbons, etc. are to be handled, however, the precision or repeatability is very much poorer, anything from 10% or worse. There are two main reasons for this. The non-wetting coating or finish is rapidly destroyed by organic solvents and, although some tips might work properly, it is likely that droplets of liquid will remain behind on the inner walls of the tip. One recommended procedure is to pre-wash the tips in the solvent to get uniform performance; this not only destroys the coating, however, but encourages droplet formation. Also, when standard tips are used for volumes between 1 μl and 20 μl, there is a comparatively large dead space above the liquid. This results in premature sample ejection, due to the build up of solvent vapour pressure in this space which forces the liquid out. Low boiling point solvents such as chloroform are particularly prone to this problem.

Choice of solvent

The problem can be alleviated to some extent by careful choice of solvent. Chloroform could be replaced, for example, by 1,2-dichloroethane which has a much lower vapour pressure. Avoidance of solvents of low boiling point reduces the problem of expulsion of the sample from the pipette tip, but does not prevent the formation of droplets within the tip.

Choice of tip type and material

The dead-space above the sample in the dart-like tip is much reduced by the use of the type of micropipette which uses fine capillary tips. The Oxford Ultramicro-sampler is an example. Use of this device completely overcomes the solvent expulsion problem. The principal disadvantage from the point of view of electrothermal atomisation with a graphite furnace is that its maximum capacity is 5 μl. This may be too little for the sensitivity of some elements. The tip material provided with this syringe is still prone to droplet formation, but this can be replaced by PTFE capillary tubing of the correct dimensions, e.g. Polypenco size TW 24, which appears to overcome the problem completely. Good precision should be attainable with tips made from this material, provided the tips are cut across at 90° to the tube axis. Chamfered tips give rise to a variable position of the meniscus with consequent loss of reproducibility.

6. Sample preparation

In many analyses, preparation will consist either of dilution of an already liquid sample or dissolution of a solid sample followed by dilution. Contamination introduced by the diluents or solvents used can be significant. The need for high quality deionised water has already been emphasized. Normal analytical reagent-grade solvents and reagents are generally not sufficiently pure and materials of a higher degree of purity, e.g. BDH 'Aristar', Merck 'Suprapur' grades or their equivalent, must be used. Sometimes solvents even in these grades give rise to a significant blank and this can only be reduced by further distillation.

Minimum quantities of acids are needed if dissolution is carried out in a PTFE pressure dissolution vessel which is capable of operating at temperatures up to 180° C. With organic matrices, however, only the special versions of this apparatus capable of withstanding very high pressures should be used. For completely inorganic dissolutions, there is also an all-PTFE vessel, which is limited to temperatures below 120° C, but which completely removes the risk of contamination from the metal outer-casing.

7. Chemical separations

The types of separation procedure described elsewhere in this book for the improvement of sensitivity and for matrix separation in flame atomic absorption analysis can, in principle, be employed for the same purposes before electrothermal atomisation.

In solvent-extraction methods only reagents and solvents of the very highest purity can be used and it is most important that blank extractions should be run wherever appropriate. Electrothermal atomisers allow a greater degree of flexibility than the flame, because the solvent does not have to be flame-compatible. The whole range of organic solvents can thus be considered. In particular, the separations using APDC and 8-hydroxyquinoline may, under the correct conditions, be more successful using chloroform than methyl isobutyl ketone which is recommended for flame work. Thus, the best solvent for a particular separation can be employed, the only limitation being imposed by the difficulties of handling organic solvents with micropipettes.

It has been confirmed that elements in compounds in organic solvents usually give the same response in electrothermal atomisation as the same element in an aqueous standard solution. This is because the solvent is removed at the drying stage and most organic complexes are converted to stable inorganic compounds at low ashing temperatures.

Ion exchange separations may also be done in preparation for electrothermal atomisation. An interesting variant on this method is where the resin itself, containing the bonded analyte element, is subjected to direct analysis in the solid phase. In one example one litre of seawater is passed through 500 mg of chitosan (a natural chelating polymer). The resin was then homogenized and 5 mg samples of this were analysed for vanadium. Response of vanadium from the resin and from aqueous standards was shown to be the same.

VIII. ATOMIC EMISSION SPECTROMETRY

The reader will recall from Chapter 1 that when a specific atom in the ground state E_0 absorbs energy in discrete amounts E it is raised to the excited state E_1. This discrete amount of energy is called a quantum.

$$E_0 + E \longrightarrow E_1 \qquad (2)$$
$$\text{ground state} \quad \text{quantum} \quad \text{excited state}$$

$$E_1 \longrightarrow E_0 + E \qquad (3)$$
$$\text{excited state} \quad \text{ground state} \quad \text{photon}$$

Expression (3) shows that when the excited atom returns to the ground state radiant energy will be given off equivalent to E. This quantum of energy is called a photon. Atoms are excited when energy is absorbed and can emit radiant energy when returning to the ground state. The photons of energy emitted are characteristic of the excited atoms in the system. This reality forms the basis of the technique of atomic emission spectrometry.

A flame emission spectrometer therefore consists of an atom source, a monochromator and detector and is therefore simpler instrumentally than the corresponding atomic absorption system. Particular developments engendered by atomic absorption have restimulated interest in flame emission spectrometry after a dormant period. Chief of these is the use of the nitrous oxide—acetylene flame which is sufficiently hot to stimulate thermal atomic-emission from a wide range of metal elements.

The technique of atomic emission does not require a hollow-cathode lamp for the analyte element and therefore an analysis may be contemplated for elements for which the laboratory does not possess a hollow-cathode lamp.

Flame emission techniques generally, require that the monochromator be more strongly resolving than in atomic absorption techniques. In other words the analyst has to use narrower slits in the monochromator. The most effective way to find the best slit to use in a particular instance is to measure the signal-to-noise ratio, that is the percent relative standard deviation for a series of measurements on a suitable standard with each of the slits that are available. The analyst may be surprised to find that in practice the best precision is very often found with extremely narrow spectral bandwidths. This is because unwanted spectral interference is filtered out from the flame.

The following elements have better detectability in simple solutions by flame emission techniques compared with atomic absorption techniques: Ca, Ba, Y, La, W, Re, Ir, In, Al, Sn, most rare earths, all alkali metals.

Chapter 4a

Water and effluents

B. J. FAREY and L. A. NELSON*
*Directorate of Scientific Services, Thames Water Authority,
Roseberg Avenue, London EC1 4TP (Gt. Britain)*

I. INTRODUCTION

A. Characteristics of water and effluents

Water is often classed as the universal solvent because most elements and their compounds dissolve to some extent in it. Further, compared to other matrices, water has a high dilution for all substances held in solution. However, waters reaching a water examination laboratory do show a wide range in quality and composition of dissolved and suspended material. Organic carbon concentrations, for example, can vary from around $100\,\mathrm{mg\,l^{-1}}$ in a crude sewage effluent to $1\,\mathrm{mg\,l^{-1}}$ in a well water used for public water supply. Similarly, a ground water from chalk strata will have a significant hardness with calcium levels at around $100\,\mathrm{mg\,l^{-1}}$ whereas soft surface-waters from some upland catchment areas in the U.K. have calcium values below $5\,\mathrm{mg\,l^{-1}}$. Trace elements in water show similar variations; strontium levels in the London chalk groundwater change by an order of magnitude from the outcrop to the basin. In addition to this trace elements are often present at relatively high levels in natural waters due to anthropogenic input; trade effluents and sewage effluents are frequently enriched in heavy metals.

B. Speciation of metals in water

Owing to the wide variation in water quality, dissolved metals are present in varying speciation. Metals frequently occur in the ionised form as hydrated cations; however, they may be present as oxyanions or complexed to varying degrees with anionic ligands. They often occur in different oxidation states depending on the redox potential of the water and, in waters with a high organic content such as crude sewage effluent, a significant fraction may be complexed with organic compounds or present as organometallic species. In waters with increased suspended solids, much of the metal may be adsorbed on the suspended material, or the metal itself may be present as an insoluble species.

---

* Present address: Institute for Marine Environmental Research, Natural Environment Research Council, Plymouth PL1 3DH, Gt. Britain.

TABLE 1

LIMITS ($mg\,l^{-1}$) FOR METAL LEVELS IN SURFACE WATER INTENDED FOR THE ABSTRACTION OF DRINKING WATER

(Offical Journal of the European Commission. Council Directive 16 June 1975.)

Metal	Treatment subsequent to abstraction					
	Simple physical, disinfection		Normal physical, chemical and disinfection		Intensive physical and chemical extended treatment and disinfection	
	Guideline	Mandatory	Guideline	Mandatory	Guideline	Mandatory
As	0.01	0.05		0.05	0.05	0.1
Ba		0.1		1		1
Be*						
Cd	0.001	0.005	0.001	0.005	0.001	0.005
Co*						
Cr (total)		0.05		0.05		0.05
Cu	0.02	0.05	0.05		1	
Fe (dissolved)	0.1	0.3	1	2	1	
Hg	0.0005	0.001	0.0005	0.001	0.0005	0.001
Mn	0.05		0.1		1	
Ni*						
Pb		0.05		0.05		0.05
Se		0.01		0.01		0.01
V*						
Zn	0.5	3	1	5	1	5

* These elements are required to be determined in water but limits have not yet been decided.

C. Public health aspects

Because many heavy metals are toxic to man, animals and plants, it is necessary to monitor continually potable water, river water and trade and sewage effluents to check that the metal levels are below the predefined safe limits. In this way, water quality is preserved and the health of the population is safeguarded. It is because of the public health aspects that toxic limits for metals in surface waters have been introduced and those set by the European Economic Community are displayed in Table 1. It can be seen that all limits are at the trace level and in order to comply with these directives there will be a need for the regular analysis of raw and potable waters for these metals. This is the function of quality-control water laboratories.

D. Atomic absorption spectrometry as applied to the analysis of waters and effluents

Atomic absorption spectrometry (AAS) is a rapid, versatile technique for the determination of metals in water. Consequently, it is highly suitable for use in a quality-control laboratory analysing large numbers of samples per day. Since water samples can be introduced directly into an atomic absorption spectrometer, often only a minimum of pretreatment is necessary. For certain classes of samples, however, a wet-ashing pretreatment may be essential. Also, in some determinations, a preconcentration method can be used to increase the sensitivity of the analytical technique. A further advantage of atomic absorption analysis is that interferences, though important, are few and well documented and therefore it is ideally applicable to the analysis of the wide range of samples encountered.

II. SAMPLING AND STORAGE

Sample collection is an important stage in the monitoring of waters for trace elements because, by transferring water from its natural environment into a sampling container, the dynamic equilibrium of the chemical and biological species may be altered. Contaminants can leach out of the vessel into the sample; and losses of analyte can occur by interaction with the container walls and also by other means. As a consequence careful consideration must be given to the particular sampling containers employed; furthermore, the addition of specific reagents is usually necessary to ensure the stability of the analyte in solution. Finally, the analyst often wishes to distinguish between dissolved and particulate metals in the sample. Because of this a filtration stage will be necessary prior to analysis and this is done immediately following sample collection. The various techniques for sample collection and storage of waters and effluents for trace element analysis are briefly described in this section. A comprehensive treatise on this subject can be found in Batley and Gardner [1].

A. Sample collection

In the collection of water and effluents for trace metal analysis, a polyethylene bucket with nylon line suffices. The contact time of the sample with the container is short and therefore the danger of contamination from the bucket is small. Even so, it is prudent to wash the bucket with 20% nitric acid followed by vigorous rinsing with distilled-deionised water prior to using it for sample collection. The container should also be rinsed with sample immediately before collection of actual sample.

B. Filtration

When the level of dissolved as opposed to total metal is required in the sample, it is necessary to filter the water to remove suspended matter. This should be done as soon as possible following sample collection. The reason for this is that in unfiltered samples contact of the dissolved fraction with particulate matter for extended periods of time is likely to lead to changes in the distribution of chemical forms of heavy metals in solution. Generally it is now accepted that the fraction which passes a 0.45 μm pore-size filter defines the dissolved metal fraction. A number of commercially available membrane filters are available for filtering waters. For trace-metal analysis of water samples, polycarbonate or cellulose ester filters under the brand names Nuclepore, Sartorius, Millipore and Gelman are commonly used. Although these filters normally do not represent a significant cause for sample contamination, filters should be washed initially with a small volume of dilute nitric acid, then with distilled-deionised water and finally with a portion of the sample to be filtered. The filtration apparatus supplied commercially with the filter should be washed likewise. Pressure filtration using compressed air or pressurised nitrogen gas is the most convenient and rapid method of filtration and can be done at the site of sampling.

C. Sample preservation

Losses of metals from dilute aqueous solution on storage are well documented. To prevent this it is usually necessary to acidify the sample after collection and filtration to pH 1. If the sample is to be analysed subsequently by flame AAS hydrochloric acid should be used (Section III.C.2); alternatively, prior to flameless electrothermal atomic absorption analysis, nitric acid should be added to preserve the sample (IV.B). The type of storage container is also important and high-density polyethylene is the preferred material for sample bottles. Here the adsorptive losses of metals appear to be lower than on glass. To avoid container contamination of the sample the container should be leached with dilute nitric acid for several days prior to use. This will remove surface contamination from the container material. Subsequent to the acid-leach, containers are washed in distilled-deionised water and then with a portion of the sample. Storage of samples for mercury analysis requires special conditions and these will be discussed later.

III. Analysis

The choice of analytical procedure for the determination of metals in water using AAS is dictated by two important factors. (a) The suspended solids and organic content of the water determines whether a pretreatment is necessary. (b) The particular metal and its level of concentration in the sample decides whether a pre-concentration technique is required. In addition, this factor may be useful in choosing between employing a flame or flameless mode of analysis.

A. Sample pretreatment

The pretreatment, where necessary, of water and effluent samples prior to metal determination usually takes the form of a wet-ashing procedure. This is required to oxidise both the particulate and dissolved organic matter and release metals combined and complexed with organic compounds into solution in the inorganic form. Therefore, wet ashing is necessary for organic-rich water samples and waters with high levels of suspended solids. This includes many sewage and trade effluents and raw waters. It should be mentioned here that a wet-ashing procedure is often essential prior to flame AAS since it is physically impossible to aspirate samples high in suspended matter directly into the flame of an atomic absorption spectrophotometer, because of possible blocking of the nebuliser. Furthermore, high levels of dissolved organic material and suspended solids in the sample may introduce a non-specific scatter interference in the analysis.

Described below are two wet-ashing techniques for water and wastewater samples prior to AAS.

1. *Nitric acid method*

Transfer 50—100 ml of the sample to a beaker and add 5 ml of concentrated nitric acid. Evaporate the sample to near dryness on a hotplate ensuring that it does not boil. Cool and add another 5 ml of concentrated nitric acid. Cover the beaker with a watchglass and return to the hotplate. Continue heating, adding additional acid as necessary until digestion is complete; this is indicated by a light-coloured residue. If this is particularly difficult to obtain, an addition of a small quantity of 100 vol. hydrogen peroxide (2—3 ml) together with the nitric acid will assist in the digestion. Following this, add 1—2 ml of concentrated hydrochloric acid and warm the beaker slightly to dissolve the residue. Wash down the beaker walls and watchglass with distilled water and filter the sample to remove silicate and other insoluble material. Adjust the volume to 50 to 100 ml prior to analysis.

2. *Sulphuric acid—nitric acid—perchloric acid method*

Take a 100 ml sample of water in a beaker. Add a crystal of sodium sulphite (this minimises the loss of hexavalent chromium as chromyl chloride on heating). Add 1 ml of concentrated sulphuric acid, heat the sample on a hot plate and evaporate until fuming. Cool. Add 25 ml of distilled water and 2 ml of concentrated nitric acid. Evaporate to near dryness on a hot plate. Cool. Add 25 ml of distilled water, 1 ml of concentrated nitric acid and 1 ml of perchloric acid. Evaporate on a hot plate to near dryness. Cool. Take up residue into 1% hydrochloric acid and filter if necessary.

It is to be noted that perchloric acid should be handled with due respect

References pp. 93—94

for its hazardous properties. These have been well documented in an article by Rooney [2]; see also Chapter 2 Section II. The essence of a successful perchloric-acid based wet-oxidation is to arrange for the redox potential of the solution to rise from a level where the most labile materials are oxidised to the final fuming perchloric acid state. In the above method this is achieved by initially carbonising the sample with sulphuric acid and subsequently oxidising the residue with nitric acid and then with a mixture of perchloric and nitric acids.

B. Preconcentration

Preconcentration of metals from water samples prior to atomic absorption spectrophotometric analysis increases the sensitivity of the analytical technique. It is only necessary when the level of metal required to be determined in the unconcentrated sample is below that which can be confidently measured by direct aspiration.

There are four widely used methods for preconcentrating trace metals from water, namely: evaporation, chelation—solvent extraction, ion-exchange and coprecipitation.

1. *Evaporation*

In this method preconcentration is achieved by evaporating the sample to low volume under a heat lamp or on a hot plate. An obvious disadvantage of this technique is that in addition to the analyte being concentrated, other species which might interfere in later analysis will be concentrated with it. The method is also subject to contamination throughout the evaporation period and furthermore, losses of volatile elements may occur. Notwithstanding, the technique has an application in water analysis where waters are relatively 'clean' with a low total dissolved solids content.

2. *Chelation—solvent extraction*

This is the most commonly used technique for the preconcentration of trace elements from water. The principle of the method is that the analyte, after being complexed with a chelating agent is extracted into a solvent. The volume reduction which results from the solvent volume being less than that of the sample gives a preconcentration factor. Furthermore, since the solvent often enhances the sensitivity of the flame atomic absorption technique, due to greater atomiser efficiency, the total increase in sensitivity is greater than that due to preconcentration alone. An additional advantage of the technique is that the analyte is removed from any interfering matrix in the sample. A useful guide to the application of chelation—solvent extraction procedures on water and effluent samples prior to atomic absorption analysis is given by Parker [3].

A variety of solvent-extraction systems have been developed and these are applicable to the determination of many elements; however, all chelation—solvent extraction schemes must satisfy a number of criteria. For effective extraction, a stable metal chelate should be formed which has low solubility in the aqueous phase but high solubility in the organic phase. The organic phase should have limited solubility in water but it must be suitable for combustion in an atomic absorption apparatus if the flame technique is to be used. As a consequence there are three restrictions placed upon organic solvents which can be burned satisfactorily in atomic absorption instruments. (a) Highly volatile solvents are unsatisfactory. These volatilise during and after nebulisation. The resulting expansion in the nebuliser causes an increase in the velocity of the gases through the burner slot and the flame is frequently extinguished by lifting off the burner head. (b) Chlorinated hydrocarbons are unsuitable. These are only partly combustible giving rise to poisonous gases such as phosgene in the flame. (c) The use of aromatic hydrocarbons as solvents is unwise. These burn with a sooty flame. The resulting reducing conditions created in the flame affect the analyte absorbance, raising the noise level and causing blockage of the burner slot.

The most effective solvents for use in atomic absorption are medium weight, low volatile aliphatics, alcohols and ketones. Frequently used solvents are methyl isobutyl ketone (MIBK) and ethyl propionate. These solvents have viscosities and surface tensions such that the efficiency of nebulisation is increased.

Dithiocarbamates are predominantly used as chelating agents in solvent extraction systems. Diethyl dithiocarbamate (HDDC) and its salts and derivatives are suitable for extraction of many elements into MIBK. Being a weak acid, HDDC is undissociated in acid solutions and is consequently unable to form chelates; as a result it is unsuitable for extractions from solutions with pH < 3. Similarly, sodium diethyl dithiocarbamate (NaDDC) is readily soluble in the aqueous phase but it is less effective below pH 3. Diethylammonium diethyl dithiocarbamate (DDDC) may be used for extractions from highly acid solutions up to pH 1. Ammonium pyrrolidine dithiocarbamate (APDC) is the most widely used chelating agent for atomic absorption. Used with MIBK it will efficiently extract a large number of metals over a wide pH range. In addition to the dithiocarbamates, 8-quinoline (8-hydroxyquinoline, oxine) has been used for extraction into ethyl propionate [4].

(i) *MIBK—APDC extraction*

Reagents. Solvent: methyl isobutyl ketone (MIBK). Chelating agent: ammonium pyrrolidine dithiocarbamate (APDC). Make up a 1% (m/v) solution of APDC in water. Prepare fresh daily.

Method (for a ca. 20 times concentration factor). Measure 200 ml of sample into a beaker, adjust the sample to the appropriate pH (see note 1) with dilute hydrochloric acid or 10% sodium hydroxide. Pour the mixture into a separating funnel. Add 4 ml of the APDC solution followed by 10 ml of MIBK. Shake for two minutes and allow the layers to separate (five minutes). Run off the aqueous layer to waste and finally draw off the organic layer into small stopped glass tubes to avoid the solvent evaporating. Aspirate into the flame of an atomic absorption spectrophotometer as soon as possible.

Note that: (a) The pH adjustment of the sample is specific to the particular element to be determined. See Table 2. (b) Standards and blanks are extracted under the same conditions as the samples. (c) It is worthwhile extracting an additional solution of the analyte into a larger volume of MIBK (50—100 ml). This MIBK solution is used specifically in optimising flame conditions and analyte sensitivity and allows the other extracted analytical standards to be used solely for calibrating the atomic absorption instrument. This analyte solution should be prepared so that the extracted optimisation solution gives a significant absorption on the spectrophotometer.

The APDC—MIBK extraction system is widely used to determine a variety of metals in water. In both the U.K. [6, 7] and the U.S.A. [8] it is the standard method for the determination of lead and cadmium in water. It is also used as a standard method [8] in the determination of hexavalent chromium. In order to determine total chromium, trivalent chromium is oxidised to hexavalent chromium by bringing the sample to the boil and adding sufficient potassium permanganate solution (0.1 N) dropwise to give a persistent pink colour while the solution is boiled for 10 min.

There have been a number of developments in the APDC—MIBK extraction system. In one of them Kinrade and Van Loon [9] used two chelating agents, APDC and DDDC. They maintained that the DDDC has a stabilising effect on all the metal complexes in the system. In this method, they adjusted the pH of 200 ml of aqueous sample with 4 ml of citrate buffer (1.2 M sodium citrate and 0.7 M citric acid) to around pH 5.0. 5 ml of the chelating solution (1% (m/v) each of APDC and DDDC in water) and 35 ml (or less) of MIBK are added and the extraction is carried out as described earlier. By this

TABLE 2

OPTIMUM pH FOR THE APDC/MIBK EXTRACTION SYSTEM [5]

Analyte	pH of sample	Analyte	pH of sample	Analyte	pH of sample
As	2—6	Fe	2—5	Pd	4—6
Bi	1—6	In	2—9	Pt	3
Cd	1—6	Mn	2—4	Se	3—6
Co	2—4	Mo	3—4	Te	4
Cr	3—9	Ni	2—4	Tl	3—10
Cu	<0.1—8	Pb	<0.1—6	V	4
				Zn	2—6

procedure, eight metals can be extracted, i.e. silver, cadmium, cobalt, copper, iron, nickel, lead and zinc.

An alternative extraction system [4] employs diphenylcarbazone, 8-quinoline and acetylacetone in ethyl propionate and using this as many as ten different metal ions (Ag^+, Al^{3+}, Be^{2+}, Cd^{2+}, Co^{2+}, Cu^{2+}, Fe^{3+}, Ni^{2+}, Pb^{2+} and Zn^{2+}) can be concentrated by a single extraction.

(ii) *Diphenylcarbazone, 8-quinoline, acetylacetone—ethyl propionate extraction [4]*

Diphenylcarbazone, 0.1 g, 8-quinoline, 0.75 g, and acetylacetone, 20 ml, are dissolved in ethyl propionate and the volume is made up to 100 ml.

To 100 ml of sample 10 ml of 1 M ammonium tartrate is added and the pH of the solution is adjusted to 6.0 ± 0.5 by adding dilute ammonium hydroxide or tartaric acid solution as necessary. Transfer the solution to a 150 ml separatory funnel and add 10 ml of extractant solution. Shake for one minute then allow to stand for three to five minutes. Withdraw the organic layer and store in stoppered tubes prior to analysis. The APHA [8] use a similar extraction technique for the determination of beryllium and aluminium in water employing a nitrous oxide flame in the subsequent atomic absorption analysis, and using MIBK as a solvent.

3. Ion exchange

Ordinary anion and cation ion-exchange resins are of limited use for the analytical concentration of trace elements from water, because of their lack of selectivity. This is especially so with strong electrolytes such as seawater. In this case the major ions; sodium, magnesium, calcium and strontium, are retained preferentially. However, the recent advent of commercial chelating resins based mainly on iminodiacetic acid-substituted cross-linked polystyrene, makes it possible to concentrate trace elements from waters. In consequence, a number of researchers have used chelating resins for trace-metal preconcentration from seawater and natural waters.

The chelating resin preconcentration method was originally developed by Riley and Taylor [10] for the determination of copper, nickel, cobalt, zinc and cadmium in seawater using Chelex-100 resin. They also investigated the uptake of trace elements from both distilled water and seawater by the chelating ion-exchange resins Chelex-100 and Permutit-S 1005. The resins retained the following elements with an efficiency of ca. 100%: Ag, Bi, Cd, Cu, In, Pb, Mo, Ni, rare earths, Re (90% only), Sc, Th, W, V, Y and Zn. Manganese was retained quantitatively only by the Chelex resin. The pH of uptake varied from 5—9 for each metal. The following elements were found to be removed with 100% efficiency from the resin by means of 2 N mineral acids: Bi, Cd, Co, Cu, In, Pb, rare earths, Sc, Th, Y and Zn. Ammonia (4 N) completely removes Mo, W, V and Re. It should be noted

References pp. 93—94

that in these experiments ionic tracers of the metals were used to test recoveries.

Subsequent work has shown problems with the chelating resin technique. These are: (a) Use of the resin in the H^+ form involves a decrease in the pH of the effluent because of the release of hydrogen ions, thus causing a deterioration in the efficiency of the uptake of metal ions on to the resin. The maximum efficiency is observed when the resin is converted to forms other than H^+ (e.g., Na^+, NH_4^+, K^+, Mg^{2+}, Ca^{2+}) [11]. (b) A significant fraction of Cu, Pb, Cd and Zn in seawater exists in a form which is not retained by the chelating resin Chelex-100 [12], even after acid digestion [11]. Figura and McDuffle [13] have likewise obtained incomplete recoveries for the uptake of Cd, Co, Cu, Ni and Zn from river water by columns of Ca-chelex.

Clearly, the preconcentration of trace metals by chelating resins remains in the development stage. The technique is not routine and has so far found little application to the analysis of waters and effluents.

4. Co-precipitation

In this technique the analyte elements are concentrated on a 'carrier precipitate' which is then dissolved in a much smaller quantity of solution. The main advantage of this method is that very high concentration factors can be achieved. There are, however, a number of drawbacks. (a) Separations are lengthy and time consuming. (b) The analytical solution after preconcentration contains a high level of carrier precipitate. This can give rise to non-specific scatter and matrix interference in the flame of the atomic absorption instrument. (c) Contamination from the carrier can be a problem.

Because of these disadvantages the co-precipitation technique is only of limited use in the analysis of waters and effluents.

A recently developed technique for concentrating trace metals from natural waters prior to atomic absorption analysis utilises an APDC co-precipitation and since it has been used successfully to determine trace metals in freshwater [14], we describe it below.

Tris(pyrrolidine dithiocarbamato-)—cobalt(III) chelate, the precipitate formed by adding ammonium pyrrolidine dithiocarbamate to cobalt(II) solutions, has been found to be a good matrix for preconcentrating lead and several other metals by co-precipitation. Concentration factors of 40 to 400 are available by the method. The analyte is co-precipitated on the Co-APDC from a litre of sample. The precipitate is filtered on a fine porosity glass sinter and redissolved in a small volume of 6 M nitric acid. The solution is then used for atomic absorption analysis.

The Co-APDC co-precipitation procedure for preconcentration has been shown to be suitable for the determination of lead in freshwater samples. Cadmium determination at concentrations ca. 0.1 mg l^{-1} in freshwater can also be performed using this procedure; at lower concentrations an appropriate recovery factor should be used. In addition, the method appears to be suitable for preconcentration of copper and nickel from freshwater. Reducing one litre of sample to 25 ml by the Co-APDC method provides a net concentration factor of 40. By adjustment of sample volume and the nitric acid solution volume, concentration factors as high as 400 can be obtained.

C. Flame techniques

1. *Preparation of standards*

Except in certain specialised cases it is a necessary part of the procedure of AAS for all absorption signals from samples to be directly compared with those from prepared standards. As a consequence, it is essential to calibrate the instrument for each element being determined. However, if the response of the instrument is known to be linear over a particular range for an element, it is then only necessary to prepare a top standard and half standard in the working range of the level of element in the sample. It is essential to prepare standards on the same day as the analyses are performed due to possible deterioration of the standards with time. Standards are commonly prepared from stock solutions at the $1{,}000\,\text{mg}\,\text{l}^{-1}$ level. These solutions are commercially available for use in AAS. Generally, standards can be prepared in one dilution with the use of micropipettes. Dilutions for the preparation of standards of varying concentrations using micropipettes are shown in Table 3. Working standards and blanks are acidified to the same extent as samples. All standards and blanks are usually prepared in glass volumetric flasks and it is good practice to keep these continually soaking in 10% nitric acid, when not in use, to avoid contamination. Prior to use they should

TABLE 3

DILUTION TABLE FOR $1000\,\text{mg}\,\text{l}^{-1}$ STANDARDS

(Concentrations in $\text{mg}\,\text{l}^{-1}$)

Pipette size (μl)		Volumetric flask size (ml)								
		10	25	50	100	200	250	500	1000	2000
5	1	0.50	0.20	0.10	0.05	0.025	0.02	0.01	0.005	0.0025
10	1	1.00	0.40	0.20	0.10	0.05	0.04	0.02	0.01	0.005
20	1	2.00	0.80	0.40	0.20	0.10	0.08	0.04	0.02	0.01
50	1	5.00	2.00	1.00	0.50	0.25	0.20	0.10	0.05	0.025
100	1	10.00	4.00	2.00	1.00	0.50	0.40	0.20	0.10	0.05
200	1	20.00	8.00	4.00	2.00	1.00	0.80	0.40	0.20	0.10
250	1	25.00	10.00	5.00	2.50	1.25	1.00	0.50	0.25	0.125
500	1	50.00	20.00	10.00	5.00	2.50	2.00	1.00	0.50	0.25
1000	1	100.00	40.00	20.00	10.00	5.00	4.00	2.00	1.00	0.50

be well washed in distilled-deionised water. Metal-free distilled-deionised water should be used in the preparation of the aqueous standards and blanks.

2. *Direct aspiration of sample*

The majority of flame analyses in water and effluents presents few problems. A pretreatment on the sample is performed only when necessary, as described earlier. Standards are prepared in the linear range of the analytical curve and blank solutions are also made up. It is preferable to acidify blanks, standards and samples to 1% with hydrochloric acid. Apart from acting as a preservative, it promotes atomisation of the analyte by forming volatile metal chlorides. The atomic absorption instrument is then set up and flame conditions and absorbances are optimised for the analyte. Following this, blanks, standards and samples are aspirated into the flame; absorbances are recorded and results calculated.

3. *Use of nitrous oxide—acetylene flame*

For the majority of elements commonly determined in water by AAS, an air—acetylene flame (2300°C) is sufficient for their atomisation. However, a number of elements are refractory and they require a hotter flame to promote their atomisation. Because of this, a nitrous oxide—acetylene flame (3000°C) is used for the determination of these elements. Refractory elements routinely determined in water are aluminium, barium, beryllium, chromium and molybdenum. Chromium shows different absorbances for chromium(III) and chromium(VI) in an air—acetylene flame [15] but use of a nitrous oxide—acetylene flame overcomes this. Barium, being an alkaline earth metal, ionises in a nitrous oxide—acetylene flame, giving reduced absorption of radiation by ground state atoms, however in this case an ionisation suppressor such as potassium should be added to samples, standards and blanks.

4. *Interferences*

The relatively simple procedure of analysing waters for trace metals using AAS is complicated by the fact that a number of metal determinations suffer from interferences. These interferences are well documented in the literature and there are now standard methods for overcoming them. In the main, interferences encountered in the atomic absorption spectrophotometric analysis of water can be divided into four categories; chemical interferences, matrix interferences, non-specific absorption and ionisation.

(i) *Chemical interferences*
This is the most common interference in flame AAS. The chemical interference prevents, enhances or suppresses the formation of ground state

atoms in the flame. In water samples, a frequent suppressive interference is produced by aluminium, silicon and phosphorus in the determination of magnesium, calcium, strontium and barium. Aluminates, silicates and phosphates are formed which are refractory in the analytical flame used. The standard procedure for overcoming this is to add a reagent that will preferentially form a compound with the interfering anion leading to the release of the analyte. Lanthanum chloride is used as such a release agent in the determination of magnesium, calcium and strontium in water.

The technique for the determination of calcium and magnesium in waters and effluents is described below and is based on that reported in HMSO [16, 17].

Reagents. 5 M Hydrochloric acid—1% (m/v) lanthanum. Lanthanum chloride solutions (10% as lanthanum) for the preparation of this reagent are available commercially.

Method. Depending on the calcium and magnesium content in the water, the analyst should dilute the sample such that the observed analyte absorbance falls on the linear portion of the calibration curve of the atomic absorption instrument. Prior to dilution, add 10 ml of the lanthanum—hydrochloric acid solution to the sample in a 100 ml volumetric flask. Dilute with water to the mark. Blanks and standards should also be prepared to the same concentration with the lanthanum chloride reagent. Having optimised the instrument conditions, samples, standards and blanks are aspirated into an air—acetylene flame.

For strontium determinations; samples, standards and blanks are made to 1% (m/v) in lanthanum to prevent the formation of refractory compounds. Analyses for strontium also suffer from ionisation interferences which are discussed later.

Another chemical interference encountered in AAS is that with chromium determinations where the chromium absorption in an air—acetylene flame is suppressed by iron, cobalt and nickel. In this instance, the iron interference can be minimised by the addition of 2% (m/v) ammonium chloride solution to the sample and standard solutions. This compound helps volatilise the chromium as chromyl chloride (CrO_2Cl).

An alternative technique for preventing the analyte forming refractory compounds in the flame is to add a compound to the sample which forms a preferential complex with the analyte. Ethylenediaminetetraacetic acid (EDTA) is often used since it complexes with the cation thus preventing its association with an anion that can lead to the formation of a refractory compound.

(ii) *Matrix interference*

This interference is caused by the physical nature of the matrix enhancing or depressing the sensitivity. For example, a crude sewage effluent may be

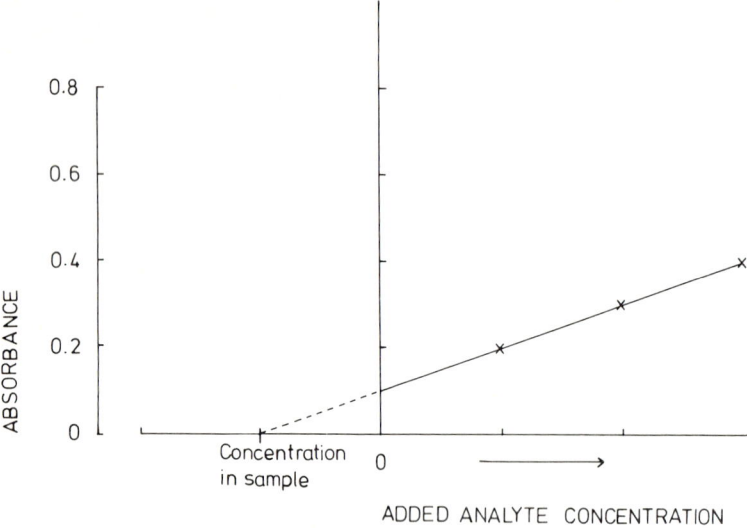

Fig. 1. Method of standard additions.

more viscous than the distilled water standard and therefore its relative flow rate into the nebuliser of the atomic absorption spectrophotometer will be slower. This will decrease the rate of atomisation of the analyte and thus lower the sensitivity. There are three techniques which can be used to overcome matrix interference. (a) The method of standard additions. In this, successive standards are prepared in the sample matrix. On drawing up the calibration curve the intercept on the concentration axis will give the analytical result. Any enhancement or depression of the signal by the matrix will be corrected for (Fig. 1). (b) Matching the matrix of the standards with that of the sample. (c) Solvent extraction or an alternative preconcentration technique which removes the cation to be determined from the interfering matrix.

(iii) *Non-specific absorption*

This is a common problem with water analysis and can give rise to an erroneously high analytical result. If the sample contains a high concentration of dissolved salts as in brackish waters or hard waters, the dried salt particles formed in the flame scatter the incident radiation from the source. Since the intensity of the transmitted radiation will be decreased, there will be an apparent increase in the absorbance signal. The non-specific scatter effect is wavelength dependent and is more pronounced at lower wavelengths. There are two ways commonly used to overcome this effect. (a) Solvent extraction to remove the element from the interfering matrix. (b) By using a deuterium background corrector.

It is standard practice to use method (b) when determining metals with absorbing wavelengths below 300 nm. A deuterium background corrector is standard equipment and is commercially available on all modern atomic absorption spectrophotometers. Alternatively, if the analyst does not possess a deuterium background corrector, the metal determination can be repeated at a nearby non-absorbing line. This will represent the scatter readings and can be subtracted from the signal obtained at the absorbing line. The resultant absorbance figure will be of the analyte only.

(iv) Ionisation interferences

This effect mostly occurs with alkali and alkaline earth metals. The low ionisation potentials of these elements cause them to be readily ionised in the flame with a resultant lowering of the population of ground state atoms and a suppression of sensitivity. The technique used to overcome this is to add an easily ionised salt such as potassium chloride to samples and standards. This ionises in preference to the analyte in the flame and enhances sensitivity. As an example, strontium, barium and aluminium are subject to ionisation in the flame. In water analyses, this is suppressed by adding potassium to the samples and standards so that the solution contains $2\,000\,\text{mg}\,\text{l}^{-1}$ potassium.

D. Specialised techniques

Mercury and arsenic and selenium and the other metalloids all show poor sensitivity when determined by conventional flame AAS. Because of this, techniques have been developed for the determination of these elements involving a specialised method of atomisation of the analyte. In this way, the sensitivity of the atomic absorption determination has been increased. These methods are now standard routine techniques in water analysis and most atomic absorption spectrophotometer manufacturers market special accessories for the determination of mercury, arsenic and selenium.

1. *Mercury by cold-vapour atomic absorption*

This is a flameless technique initially developed by Hatch and Ott [18]. The inorganic mercury in the sample is reduced with stannous chloride solution in a reaction flask. Any inorganic mercury ion in the sample is thus converted to metallic mercury which is volatilised and passed into an absorption cell using nitrogen, argon or compressed air as the carrier gas. Light from a mercury hollow-cathode lamp at wavelength 253.7 nm is diverted through the axis of the cell and absorption of radiation by the mercury atoms in the cell is thus proportional to the population of atoms present. The process is straightforward and simple to operate. A detailed description on the operation of the method can be obtained from the appropriate manufacturer's handbook.

There are, however, interferences in the method which should be mentioned.

(a) Water vapour absorbs radiation at the 253.7 nm resonance line of mercury; for this reason, passage of water vapour into the absorption cell must be avoided. This is done by the use of drying agents placed in the gas line between the reaction vessel and the absorption cell or, alternatively, by heating the cell with an incandescent lamp.

(b) Sulphide ion interferes in the stannous chloride reduction and the subsequent volatilisation of mercury. This can be overcome by using a suitable pretreatment on the sample.

(c) Some volatile organic substances also absorb in the ultraviolet region and such absorption can constitute a significant error in the analysis. In order to correct for this error it is necessary to run a sample analysis omitting the stannous chloride reduction. If a signal is observed this will indicate organic interference. To correct for the interference, this absorbance is subtracted from the absorbance value following sample reduction. Alternatively, the deuterium background correction lamp can be used to compensate for the organic interference.

In addition to analytical interferences, there are two important problems with mercury determinations in water.

(i) *Mercury loss from aqueous solutions on storage*

A very rapid rate of loss of mercury from aqueous solutions can occur after water samples have been collected. This is well documented and much work has been done in investigating the loss mechanism. In consequence, various preserving agents for mercury in aqueous solution have been recommended. It is agreed that low pH values, high ionic strengths and oxidising environments help in keeping mercury in solution [19]. An effective reagent developed by Feldman [20] for preserving mercury in dilute solution is nitric acid and potassium dichromate. Results have shown that $\mu g\, l^{-1}$ mercury solutions treated with this reagent and stored in glass containers are stable for several months. Samples and standards are prepared to 5% in nitric acid and 0.01% in potassium dichromate. It is essential to store these solutions in glass containers since mercury has been found to diffuse through polyethylene [21]. Also volatile organics can leach out of plastic containers and these will interfere with the cold-vapour detection.

(ii) *Pretreatment of samples for mercury analysis*

A characteristic of the cold-vapour atomic absorption technique is that only inorganic mercury in the sample is measured. The reason for this is that the stannous chloride solution does not reduce mercury bonded to organic compounds to metallic mercury. As a consequence, organomercurials must be converted to inorganic mercury in solution prior to reduction if a total mercury determination is required. This is important for water samples since mercury in natural waters is present to some extent as stable organomercury associations. Various procedures have been reported for breaking down organomercurials in waters and effluents and most of these methods require a vigorous oxidising procedure. A sulphuric acid—permanganate digestion has been developed [22]. Organomercury associations can also be broken down by ozonisation of the sample [23] or through ultraviolet irradiation [24]. Listed below are two techniques for the pretreatment of water samples prior to the cold-vapour atomic absorption determination for mercury.

(a) Persulphate—permanganate oxidation. This technique is based on that reported by Kopp et al. [25].

Reagents. Sulphuric acid (18 N), nitric acid (sp. gr. 1.42), potassium permanganate 5% (m/v) solution, potassium persulphate 5% (m/v) solution, sodium chloride 12% (m/v)—hydroxylamine sulphate 12% (m/v) solution.

Method. Transfer 100 ml of sample to a 300 ml glass flask. Add 5 ml of sulphuric acid and 2.5 ml of nitric acid, mixing after each addition. Add 15 ml of potassium permanganate solution to each sample bottle. Shake and add additional portions of permanganate solution until the purple colour persists for at least 15 min. Add 8 ml of potassium persulphate to each bottle and heat for two hours in a water bath at 95°C. Cool and add 6 ml of sodium chloride—hydroxylamine sulphate solution to reduce the excess of oxidant prior to analysis. Standards and blanks should be treated in the same way as samples.

Notes. The method removes the sulphide interference from the cold-vapour mercury signal. Concentrations of sulphide as high as 20 mg l^{-1} S^{2-} (as Na_2S) do not interfere with the recovery of inorganic mercury added to distilled water. However, the oxidation technique suffers from chloride interference. If chloride is present in the sample it utilises oxidant and is oxidised to chlorine which interferes with the cold-vapour detection by absorbing radiation at the same wavelength as mercury.

(b) Bromination pretreatment [26, 27]

Reagents. Potassium bromate (0.1 N)—potassium bromide (1% m/v) solution: dissolve 2.784 g anhydrous potassium bromate in distilled water; add 10 g potassium bromide and dilute to one litre with distilled water. Hydrochloric acid (sp. gr. 1.18). Sodium chloride (12% m/v)—hydroxylammonium chloride (12% m/v) solution.

Method. Hydrochloric acid (5 ml) is added to 50 ml of the water solution in a 100 ml calibrated flask, an aliquot (1—3 ml) of the bromate—bromide solution is added and the flask is shaken. Following an allowed reaction time, the bromination is terminated by removing the excess of bromine by the addition of 1—2 drops of hydroxylamine reagent. The solution is then made up to the 100 ml mark with distilled water. Standards and blanks are treated in the same way as samples.

Notes. Using 2 ml of bromate—bromide reagent and 5 min reaction time at room temperature, quantitative recoveries of organically bound mercury from natural waters and effluents have been obtained. In addition, the bromination pretreatment removes the sulphide interference from the cold-vapour mercury signal; following bromination with 3 ml of reagent, sulphide concentrations as high as 24 mg l^{-1} S^{2-} (as Na$_2$S) have been shown not to interfere with the recovery of inorganic and organically bound mercury in distilled water. Unlike the persulphate—permanganate pretreatment, the bromination procedure does not suffer from chloride interference.

2. *Arsenic and selenium by hydride generation*

The analysis of arsenic and selenium by conventional flame atomic absorption presents problems. The difficulty arises from the location of their resonance lines, 193.7 nm and 196.0 nm, respectively, in the low ultraviolet region of the spectrum where acetylene flame absorption is severe. To overcome this, an argon—hydrogen entrained-air flame may be employed. However, the interferences associated with the use of this low-temperature flame are marked. Alternatively, arsenic or selenium may be separated from the sample matrix prior to the determination. This is done by converting the analyte elements into volatile hydrides using a suitable reductant. The hydride is then passed into an argon—hydrogen entrained-air flame or alternative atomiser for atomic absorption measurements. Varieties of this technique have been in use for several years and all of the major atomic absorption instrument manufacturers now offer equipment permitting modification of standard instruments for these analyses. In addition to measuring arsenic and selenium, the hydride generation method has been developed for the determination of antimony [28], tellurium, bismuth, germanium, tin and lead [29].

Any hydride atomic absorption analytical method can be thought of as being composed of three steps.

(a) Generation of the hydride. In this, either a zinc metal or a homogeneous sodium borohydride reduction is used. In each case the sample is acidified prior to reduction. For arsenic and selenium determinations using the zinc metal reduction technique, Caldwell et al. [3] recommended acidifying 25 ml of water sample with 20 ml of concentrated hydrochloric acid and 5 ml of 18 N sulphuric acid before adding the zinc metal slurry. On the other hand, Thompson and Thomerson [29] prepared solutions in 1.5 M hydrochloric acid for arsenic reduction to arsine using a zinc column. They also found that the instrument response to arsenic, bismuth, germanium, antimony, selenium and tellurium when using sodium borohydride was not very dependent on the hydrochloric acid concentration (1 to 4 M).

(b) Transfer of the hydride to the atomiser. Here either stop-flow or continuous flow techniques are used.

(c) Decomposition of the hydride to gas-phase metal atoms within the optical axis of the atomic absorption spectrometer. There are a number of modifications of the atomisation technique, one of which involves passing the hydride into a heated silica tube where it decomposes into analyte atoms [29]. This avoids dilution of the generated hydrides by the burner gases and also allows the use of an air—acetylene flame since the resonance line does not pass directly through the flame.

(i) *Oxidation state of the element*

The magnitude of the atomic absorption signal in the hydride generation technique is dependent on the oxidation state of the element in the sample. The absorbances from an arsenic(V) solution have been found to be lower than that from identical concentrations of arsenic(III); similarly solutions of selenium(VI) and tellurium(VI) gave a negligible response compared with equivalent amounts of selenium(IV) and tellurium(IV) [29]. The reason for the depression of the arsenic signal from arsenic(V) solutions has been suggested by Siemer and Koteel [31] to be due to a slower rate of reduction of arsenic(V) compared with arsenic(III). As a consequence both arsenic and selenium should be fully converted to the (III) and (IV) oxidation states, respectively, in the sample prior to hydride generation. In order to do this a pre-reductant is used. If zinc is employed as the reducing agent in the hydride generation, the prior addition of tin(II) chloride and potassium iodide to the sample is recommended for the pre-reduction of arsenic to its lowest oxidation state [30]. Similarly, tin(II) chloride is used for the pre-reduction of selenium(VI) to selenium(IV).

Accordingly, in arsenic and selenium determinations the following conditions are employed for pre-reduction. For arsenic determinations; to 25 ml of acidified sample or standard, 1 ml of 20% potassium iodide and 0.5 ml of 100% tin(II) chloride are added. At least 10 min are allowed for the arsenic to be reduced from the (V) oxidation state to the (III) oxidation state. For selenium determinations; to 25 ml of acidified sample or standard, 0.5 ml of 100% tin(II) chloride are added. At least 10 min are allowed for the selenium to be reduced to its lowest oxidation state. Note that the tin(II) chloride pre-reduction is not compatible when the sodium borohydride reduction is used in the hydride generation technique. Siemer and Koteel [31] have reported that the pre-reduction step requiring addition of $KI/SnCl_2$ solution in the arsenic determination is undesirable because of a definite matrix effect attributable to the tin(II) ion. They suggest the use of iodide ion in highly acidified solutions as a suitable pre-reductant. Ascorbate is also added to prevent the inevitable air or iron(III) oxidation of the iodide ion. 0.5 ml of the 1 M potassium iodide 1—10% sodium ascorbate solution is added to the acidified sample (3.7 M hydrochloric acid for 3.5 min pre-reduction time). Following this, arsenic is determined by hydride generation using sodium borohydride reduction.

(ii) *Pretreatment*

Since only inorganic arsenic and selenium can be reduced to the corresponding hydride, organic species of these elements should be converted

References pp. 93—94

to the inorganic form prior to determination of the total element by hydride generation. Caldwell et al. [30] have recommended a method for converting organic arsenic to inorganic arsenic. The technique involves a nitric acid—sulphuric acid—perchloric acid oxidation and is used where total arsenic levels in water are required [8]. The method is as follows. To 50 ml of sample, 10 ml of concentrated nitric acid and 12 ml of 18 N sulphuric acid are added. This mixture is evaporated to SO_3 fumes. Oxidising conditions must be maintained at all times to avoid the loss of arsenic. This is accomplished by adding small amounts of nitric acid whenever the red-brown NO_2 fumes disappear. The solution is slightly cooled, 25 ml of distilled water and 1 ml of perchloric acid are added and again the mixture is evaporated to SO_3 fumes. The solution is cooled, 40 ml of concentrated hydrochloric acid is added, and volume is brought to 100 ml with distilled water.

IV. FLAMELESS ATOMIC ABSORPTION

A. Application to waters and effluents

Flameless atomic absorption using an electrothermal atomiser is essentially a non-routine technique requiring specialist expertise. It is slower than flame analysis; only 10—20 samples can be analysed in an hour; furthermore, the precision is poorer (1—10%) than that for conventional flame atomic absorption (1%). The main advantage of the method, however, is its superior sensitivity; for any metal the sensitivity is 100—1000 times greater when measured by the flameless as opposed to the flame technique. For this reason flameless atomic absorption is employed in the analysis of water samples where the flame techniques have insufficient sensitivity. An example of this is with the elements barium, beryllium, chromium, cobalt, copper, manganese, nickel and vanadium, all of which are required for public health reasons to be measured in raw and potable waters (section I.B). Because these elements are generally at the $100 \mu g l^{-1}$ level and less in water, their concentration is below the detection limit when determined by flame atomic absorption; as a result, an electrothermal atomisation (ETA) technique is often employed for their determination.

Apart from its high sensitivity, the ETA method has a number of additional advantages in the analysis of water and effluents for metals. (a) Only very small samples (1—100 μl) are required for the analysis. (b) Since the sample is dried and ashed in the electrothermal atomiser, pretreatment procedures prior to analysis are often not required. It is possible to introduce waters containing a high level of organic matter directly into the atomiser.

B. Methodology

Since metals are determined at the $\mu g\,l^{-1}$ level in water using the electrothermal atomiser, contamination is a potential problem. Consequently, cleanliness and use of ultra-pure reagents and distilled-deionised water is essential.

Following sample collection, samples should be made 1% (v/v) in nitric acid. In addition, although most commercially available electrothermal atomisers have facilities for ashing the samples before atomisation, a pretreatment is sometimes necessary on organic and turbid waters. This ensures homogeneity of the analyte in solution and reduces matrix interference. A general pre-treatment prior to sample injection into the atomiser is as follows. 25 ml of sample are measured into a 100 ml beaker. One ml of 30% hydrogen peroxide and 100 μl of concentrated nitric acid are added. The sample is placed on a hot plate and digested at 75°C for 30 min, it is then removed and allowed to cool to room temperature. The sample is diluted to 25 ml.

Electrothermal atomisers fall into two classes, i.e. filaments and furnaces. The former category includes all devices in which an electrically heated filament, rod, strip or boat is used and where the atomic vapour passes into an unconfined volume above the viewing area; on the other hand, furnaces usually consist of an electrically heated carbon tube into which the sample is injected. The optical axis of the hollow-cathode lamp light beam passes through the centre of this tube. Electrothermal atomisers are connected to a programmable power supply such that the sample can be dried, ashed and atomised at preset temperatures and times.

For injection of the water sample into the atomiser, micropipettes are used; these are now commercially available and commonly specified to a 1% accuracy. Pipette tips are known to be contaminated with Fe, Zn and Cd, thus they should be soaked in 10% nitric acid and then washed in distilled-deionised water and sample prior to use. Accurate, precise pipetting and the correct adjustment of the drying, ashing and atomisation programme are essential factors required for a successful flameless atomic absorption analysis. When pipetting the sample, the water droplet must be positioned reproducibly on the filament or in the furnace and it should be of an optimum size such that it does not run or spit during heating. If this happens, irreproducible absorption peaks may result.

The purposes of the heating programme are as follows:
(a) Drying the sample: this should be done so that there is a gradual removal of water. Any sudden evaporation of water will displace the sample and poor precision may ensue in the determination. The drying process can either be monitored visually or through the use of a deuterium hollow-cathode lamp.
(b) Sample ashing: here, the sample matrix is either completely destroyed and removed or modified. One of the purposes of this is to avoid subsequent non-specific absorption and matrix interference in the analyte signal. Similar

to the previous process, ashing should be accomplished with minimum disturbance to the sample. Volatilisation of the analyte must be avoided at this stage.
(c) Atomisation of the analyte. Here, complete atomisation of the metal should be ensured to give maximum sensitivity and absence of "memory" effects.

Subsequent to optimisation of the heating programme, the instrument should be calibrated with aqueous standards. When a linear and reproducible calibration curve is obtained, a series of standard additions on the samples should be performed in order to elucidate whether a matrix interference is operating. This is evident when the analyte absorbance is enhanced or depressed in the sample matrix compared to the pure standard. Often however, interference is indicated by the shape of the analyte absorbance peak differing in the sample relative to that in the standard or when the analyte absorbance in the sample is irreproducible or spurious.

C. Interferences

These are often encountered in flameless atomic absorption analysis and in many cases the interfering mechanism is not clearly understood. There is much literature on the subject, however, and listed below are the types of interference most commonly met with in the ETA analysis of water.

(a) Interferences of the inter-element type. These are notable in lead analysis where the presence of variable amounts of cations and anions inhibit the atomisation process resulting in variable lead sensitivity [32].
(b) The loss of analyte as volatile halide during ashing in the presence of high salt concentrations. This will cause a reduction in the analyte absorbance and thus a loss in sensitivity.
(c) Non-atomic absorption. This is magnified for flameless as opposed to flame spectrophotometry [33]. It is due to light scattering by solid particulate species in the electrothermal atomiser and also to molecular absorption. The latter is associated with matrix species being vaporised along with the analyte atomic species and absorbing a portion of the analyte atomic resonance line emitted from the hollow-cathode lamp.
(d) In graphite rod and furnace atomiser systems, there is a tendency for some metals to react with the graphite carbon to form refractory carbides. This suppresses their atomisation and as a result the analyte absorbance is reduced.

Both chemical and instrumental means have been developed for overcoming interferences in flameless atomic absorption and these are discussed below.

(a) Background correction. For non-atomic absorption below 350 nm a continuum lamp can be used for subtracting background absorbance at the resonance wavelength of the analyte. Modern background correction lamps can correct for over two absorbance units [33]. As an alternative method, a non-absorbing line close to the analytical line can be used to measure non-specific absorption. The problem with this however, is that when the non-specific absorption is due to molecular absorption by alkali metal halides, it has been found to be strongly wavelength dependent. The steep slope of the molecular absorption bands make it very difficult to find a non-absorbing line which is close enough to the resonant line to achieve an accurate correction [34]. As a consequence a continuum source should be used wherever possible for eliminating non-specific absorption interference. Notwithstanding, it has recently been shown [35] that the conventional deuterium background correction is inadequate for removing background absorbance from the analytical signal. In this instance it was found that the apparent concentration of chromium in urine was highly dependent on the amount of background present during atomisation. For this reason, where there is a possibility of background interference in the analyte signal, it should be minimised as well as being corrected for, in order to obtain the best results.

(b) Method of standard additions. Prior to embarking on a series of analyses, the results of the determinations found using the direct comparison method should be compared to those obtained using the method of standard additions. This will indicate whether there is a matrix interference acting on the analyte signal. The standard addition only corrects for matrix enhancement or suppression of the analyte signal, it does not correct for non-specific absorption (Fig. 2). Often however, a matrix interferent does not act uniformly on the analyte absorbance at atomisation [36] and in this instance the method of additions is not applicable. Moreover, the additions technique is time consuming and lengthy. As a consequence it often is not suitable for routine application.

(c) Peak-area measurement. The interferent in the sample acting on the analyte signal may serve to delay or retard atomisation, in which case an equal amount of analyte will be atomised on each determination, but the atomisation will proceed at different rates in different matrices. Clearly, in this case, the area of the absorbance signal will be a more reproducible measurement than the peak height. This was shown by Aldous et al. [37] when measuring lead in potable waters using a Delve's cup type atomiser.

(d) Matrix modification. This is one of the principal techniques of interference removal in flameless atomic absorption. In this, the sample is treated with a reagent such that during the heating programme in the atomiser, the sample matrix is modified with the result that a more efficient atomisation is ensured. A general application of this is with samples containing high chloride concentrations where there is a danger of the analyte being lost as a volatile metal halide during the ashing process. To prevent this, oxyacids are added to all samples and standards prior to analysis. In this case, atomisation proceeds via the oxide [38]. The use of oxyacids will also encourage

References pp. 93—94

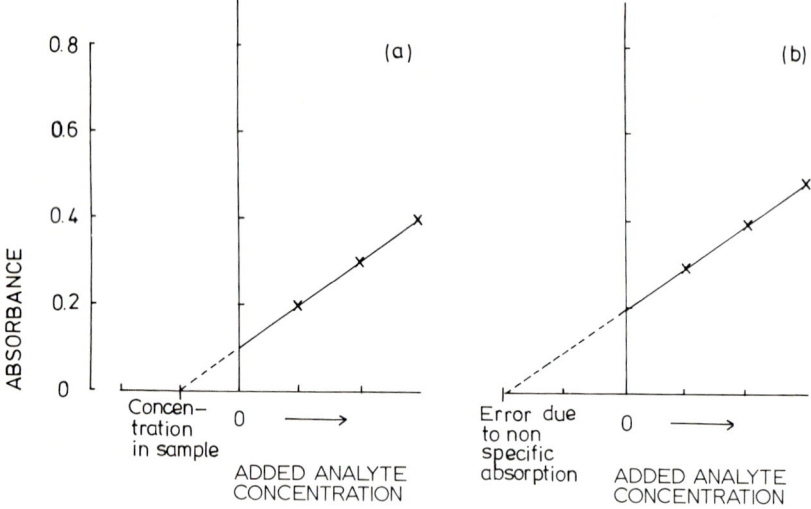

Fig. 2. (a) Method of standard additions. (b) Error introduced due to non-specific absorption.

the formation of alkali oxyacid salts from alkali salts in the sample. Alkali salts can give rise to serious background interference and it has been shown [34] that whereas alkali metal oxyacid salts when completely atomised on a carbon rod, showed little molecular absorption between the wavelengths 200 to 380 nm, alkali metal halides gave broad bands of molecular absorption. As an example of metal halide interference removal, Hodges [32] found that 0.1 mg l^{-1} lead standards with added calcium chloride, magnesium chloride and sodium sulphate showed no interference when made to 2% with orthophosphoric acid. Furthermore, Fordham [39] showed in the analysis of stream waters for cadmium, lead and zinc where many samples exhibited high non-atomic absorption, that the background was reduced considerably by a nitric acid digestion within the furnace.

Finally, Czobik et al. [4] reported that in the analyte—interferent systems: Pb—$CdCl_2$, Sn—$CdCl_2$, Sn—$CuCl_2$, Zn—$PbCl_2$, Zn—$CuCl_2$, and Zn—$CdCl_2$, sulphuric acid and phosphoric acid successfully removed the interferences. Nitric acid was found to be less effective although it removed the interference in the Pb—NaCl system.

Since it has generally been shown that oxyacids aid in removing interferences in flameless atomic absorption, it is recommended that all samples and standards are prepared in 1% nitric acid prior to flameless analysis.

There are alternative ways of modifying the matrix to remove the interference. Organic compounds have been shown to be effective. Regan and Warren [41] studied potential interferences on lead determinations in water. 100 mg l^{-1} of calcium, magnesium, strontium and barium were added to lead solutions in hydrochloric acid and nitric acid. The most severe interferences resulted from magnesium in hydrochloric acid. The authors found that additions of 1% ascorbic acid to each of the interference solutions reduce all interferences on lead to less than 7%. Similarly, Fuller [42] found that 1% oxalic acid removed the interference of 0.1% $MgCl_2$ on the lead absorption signal.

In saline waters, salt produces a large background signal unless it is driven off during the ashing stage. However, many metals volatilise at temperatures lower than that which will drive off the sodium chloride. Ediger et al. [43] and Ediger [44] proposed the addition of ammonium nitrate to convert the sodium chloride to products which can be driven off in the ashing step below 500°C. They pipetted 20 μl of a 50% ammonium nitrate solution together with 50 μl of a seawater sample into the graphite furnace. The combined solutions were dried for 30 s at 120°C, charred for 60 s at 1000°C and atomised for 10 s at 2500°C. The method was found to be successful in the determination of copper. Manning and Slavin [45] evaluated the effect of several matrix modifiers on the $MgCl_2$ interference in the determination of lead. 2% (m/v) orthophosphoric acid, 5% (m/v) ammonium nitrate, 5% (m/v) oxalic acid and 5% (m/v) sucrose were tested. Out of these ammonium nitrate was found to be the most effective in removing the interference.

Using the ETA technique for the determination of volatile elements there is a danger of losing the analyte in the ashing step of the heating programme prior to atomisation. In this instance reagents can be added to the sample to render the analyte element more refractory. An example of this is the addition of 0.5% (m/v) $Ni(NO_3)_2$ to samples for arsenic and selenium analysis enabling ashing temperatures of 1100°C to be used as opposed to the lower temperatures of 600°C and 400°C respectively. Also when samples for lead analysis are made 2% (m/v) in orthophosphoric acid ashing temperatures up to 1000°C can be used without loss of lead [32]. Similarly, ammonium phosphate or ammonium sulphate added to solutions (1% m/v) for cadmium assay enables the ashing temperature to be raised from 250°C to 800°C. In mercury determinations, using a heated graphite atomiser, it was found by Issaq and Zielinski, Jr. [46] that there was a fifty-fold signal enhancement when hydrogen peroxide was added (1% m/v) to the solutions. They postulated the formation of a stable mercury peroxy complex restricting mercury vaporisation losses.

In flameless atomic absorption the analyte often tends to react with the graphite furnace or rod to form carbides. In such cases atomisation is suppressed. 'Release agents' are used to react preferentially with the graphite 'releasing' the analyte on atomisation. An application of this is in the determination of aluminium, barium, beryllium, silicon and tin. A large enhancement of the signal has been observed [47] when calcium (as the nitrate) is added to the analytical solutions. This has been suggested as due to a reduction in the formation of carbide in the presence of calcium. A calcium level of 1000 to 2000 mg l^{-1} in the solutions has been reported as the optimum in most cases.

References pp. 93—94

(e) *Pretreatment of the graphite furnace or rod.* In a number of cases workers have pretreated the graphite rod or furnace with 'release agents' prior to injecting the sample. Runnels et al. [48] treated the graphite furnace with a suitable carbide forming element such as lanthanum and zirconium. This was found to eliminate the interference in the determination of beryllium, chromium, manganese and aluminium in petroleum. Similarly, Thompson et al. [49] used lanthanum for treating graphite furnaces. They injected 15 µl of a 20% (m/v) lanthanum nitrate hexahydrate solution into the graphite tube and carefully evaporated the drop to dryness by slowly increasing the voltage of the evaporation stage over a period of 90 s. The tube was then passed through the dry, ash and atomisation stages of the heating programme. This procedure was then repeated for a total of three 15 µl injections and it was found to reduce the matrix suppression effect to less than 7% for cadmium and lead in natural waters.

Alternatively, molybdenum has been used for furnace pretreatment. Hodges [32] pretreated the graphite cylinder by firing ten times with 1% ammonium molybdate. This removed the background signal in the determination of lead in urine. Samples were previously acidified with 2% orthophosphoric acid. Also, Manning and Slavin [45] reported that by coating the interior of the graphite tubes with molybdenum, the interference of 1% NaCl or $MgCl_2$ on lead determinations was removed. This was after 4% ammonium nitrate had been added as a matrix modifier. In their treatment, a molybdenum solution was prepared by dissolving 9.2 g of $(NH_4)_6 \cdot Mo_7O_{24} \cdot 4H_2O$ in 40 ml of 5% NH_4OH and diluting to 50 ml in a volumetric flask. This results in approximately 10% Mo (m/v). 25 µl of this solution are introduced into the furnace and the tube is fired. This is then repeated for three sequences.

(f) *Chelation and solvent extraction of analyte prior to injection and atomisation.* One of the most effective means of minimisation of interference is to remove the analyte from the sample matrix. Methods of chelation and solvent extraction described earlier (Section III.C.2) can be used and the extract injected into the atomiser. Owing to the small quantity of extract required to be injected, the whole procedure may be performed on a micro scale. Furthermore, unlike the flame, there is no restriction on the solvents which can be used.

Pellenbarg and Church [50] employed a variation of this technique. They extracted estuarine water samples using the APDC—MIBK procedure. The organic extract was then evaporated to dryness and the residue wet ashed with 4.0 ml portions of concentrated nitric acid. Subsequently the nitric acid was evaporated and the moist residue made up to 5.0 ml with 0.1 M nitric acid. Copper was determined in this solution by injection into a carbon-rod atomiser.

The authors thank Mr. R. Atkins, Directorate of Scientific Services, Thames Water Authority for his assistance in obtaining the literature necessary in the preparation and Miss M. Saunders and colleagues of Metropolitan Water Division, also of Thames Water Authority, for typing the manuscript.

REFERENCES

1. G. E. Batley and D. Gardner, Water Res., 11 (1977) 745—756.
2. R. C. Rooney, IL Circular No. 4, June 1978, Instrumentation Laboratory (U.K.) Ltd.
3. D. R. Parker, Water Analysis by Atomic Absorption Spectroscopy, Varian Techtron Pty. Ltd., Springvale, Australia, 1972.
4. S. L. Sachdev and P. W. West, Environ. Sci. Technol., 4 (1970) 749—751.
5. C. E. Mulford, At. Absorpt. Newsl., 5 (1966) 88—90.
6. Lead in Potable Waters by Atomic Absorption Spectrophotometry, HMSO, London, 1977.
7. Cadmium in Potable Waters by Atomic Absorption Spectrophotometry, HMSO, London, 1977.
8. Methods for the Examination of Water and Wastewater, American Public Health Association, Washington, DC, 1976, pp. 144—162.
9. J. D. Kinrade and J. C. Van Loon, Anal. Chem., 46 (1974) 1894—1898.
10. J. P. Riley and D. Taylor, Anal. Chim. Acta, 40 (1968) 479—485.
11. M. I. Abdullah, O. A. El-Rayis and J. P. Riley, Anal. Chim. Acta, 84 (1976) 363—368.
12. T. M. Florence and G. E. Batley, Talanta, 23 (1976) 179—186.
13. P. Figura and B. McDuffle, Anal. Chem., 49 (1977) 1950—1953.
14. K. V. Krishnamurty and M. M. Reddy, Anal. Chem., 49 (1977) 222—226.
15. K. C. Thompson, Analyst, 103 (1978) 1258—1262.
16. Calcium in Raw and Potable Waters by Atomic Absorption Spectrophotometry, HMSO, London, 1977.
17. Magnesium in Raw and Potable Waters by Atomic Absorption Spectrophotometry, HMSO, London, 1977.
18. W. R. Hatch and W. L. Ott, Anal. Chem., 40 (1968) 2085—2087.
19. J. Carron and H. Agemian, Anal. Chim. Acta, 92 (1977) 61—70.
20. C. Feldman, Anal. Chem., 46 (1974) 99—102.
21. M. H. Bothner and D. E. Robertson, Anal. Chem., 17 (1975) 592—595.
22. S. H. Omang, Anal. Chim. Acta, 54 (1971) 415—420.
23. L. Lopez-Escobar and D. N. Hume, Anal. Lett., 6 (1973) 343—353.
24. H. Agemian and A. S. Y. Chau, Anal. Chem., 50 (1978) 13—16.
25. J. F. Kopp, M. C. Longbottom and L. B. Lobring, J. Am. Water Works Assoc., (1972) 20—25.
26. B. J. Farey, L. A. Nelson and M. G. Rolph, Analyst, 103 (1978) 656—660.
27. B. J. Farey and L. A. Nelson, Anal. Chem., 50 (1978) 2147—2148.
28. J. Y. Hwang, P. A. Ullucci, C. J. Mokeler and S. B. Smith Jr., Am. Lab., (1973) 43—47.
29. K. C. Thompson and D. R. Thomerson, Analyst, 99 (1974) 595—601, 1182.
30. J. S. Caldwell, R. J. Lishka and E. F. McFarren, J. Am. Water Works Assoc., (1973) 731—735.
31. D. D. Siemer and P. Koteel, Anal. Chem., 49 (1977) 1096—1099.
32. D. J. Hodges, Analyst, 102 (1977) 66—69.
33. J. Y. Hwang and G. P. Thomas, Am. Lab., (1974) 55—63.
34. B. R. Culver and T. Surles, Anal. Chem., 47 (1975) 920—921.
35. B. E. Guthrie, W. R. Wolf and C. Veillon, Anal. Chem., 50 (1978) 1900—1902.

36 W. M. Barnard and M. J. Fishman, At. Absorpt. Newsl., 12 (1973) 118—124.
37 K. M. Aldous, D. G. Mitchell and F. J. Ryan, Anal. Chem., 45 (1973) 1990—1993.
38 J. M. Ottoway, Proc. Anal. Div. Chem. Soc., 13 (1976) 185—192.
39 A. W. Fordham, J. Geochem. Explor., 10 (1978) 41—51.
40 E. J. Czobik and J. P. Matousek, Anal. Chem., 50 (1978) 2—10.
41 J. G. T. Regan and J. Warren, Analyst, 101 (1976) 220—221.
42 C. W. Fuller, At. Absorpt. Newsl., 16 (1977) 106—107.
43 R. D. Ediger, G. E. Peterson and J. D. Kerber, 13 (1974) 61—64.
44 R. D. Ediger, At. Absorpt. Newsl., 14 (1975) 127—130.
45 D. C. Manning and W. Slavin, Anal. Chem., 50 (1978) 1234—1238.
46 H. J. Issaq and W. L. Zielinski, Jr., Anal. Chem., 46 (1974) 1436—1438.
47 K. C. Thompson, R. G. Godden and D. R. Thomerson, Anal. Chim. Acta, 74 (1975) 289—297.
48 J. H. Runnels, R. Merryfield and H. B. Fisher, Anal. Chem., 47 (1975) 1258—1263.
49 K. C. Thompson, K. Wagstaff and K. C. Wheatstone, Analyst, 102 (1977) 310—313.
50 R. E. Pellenbarg and T. M. Church, Anal. Chim. Acta, 97 (1978) 81—86.

Chapter 4b

Marine analysis by atomic absorption spectrometry

HIROKI HARAGUCHI and KEIICHIRO FUWA
*Department of Chemistry, Faculty of Science, University of Tokyo,
Bunkyo-ku, Tokyo 113 (Japan)*

I. INTRODUCTION

In marine chemistry, concentration, chemical state, material balance and cycle of both elements and compounds are the main subjects of study for describing the oceans. The oceans, which cover about 71% of the earth's surface, form a complex multidimensional system with varied constructions and materials. Thus, analyses of the components of seawater, marine plants and animals, and sediments give the bases for the interpretation of material balances and geochemical phenomena in oceans.

The development of marine chemistry has depended upon proper analytical techniques and sampling devices. In the 18th and 19th centuries, Boyle and Lavoisier, who were the early pioneers in marine chemistry, investigated the various salts in seawater. Dittmar examined the components of sea water using gravimetric and volumetric analytical techniques. Before Dittmar's work, K, Na, Ca, Mg, S, Cl, Br, B, Sr and F had been found in seawater. Dittmar noted that the ratios of the major constituents are almost constant, while the total salt concentration is variable [1].

Various spectroscopic techniques such as flame photometry, emission spectroscopy, atomic absorption spectrometry, spectrophotometry, fluorimetry, X-ray fluorescence spectrometry, neutron activation analysis and isotope dilution mass spectrometry have been used for marine analysis of elemental and inorganic components [2]. Polarography, anodic stripping voltammetry and other electrochemical techniques are also useful for the determination of Cd, Cu, Mn, Pb, Zn, etc. in seawater. Electrochemical techniques sometimes provide information on the chemical species in solution.

Each spectroscopic method has a characteristic application. For example, flame photometry is still applicable to the direct determination of Ca and Sr, and to the determination of Li, Rb, Cs and Ba after preconcentration with ion-exchange resin. Fluorimetry provides better sensitivities for Al, Be, Ga and U, although it suffers from severe interference effects. Emission spectrometry, X-ray fluorescence spectrometry and neutron activation analysis allow multielement analysis of solid samples with pretty good sensitivity and precision, and have commonly been applied to the analysis of marine organisms and sediments. Recently, inductively-coupled plasma (ICP)

References pp. 119—122

emission spectrometry has been increasingly used as a method for simultaneous multielement analysis [3, 4].

Atomic absorption spectrometry (AAS) is a relatively new analytical technique among the spectroscopic methods. As described in the previous chapters, AAS gives high sensitivity, precision and accuracy along with experimental convenience and a wide instrumental availability. Therefore, this technique has been extensively employed for the analysis of marine samples. However, the elemental contents of marine samples are generally very low, and suitable preconcentration procedures are required. Recent development of graphite-furnace techniques and gas generation techniques has extended the applicability of AAS to marine analysis. The determination of Cd, Cu, Ni, Pb, Hg, As, Sb, Se, Sn and Te has become much easier as a result of the development of these techniques.

In analysis of marine samples by AAS problems are encountered in sampling procedure, sample storage, sample treatment and measurement procedures. Analytical difficulties arise from the low concentration of most elements and complex matrices in marine samples. In this chapter, a general discussion on marine analysis by AAS will be provided in terms of seawater, marine organisms and sediments.

II. SEAWATER

Seawater contains about 3.5% salts, in which the content of sodium chloride is about 80%. The concentration of dissolved salts as well as temperature and pressure influence the physical properties of seawater. The total salt concentration is usually called "salinity". Salinity is generally measured by the electrical conductivity or determination of chloride content. At present, salinity(S) is defined as: $S = 1.80655$ Cl (Cl is the concentration of chloride in seawater) [5]. Dissolved oxygen and silica are usually measured as additional parameters to characterize seawater. The concentrations of nitrogen and phosphorus are the indices of nutrients and measure the fertility and production of the oceans.

The average elemental composition of seawater is shown in Table 1 [6]. The elements which exist at more than $1\,\mu g\,ml^{-1}$ are the major elements, and those less than $1\,\mu g\,ml^{-1}$ are the trace elements. Of these, only twelve elements are present at a concentration greater than $1\,\mu g\,ml^{-1}$, and most other elements range from $0.5\,\mu g\,ml^{-1}$ to much less than $1\,ng\,ml^{-1}$ in content.

The concentrations of trace elements in seawater varies geographically, spatially, and in depth [7]. Such variations are generally caused by the biological activities and physicochemical processes in the oceans. Thus, readers should note that the concentrations of trace elements given in Table 1 are not representative of those for all oceans. However, the extremely low levels of elements in concentrated salt matrix make accurate and precise analysis of seawater difficult. In recent interlaboratory investigations of seawater analysis by skillful marine or analytical chemists, a wide range of

TABLE 1

ELEMENTAL COMPOSITION OF SEAWATER

Element	Concentration ($\mu g\ ml^{-1}$)	Element	Concentration ($\mu g\ ml^{-1}$)
Cl	19000	As	0.003
Na	10500	Sn	0.003
Mg	1350	U	0.003
S	885	Mn	0.002
Ca	400	V	0.002
K	380	Ti	0.001
Br	65	Ce	0.0004
C	28	Ag	0.0003
Sr	8.1	Y	0.0003
B	4.6	Co	0.00027
Si	3	Cd	0.00011
F	1.3	W	0.0001
N	0.5	Se	0.00009
Li	0.18	Ge	0.00007
Rb	0.12	Cr	0.00005
P	0.07	Th	0.00005
I	0.06	Ga	0.00003
Ba	0.03	Pb	0.00003
Al	0.01	Hg	0.00003
Fe	0.01	Bi	0.000017
Zn	0.01	Au	0.000011
Mo	0.01	Ra	6×10^{-11}
Ni	0.0054	Ru	6×10^{-16}
Cu	0.003		

H. J. M. Bowen, Trace Elements in Biochemistry, Academic Press, New York, 1966.

analytical values were reported for some trace elements [8, 9]. Chakrabarti et al. [10] summarized the problems in seawater analysis as follows: (a) Problems of sampling, sample collection, filtration, and storage. (b) Preconcentration and separation. Determination of trace metals in marine samples invariably involves some preconcentration which can cause considerable contamination. (c) Contamination from reagents, solvents, laboratory ware and the laboratory environment. (d) Inadequate appreciation of the limitations of the analytical technique (or techniques) used in the determination. (e) Lack of adequate tests of techniques, methods and procedures. (f) Lack of assessment of the reliability of techniques, methods and procedures by extensive inter- and intra-laboratory intercalibration studies. (g) Lack of adequate standard reference materials for calibration purposes, and a standard reference sample that could be certified as blank, i.e., a seawater facsimile.

In order to obtain reliable analytical values, all the items pointed out above should be carefully examined at each step throughout the overall experimental procedures. Sampling, sample collection, filtration and storage

References pp. 119—122

are primarily important before instrumental analysis. As summarized in Table 2, flame and graphite-furnace atomic absorption spectrometry provide quite good sensitivities [11, 12], where the concentration ranges of trace elements are also given [13]. For most trace elements, however, the sensitivities (or detection limits) are not high enough to determine them directly without preconcentration. Great care should be taken in the preconcentration step, where contamination from water, chemicals, solvents, glassware and the laboratory environment is often easily introduced.

At the sampling step, contamination from sampling bottles and hydrowire should be minimized [10]. Old types of sampling devices such as Nansen and Knudsen are constructed mostly from brass, which may cause contamination of Cu and Zn into seawater. The recently developed Niskin plastic sampling bottle provides less contamination from the sampler walls. However, the materials used for the end-caps or springs, rubber or Teflon-coated stainless steel, sometimes release significant amounts of Ba, Cu, Sb and Zn [14].

Bruland and Franks [15] compared the sampling methods using the Teflon-coated 30-1 PVC ball-valve samplers (Go-Flo, General Oceanics) and the CIT deep-water sampler, designed and constructed by Schaule and Patterson. Careful sample handling and processing were carried out on board ship inside a modular Porta-lab equipped with a positive pressure filtered air supply and specially designed for trace-metal analysis. The seawater samples were not filtered, so as to avoid contamination from the filter material, since the primary purpose of their study was the comparison of the sampling techniques. The collected seawater samples were analysed by solvent-extraction flameless AAS which is described in detail in section II.B.1. Furthermore, they investigated the vertical profiles of Cu, Cd, Zn and Ni in terms of the seawater samples collected with the Go-Flo and CIT samplers. The results for comparison of profiles obtained with the Go-Flo and CIT samplers are shown in Fig. 1 [15]. The comparative results shown in Fig. 1 demonstrate the consistency between the Go-Flo and CIT samplers for four trace metals in deep seawater down to 3000 m. Bruland and Franks noted [15] that the samples at depths of 630 m and 1560 m collected with the CIT sampler were contaminated because of a faulty seal. Great care was taken in terms of chemicals, vessels, cleaning and preconcentration techniques, and analytical methods. Consequently, it is clearly very difficult to obtain consistent analytical results for ultra-trace metals in seawater. Concerning sampling techniques and contamination from samplers, see ref. 16.

Suspended particulate matter and marine organisms are present in seawater. In seawater analysis, it is desirable to separate the particulate and dissolved fractions, and to analyse them separately. A comparison of analytical results for filtered and unfiltered seawater is shown in Table 3 [15], where upper and lower limits of particulate metal concentrations are also given. As can be seen in Table 3, the contents of particulate trace metals are less than 1% of the total metal concentration, which is generally within

TABLE 2

THE DETECTION LIMITS OF FLAME AND GRAPHITE FURNACE ATOMIC ABSORPTION SPECTROMETRY AND TRACE ELEMENT CONCENTRATION IN SEAWATER

Element	Detection limit[a]		Trace element concentration[d] ($\mu g\,l^{-1}$)	
	FAA[b] ($\mu g\,ml^{-1}$)	GFAA[c] (ng)		
Ag	1	0.0001	0.1 —	0.7
Al	30	0.001	0 —	7
As	30	0.008	0.8 —	8.0
Au	20	0.001	0.004 —	0.027
B	2500	0.02		
Ba	20	0.006	4 —	20
Be	2	0.00003	0.001 —	0.03
Bi	50	0.004	0.001 —	0.04
Ca	1	0.0004		
Cd	1	0.00008	0.07 —	0.71
Co	2	0.002	0.078 —	0.34
Cr	2	0.002	0.1 —	0.5
Cs	50	0.004	0.28 —	0.5
Cu	1	0.0006	0.6 —	20
Fe	4	0.01	0.1 —	61.8
Ga	50	0.001	0.030 —	0.037
Hg	500	0.02	0.01 —	0.075
In	30	0.0004	0.0001—	0.0012
Li	1	0.003	173 —	183
Mn	0.8	0.0002	0.2 —	8.6
Mo	30	0.003	2.1 —	18.8
Ni	5	0.009	1.1 —	4.0
Pb	10	0.004	0.02 —	0.04
Pd	10	0.004	—	
Pt	5	0.01	—	
Rb	5	0.001	119 —	125
Rh	20	0.008	—	
Sb	30	0.005	0.01 —	0.4
Se	100	0.009	0.052 —	0.11
Si	100	0.000005	—	
Sn	0.05	0.02	0.008 —	0.04
Te	50	0.001	—	
Ti	90	0.04	1.9	
Tl	20	0.001	0.0101—	0.019
V	20	0.003	0.2 —	4
Zn	1	0.00003	1.0 —	50

[a] FAA = flame AAS, GFAA = graphite furnace AAS.
[b] All values cited from Winefordner et al. [11].
[c] All values cited from L'vov [12].
[d] The values cited from Brewer [13].

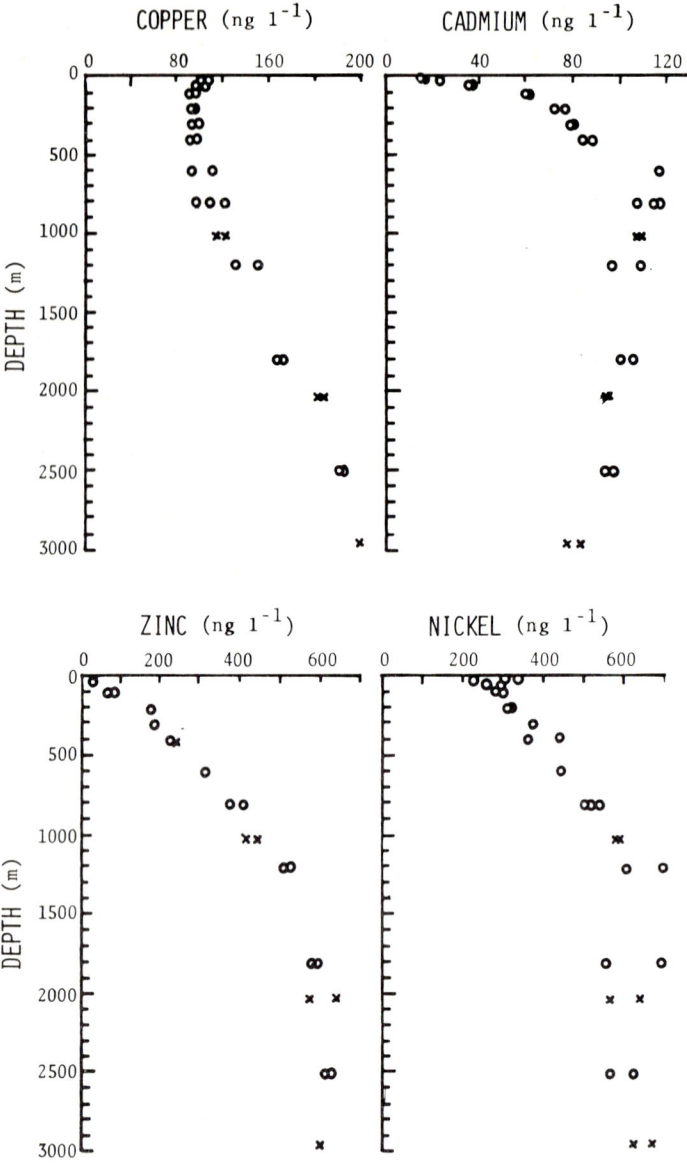

Fig. 1. Comparison of profiles obtained with Go-Flo (O) and CIT (×) samplers. Data points represent replicate extraction values. (From ref. 15.)

the range of estimated precision. Thus, filtering deep seawater does not cause significant differences. However, the ranges of particulate values at 25 m are 1—9, 1.5—4, 3—27 and 0.2—2% of total Cu, Cd, Zn and Ni, respectively, as calculated from the data for the unfiltered and particulate samples in Table 3. Therefore, differences due to filtration should be taken into account

TABLE 3

COMPARISON OF FILTERED AND UNFILTERED SEAWATER SAMPLES ANALYSED BY THE SOLVENT EXTRACTION—ATOMIC ABSORPTION METHOD[a]
(All values in ng l^{-1})

Sample	Copper	Cadmium	Zinc	Nickel
25 m: unfiltered	110, 109	30.9, 31.5	24, 28	332, 320
filtered	104, 111	28.8, 25.0	22, 31	325, 321
particulate	1—9	0.4—1.0	1—7	1—6
1800 m: unfiltered	173, 170	107, 102	579, 588	556, 693
filtered	158, 162	107, 109	525, 565	603, —
particulate	1.5—3	0.05	1.5—2	5—10

[a] Ref. 15.

for coastal and surface seawater. In addition, contamination from filter materials and filtration apparatus and adsorption of trace metals on the filters, which may produce appreciable error, have to be checked carefully before the analysis of seawater samples. Significant contamination of seawater with Cu [17], Fe and Mn [18], and adsorption of Hg [14] have been reported.

A. Storage of seawater samples

The acidification of seawater samples is required for the storage of seawater in order to prevent the precipitation, adsorption of trace metals on the container walls or volatilization loss from solution. Generally, nitric acid is added to seawater to bring the pH of the solutions to below 1.5. The addition of acids immediately after collection on board ship may be preferable for sample storage rather than addition before analysis in the laboratory.

Many workers have investigated the significant loss of trace metals from water samples upon storage [19—26]. The extent of loss varies with the pH, the length of storage period, the types of containers, and the concentration of the metals. Eichholz et al. [19] found loss of trace metals from water samples when stored in polyethylene and borosilicate (Pyrex) glass containers. Robertson investigated losses of Ag, Co, Fe, In, Sc and V from seawater stored in Phnex and polyethylene containers using a neutron activation technique was applied for monitoring, and found significant losses of these metals [20]. Recently, Subramanian et al. [26] studied losses of eleven trace metals (Ag, Al, Cd, Co, Cr, Cu, Fe, Mn, Ni, Pb, Sr, V and Zn) as a function of contact time (days), at various solution pH's (1.5—8.0), where water samples were stored in Pyrex glass, Nalgene linear polyethylene, and Teflon containers. Graphite furnace AAS was used for monitoring of metal contents in the solutions. The percent losses of Cd and Pb from synthetic and river-water samples stored in Pyrex glass and polyethylene containers at various pH values as a function of contact time (for 30 days) are shown in Figure 2. From these studies, it was found that the acidification to pH 1.5 with nitric

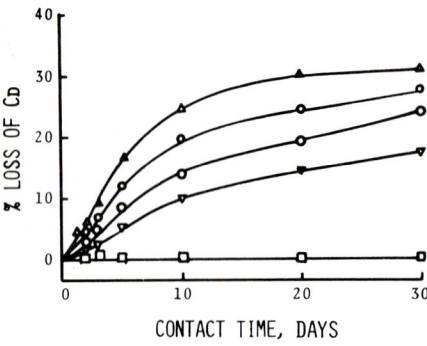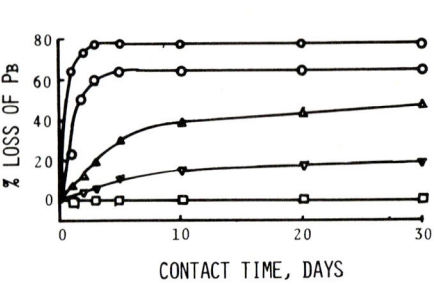

Fig. 2. (a) Loss of cadmium from Pyrex glass and Nalgene containers. Rideau River water sample (0.12 μg l^{-1} Cd); Pyrex glass: (□) pH 1.6, 2.5, 4.0; (●) pH 6.0; (△) pH 8.0. Nalgene: (□) pH 1.6, 2.5, 4.0, 6.0, 8.0. Synthetic water sample (0.10 μg l^{-1} Cd); Pyrex glass: (□) pH 1.6, 2.5, 4.0; (▽) pH 6.0; (○) pH 8.0. Nalgene: (□) pH 1.6, 2.5, 4.0, 6.0, 8.0 (b) Loss of lead from Pyrex glass and Nalgene containers. Rideau River water sample (4 μg Pb l^{-1}); Pyrex glass and Nalgene containers: (□) pH 1.6; (△) pH 2.5, 4.0; (○) pH 6.0, 8.0. Synthetic water sample (5 μg Pb l^{-1}); Pyrex glass and Nalgene containers: (□) pH 1.6, (▽) pH 2.5, 4.0; (●) pH 6.0, 8.0. (From ref. 26.)

acid and storage in a Nalgene container is the most effective way of minimizing the loss of trace metals from natural waters.

The loss of mercury from water samples on storage has been shown to be a serious problem by many workers [24—26]. These losses of mercury are caused by rapid adsorption on container walls [25, 29] and reduction of mercury to the atomic state followed by volatilization from solution [29]. Lo and Wai reported that 81% of mercury in untreated samples was lost to the walls of the polyethylene containers and the remaining 19% was volatilized to the atmosphere [29]. Bothner and Robertson observed mercury contamination of seawater samples due to the diffusion of mercury vapor from the laboratory into the polyethylene containers [31].

Since polyethylene is one of the most troublesome materials with regard to losses of mercury, Pyrex glass container should be used with some preservative for the storage of mercury-containing seawater samples [25, 31]. Work has still to be done on preservatives. Feldman proposed the addition of 5% (v/v) HNO_3 and 0.05% $K_2Cr_2O_7$ in the case of a polyethylene container or of 0.01% $K_2Cr_2O_7$ for a borosilicate glass container [25]. He also recommended that HNO_3 and/or $K_2Cr_2O_7$ should be added to the containers before collection of the samples in order to stabilize the mercury immediately [25]. Dokiya et al. [33] showed that L-cystein with dilute HCl is a good preservative for mercury even at the 1 ng ml^{-1} level in water samples. Matsunaga et al. [34] acidified seawater samples to 0.4 N with H_2SO_4 and stored them for more than three weeks at room temperature. Using cold-vapor AAS they noted that the mercury concentration increased gradually during the storage, and became almost constant after three weeks. Considering the contamination introduced by the preservatives, the addition of H_2SO_4 appears to be the most effective way of storing seawater samples for mercury analysis.

B. Preconcentration

The detection limits obtained by flame and graphite furnace AAS and the concentration levels of the elements in seawater are summarized in Table 2. In general, graphite furnace AAS provides better sensitivities for many elements than the flame technique. Even so, AAS sensitivity is insufficient for the direct determination of most ultra-trace elements. Furthermore, concentrated salts and undissolved particulates cause severe interferences with the determination of trace elements by AAS. Therefore, it is necessary to concentrate the analytes before the determination, and, if possible, to separate the analytes from dissolved major constituents and particulates. Solvent extraction, coprecipitation and ion-exchange techniques are the most widely used techniques for the preconcentration of seawater. In the following sections, these techniques will be reviewed. It should be noted here that the efficiencies of the recovery of the analytes as well as the contamination from reagents and solvents must be carefully examined when the preconcentration techniques are applied. Chakrabarti et al. [10] have summarized the work on the application of preconcentration techniques to marine analysis by AAS. Hence, only some representative applications will be introduced hereafter.

1. *Solvent extraction*

Solvent extraction is one of the methods widely used for concentration and separation. Most heavy metals are extracted with chelating reagents into organic solvents [35]. Some chelating agents commonly used in atomic absorption spectrometry are shown in Fig. 3. APDC and DDC are most commonly used in AAS. Solvents such as ketones, esters, ethers, alcohols, and other oxygen-containing hydrocarbons are suitable for the flame atomic absorption technique. Of these solvents, MIBK (methyl isobutyl ketone, 4-methylpentane-2-one; $CH_3-CH-H_2COCH_3$ | CH_3) is most widely used together with chelating agents such as APDC, DDC and oxine. Solvent extraction has the following advantages for AAS; (i) concentration of trace elements, (ii) enhancement of atomic absorption signals, (iii) separation of trace metals from interfering components. Details of the solvent extraction technique in atomic absorption spectrometry can be found in the book by Cresser [35].

Brooks et al. [36] pioneered solvent extraction for seawater with an APDC—MIBK extraction technique for the determination of six elements in seawater by flame AAS. However, the metal complexes extracted into

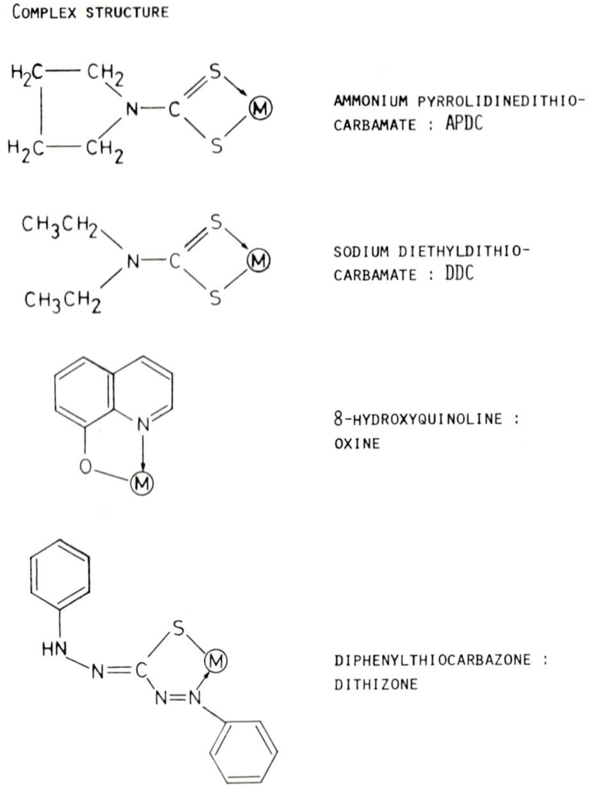

COMPLEX STRUCTURE

AMMONIUM PYRROLIDINEDITHIO-
CARBAMATE : APDC

SODIUM DIETHYLDITHIO-
CARBAMATE : DDC

8-HYDROXYQUINOLINE :
OXINE

DIPHENYLTHIOCARBAZONE :
DITHIZONE

Fig. 3. Typical chelating agents commonly used in atomic absorption spectrometry.

MIBK were unstable (decomposing within one day), which makes the technique inconvenient for routine analysis.

Recently, Jan and Young [37] investigated an APDC—MIBK extraction of trace metals (Ag, Cd, Cr, Fe, Ni, Pb and Zn) followed by back extraction with 4 N nitric acid, and analysed them by flameless AAS. They also compared their results with those obtained by the Chelex-100 cation-exchange [38] and APDC—MIBK single extraction methods. The method proposed by Jan and Young offers the advantages in consumption of the small sample volumes, the better recovery efficiency, and the stability of the metal complexes in the acid extracts over the ion-exchange and APDC—MIBK single extraction methods.

The procedure of MIBK—nitric acid successive extraction employed by Jan and Young [37] is as follows. For the APDC—MIBK extraction, a 200 ml seawater sample in a Teflon beaker containing 2 ml of 1% APDC is heated to incipient boiling at a pH of about 4. After cooling to room temperature, 7 ml of MIBK is added to the sample, and the mixture is transferred into a polyethylene bottle and shaken for 25 min on a mechanical shaker. After allowing the layers to separate in a separatory funnel for 20 min, the organic layer is collected into a polyethylene bottle.

For the acid back extraction, 5 ml of 4 N HNO_3 is pipetted into the MIBK extract obtained from the previous step, and the mixture is shaken for 20 min. After transferring the mixture into a Teflon separatory funnel and standing for 20 min, the acid layer is drained into a polyethylene bottle and preserved in a refrigerator until analysed.

TABLE 4

EXPERIMENTAL CONDITIONS AND DETECTION LIMITS IN FLAMELESS ATOMIC ABSORPTION SPECTROMETRIC ANALYSIS OF SEAWATER EXTRACTS[a]

	Ag	Cd	Cr	Cu	Fe	Ni	Pb	Zn
Wavelength (nm)	328.1	228.8	357.9	324.7	248.3	232.0	217.0	213.9
Lamp current (mA)	3	5	5	3	5	5	5	5
Spectral slit (nm)	0.5	0.5	0.2	0.5	0.2	0.2	1.0	0.5
Dry[b] (50 s)	3.5	3.5	3.5	3.5	3.5	3.5	3.5	3.5
Ash[b] (20 s)	5.0	4.0	6.0	6.0	6.0	6.0	4.0	4.0
Atomize[b] (2 s)	6.5	7.0	7.5	7.0	8.0	8.0	6.5	6.5
N_2 gas (l min^{-1})	4	4	4	4	4	4	4	4
H_2 gas (l min^{-1})	—	—	1	—	—	—	—	—
Detection limit ($\mu g\,l^{-1}$)	0.02	0.003	0.05	0.02	0.20	0.10	0.03	0.03

[a] Ref. 37. Analysed by Varian-Techtron atomic absorption spectrometer (Model AA-6) with a carbon rod atomizer (Model 63).
[b] Arbitrary dial settings of temperature on the M-63 power supply.

The experimental conditions and detection limits of flameless AAS for eight metals in seawater extracts obtained by Jan and Young are summarized in Table 4, where a Varian-Techtron AA-6 atomic absorption spectrometer equipped with a carbon-rod atomizer (Model 63) was used with a background corrector, and 2.5 µl of treated sample solution was injected into the graphite tube [37]. The analytical results of trace metals in seawater using MIBK single extraction and MIBK—HNO_3 successive extraction methods are compared in Table 5. As can be seen from Table 5, the precision of the MIBK—HNO_3 successive extraction method is improved, in addition to the stability of the extract solution, compared to that of the MIBK single extraction method. In another experiment, they obtained mean relative standard deviations ranging from 18 to 25% for those metals present below 1 $\mu g\,l^{-1}$. The results for the recovery test obtained by the method mentioned above are also summarized in Table 6 [37]. Generally, the recoveries are not enough for all the elements, especially for silver. This indicates that the solvent extraction at pH 4 is not suitable or optimal for these elements. It is necessary to examine the optimum pH for each element in order to improve the efficiency of the solvent extraction technique.

Bruland and Franks [15] developed a more efficient solvent extraction method for Cu, Cd, Zn and Ni, using a double extraction technique with

References pp. 119—122

TABLE 5

TRACE METALS ANALYSIS OF SEAWATER USING MIBK SINGLE EXTRACTION AND MIBK—HNO$_3$ SUCCESSIVE EXTRACTION METHODS [37]

Method	Concentration (μg l^{-1})					
	Ag	Cr	Cu	Fe	Ni	Pb
MIBK—HNO$_3$ extraction	<0.02	0.14	0.78	3.28	0.59	0.29
	<0.02	0.17	0.73	3.57	0.59	0.25
	<0.02	0.16	0.82	3.51	0.59	0.28
	<0.02	0.12	0.82	3.10	0.59	0.31
Mean	<0.02	0.15	0.79	3.37	0.59	0.28
SD[a]	—	0.022	0.043	0.22	0.00	0.025
% RSD[b]	—	7.3	5.4	6.5	0.0	8.9
MIBK single extraction	<0.01	0.11	0.95	3.40	0.51	0.32
	<0.01	0.15	0.75	3.47	0.54	0.40
	<0.01	0.15	0.67	3.15	0.54	0.34
	<0.01	0.19	0.84	4.05	0.56	0.43
Mean	<0.01	0.15	0.80	3.52	0.54	0.37
SD	—	0.033	0.120	0.380	0.021	0.051
% RSD	—	22	15	11	3.9	14

[a] SD = standard deviation.
[b] RSD = relative standard deviation.

TABLE 6

RECOVERY OF METALS FROM SPIKED SEAWATER OBTAINED BY SOLVENT EXTRACTION—ATOMIC ABSORPTION SPECTROMETRY [37]

Element	Added[a] (μg)	Found[b] (μg)		Av. recovery (%)
		Mean	SD[c]	
Ag	0.037	0.007	0.0007	19
	0.185	0.024	0.0006	13
Cd	0.020	0.016	0.0025	80
	0.100	0.081	0.0085	81
Cr	0.025	0.020	0.0026	80
	0.125	0.117	0.0279	94
Cu	0.100	0.103	0.0219	103
	0.500	0.373	0.0528	75
Fe	0.100	0.069	0.0095	69
	0.500	0.381	0.0796	76
Ni	0.100	0.080	0.0081	80
	0.500	0.391	0.0099	78
Pb	0.050	0.035	0.0023	70
	0.250	0.163	0.0329	65
Zn	0.050	0.059	0.0064	118
	0.250	0.192	0.0439	77

[a] Filtered island control seawater (200 ml) spiked with metal standard solution.
[b] After correcting for concentration measured in unspiked sample.
[c] SD = standard deviation, $n = 3$.

APDC/DDDC (diethylammonium diethyldithiocarbamate) into chloroform and back-extraction into 7.5 N HNO_3. As this method is a great improvement compared with other methods the experimental procedure is now described in detail.

Approximately 250 g of acidified seawater was taken into a 250 ml Teflon separatory funnel and buffered to about pH 4 with 2 ml of ammonium acetate; 1 ml of a solution containing 1% (w/v) each of APDC and DDDC (stabilized in a 1% ammonium hydroxide solution and purified by chloroform extraction) and 8 ml of chloroform were added, and the mixture shaken vigorously for 2 min. After 5 min of phase separation, the chloroform fraction was drained into a 125 ml Teflon separatory funnel, and 4 ml of 7.5 N nitric acid added to degrade the dithiocarbamates. An additional 6 ml of chloroform was poured into the original seawater sample for a second extraction of the chelated species, as well as for the rinsing of the solution and the separatory funnel. After phase separation, the chloroform was combined with the first fraction and shaken vigorously for 2 min. The phases were allowed to stand 5 min for separation, then the chloroform phase was discarded.

The 4 ml acid phase containing the back-extracted metals was then drained into a 10 ml fused quartz beaker, and 2 ml of 7.5 N nitric acid was used to rinse the 125 ml separatory funnel and stem. This rinse was combined with the acid in the quartz beaker. The 6 ml back-extract was evaporated to dryness, and the residue was further oxidized by 250 μl or 500 μl of concentrated nitric acid. The residue in the quartz beaker was redissolved in warm 1 N nitric acid and quantitatively transferred with 250 μl rinses (ca. 1.25 ml total) to a 1.8 ml polyvial. This solution was provided for the analysis by furnace AAS.

After the completion of the procedure above described, the concentration factor was 200-fold, and alkali and alkaline earth elements, which often provide spectral interferences in furnace AAS, could be removed. The overall recoveries were examined by using seawater samples spiked with ^{115m}Cd, ^{64}Cu and ^{65}Zn, and were 98.9 ± 0.7% (n = 10) for Cd, 97.9 ± 1.2% (n = 3) for Cu and 99.0 ± 0.8% (n = 5) for Zn. The experimental results obtained by Bruland and Franks are shown in Figs. 1 and 4 and Table 3.

Cr(VI) is a toxic element, and its environmental pollution should be monitored even in seawater. CrO_4^{2-} is a stable chemical species in seawater, while Cr(III) also exists in relatively high amounts. Therefore, a separation of Cr(VI) from Cr(III) is necessary in the analysis of Cr(VI). For this purpose, the solvent extraction technique can also be used, being followed by atomic absorption analysis. Many workers have investigated the solvent extraction of total Cr in seawater, where Cr was extracted with acetylacetonate, DDC, APDC and analysed by AAS [37—42]. Hiiro et al. examined in detail the separation of Cr(VI) from Cr(III) in seawater [42]. The effect of pH values on extraction of Cr(VI) is shown in Fig. 5. Cr(VI) is most effectively extracted near pH 5, while Cr(III) is increasingly extracted above pH 4. Therefore,

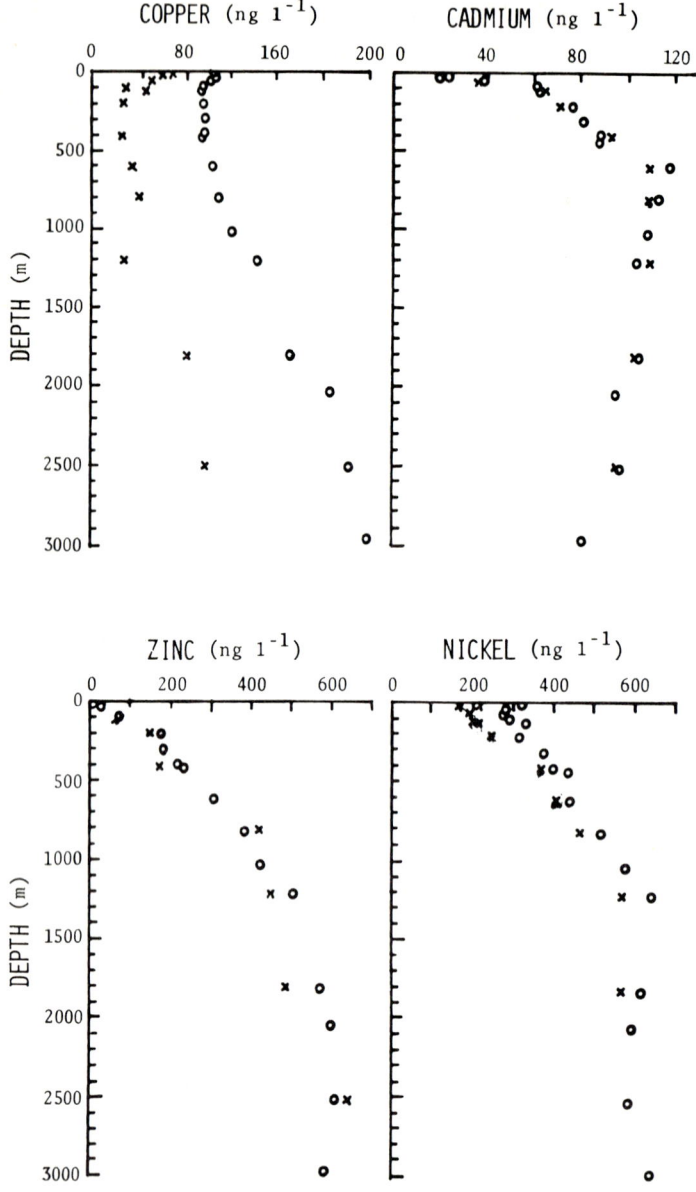

Fig. 4. Comparison of profiles from solvent extraction (○, values expressed as average of two or more replicates) and Chelex-100 (×) techniques. (From ref. 15.)

Cr(VI) must be extracted below pH 4. More than 2×10^{-3} M DDC is desirable for the extraction of Cr(VI), as shown in Fig. 5. From these results, Hiiro et al. proposed the following procedures for the determination of Cr(VI): 100 ml seawater is taken into a separatory funnel, and 2 ml of 1 M

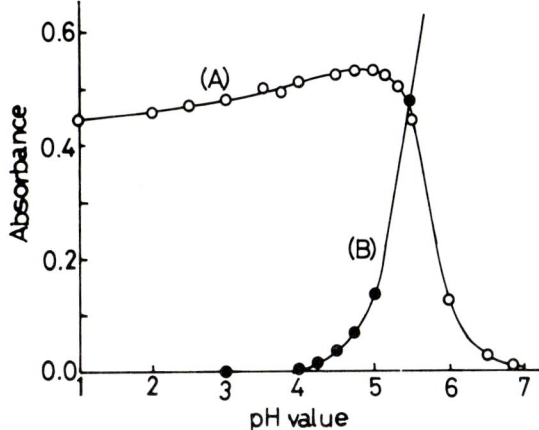

Fig. 5. Effect of pH value on the extraction of Cr(VI) using the flame method. (A) Cr(VI): 0.5 ppm; (B) Cr(III): 20 ppm; V_{aq}: 100 ml (From ref. 42.)

CH_3COOH-CH_3COONa(3:1) buffer solution (pH 4) and 2 ml of 0.1 M DDC solution are added into the funnel. Then, 10—30 ml MIBK is added to the sample and shaken for about 3 min. After the separation of the aqueous and organic layers, the organic layer is used for the atomic absorption analysis at 357.9 nm. The sensitivity (1% absorption) of this method is ca. 0.02 ng ml^{-1} by furnace AAS and 0.4 ng ml^{-1} by flame AAS.

The oxine—MIBK system has been applied to the determination of Fe in seawater by flame AAS [43, 44].

2. Coprecipitation

In the coprecipitation method, trace metals are concentrated by adsorbing onto the surface of a precipitate. As the precipitate, hydroxides of Fe(III), Mn(IV), Al(III), Bi(II), and Zr(IV) and sulfides of Co(II), Pb(II) and Fe(II) are frequently used. In AAS the use of hydroxide precipitates is more popular than that of sulfides. Coprecipitation using the hydroxide precipitates is generally unspecific, and many trace metals may be concentrated simultaneously. However, some limitations are often found in the coprecipitation technique [10].

Contamination of the analytes from the carriers (the precipitates) should be first examined, and the blank test carried out carefully. Great care should also be taken in terms of the recoveries of the analytes, because the procedures in the coprecipitation are sometimes time-consuming and irreproducible. Some efficiencies of recovery for Zr(IV) coprecipitation along with the determined values of trace elements in seawater are summarized in Table 7, where inductively-coupled plasma (ICP) emission spectrometry was applied for the simultaneous multielement analysis [45]. In this experiment, 10 mg of Zr(IV) was added to 1 l of seawater, the precipitation made

TABLE 7

DETERMINATION OF TRACE ELEMENTS IN SEAWATER AND RECOVERIES USING THE ZIRCONIUM-COPRECIPITATION TECHNIQUE

Element	Concentration (ng ml^{-1})	Recovery (%)[a]	Element	Concentration (ng ml^{-1})	Recovery (%)[b]
Cd	0.11 ± 0.07	50	Al	10.3 ± 0.1	105
Co	0.40 ± 0.09	80	As	3.5 ± 0.2	100
Cr	0.38 ± 0.02	9	Fe	8.1 ± 0.7	111
Cu	0.91 ± 0.22	95	La	1.3 ± 0.2	96
Mo	0.78 ± 0.17	20	Mn	7.4 ± 0.1	98
Ni	1.4 ± 0.6	71	Pb	4.5 ± 0.9	95
Ti	0.38 ± 0.04	99	Sb	3.0 ± 1.0	106
Y	0.13 ± 0.05	100	V	1.8 ± 0.1	91
			Zn	3.2 ± 1.9	100

[a] Recovery of 2.5 µg of each spiked element.
[b] Recovery of 5.0 µg of each spiked element.

at pH 9.2, and the total solution was made up to 50 ml after dissolving the precipitate with nitric acid solution. As can be seen from the results, the recoveries of Cd, Ni, Co, Mo and Cr are not satisfactory, probably due to insufficient precipitation of hydroxides of Cd, Ni and Co at pH 9.2 and insufficient adsorption of Mo and Cr because of their soluble anion forms.

Satō and Saitoh determined As, Cr and Pb in seawater by graphite-furnace AAS with the Zr-coprecipitation technique [46]. They added 1 ml of 0.13 M ZrOCl$_2$ solution (about 10 mg of Zr) into 1 l seawater, and adjusted to pH 9 with ammonia solution. The precipitate was filtered with No. 5 filter paper, washed with 2% ammonia solution, and dissolved with hot 2 N hydrochloric acid. The final solution volume was made up to 50 ml. The treated solution (50 µl) was subjected to graphite-furnace AAS (spectrophotometer; Hitachi Model 207, Graphite furnace; Perkin-Elmer Model HGA-70). The recoveries for As, Cr and Pb were almost 100%, when the samples containing only a few nanograms were analysed. The interference effects of the 18 kinds of metal ions and anions examined were almost negligible for concentrations upto 1000—10000 fold. Satō and Saitoh [46] also suggested this method might be applicable to Cd^{2+}, Co^{2+}, Cu^{2+}, Fe^{3+}, Al^{3+}, Mn^{2+}, Ni^{2+}, Sb^{3+}, Sb^{5+}, Se^{4+}, Se^{6+}, Zn^{2+}, V^{5+}, and so forth. Other applications of coprecipitation techniques can be found in the references [47—50].

The coprecipitation method combined with colloid flotation using stearylamine, sodium oleate, etc. has been used to preconcentrate analytes in seawater [49, 50].

3. Ion exchange

Various ion-exchange procedures have been developed in analytical chemistry. Riley has published a detailed review of various ion-exchange techniques for concentration of trace metals in seawater [14]. The extremely high content [12] of sodium ion in seawater makes cation-exchange resins of little use. Anion-exchange resins are useful for elements existing as anions, especially chloro-complexes in seawater. Although Brooks [52] reported large enrichment factors for Au, Bi, Cd and Tl with Amberlite IRA-100 resin, there have not been many applications of anion-exchange resins to seawater analysis [53, 54].

Recently, the use of chelating exchange resins (e.g., Chelex-100) in preconcentration of trace metals in seawater has become popular [15, 37, 55—61]. Chelex-100 is a copolymer of vinylbenzyliminodiacetic acid, styrene and divinylbenzene, and shows strong chelating activity for various trace metals [12, 55, 56].

Bruland and Franks [15] investigated the determination of Cu, Cd, Zn and Ni in seawater by graphite-furnace atomic absorption spectrometry, carrying out a strict comparison of sampling techniques and preconcentration techniques (solvent extraction and chelating resin methods) [15].

The concentration technique with the Chelex-100 resin can be summarized as follows. The resin (Na form; 100—200 mesh) was cleaned with 6 N HCl by daily suspension, decantation and addition of fresh acid for one week; it was then rinsed with distilled water and 2 N HNO_3, and finally cleaned with distilled water. The resin was changed to the ammonia form and 7.5 ml of the resin was packed in disposable polyethylene columns. The excess of ammonia was washed with distilled water to reduce the pH of the effluent to below 9. The concentration procedure of seawater with the Chelex resin was performed immediately after sample collection on shipboard. Seawater (4 l) was taken from the samplers into acid-cleaned polyethylene bottles. It was pumped through Teflon tubing directly into the resin column, where a peristaltic pump was used to maintain the flow rate at 3.5—4.5 ml min^{-1}. Upon completion of pumping, the column was capped at both ends, covered with polyethylene bags, and frozen for storage. After return to the lab, the resins were thawed, washed with distilled water to remove excess of salts, and eluted with 30 ml of 2 N HNO_3 into polyethylene bottles. The concentration factor obtained by this concentration procedure was about 120-fold. The recoveries of trace metals for this Chelex procedure were examined with ^{65}Zn- and ^{115m}Cd-spiked seawater samples and found to be almost 100%.

The treated effluents were analysed by graphite-furnace AAS. Using the concentration procedures described above, large amounts of Mg and Ca were eluted in the final step; Na and K were also eluted in the final step. These latter components interfered with the determination of Cd, Cu, Pb and Zn by AAS. Therefore, the standard solutions with matched matrices (ca. 188 μg ml^{-1}; Mg 1700 μg ml^{-1}; Na 150 μg ml^{-1} and K 20 μg ml^{-1}) were prepared for the analysis.

Bruland and Franks [15] compared the results obtained by Chelex-100 concentration with those by solvent extraction (a double extraction with

APDC/DDDC into chloroform and back-extraction into 7.5 N HNO_3). Vertical profiles obtained by both techniques are shown in Figure 4. The profiles for cadmium and zinc show good agreement between the solvent extract and Chelex results, although the Chelex results are somewhat lower than those by solvent extraction concentration. The values for nickel were also lower when the Chelex technique was employed. The mean difference between the two techniques was $65 \pm 35 \, \text{ng l}^{-1}$.

A large discrepancy between the two concentration techniques was found for the copper results. The average difference was $72 \pm 27 \, \text{ng l}^{-1}$. Bruland and Frank analysed the Chelex column effluent by the solvent extraction technique, and found 63 and $135 \, \text{ng l}^{-1}$ as the copper content for the samples at 25 m and 2500 m, respectively. These values are almost equal to the difference between the Chelex and solvent extraction results. Therefore, they concluded that about 60% of the copper in seawater (unfiltered and unacidified) is not removed by the Chelex technique. As Riley et al. suggested, copper in seawater is not liberated by the Chelex resin because of association with colloids and fine particulates [62]. In order to avoid this error, acidification and heating of seawater is necessary prior to the Chelex treatment. According to the results of Bruland and Franks, acidification and storage followed by solvent extraction appears to be superior to the Chelex resin concentration for the quantitative determination of copper in seawater [15]. Similar problems have been pointed out by Eisner and Mark, Jr. [63] and Florence and Batley [64].

Muzzarelli and Marinelli [65—67] proposed the use of a natural chelating polymer, chitosan, which is de-acetylated chitin containing glucosamine and *N*-acetyl glucosamine. Sugawara et al. [68] have developed another promising chelating agent, controlled pore glass (GPC) immobilized chelate(s). The properties and feasibilities of the GPC-immobilized chelates have been summarized by Chakrabarti et al. [10].

4. *Other preconcentration techniques*

Solvent extraction, coprecipitation and ion-exchange techniques are the main concentration methods used for seawater analysis. Other interesting concentration techniques, such as electrodeposition, amalgam trap (for mercury), a cold trap-vaporization system for hydride generation, and recrystallization, are often used by marine and analytical chemists. The first three methods are briefly reviewed here.

Electrodeposition is a unique concentration technique, because the separation of trace metals from interfering matrix species can be easily carried out at the same time. Thus, the detection limits of AAS can be improved remarkably without further chemical preconcentration steps and spectral and chemical interferences due to major components in seawater can be also eliminated easily. Thus, many applications of the technique to seawater analysis by AAS have been made, especially in flameless AAS [69—78]. The

HMDE (hanging mercury drop electrode) [71, 72], gold-foil [73], copper-wire [74], tungsten-wire [75, 76] and pyrolytic graphite-coated tube [78] have been used as the electrodes for electrochemical deposition, and successfully applied to the determination of Cu, Cd, Pb, Zn, Hg and so forth. In atomic absorption analysis the electrodes are usually heated directly for atomization of metals.

Flame AAS is not sufficiently sensitive for the determination of mercury. Thus, flameless AAS employing the cold-vapor generation technique [79] has been developed to improve the analytical sensitivity for mercury. In the cold-vapor atomic absorption technique, a long tube absorption cell is usually used, and a detection limit of the order of sub-ppb can be obtained. Recently, Tanabe et al. [80] reported a detection limit of $0.002 \, \text{ng} \, \text{l}^{-1}$ using the 185.0 nm atomic line instead of 253.7 nm. The concentration of mercury in seawater is, however, much lower than the detection limit obtained by such a cold-vapor technique [81]. A few $\text{ng} \, \text{l}^{-1}$ of mercury in seawater have been reported by some workers [82, 83], and a sub-ppt level of organic mercury has been reported [84]. Thus, proper preconcentration of mercury in seawater should be carried out before atomic absorption analysis. Among others, gold-foil [73], gold-wire [85, 86], gold-sponge [87], silver metal particles [83, 88], a cold trap cooled with liquid nitrogen [89] have been used for the trap-concentration of mercury generated from seawater by proper reduction. These metallic or cold traps are generally heated up to about 400—500°C immediately after the concentration process. The quantitative determination of mercury at the sub-ppt level can be achieved using preconcentration techniques. It should, however, again be stressed that the preservation of mercury during the storage of samples is a more critical factor than the analytical procedures. Therefore, immediate analysis on the shipboard may be desirable [83]. The present authors and co-workers have developed a sensitive non-dispersive vacuum-ultraviolet (185.0 nm) mercury atomic absorption analyser [83, 90], which is a compact instrument easily set even in the shipboard laboratory.

Owing to the development of the hydride generation technique, the analytical sensitivities for As, Sb, Sn, Te, Bi and Pb have been improved significantly, to the ppb or sub-ppb level. The use of sodium borohydride ($NaBH_4$), which is strongly reducing, has led to successful improvement of sensitivity and analytical convenience. Recent topics of interest in this field include species analysis of inorganic and organic compounds. First, Braman and Foreback [91] investigated sequential volatilization technique for arsines as a result of their interest in environmental chemistry. Arsenic compounds are toxic and they are used as silvicides and pesticides [91]. Talmi and Bostick [92] used gas-chromatographic separation of sodium borohydride reduced arsenic compounds with a microwave discharge emission spectrometric detection system [92]. Furthermore, Edmonds and Francesconi [92] and Andreae [94] used the separation technique proposed by Braman and Foreback [91], and detected arsines by atomic absorption spectrometric methods.

References pp. 119—122

TABLE 8

THE PROPERTIES OF ARSENIC COMPOUNDS AND ARSINES [94]

Molecules	pK_a	Reduction pH	Reduction product	B.p. (°C)
As(III), arsenous acid, $HAsO_2$	9.23	4	AsH_3	−55
As(V), arsenic acid, H_3AsO_4	2.25	1—2	AsH_3	−55
Monomethylarsonic acid, $CH_3AsO(OH)_2$	2.60	1—2	CH_3AsH_2	2
Dimethylarsenic acid, $(CH_3)_2AsO(OH)$	6.19	1—2	$(CH_3)_2AsH$	35.6
Trimethylarsine[a], $(CH_3)_3As$	—	1—4	$(CH_3)_3As$	70
Phenylarsonic acid, $C_6H_5AsO(OH)_2$	3.59	1—2	$C_6H_5AsH_2$	148

[a] Or its oxidized form, presumed to be $(CH_3)_3AsO$.

The properties of arsenic compounds and their reduced forms are summarized in Table 8 [94]. As can be seen, various forms of arsines show different boiling points. Thus, these arsines can be trapped and concentrated in a liquid-nitrogen cold trap, and then separated sequentially by heating the trap. The stripping and trapping efficiencies of arsines were examined by Andreae [94] and found to be more than 90%. Andreae determined the arsines with an atomic absorption detector using a hydrogen-rich hydrogen/air flame. The detection limit at 197.3 nm was $0.5 \, \text{ng} \, l^{-1}$. The determined values of arsenic compounds in natural waters, including seawater, are shown in Table 9. The concentrations of As(VI) are higher compared with other arsenic compounds. Since the use of MMAA and DMAA as pesticides is increasing, the concentrations of these compounds should be monitored in relation to environmental pollution of the oceans. Braman and Tompkins [95] reported the determinations of inorganic and organic tin compounds in seawater by means of a technique similar to that for arsenic, although they used a flame emission spectrometric method in a hydrogen/air flame.

III. MARINE ORGANISMS

The study of elemental compositions of marine organisms (animals and plants) is an important research subject of marine chemistry. Generally, elemental concentrations in the organisms are much higher than the abundances of seawater [6] (see also Tables 1 and 2). This phenomenon is usually interpreted as accumulation of elements from seawater. The marine organisms take up nutrient and trace elements from seawater, and concentrate them inside their bodies. The concentration of trace elements in organisms is sometimes anomalously high. The concentration of elements from seawater is defined as the concentration factor [6]; it may be as large as $10—10^6$ [6, 96]. According to references, the elements concentrated are sometimes specific for the species of the organisms, and sometimes not

TABLE 9

CONCENTRATIONS OF ARSENIC SPECIES IN NATURAL WATERS (ng As ml^{-1})

Locality and sample type	As(III)	As(V)	MMAA[a]	DMAA[b]
Seawater, Scripps Pier, La Jolla, Calif.				
5 Nov. 1976	0.019	1.75	0.017	0.12
11 Nov. 1976	0.034	1.70	0.019	0.12
Seawater, San Diego Trough				
Surface	0.017	1.49	0.005	0.21
25 m below surface	0.016	1.32	0.003	0.14
50 m below surface	0.016	1.67	0.003	0.004
75 m below surface	0.021	1.52	0.004	0.002
100 m below surface	0.060	1.59	0.003	0.002
Sacramento River, Red Bluff, Calif.	0.040	1.08	0.021	<0.004
Owens River, Bishop, Calif.	0.085	42.5	0.062	0.22
Colorado River, Parker, Ariz.	0.114	1.95	0.063	0.051
Colorado River, Slough near				
Topock, Calif.	0.085	2.25	0.13	0.31
Saddleback Lade, Calif.	0.053	0.020	<0.002	0.006
Rain, La Jolla, Calif.				
10 Sept. 1976	<0.002	0.180	<0.002	0.024
11 Sept. 1976	<0.002	0.094	<0.002	<0.002

[a] Monomethylarsonic acid.
[b] Dimethylarsenic acid.

specific [97]. Therefore, the concentrations of elements may be primarily linked to the food chain [6, 98].

On the other hand, marine organisms influence elemental compositions in seawater. They concentrate the elements from seawater and return them to the ocean after death. Thus, the organisms play an important role in the movement, transfer, and circulation of elements from surface to deep waters or from seawater to organic matter. For the last 20 years AAS has been extensively applied to the analysis of marine organisms. When AAS is employed, the storage of organisms and sample treatments (digestion by ashing, wet acidic digestion, lyophilization, or low-temperature plasma ashing) must be carefully investigated if accurate and precise analytical data is to be obtained. The following three methods are usually employed as pretreatment of marine organisms for storage: (A) freezing the wet sample, (B) drying the wet samples in an oven at 80—120°C to remove water, and (C) lyophilization. The drying and lyophilization methods may be more convenient than freezing, because the dried samples are easily stored without freezing boxes or rooms.

Samples dried in an oven can be used for the determination of most elements except for mercury which is extremely volatile. Therefore, it is most important to know whether any loss of mercury occurs during the drying or lyophilization process. Pillay et al. [99] reported considerable losses of mercury from plankton/algae during low-temperature oven drying

TABLE 10

LOSS OF MERCURY FROM PLANKTON AND SEDIMENT SAMPLES DURING LOW-TEMPERATURE OVEN DRYING [99]

Sample identification[a]	Initial levels of mercury (ppm)	Loss of mercury (%)
A. Plankton/algae		
PL—Bx	17.86	51.1
PL—By	17.86	71.7
PL—Bz	17.86	60.6
B. Sediment/silt		
S/S—EA	2.25	23.6
S/S—EB	2.25	12.4
S/S—EC	2.25	12.4

[a] Drying for 50 h in a laboratory oven at 60°C.

prior to neutron activation analysis and cold-vapor atomic absorption spectrometry. The results are shown in Table 10 along with the results for sediment. The samples were dried in the laboratory oven at 60°C for 50 h or until they attained constant weight. Significant losses of mercury were found even for low-temperature drying. Thus, oven-drying of marine samples is not suitable for mercury analysis.

The possible loss of mercury from biological materials upon lyophilization has been also studied by several workers. Some workers have observed losses of mercury [99, 100], and others no losses from fish samples and animal organisms [101, 102]. Recently, Ramelow and Hornung [103] investigated, using an AAS technique, mercury losses upon lyophilization of fish and mussel samples, carefully comparing results with the analyses of fresh wet-samples. In their experiment, fresh fillets of single or several individual specimens were homogenized in a blender, spread inside small glass bottles and frozen overnight in a freezer. Before placing the samples in the lyophilizer, they were frozen at $-70°C$ in a slurry of ethanol and dry ice, and then left in the lyophilizer for about 24 h. The mussel samples were prepared by grinding the soft parts of several individual specimens in an agate mortar until visibly homogeneous. Then, they were treated in the same manner as the fish samples. The fresh fish and mussel samples were prepared for analysis by digesting amounts of wet tissue (0.3—1 g) in Teflon-lined, high-pressure decomposition vessels. The homogenized samples were digested in the same manner as fresh samples. The analytical results are summarized in Table 11 [103], where the concentration of mercury in the lyophilized samples has been converted to wet weight. The average water losses of lyophilized samples were 75.9% for fish and 74.3% for the mussel. The results in Table 11 show no significant losses of mercury from the fish

TABLE 11

A COMPARISON OF MERCURY CONCENTRATION IN BIOLOGICAL SAMPLES USING WET AND LYOPHILIZED SAMPLES [103]

Organisms	Hg concentration (μg g^{-1} wet weight)					
	Wet sample			Lyophilized sample		
	No. of samples	Range	Average ± s.d.	No. of samples	Range	Average ± s.d.
PISCES (FISH)						
Synodontidae						
Saurida sp.	9	0.05–0.65	0.24 ± 0.19	9	0.07–0.70	0.27 ± 0.20
Darangidae						
Trachurus sp.	3	0.06–0.19	0.12 ± 0.07	3	0.07–0.19	0.12 ± 0.06
Mullidae						
Mullus sp. 1	25	0.02–0.35	0.13 ± 0.06	25	0.03–0.23	0.15 ± 0.05
Mullus sp. 2	1	0.09	—	1	0.13	—
Sparidae						
Boops sp.	2	0.07–0.08	0.08 ± 0.007	2	0.08–0.08	0.08 ± 0
Pagellus sp.	6	0.07–0.28	0.17 ± 0.07	6	0.08–0.28	0.18 ± 0.08
Sphyraenidae						
Sphyraena sp.	2	0.08–0.16	0.12 ± 0.06	2	0.08–0.17	0.12 ± 0.06
Triglidae						
Chelidonichthys sp.	1	0.09	—	1	0.13	—
MOLLUSCA						
Donacidae						
Donax sp.	8	0.07–0.80	0.38 ± 0.29	8	0.08–0.93	0.37 ± 0.26

samples upon lyophilization. As can be seen from the above results, lyophilization is more efficient than oven drying. However, when a large number of samples are analysed, lyophilization is not practically convenient. In such cases, oven drying may be chosen when mercury analysis is not required.

IV. SEDIMENT

The main components of marine sediments are inorganic aluminosilicate minerals which are usually accumulated on the sea floor by river and other geological activities, and also skeletons and shells of marine organisms (mainly calcium carbonate and silica) [2]. Of course, some metal salts or particulates which precipitate from seawater form new minerals, e.g. manganese nodules [2]. The chemical compositions of the three principal types of sediments in the ocean are shown in Table 12 [105]. Most of the sediments found in the deep-sea floor are mixtures of these three principal minerals. Study of the sediments in the oceans and seashores can provide important data related to geochemical, oceanographical or biological circulation and deposition of elements, formation and distribution of marine sediments, and exploitation of marine resources.

The major elements of marine sediments are still determined by classical gravimetric and volumetric methods. Atomic absorption spectrometric methods have been applied to sediment analysis, although emission spectrometry, X-ray fluorescence spectrometry and neutron activation analysis

TABLE 12

CHEMICAL COMPOSITION OF THREE PRINCIPAL SEDIMENTS IN THE OCEAN (wt.%) [104]

Composition	Red clay	Calcareous ooze	Siliceous ooze
SiO_2	53.93	24.23	67.36
TiO_2	0.96	0.25	0.59
Al_2O_3	17.46	6.60	11.33
Fe_2O_3	8.53	2.43	3.40
FeO	0.45	0.64	1.42
MnO	0.78	0.31	0.19
CaO	1.34	0.20	0.86
MgO	4.35	1.07	1.71
Na_2O	1.27	0.75	1.64
K_2O	3.65	1.40	2.15
P_2O_5	0.09	0.10	0.10
H_2O	6.30	3.31	6.33
$CaCO_3$	0.39	56.73	1.52
$MgCO_3$	0.44	1.78	1.21
C	0.13	0.30	0.26
N	0.016	0.017	—

are also used because of their advantages of direct multielement analysis [2]. Chakrabarti et al. [10] have summarized the use of flames and graphite-furnace AAS. Atomic absorption methods as well as classical analytical methods require acidic wet digestion or fusion with adequate salts. These sample treatments are not easy, particularly when trace elements are analysed. In addition, variable and complex compositions of major elements cause severe matrix interference with the determination of trace metals in marine sediments [105—107].

V. CONCLUSION

This chapter has briefly described the present status of AAS applied to marine analysis, specifically seawater, marine organisms and sediments. Atomic absorption spectrometry has been increasingly applied to the compositional analysis of marine samples. Problems have been, and will be met in each experimental step, sampling, storage and standardization of analytical methods, even though AAS is one of the best suited, most sensitive, precise and selective methods for the purpose. The oceans hold limitless future resources. Therefore, accurate data related to the oceans from the aspects of geology, chemistry, physics and biology is required for future exploitation. Atomic absorption spectrometry will no doubt contribute to the future development of marine science and the battle against further destructive pollution.

REFERENCES

1. E. C. Goldberg, W. S. Broecker, G. M. Gross and K. K. Turekian, Radioactivity in the Marine Environment, National Academy of Sciences, Washington, DC, 1971.
2. G. Thompson, in E. L. Grove (Ed.), Applied Atomic Spectroscopy, Vol. 1, Plenum Press, New York, 1978, pp. 273—300.
3. V. A. Fassel and R. N. Kniseley, Anal. Chem., 46 (1974) 1110A, 1155A.
4. V. A. Fassel, Science, 202 (1978) 183.
5. T. R. S. Wilson, in J. P. Riley and G. Skirrow (Eds.), Chemical Oceanography, 2nd ed., Vol. 1, Academic Press, New York, 1975.
6. H. J. M. Bowen, Trace Elements in Biochemistry, Academic Press, New York, 1966.
7. E. A. Boyle, F. Sclater and J. M. Edmond, Nature, 263 (1976) 42.
8. P. G. Brewer, N. Frew, N. Cutshall, J. J. Wagner, R. A. Duce, P. R. Walsh, G. L. Hoffman, J. W. R. Dutton, W. F. Fitzgerald, C. D. Hunt, D. C. Girvin, P. G. Clem, C. Patterson, D. Settle, B. Glover, B. J. Presley, J. Trefry, H. Windom and R. Smith, Mar. Chem., 2 (1974) 69.
9. K. Sugawara, Deep-Sea Res., 25 (1978) 323.
10. C. L. Chakrabarti, K. S. Subramanian and T. Nakahara, The Application of Atomic Absorption Spectrometry to the Analysis of Trace Metals in Non-biological Marine Samples, Rep. 6 (NRCC No. 17530), National Research Council Canada, 1978.
11. J. D. Winefordner, J. J. Fitzgerald and N. Omenetto, Appl. Spectrosc., 29 (1975) 369.
12. B. V. L'vov, Atomic Absorption Spectrochemical Analysis, translated by J. H. Dixon, Adam Hilger, London, 1970.

13 P. G. Brewer, in J. P. Riley and G. Skirrow (Eds.), Chemical Oceanography, Vol. 1, 2nd edn., Academic Press, New York, 1975.
14 J. P. Riley, in J. P. Riley and G. Skirrow (Eds.), Chemical Oceanography, Vol. 3, 2nd edn., Academic Press, New York, 1975.
15 K. W. Bruland and R. Franks, Anal. Chim. Acta, 105 (1979) 233.
16 D. A. Segar and G. A. Berberian, in T. R. P. Gibb, Jr. (Ed.), Analytical Methods in Oceanography, Adv. Chem. Ser. No. 147, American Chemical Society, Washington, DC, 1975.
17 K. T. Marvin, R. P. Proctor and R. A. Neal, Limnol. Oceanogr., 15 (1970) 320.
18 W. R. Hirsbrunner and P. J. Wangersky, Mar. Chem., 3 (1975) 55.
19 G. G. Eichholz, A. E. Nagel and R. B. Hughes, Anal. Chem., 37 (1965) 863.
20 D. E. Robertson, Anal. Chim. Acta, 42 (1968) 533.
21 R. A. Durst and B. T. Duhart, Anal. Chem., 42 (1970) 1002.
22 W. G. King, J. M. Rodriguez and C. M. Wai, Anal. Chem., 46 (1974) 771.
23 R. D. Edigar, At. Absorpt. Newsl., 12 (1973) 151.
24 R. V. Coyne and J. A. Collins, Anal. Chem., 44 (1972) 1093.
25 C. Feldman, Anal. Chem., 46 (1974) 99.
26 K. S. Subramanian, C. L. Chakrabarti, J. E. Sueiras and I. S. Maines, Anal. Chem., 50 (1978) 444.
27 S. Shimomura, Y. Nishimura and Y. Tanse, Bunseki Kagaku, 17 (1968) 1148; 18 (1969) 1072.
28 L. C. Bates, Radiochem. Radioanal. Lett., 6 (1971) 139.
29 J. Lo and C. Wai, Anal. Chem., 47 (1975) 1869.
30 S. H. Omang, Anal. Chim. Acta, 53 (1971) 415.
31 M. Bothner and D. Robertson, Anal. Chem., 47 (1975) 592.
32 K. I. Mahan and S. E. Mahan, Anal. Chem., 49 (1977) 662.
33 Y. Dokiya, H. Ashikawa and K. Fuwa, Spectrosc. Lett., 7 (1974) 551.
34 K. Matsunaga, M. Nishimura and S. Konishi, Nature, 258 (1975) 224.
35 M. S. Cresser, Solvent Extraction in Flame Spectroscopic Analysis, Butterworths, London, 1978.
36 R. P. Brooks, B. J. Presley and I. R. Kaplan, Talanta, 14 (1967) 809.
37 T. K. Jan and D. R. Young, Anal. Chem., 50 (1978) 1250.
38 J. P. Riley and D. Taylor, Anal. Chim. Acta, 40 (1968) 470.
39 B. Belaughter, At. Absorpt. Newsl., 4 (1968) 273.
40 Y. K. Chau, S. S. Sim and Y. H. Wong, Anal. Chim. Acta, 43 (1968) 13.
41 T. R. Gilvert and A. M. Clay, Anal. Chim. Acta, 67 (1973) 289.
42 K. Hiiro, T. Owa, M. Takaoka, T. Tanaka and A. Kawahara, Bunseki Kagaku, 25 (1976) 122.
43 J. L. Jones and R. D. Eddy, Anal. Chim. Acta, 43 (1968) 165.
44 K. Hiiro, T. Tanaka and T. Sawada, Bunseki Kagaku, 21 (1972) 635.
45 H. Haraguchi, Y. Nojiri, M. Matsui and K. Fuwa, unpublished data.
46 A. Satō and N. Saitoh, Bunseki Kagaku, 25 (1976) 663.
47 Y. K. Chau and P. Y. Wong, Talanta, 15 (1968) 867.
48 T. Owa, K. Hiiro and T. Tanaka, Bunseki Kagaku, 21 (1968) 878.
49 N. Rothstein and H. Zeitlin, Anal. Lett., 9 (1976) 461.
50 M. Hiraide, Y. Yoshida and A. Mizuike, Anal. Chim. Acta, 81 (1976) 185.
51 E. A. Boyle and J. M. Edmond, Anal. Chim. Acta, 91 (1977) 189.
52 R. R. Brooks, Analyst (London), 85 (1960) 745.
53 A. D. Matthews and J. P. Riley, Anal. Chim. Acta, 51 (1970) 455.
54 K. Hiiro, A. Kawahara and T. Tanaka, Bunseki Kagaku, 22 (1973) 1210.
55 J. P. Riley and D. Taylor, Anal. Chim. Acta, 40 (1968) 175.
56 J. P. Riley and D. Taylor, Anal. Chim. Acta, 40 (1968) 479.
57 H. L. Windom and R. G. Smith, Deep-Sea Res., 19 (1972) 727.
58 J. P. Riley and D. Taylor, Deep-Sea Res., 19 (1972) 307.

59 A. Sato, T. Oikawa and N. Saitoh, Bunseki Kagaku, 24 (1975) 584.
60 C. H. Van der Weijden, Chem. Geol., 18 (1976) 65.
61 R. Frache, F. Baffi, A. Dadone and G. Zanicchi, Mar. Chem., 4 (1976) 365.
62 M. I. Abdullah, O. A. El-Rayis and J. P. Riley, Anal. Chim. Acta, 84 (1976) 363.
63 U. Eisner and H. B. Mark, Jr., Talanta, 16 (1969) 27.
64 T. M. Florence and G. E. Batley, Talanta, 23 (1976) 179.
65 R. A. A. Muzzarelli and M. Marinelli, Anal. Chim. Acta, 64 (1973) 371.
66 R. A. A. Muzzarelli and M. Marinelli, Anal. Chim. Acta, 69 (1974) 35.
67 R. A. A. Mazzarelli and M. Marinelli, Anal. Chim. Acta, 70 (1974) 283.
68 K. F. Sugawara, H. H. Weetall and G. D. Schucker, Anal. Chem., 46 (1974) 489.
69 H. Brandeuberger and H. Bader, At. Absorpt. Newsl., 6 (1967) 101.
70 C. Fairless and A. J. Bard, Anal. Lett., 5 (1972) 433.
71 C. Fairless and A. J. Bard, Anal. Chem., 45 (1973) 2289.
72 F. O. Jensen, J. Dolezal and F. J. Langmyhr, Anal. Chim. Acta, 72 (1974) 245.
73 J. Olafsson, Anal. Chim. Acta, 68 (1974) 207.
74 M. P. Newton and D. G. Davis, Anal. Lett., 8 (1975) 729.
75 W. Lund and B. V. Larsen, Anal. Chim. Acta, 70 (1974) 299.
76 W. Lund and B. V. Larsen, Anal. Chim. Acta, 72 (1974) 57.
77 Y. Thomassen, B. V. Larsen, F. J. Langmyhr and W. Lund, Anal. Chim. Acta, 83 (1976) 103.
78 G. E. Batley and J. P. Matousek, Anal. Chem., 49 (1977) 2031.
79 W. R. Hatch and W. L. Ott, Anal. Chem., 40 (1968) 2085.
80 K. Tanabe, J. Takahashi, H. Haraguchi and K. Fuwa, Anal. Chem., 52 (1980) 453.
81 S. Yamazaki, Y. Dokiya, T. Watanabe and K. Fuwa, Ecotox. Environ. Safety, 2 (1978) 1.
82 K. Matsunaga, M. Nishimura and S. Konishi, Nature, 258 (1975) 224.
83 J. Takahashi, H. Haraguchi and K. Fuwa, Chem. Lett., (1981) 7.
84 M. Fujita, K. Iwashima, I. Fukuoka, E. Takabatake and N. Yamagata, Suishitsu Odaku Kenkyu, 1 (1978) 133.
85 R. A. Carr, J. B. Hoover and P. E. Wilkniss, Deep-Sea Res., 19 (1972) 747.
86 S. Nishi, Y. Horimoto and N. Nakano, Bunseki Kagaku, 23 (1974) 386.
87 D. H. Anderson, J. H. Evans, J. J. Murphy and W. W. White, Anal. Chem., 43 (1971) 1511.
88 M. Nishimura, K. Matsunaga and S. Konishi, Bunseki Kagaku, 24 (1975) 655.
89 W. F. Fitzgerald, W. B. Lyons and C. D. Hunt, Anal. Chem., 46 (1974) 1882.
90 H. Haraguchi, J. Takahashi, K. Tanabe, Y. Akai, A. Homma and K. Fuwa, Bunseki Kagaku, 29 (1980) 348.
91 R. S. Braman and C. C. Foreback, Science, 182 (1973) 1247.
92 Y. Talmi and D. T. Bostick, Anal. Chem., 47 (1975) 2145.
93 J. S. Edmonds and K. A. Francesconi, Anal. Chem., 48 (1976) 2019.
94 M. O. Andreae, Anal. Chem., 49 (1977) 820.
95 R. S. Braman and M. A. Tompkins, Anal. Chem., 51 (1979) 12.
96 V. T. Bowen, J. S. Olsen, C. L. Osterberg and J. Ravera, Radioactivity in the Marine Environment, National Academy of Science, Washington, DC, 1971.
97 D. B. Carlisle, Nature, 181 (1958) 922.
98 G. D. Nicholls, H. Curl and V. T. Bowen, Limnol. Oceanogr., 5 (1960) 472.
99 K. K. Pillay, C. C. Thomas, Jr., J. A. Sondel and C. M. Hyche, Anal. Chem., 43 (1971) 1419.
100 R. Litman, H. L. Finston and E. T. Williams, Anal. Chem., 47 (1975) 2364.
101 P. D. LaFleur, Anal. Chem., 45 (1973) 1534.
102 M. Friedman, E. Miller and J. Tanner, Anal. Chem., 46 (1974) 236.
103 G. Ramelow and H. Hornung, At. Absorpt. Newsl., 17 (1978) 59.
104 S. K. El Wakeel and J. P. Riley, Geochim. Cosmochim. Acta, 25 (1961) 110.

105 K. W. Bruland, K. Bertine, M. Koide and E. D. Goldberg, Environ. Sci. Technol., 8 (1974) 425.
106 V. Talbot, R. J. Magee and M. Hussain, Mar. Pollut. Bull., 7 (1976) 53.
107 R. T. T. Rantala and D. H. Loring, At. Absorpt. Newsl., 14 (1975) 117.

Chapter 4c

Analysis of airborne particles in the workplace and ambient atmospheres

THEO J. KNEIP and MICHAEL T. KLEINMAN
New York University Medical Center, Institute of Environmental Medicine, 550 First Avenue, New York, NY 10016 (U.S.A.)

I. INTRODUCTION

This chapter presents a rationale for the use of atomic absorption spectrometry (AAS) for trace element analysis of air samples, and a comparison with other analytical methods currently in use. Sampling techniques, sample preparation and analytical methods, and applications to workplace and ambient atmospheres are also discussed. Step-by-step procedures will be given which can be used to analyze air-filter samples for a broad spectrum of possible analytes.

Before the 1960's, the analysis of toxic elements in airborne materials employed separations and colorimetric determination for single-element problems, or spectrographic methods for multielement, multisample studies. Variable matrices in most aerosols sampled had prevented sensitive, but interference-prone, flame-emission methods from attaining much usage. The increased concern over the environmental effects of toxic elements in the late 1960's resulted in a need for greater sensitivity and ease of operation in measurements of these elements. The many laboratories with increased responsibilities found AAS most useful because of its accuracy, sensitivity, and relative lack of matrix effects, plus the low cost of the equipment.

By the late 1960's a number of laboratories were using the method. Among others, Kneip et al. [1] reported AAS analyses of matter collected on filters from urban and rural air. Thompson et al. [2] reported the use of AAS in a study of sample preparation methods used in the continuing development of the multielement methods in use by the United States Environmental Protection Agency.

The more recent success achieved with Delves cup and heated graphite sources for microsample applications in AAS has widened the applicability of AAS to include major-element analysis of microsamples and trace-element analysis of macrosamples. The range of sensitivities now available affords relatively easy analysis of samples of airborne materials over the full range of concentrations important in studies of air pollution and occupational health problems.

References pp. 137—138

A. Comparison of multielement methods

Despite the availability of neutron activation, X-ray fluorescence and spectrographic multielement methods, atomic absorption continues to be very important. It has advantages of versatility and ease of calibration for laboratories with single element or variable analytical requirements. The high capital cost of the equipment makes the other methods competitive only for multielement, multisample programs.

Even in relatively large programs, few laboratories will justify the initial expense and calibration effort required for development of the emission spectrographic method. As reported by Scott et al. [3], sample preparation will generally not differ significantly from that required for AAS. Instrumental neutron activation analysis (INAA) is only attractive where a reactor is already available, and multielement analysis by this technique requires the use of high resolution Ge(Li) crystals and multiple irradiations for elements with differing activation product half-lives. The key elements, cadmium, nickel and lead still require analysis by AAS because of limitations of the INAA method [4].

The X-ray fluorescence method normally affords adequate sensitivity for most elements of interest. However, overlapping X-ray emission lines may limit the use of this method. For instance, a lead line seriously interferes in the analysis of arsenic, and some transition elements may have overlapping spectra. Because of difficulties in calibration for "thick" samples, particle deposits on filters must be kept to minimal values of mass per unit area. For urban atmospheres, sampling is limited to about a one to six-hour period by the combination of detection limit and sample loading requirements. The elimination of sample preparation makes this method attractive despite relatively high capital costs. Sample mass loading limitations are even more severe for X-ray methods using particle excitation, such as with protons, than with the photon excitation fluorescence technique.

II. APPLICATIONS

For purposes of convenience, atmospheres of interest have generally been classified as workplace (occupational) or ambient. Some studies have been performed on indoor air quality, but indoor aerosols have generally not been analyzed other than for occupational exposures.

The objectives of studies of ambient and workplace air differ considerably. Ambient studies have been carried out to provide general knowledge, observe differences between locations, relate trace-element concentrations to sources, and seek correlations to other air pollution indices or to human health studies. Workplace measurements, on the other hand, are made to assure control of toxic materials at concentrations safe for the workers, as well as to seek better data on health effects. In the United States acceptable levels are defined by the Occupational Safety and Health Administration of the U.S. Department of Labor. These levels can be changed to higher or lower values as new studies are performed and better data become available for relating exposures and workers' health.

A. Flame AAS

Flame AAS methods have been applied to analyses of metals in ambient air [1—8]. If sufficiently large samples are taken, a wide range of elements can be determined.

Several methods are in general use for introduction of a material into the flame. These include solution aspiration, gas (hydride) evolution and entrainment into the flame, and direct introduction of solid substrates. All three have been used for flame-AAS analysis of trace elements from air sampling; however, most work is carried out by use of the first two methods. In each case a suitable solution is first obtained by the dissolution of the elements of interest from the sample. Sample dissolution will be discussed in greater detail in a following section.

The topic of interferences in AAS analyses is dealt with in Chapter 3 of this book. In general it is wise to take the precaution of checking for sample matrix interferences using the method of standard additions, to make general use of background correction techniques unless proven unnecessary, and to match closely the matrices of samples and standards. These precautions will limit the likelihood of errors due to a variety of potential interferences.

Solution aspiration rates, fuel and oxidant mixtures, gas flow rates, burner choice, matrix effects and interelement interferences must all be taken into account when using flame AAS. While optimal choices for the above parameters vary from instrument to instrument, recommendations which afford reasonable starting points for operation have been published by both the Intersociety Committee for Methods of Air Sampling and Analysis (ISC) [7] and the National Institute of Occupational Safety and Health (NIOSH) [8]. These recommendations were the result of ISC efforts supported by both the United States Environmental Protection Agency and NIOSH.

B. Electrothermal AAS

Electrothermal atomizer (ETA) methods are often orders of magnitude more sensitive for a given element than corresponding flame-AAS techniques. These ETA techniques will find great application to situations in which ultra-low ($pg\,m^{-3}$) concentrations are important, or where total sample size is severely limited. Specific instances where ETA will be advantageous are in size classification studies of aerosol particles using multistage impactor samplers which typically segregate aerosols into four or more size fractions and collect at rates of a few liters per minute. Even when high-volume impactor collectors are used in relatively dirty ambient atmospheres, there

References pp. 137—138

are often one or more sample size-fractions which contain a small percentage of the total collected mass.

Studies of variations of the time course of concentration for trace elements can also benefit from the high sensitivity of ETA techniques. Until recently, these applications were in the province of techniques such as neutron activation or X-ray fluorescence, which require the use of high-cost irradiation, counting and computing facilities.

The new, commercially available heated graphite ETA devices are available at much lower cost than XRF devices, and will certainly encourage expansion of the application of low total mass collection type studies. These heated graphite rods or furnaces are the type of ETA devices which one is most likely to encounter in general air pollution applications and most of this chapter's comments will be directed toward them; however, the comments will also be pertinent to special purpose devices such as heated foil and wire atomizers. The reader is directed to Chapters 2 and 3 and to Siemer [9] for general descriptions of ETA devices, their applications, their characteristics and their limitations. Bancroft [10] has reviewed ETA applications to the ultratrace determination of metals.

III. GENERAL METHODS

In every analytical problem a unified approach must be taken to obtain results of satisfactory accuracy. Considerations of interferences due to contamination are of the utmost importance in the analysis of trace elements from air samples. In many cases attention must be given to selection of high-purity acids and other reagents, and particularly to the purity requirements for collection media such as filters.

The best choice of sampling medium depends on the types of substances to be analyzed. It is becoming increasingly rare for an investigator to have a single substance which is of interest to him. For many trace elements, filters such as Gelman Spectrograde, Whatman EPM 1000 or S and S No. IHV are recommended because of their high collection efficiency, relatively low contamination background, and good handling properties. If, however, anionic species such as nitrate or sulfate are to be analyzed in the same sample, prewashed quartz filters may be preferable [11, 12]. Teflon-coated glass-fiber filters have also been used in such studies.

Choice of a dissolution method also depends on the range of substances to be analyzed, but in some specific instances the chemical form of the element is of even more importance. For example, particles formed at high temperature usually contain refractory oxides and silicates which may require fusion with either an acidic or basic flux, or hydrofluoric acid treatment to remove silica, or a combination of these techniques.

It is recommended that in every new problem a careful study is made of responses for standard additions of the element in question using real samples. In the application of some techniques to samples with widely varying compositions, matrix effects may require standard additions for every sample.

A. Sampling

The relative sensitivity of various AAS techniques must be taken into account if samples are to be collected with foresight. By first deciding upon the lowest atmospheric concentration which the method is required to detect and comparing this with known detection limits, an air sample can be collected which contains a sufficient mass of the desired analyte(s). We detail these considerations in Sections IV and V for ambient and workplace atmospheres, respectively.

The use of appropriate sampling and collection methods is obviously critical to the accuracy, precision and interpretability of the results. It is beyond the scope of this chapter to cover the details of such methods. The reader is cautioned that sample collection is not trivial, but has been thoroughly investigated; the reader must understand both the principles and practice of sampling if satisfactory results are to be obtained. Sampling of aerosols, whether solid particles, liquid droplets or both is performed most often by total collection using one of a variety of filters. Collection of size-fractionated particles is performed with devices which use electrostatic precipitation, impaction, centrifugation or diffusion for separation of sizes. Gas sampling methods generally involve absorption in solution, adsorption on solids, condensation, or primary collection as a gas in a container. Reviews of the theories, principles and practices of sampling for aerosols are published every five years in Air Sampling Instruments [5], and are also reviewed in the publications of the Intersociety Committee [7] and the American Society for Testing and Materials (ASTM) [6]. Recently Corn and Esman [13] have reviewed this information and have discussed reasons for particular choices between various techniques.

The information in Table 1 indicates recommended sampling media, general analytical conditions and estimated sensitivities for detection of 37 elements. These data can be used as guides in designing a sampling procedure for a specific application.

Atomic absorption techniques are less prone to interferences than other forms of spectroscopy and samples collected for air pollution analysis generally do not require exotic treatments to remove interferences. Background corrected instruments, use of linear response regions or correction of curvature and care regarding ionization or chemical interferences alleviate most problems. The reader should consult the general chapter on interferences in this volume (Chapter 3). The concentrations of constituents sought are usually higher than those of interfering elements. Those elements of greatest interest, such as lead, cadmium and mercury, are either free of serious interferences or have well understood and easily controlled problems.

References pp. 137—138

TABLE 1

RECOMMENDED SAMPLING MEDIA, SOLUTION CONDITIONS, AND ESTIMATED SENSITIVITIES

Element	Sampling media[a]	Solution conditions			Estimated[e] sensitivities	
		Ionization buffer[b]	Release agent[c]	Other[d]	Flame (ng ml^{-1})	ETA (pg)
Ag	ABC			a	36	0.2
Al	B	X	X	b	760	2.7
As[b]	ABC			c	1	35
Au	ABC			c	—	5
Ba	ABC	X	X		200	7.0
Be	AB*C			b, d	17	1.2
Bi	ABC				220	7.0
Ca	ABC	X	X		20	0.2
Cd	ABC				11	0.2
Co	ABC			b	66	3.8
Cr	ABC			b	55	5.1
Cu[g]	ABC				40	3.3
Fe	ABC			b	62	1.7
Hg	D[h] E[i]				(20 ng m^{-3} [h])	260
In	ABC				380	20
K	AB*C	X			10	—
Li	ABC	X			11	11
Mg	ABC	X	X		3	0.06
Mn	ABC				25	0.5
Mo	ABC			e	600	7.0
Na	B*C	X			3	—
Ni	ABC				66	5.2
Os					—	270
Pb[j]	ABC				110	3.0
Pd	ABC		X		—	12
Pt	ABC		X		—	82
Rb	ABC	X			42	—
Sb[f]	ABC				10	16
Se[f]	ABC				1	30
Si	B				—	18
Sn	ABC	X	X		—	36
Sr	ABC	X	X		44	3.3
Te	ABC	X	X		200	7
Ti	B	X	X		—	1000
Tl	ABC	X			280	3.0
V	ABC	X	X	b, e	880	38
Zn	ABC		X		9	0.1

[a] *Indicates preference, if any. A, Glass fiber. B, Membrane. C, Quartz fiber. D, Silver wool (cold vapor method). E, Hopcalite (graphite furnace-ETA). The use of filter types such as cellulose fiber or Nuclepore requires unusual care to avoid significant losses of small particles due to penetration of the filter [14]. Many elements may be effected by excessive dissolved silica when glass fiber filters are used unless the precautions noted in the procedure described in Section III. B. 1 are carefully followed.

The addition of an easily ionized substance such as cesium, which is usually not an analyte of interest in air pollution work, can reduce problems in alkali metal determination. A releasing agent, such as lanthanum, is useful for overcoming problems in analyzing for alkaline earth elements, and rare earth elements. Information on the use of these additives is given for specific cases in Table 1.

B. Sample preparation

Sample preparation is generally the simplest available procedure. Acid dissolution, followed by filtration or centrifugation is often sufficient to provide a solution suitable for aspiration. In some cases it may be necessary to separate chemically interfering and sought elements. In general, when analyzing air-filter samples, separations are not required. Other elements are easily lost at even moderate temperatures. One must often vary the initial sample treatment to account for the known chemical behavior of the elements sought. This is particularly true for volatile elements such as mercury, selenium, and arsenic which are readily lost when insufficient care is taken during sample preparation. Losses in preparation have also been reported for cadmium, copper, lead and zinc [2]. Sources of error due to chemical interferences from sample preparation are also suggested in Table 1.

1. *Dissolution method*

In most situations encountered in analyzing filters collected in either the workplace or the ambient environment, sample dissolution with nitric acid or a mixture of nitric and perchloric acids is adequate to obtain quantitative results. Since use of perchloric acid requires special fume hoods and is recommended only for chemists well trained in safe methods for use of this

---

[b] Add 1000 μg Cs ml^{-1} to standards and samples.

[c] Add 1000 μg La ml^{-1} to standards and samples.

[d] a, Do not use halide acids. b, Some compounds of these elements may be difficult to dissolve with oxidizing acids (the recommended method uses HNO$_3$). c, Maintain a minimum of 1% v/v HNO$_3$, analyze in 4 hours maximum. d, Final solution should be diluted in HCl. e, Add 1000 μg Al ml^{-1} to standards and samples.

[e] Estimates based on conservative values for concentrations or amounts of analytes which produce 1% reduction in incident light, i.e., 0.0044 absorbance. ETA estimates based on 10 μl injected from a total volume of 10 ml.

[f] Measured by hydride evolution for the flame method.

[g] Copper brushes on sampler motors can cause a serious interference.

[h] Elemental mercury measured by the cold vapor method.

[i] Carbon rod or graphite furnace following collection on Hopcalite and acid leach.

[j] EDTA addition can prevent anion interferences other than chloride, phosphate and peroxydisulfate.

References pp. 137—138

chemical, we suggest use of the following nitric acid extraction procedure which has been partially based on methods of the Intersociety Committee [7] and NIOSH [8].

(i) *Nitric acid extraction*

The samples, including clean blanks (minimum of 1 filter blank for every 10 filter samples), are transferred to clean 125-ml Phillips or Griffin beakers and sufficient concentrated HNO_3 is added to cover the sample (about 50 ml). Each beaker is covered with a watch glass and heated on a hot plate (140°C) in a fume hood until a slightly yellow solution is produced. Approximately 30 min of heating will be sufficient for most air samples. Membrane and cellulose filters will completely dissolve with this treatment. For glass or quartz fiber filters, bring the sample to near boiling for 30 min, cool and decant the solution into a clean 250 ml beaker. Add a second 50 ml of HNO_3 and repeat the boiling and decantation, combining the extracts after cooling. A final extraction with boiling water is performed, the extracts combined and reduced to a very low volume ($\sim 1-2$ ml). A brownish yellow solution is indicative of residual organic compounds. Subsequent additions of HNO_3 may be needed completely to destroy high concentrations of organic material, and a few drops of H_2O_2 can be added to accelerate ashing.

Once ashing is complete, as indicated by a clear, colorless to straw colored solution, reduce the volume to about 1 ml, add 2—3 ml of 0.5% (v/v) HNO_3 and allow to stand covered overnight. This will allow silica to precipitate. Filter if necessary, washing with 0.5% (v/v) HNO_3, quantitatively collecting the filtrate and wash solutions in a 25 ml volumetric flask. If any elements are being determined which require an ionization buffer, 0.5 ml of 50 mg Cs ml^{-1} is added to the volumetric flask. If any elements requiring a releasing agent are being determined, 0.5 ml of 50 mg La ml^{-1} is added to each volumetric flask. The samples are then diluted to volume with water.

(ii) *Perchloric acid extraction*

Alternatively, the following perchloric acid procedure can be used. Perchloric acid can be used on inert (glass or quartz fiber) filters or with filters where a pretreatment has oxidized most of the organic material. In no event should reactive organic compounds be in contact with fuming (hot, concentrated) perchloric acid, nor should perchloric acid be fumed without an adequate fume trapping system or the use of a perchloric acid hood. This procedure is *not* suitable for preparation of samples to be analyzed by ETA. For proper safety precautions refer to "Perchloric Acid and Perchlorates" by Schilt [15].

The method cited here has been in routine use since 1972 without incident. The only major precaution has been extreme care in preventing any sample from evaporating to dryness. All digestions are done in a perchloric acid hood.

Place the filter (glass or quartz fiber) or filter section in a clean 250 ml

Griffin beaker. Add 50 ml of a 5% v/v solution of $HClO_4$ in concentrated HNO_3. Add a minimum amount of additional HNO_3 as required, to cover the sample. Heat to near boiling for one hour with repeated stirring. Decant the acid to a clean 250 ml beaker. Repeat this acid extraction and decant again to the same beaker. Extract the residue with 50 ml of boiling 50% HNO_3 and decant, then extract with 50 ml of boiling deionized water and decant, combining all the extracts in the same beaker.

Evaporate the accumulated extracts without boiling until the volume is reduced to approximately 20—25 ml. Increase the rate of heating to bring the solution to near boiling; a change should occur from a dark brown to black through orange to light yellow color as fumes of $HClO_4$ begin to appear. The dark coloration appears to be due to suspended solid material and the color change precedes the appearance of the fumes. Continue fuming until the solution appears clear and colorless (usually less than 5 ml final volume). Do not let the sample go to dryness!

Cool and filter with suction through a 2.4 cm glass fiber filter (Reeve Angel 934AH), into a 25 ml volumetric flask. Dilute to volume with 5% HNO_3. Prepare standards in a solution of 5 ml of $HClO_4$ diluted to 25 ml with 5% HNO_3.

The authors do *not* recommend that anyone attempt the use of perchloric acid digestions unless previously trained by an experienced chemist and fully aware of all necessary safety precautions.

2. Atomization conditions

Each element measured by flame AAS will require from 0.5 to 2 ml of solution depending on the equipment and technique in use. Ambient studies often involve determination of 8—15 elements, and those elements presented at high concentration may be analyzed in a diluted aliquot of the original analyte solution. The number of elements which can be measured in a given sample will be dependent on the composition of the actual atmosphere sampled.

Substantially less sample solution (1—100 μl) is required for each element measured by ETA. This feature may make ETA an attractive alternative to flame methods when multielement studies are planned. The automatic sample injectors and improved power supply modules available with the current generation of NFAA devices have combined to make ETA less of an art than it was previously.

Recommended conditions for flame and approximate values for ETA (graphite rod, etc.) atomizers are given in Table 2 for a number of elements important with regard to air pollution studies. Conditions are included in the table for the flame system used when hydrides of arsenic, antimony and selenium are generated and passed through the flame. Burrel [16] discusses generation of metal hydrides and cold-vapor mercury evolution techniques in great detail.

TABLE 2
TYPICAL OPERATING CONDITIONS FOR THE ANALYSIS OF METALS

Element	Analytical wavelength (nm)	AAS Spectral bandpass (nm)	Flame conditions[a]	ETA Graphite furnace conditions[b]						
				Dry		Ash		Atomize		
				Temp. (°C)	Time (s)	Temp. (°C)	Time (s)	Temp. (°C)	Time (s)	
Ag	328.1	1.0	1	90	45	600	35	1400	5	
Al	309.3	1.0	4	100	45	1000	35	2500	5	
As	193.7	1.0	5	100	45	500	15	1800	10	
Au	242.8	1.0	1	100	45	600	40	1700	10	
Ba	553.5	0.5	3	100	45	1000	60	2800	10	
Be	234.9	1.0	3	100	45	1000	40	2500	5	
Bi	223.1	0.3	1	100	45	400	40	1600	10	
Ca	422.7	1.0	1	100	45	525	30	2200	5	
		1.0	3(a)							
Cd	228.8	1.0	1	100	45	225	20	1000	10	
Co	240.7	0.3	1	100	45	700	40	2200	10	
Cr	357.9	0.5	4	100	45	1000	30	2200	5	
			1(b)							
Cu	324.7	1.0	1	100	45	600	30	1800	5	
Fe	248.3	0.3	1	100	45	800	30	2300	5	
Hg	253.7	1.0	6	100	45	200	20	1000	5	
In	303.9	0.15	1	100	45	600	20	2200	10	
K	766.5	1.0	1	100	45	825	40	2000	10	
Mg	285.2	1.0	1(c)	100	45	875	40	2100	10	
		1.0	4							
Mn	279.5	0.5	1	100	45	500	40	2200	5	
Mo	313.3	0.5	4	100	45	1100	35	2800	10	
Na	589.0	0.5	1	100	45	800	40	2000	5	
Ni	232.0	0.15	1	100	45	550	30	1950	10	
Os	290.9	0.3	4	100	45	500	20	2800	15	

TABLE 2 (continued)

Pb	217.0	1.0	1	100	45	400	30	1800	5
	283.3	0.5	1						
Pd	247.6	0.3	1	100	45	875	40	2800	10
Pt	265.9	0.5	1(d)	100	45	875	40	2800	10
Sb	217.6	0.3	5	100	45	400	30	2250	5
Se	196.0	1.0	5	100	45	875	40	2250	5
Si	251.6	0.3	4	100	45	1000	35	2800	10
Sn	235.5	0.5	4	100	45	550	40	2500	10
Sr	460.7	0.5	3	100	45	875	30	2500	10
Te	214.3	0.5	1	100	45	875	45	2250	5
Ti	364.3	0.3	4	100	45	1100	35	2800	10
Tl	276.8	1.0	1	100	45	300	30	1800	10
U	358.5	0.15	4(a)	100	45	250	20	2800	15
V	318.5	0.5	4	100	45	625	30	2800	10
Zn	213.9	1.0	1	100	45	400	30	1500	10

a 1, Air–C_2H_2 (oxidizing): fuel lean – Blue Flame.
 2, Air–C_2H_2 (reducing): fuel rich – Yellow Flame.
 3, N_2O–C_2H_2 (oxidizing): fuel lean – 5 mm red "feather".
 4, N_2O–C_2H_2 (reducing): fuel rich – 20 mm red "feather".
 5, H_2–Ar (entrained air): evolution of metal as hydride.
 6, None: cold vapor evolution, absorbance cell used.
 (a) All samples and standards should contain ionization buffer.
 (b) Prone to interference in presence of Cu, Ba, Al, Mg and Ca.
 (c) Addition of releasing agent is necessary if Si or Al is present.
 (d) Releasing agent required (2 mg ml^{-1} final concentration).

b The time and temperature settings in this table reflect average values for a carbon-rod type furnace, and are presented a typical values. The operating parameters for a specific instrument should be used in preference to these estimates.

IV. AMBIENT AIR

Using the estimates of typical ambient concentrations shown in Table 3 and the data for various elements shown in Table 1, we have calculated the volume of air which must be sampled to obtain reliable measurements for a given element. The typical airborne concentrations reported here can be compared with the lists of potentially interferring elements in Chapter 3 to determine the likelihood of a problem.

The ranges of element concentrations in ambient air are based on data from the National Air Sampling Network [17, 18], as well as a review by Rahn [19]. Sensitivity values expressed as the amount of sample which absorbs one percent of the incident light (or produces 0.0044 absorbance units), were averages compiled from the operating manuals of currently used flame AAS and graphite rod or furnace type ETA analyzers. (Reported sensitivities ranged over a factor of 10). We have, on occasion, spot checked these values, and found that the reported values generally agree with the measured sensitivities and indeed may be on the conservative side.

The recommended sample sizes were estimated by dividing the sensitivity by the lower end of the ambient concentration range, and assumes that the sample is diluted to a final volume of 25 ml after dissolution. For ETA devices, which can accept sample injections of different sizes, an injection of $10\,\mu l$ was assumed since this can be accommodated by all of the models.

V. WORKPLACE ATMOSPHERES

The elements of concern in workplace exposures are listed in Table 4. The combination of a 10 ml final volume of analyte solution and instrumental sensitivity for a $10\,\mu l$ injection may be used to guide selection of the sampling volume desired. Workplace sampling often involves study of a specific element known to be in use and established as hazardous. In general this offers the opportunity for somewhat simpler conditions for sampling and analysis. Concentrations are generally higher than in ambient air, interferences are less likely and analyses for multiple elements are generally not needed.

The standard personal sampler operates at a flow rate of about $2\,l\,min^{-1}$ and newer models will sample at up to $4\,l\,min^{-1}$. Thus, in a two-hour period, 240 to 480 l or about 0.25 to $0.5\,m^3$ can be collected. We have used a $0.25\,m^3$ sample volume to estimate the detection limits for metals in air. The dissolution method in Section II. B. 1 should be modified for analysis of workplace samples; use a 10 ml volumetric flask for the final volume, and additions of 0.2 ml of the ionization buffer and releasing agent solutions when required.

The elements surveyed in Table 4 could all be measured at their Threshold Limit Value (TLV) by the methods suggested in this chapter. If the TLV is

TABLE 3

AMBIENT AIR DATA

Element	Typical ambient concentrations (ng m^{-3})	Recommended sample volume ( ) in m^3 to measure ng m^{-3}	
		Flame	ETA
Ag	—	90(10)	1(1)
Al	1700	190(100)	1(10)
As	4—750	250(10)[a]	10(10)
Au	—	—	2(10)
Ba	30	50(100)	2(10)
Be	4—10	425(1)	3(1)
Bi	<1—50	550(10)	2(10)
Ca	770	50(100)	1(1)
Cd	4—450	280(1)	1(1)
Co	4—40	165(10)	1(10)
Cr	4—450	140(10)	2(10)
Cu	1—10,000	100(10)	1(10)
Fe	100—22,000	155(10)	4(1)
Hg	~1—3	150(20)[b]	65(10)[c]
In	—	95(100)	6(10)
K	500	25(10)	—
Li	—	43(10)	3(10)
Mg	440	75(1)	1(0.1)
Mn	10—10,000	65(10)	1(1)
Mo	<1—800	1500(10)	2(10)
Na	470	75(1)	—
Ni	<1—700	165(10)	2(10)
Os	—	—	65(10)
Pb	4—5000	275(10)	1(10)
Pd	—	—	3(10)
Pt	—	—	20(10)
Rb		105(10)	—
Sb	<1—200	250(100)[a]	4(10)
Se	3.5	250(10)[a]	8(10)
Si	—	—	6(10)
Sn	10—200	—	10(10)
Sr	—	110(10)	1(10)
Ti	10—1100	—	25(100)
Tl	—	70(100)	1(10)
V	<1—2200	220(100)	10(10)
Zn	10—10,000	22(10)	2(0.1)

[a] Measured by the hydride evolution method.
[b] Elemental mercury by the cold vapor method.
[c] Measurement by ETA after Hopcalite absorber collection and acid leach [20].

reduced, increasing the sample size or decreasing the dilution volume may bring the final solution concentration into the required concentration range. The use of a higher capacity pump or longer sampling times to increase sample size is recommended.

References pp. 137—138

TABLE 4

WORKPLACE ATMOSPHERE RELATIONSHIPS

Element	TLV[a] (μg m^{-3})	Estimated detection limits[b]	
		Flame (μg m^{-3})	ETA (ng m^{-3})
Ag	10 (metal and soluble compounds)	1.4	0.8
Al	NL[c]	30	11
As	50 (arsenic trioxide production)	0.04[d]	140
	500 (arsenic and compounds as As)		
Ba	500 (soluble compounds)	8.0	28
Be	2	0.7	4.8
Bi	NL[c]	8.8	28
Ca	2 000 (CaO)	0.8	0.8
Cd	50 (metal dust, soluble salts, CdO fume)	0.44	0.8
Co	50 (metal fume and dusts)	2.6	15
Cr	500 (soluble chromic, chromous salts)	2.2	20
	100 (chromic acid and chromate, as CrO_3)		
	100 (chromite ore processing as CrO_3)		
Cu	1 000 (dusts and mists)	1.6	13
	200 (fume)		
Fe	5 000 (iron oxide fume, as iron oxide)	2.5	6.8
	1 000 (soluble compounds)		
Hg		0.02[e]	1 040[f]
In	100 (metal and compounds)	15	80
K	2 000 (as KOH)	0.4	—
Li	25 (as LiOH)	0.44	44
Mg	10 000 (as MgO fume)	0.12	0.24
Mn	5 000 (metal and compounds)	1.0	2.0
Mo	5 000 (soluble, as Mo)	24	28
	10 000 (insoluble, as Mo)		
Na	2 000 (as NaOH)	0.12	—
Ni	100 (metal and soluble compounds)	2.6	21
Pb	50 (inorganic compounds fumes and dusts)	4.4	12
Pd	NL[c]	—	48
Rb	NL	1.7	—
Sb	50 (antimony trioxide production)	0.4[d]	64
	500 (antimony trioxide handling and use, as Sb)		
Se		0.04[d]	120
Si	10 000	—	72
Sr	NL[c]	1.8	13
Tl	100 (soluble compounds)	11	12
V	500 (V_2O_5 dust)	35	152
	50 (V_2O_5 fume)		
Zn	5 000 (ZnO fume)	0.36	0.4

VI. CONCLUSIONS

The versatility of the atomic absorption method has long been obvious. The applications in air sampling and analysis treated here represent only those methods already widely accepted as reproducible in many laboratories. Developments in this field are occurring with great rapidity, and futher methods with simpler, more direct sample handling techniques are expected to cover an even greater range of applications.

REFERENCES

1. T. J. Kneip, M. Eisenbud, C. D. Strehlow and P. C. Freudenthal, J. Air Pollut. Control Assoc., 20(3) (1970) 144—159.
2. R. J. Thompson, G. B. Morgan and L. J. Purdue, At. Absorpt. Newsl., 9 (1970) 53.
3. D. R. Scott, D. C. Hemphill, L. F. Halhoke, S. J. Long, W. A. Laseke, L. J. Pranger and R. J. Thompson, Environ. Sci. Technol., 10 (1976) 877—880.
4. G. S. Kowalczyk, C. E. Choquette and G. E. Gordon, Atmos. Environ., 12 (1978) 1143—1153.
5. American Conference of Government Industrial Hygienists, Air Sampling Instruments for Evaluation of Atmospheric Contaminants, 5th edn., ACGIH, Cincinnati, OH, 1978.
6. American Society for Testing and Materials, 1978 Annual Book of ASTM Standards, Part 26. Gaseous Fuels; Coal and Coke; Atmospheric Analysis. ASTM, 1916 Race Street, Philadelphia, PA, 19103, 1978.
7. Intersociety Committee, Methods of Air Sampling and Analysis, 2nd edn., M. Katz (Ed.), American Public Health Association, 1015 18th Street, N.W., Washington, D.C. 20036, 1977.
8. U.S. Dept. of Health, Education and Welfare, NIOSH Manual of Analytical Methods, 2nd edn., Vols. 1—4. U.S. Government Printing Office, Washington, D.C. 20402, DHEW(NIOSH) Publ. No. 77-157-A, April, 1977.
9. D. D. Siemer, Environ. Sci. Technol., 12 (1978) 539—543.
10. M. Bancroft, in T. Y. Toribara (Ed.), Environmental Pollutants: Detection and Measurement, Plenum Press, New York, 1978.
11. R. L. Tanner and L. Newman, J. Air Pollut. Control Assoc., 26 (1976) 737—747.
12. R. L. Tanner, W. H. Marlow and L. Newman, Environ. Sci. Technol., 13 (1979) 75—78.
13. M. Corn and N. A. Esman, Am. Lab., (July, 1978) 13—26.
14. M. Lippmann, Filter Media for Air Sampling, in: Air Sampling Instruments for Evaluation of Atmospheric Contaminants, 5th edn., American Conference of Government Industrial Hygienists, Cincinnati, OH, 1978.

---

[a] TLV, Threshold Limit Value [21].

[b] For 0.25 m^3; flame, final volume 25 ml; ETA, final volume 10 ml, aliquot 10 µl.

[c] NL, no limit.

[d] By hydride evolution.

[e] By the cold vapor method.

[f] After Hopcalite absorber collection, acid leach.

15 A. A. Schilt, Perchloric Acid and Perchlorates, The G. Frederick Smith Chemical Company, 867 McKinley Ave., Columbus, OH 43223, 1979.
16 D. C. Burrel, Atomic Spectrometric Analysis of Heavy-Metal Pollutants in Water, Ann Arbor Science Publishers, Ann Arbor, MI, 1974.
17 T. B. McMullen and R. B. Faoro, J. Air Pollut. Control Assoc., 27 (1977) 1198—1202.
18 D. M. Bernstein and K. A. Rahn, Trace Element Concentrations as a Function of Particle Size, in T. J. Kneip and M. Lippmann (Eds.), The New York Summer Aerosol Study, 1976. Annals of the New York Academy of Sciences, Vol. 322, The New York Academy of Sciences, New York, NY 10021, 1979.
19 K. A. Rahn, The Chemical Composition of the Atmospheric Aerosol, Technical Report, Graduate School of Oceanography, University of Rhode Island, Kingston, RI, July, 1976.
20 R. E. McCullen and M. T. Michaud, Am. Ind. Hyg. Assoc. J., 39 (1978) 684—688.
21 American Conference of Government Industrial Hygienists, TKVs^R Threshold Limit Values for Chemical Substances in Workroom Air Adopted by ACGIH for 1978. ACGIH, Cincinnati, OH, 1978.

Chapter 4d

Application of atomic absorption spectrometry to the analysis of foodstuffs

M. IHNAT

Chemistry and Biology Research Institute, Agriculture Canada, Ottawa, Ontario (Canada)

I. INTRODUCTION

A. Literature on applications of atomic absorption spectrometry to food analysis

Since the original independent proposals by Walsh [1] and Alkemade and Milatz [2, 3] a quarter of a century ago, regarding the applications of atomic absorption spectrometry (AAS) to chemical analysis, the technique has become firmly and extensively established. The AAS attributes of relative simplicity of operation, high element specificity and detectivity, adaptability to about 70 elements, and low to moderate cost of basic instrumentation have resulted in an exponential rise in interest in the technique with a concomitant innundation by publications appearing annually dealing with applications of AAS to the determination of primarily inorganic but occasionally organic determinands in a vast array of materials. In spite of this situation, however, and contrary to statements that "atomic absorption techniques are now so common place that minor innovations hardly merit publication" [ref. 4, cf. also ref. 5a] and ". . . comprehensive compendia of methods containing proven procedures are nowadays available" [6a], the literature does not abound with details (certainly not concerning approved methodology) in respect of the application of atomic spectrometry to the analysis of foodstuffs.

Several compendia of approved (official, quasi-official, standardized, recommended, etc.) methods, detailing analytical methodologies regarding flame atomic spectrometry and organic matrices are available to the food analyst. Methods for the determination of several of the more common elements in a number of foods and related materials are given in the most recent editions of the authoritative and comprehensive methods manuals of the Association of Official Analytical Chemists (AOAC) [7d—r] and the Society for Analytical Chemistry (SAC) [8d—f] and in supplements [9—20]. Included in the AOAC publications are procedures for the indirect determination of oxalic acid in canned vegetables [7p] and dibutyltin dilaurate in feeds [15]. In addition to providing details on standardized methods of analysis, the SAC publication contains an extensive bibliography [8f] on official, standardized and recommended methods of analysis found in other literature. A third source of "official" methods of analysis is technical bulletin 27 of the Ministry of Agriculture, Fisheries and Food (UK) [21]. Although collectively these three manuals probably represent the most comprehensive documentation of validated flame atomic absorption spectrometric methods of food analysis, shortcomings are apparent in that AAS methods are described for only 5—15 elements. Methods may also be found to an even more limited degree, in other manuals and reports [22d, e, 23d, 24—28]. The dearth of validated atomic spectrometric methods results from the apparent slow acceptance of this technique by analysts in food-science fields and perhaps the necessary slow progression in the validation and acceptance deliberations of organizations. The American Association of Cereal Chemists' AAS method for iron in cereals issued in 1974 [22d] is scheduled to be dropped unless there is a response from users wishing to have it retained! At the time of writing, no AAS methods for foods have been issued by the International Dairy Federation (IDF), the International Association for Cereal Chemistry (ICC) or the International Organization for Standardization (ISO). However, an ICC study group on minerals and trace elements is currently active [29] and it remains to be seen what methodologies result.

Several other monographs have been published depicting step by step analysis procedures with atomic spectrometric measurement. The volumes specializing in food analysis, generally presenting commendable treatments and discussion of methodology are, however, weak on AAS applications. Thus, methods are outlined by Lees [30d—g] only for Cd, Pb, Ca, Cu and Mg, and by Pearson [31c] only for Ca, Cd, Cu and Hg. In spite of their prefatory intention "to give proper representation to the newer techniques such as ... atomic absorption spectrophotometry", Hart and Fisher [32] devote only one page to a discussion of AAS; furthermore, an AAS procedure is given for only one element in one food matrix — copper in milk! In light of today's analytical technology, the dithizone and Nessler tube colorimetric procedures for mercury and lead, respectively, and a volumetric approach to the estimation of zinc reported by Lees [30h] add an aura of historical perspective to his handbook. A multi-authored volume on AAS edited by Pinta (ref. 33, a 1975 English translation of the 1971 French edition) presents an excellent chapter [34] on analysis of vegetable matter, including discussion and description of methods of sample treatment, and determination of major and trace elements. A further chapter [35, 36] dispenses with foods per se in one page, and devotes seven pages to the measurement of elements in wines. Chemical Analysis of Ecological Materials by Allen et al. [37d] documents, for a variety of materials, procedures which could be adapted to food commodities. AAS procedures are given for the measurement of Al, Ca, Cu, Fe, Mg, Mn, Sb, Cd, Cr, Pb, Hg, Ni and Zn, together with some discussion of the technique. Several other books devoted to principles and techniques of food analysis via a non methods-manual approach, have only briefly touched upon AAS. Joslyn [38] remarks on AAS in a footnote to the preface, mentions it in passing in the text, and presents essentially no treatment of either the technique or inorganic food components. Both MacLeod [5b] and Pomeranz and Meloan [39a, b] present elementary pedagogical treatments of principles of the technique, with the former including a section on analysis of foodstuffs by flame techniques.

Additional discussion, reviews and guides to the literature regarding AAS applications in the food and allied fields are provided by the numerous monographs [6, 40—53], and reviews [54—61] on AAS. Koirtyohann and Pickett present the most detailed discussions on analysis of foods [52a] and agricultural materials [52b], whereas more limited treatments are presented by Ramirez-Muñoz [44], Slavin [43a], Price [42, 50], Welz [6b] and Kirkbright and Sargent [51] with the inclusion of bibliographic surveys. In the first monograph on the newer technique of electrothermal atomization atomic absorption spectrometry, (EAAS), Fuller [53a] devotes four pages to applications to plant and food matrices. Good and interesting discussions in respect of biological applications are also given in some of the monographs referred to above. For an exposure to in-depth treatments of the principles and applications of flame AAS, the reader is referred to the excellent, comprehensive treatises edited by Mavrodineanu [62], and Dean and Rains [63], and written by Kirkbright and Sargent [51]. Excellent continuation reviews are the Annual Reports on Analytical Atomic Spectroscopy published by the Chemical Society, and the applications reviews published biennially in odd years in Analytical Chemistry. A recent review by Crosby [60] presents a good overview of the determination of metals in foods by a variety of techniques, including AAS. An FAO report by Schuller and Egan [58] reviews methods of trace analysis, including AAS, with reference to the determination of Cd, Pb, Hg and methylmercury compounds in foods. Several other reviews published in the 1970's [54—57, 59, 61] present additional but at times abbreviated treatment. No single recent review has yet appeared covering comprehensively the application of AAS to food analysis.

B. Scope of chapter

An analytical chemical procedure is composed of a series of operations applied in sequence, leading from the material of interest to a statement regarding its chemical composition in respect of a chemical constituent or constituents of interest. The scheme of analysis may be subdivided into the five main component activities of (a) sampling, (b) sample treatment, (c) analyte separation and manipulation, (d) analyte measurement and (e) expression and interpretation of results. Each of these is integral to the overall analytical effort and cognisance of the respective influences on data quality is essential for the generation of useful and reliable analytical information. Sampling includes the broad domain of sample collection, preservation, storage, transmission, removal of extraneous and foreign matter,

References pp. 201—210

reduction in particle size and mixing, subsampling of the bulk sample and other required procedures to yield the analytical sample. It includes considerations of conceptually correct sample type (e.g. edible portion of the foodstuff or the commodity as grown or purchased, the entire item or the portion in which the analyte is concentrated), representative (statistically sound) sampling, sampling and processing apparatus, and storage containers and contamination therefrom, and delay before analysis. Sample treatment in the context of this chapter pertains to the bringing into solution of the analyte of interest (inorganic) effected by simple extraction with aqueous solutions or partial dissolution of the organic matrix, or most commonly by the complete dissolution and removal of organic matrix by wet digestion or dry ashing/acid dissolution. The process of analyte separation and manipulation is concerned with the isolation of the analyte using techniques such as chelation—solvent extraction, co-precipitation and ion exchange to present a higher analyte concentration and simpler matrix to the measuring device, and involves associated manipulations including matrix modification when necessary. Detection and quantitative measurement of analyte concentration in the final analytical solution covers the analyte measurement step. Finally, how the analytical results are expressed depends on the objective of the exercise, e.g.: presenting the concentration of the analyte as element or oxide on a fresh weight, moisture-free weight, ash, or other weight basis; data interpretation again relates to the study objectives, and could include consideration of analytical methodological and sample population accuracies and variances.

It is with the topic of analyte determination in foods by the technique of analytical AAS that this chapter is concerned. Analyte quantitation (d above) by this technique is thus the main thrust of this treatment, but of necessity, the intimately related procedures of sample treatment (b) and analyte separation and manipulation (c) will also be discussed insofar as they bear on quantitative measurement by AAS. Food for human consumption is the main concern of this chapter. Peripheral discussion, however, of allied commodities such as plants and animal feedstuffs, is included to make the treatment more comprehensive, especially in areas where there is a dearth of publications relating to food-analysis applications of atomic spectrometry. For detailed accounts of methodologies bearing on such related materials, the reader is referred to the other chapters in this volume.

The usefulness of a methodology manual is in no small way directly related to the reliability of methods described. It is therefore preferable that methods selected for inclusion therein be tested and satisfactory performance demonstrated. This criterion was assumed to be met by selecting methods from the following sources, listed in order of estimated decreasing reliability: official methodology manuals with method subjected to interlaboratory validation > other methodology manuals > literature and independently confirmed by other laboratories > unconfirmed reports in the literature. Due to the relatively small number of validated AAS methods, and the requirements of analysts to deal with an increasing number of elements, several "non-validated" procedures are included in an attempt to fill the gap. Elements are discussed in order of their sequence in the periodic table. References listed do not provide a comprehensive coverage and reflect only a relevant sampling of the literature.

C. Elements of interest

The occurrence of elements in foodstuffs is a function of the biological roles played by the elements in the structure and physiology of the food tissue, and adventitious contamination during growth, processing and preparation. Of the at least 90 naturally occurring elements, 26 are known to be essential for animal life; in addition, boron has been demonstrated essential for the higher plants. Of the 11 essential major elements (H, C, N, O, Na, Mg, P, S, Cl, K, Ca) and 16 trace elements (B, F, Si, V, Cr, Mn, Fe, Co, Cu, Zn, As, Se, Br, Mo, Sn, I), the following are usually determined or determinable directly by AAS: Na, Mg, K, Ca, B, Si, V, Cr, Mn, Fe, Co, Cu, Zn, As, Se, Mo and Sn. A large number of other elements occur in foodstuffs as environmental contaminants, occasionally at levels toxic to the food organism or to the user of the food product. Cd, Hg and Pb (all measurable by AAS) are three elements of current interest in this context (there is, however, recent evidence that cadmium may be essential [64b]! At excessive levels, even nutritionally essential elements may exhibit toxicological character, and elements such as arsenic and tin are usually looked upon as toxic contaminants.

To give the analyst some idea of the elemental content to be expected in foodstuffs, a listing of estimated typical ranges of some of the more important elements in 12 different classes of food is presented in Table 1. The data referring to total concentrations in edible food portions were distilled from the extensive compilations on Na, K, Mg, Ca, Fe and Zn in a vast variety of foods, contained in USDA Handbooks 8, 8-1 and 8-2 [65—67], the compilations for 12 trace elements authored by Schlettwein-Gsell and Mommsen-Straub [68, 69], and a number of other publications referred to in the Table [64a, 65—119]. Selection of elements considered in this chapter relates to Table 1.

II. SAMPLING AND PREPARATION OF ANALYTICAL SAMPLES

A. Sampling

However good may be the methodology for the determination of a specific component in a sample, the result will be of little value if the sample analyzed does not represent the bulk sample and if the bulk sample does not reflect the original sample and the objective of the study. The analyst and food scientist thus must be cognizant of several facets relevant to the

References pp. 201—210

TABLE 1

COMPOSITION (mg kg^{-1}) OF FOODSTUFFS IN RESPECT OF ELEMENTS DETERMINABLE BY ATOMIC SPECTROMETRY[a]

Food Class	Na	K	Mg	Ca	B	Al	V
Cereal products	6— 6700	90— 5200	20—6900	40— 5000	0.03 — 17	0.4 — 300	0.002 —1
Dairy products	130—19000	320—21000	10—2300	240—21000	0.002— 8	0.1 — 700	0.00005—0.1
Eggs and egg products	490—12000	830—11000	40— 720	90— 2800	0.01 — 1	0.2 — 2	—
Meat and meat products	500— 1800	1400— 4500	70—1000	20— 200	0.1 — 3	0.5 — 10	0.001 —0.05
Fish and marine products	400—81000	800— 5400	90—1000	100—60000	0.1 — 5	0.4 — 8	0.002 —0.05
Vegetables	10— 4600	950—19000	14—2800	140— 2600	0.3 —125	0.02— 50	0.0008 —6
Fruit and fruit products	10— 340	350—16000	5—1000	30— 1300	0.05— 40	0.1 — 20	0.001 —0.06
Fats and oils	0— 9900	0— 230	1— 480	0— 200	0.3 — 6	0.7 — 3	0.001 —0.6
Nuts and nut products	10— 40	4500— 7700	330—4100	60— 2300	4 — 40		0.06 —0.2
Sugar and sugar products	10— 2300	10— 6200	2—2600	5— 2300	0.05— 8	0.6 — 110	—
Beverages	0— 720	0—33000	0.3—6400	0— 1800	0.02— 25	0.02—1300	0.00009—0.3
Spices and Condiments	50—10000	730—47000	100—6900	800—21000	0.5 — 95	20 —1000	0.1 —1

Food class	Cr	Mn	Fe	Co	Ni	Cu	Zn
Cereal products	0.01 — 2	0.5 —260	2 — 94	0.0001—1.2	0.08 —3.4	0.2 — 54	0.6 — 210
Dairy products	0.01 — 2	0.005— 7	1 — 21	0.0005—0.07	0.004—1.4	0.03 — 15	0.2 — 105
Eggs and egg products	0.2 — 0.5	0.08 — 2	0.3 — 108	0.005 —0.1	0.03 —0.4	0.05 — 2.4	0.1 — 62
Meat and meat products	0.06 — 0.4	0.09 — 5	5 — 65	0.005 —0.7	0.02 —0.9	0.1 —100	1 — 500
Fish and marine products	0.01 — 0.4	0.02 — 37	4 — 36	0.007 —2	0.02 —1.7	0.1 —140	2 —1600
Vegetables	0.002— 0.7	0.04 —150	6 — 96	0.002 —5	0.01 —7	0.02 — 18	0.02— 280
Fruit and fruit products	0.01 — 0.9	0.08 — 45	0.5 — 30	0.001 —0.3	0.06 —0.6	0.04 — 10	0.05— 22
Fats and oils		0.01 — 36	0.1	0.04 —0.6	0.01 —2.3	0.002— 7	0.02— 30
Nuts and nut products		0.4 — 42	22 — 60	0.04 —0.6	0.05 —1.3	0.2 — 70	0.5 — 145
Sugars and sugar products	0.02 — 0.4	0.1 — 3	0.7 — 50	0.007 —0.1	0.03	0.1 — 14	0.2 — 8
Beverages	0.003— 0.8	0.1 —900	0.09— 56	0.03 —0.9	0.01 —8	0.002— 50	0.02— 54
Spices and condiments	0.01 —10	0.5 —410	30 —1200	0.7 —1.2	0.4 —4	0.2 — 24	0.7 — 125

TABLE I (CONTINUED)

Food class	As	Se	Mo	Cd	Sn	Hg	Pb
Cereal products	0.01 — 0.5	0.002 — 4	0.06 —6	0.005–0.9	3 — 30	0.002 —0.13	0.05 —0.8
Dairy products	0.004— 1.5	0.003 — 0.2	0.00002–0.5	0.002–0.01	—	0.0006–0.3	0.001–0.8
Eggs and egg products	0.002— 0.2	0.02 — 1	0.1 —0.8	<0.01 —0.03	—	0.002 —0.07	0.03 —0.3
Meat and meat products	0.01 — 0.05	0.004 —12	0.07 —4	0.02 —0.5	—	0.0008–0.4	0.05 —1.2
Fish and marine products	0.1 —60	0.1 — 9	0.03 —0.6	0.01 —5	0.5— 5	0.002 —2	0.03 —1.7
Vegetables	0.000— 0.1	0.001 —12	0.004 —6	0.001–0.1	—	0.001 —0.06	0.01 —1
Fruit and fruit products	0.02 — 0.2	0.01 — 0.2	0.01 —0.8	0.01 —0.05	8 —130	0.001 —0.04	0.01 —1
Fats and oils	0.02 — 0.3	(0.25)	0.2 —1	0.01 —0.1	—	0.001 —0.005	0.02 —1
Nuts and nut products	—	0.02 — 8	0.2 —0.6	—	—	—	0.05 —0.2
Sugar and sugar products	0.01 — 0.08	0.003 — 0.3	0.08 —0.9	0.005–0.03	—	0.0007–0.03	0.01 —0.8
Beverages	0.01 — 0.08	0.0006– 0.25	0.00003–0.2	0.01 —0.1	0.2– 50	<0.001 —0.5	0.01 —0.7
Spices and condiments	—	0.006 — 2	0.2	(0.8)	—	—	(3)

[a] Estimated typical total concentration ranges in the edible portion on a fresh, dry, processed or prepared weight basis; adapted from data reported in refs. 64a, 65—119. Data in parentheses reflect limited information.

investigation toward which the analytical information makes a contribution, when conducting bulk sampling, selecting a food component for analysis, and reducing the bulk sample to a laboratory sample on which the actual analytical determinations are to be made. Thus, if the objective of the study is to measure the average level of manganese in the fruit of an apple orchard, then all the apples collected would be reduced to one convenient homogeneous laboratory sample for analysis. If, on the other hand, one is interested in the variability of manganese among the individual fruits, analysis of the individual fruits would be indicated. Similarly, attention must be paid to over variables having a possible bearing on concentrations of the analyte of interest: food component (e.g. tissue type, edible part or as purchased), growing and processing conditions, time of sampling (e.g. diurnal, seasonal and inter-year variation in chemical composition), length and conditions of storage, etc.

Deliberations on sampling are beyond the scope of this treatment. For coverage of this topic, the reader is referred to several selected references. Basic sampling concepts and practices have been treated by Benedetti-Pichler [120], Walton and Hoffman [121] and Bicking [122], whereas more detailed delineations of sampling procedures in respect of foodstuffs are contained in various publications on food analysis and pronouncements from specialized organizations [7a, 8a, 22a, 23a, 30a, 37a, 38a, 39c, 123a, 124a, 125a, 126, 127, 128a].

B. Preparation of analytical samples

The bulk sample initially collected and consisting of a large, most likely heterogeneous mass of perhaps tens of kilograms, must be reduced in size prior to analysis. The preparation of an analytical sample (sub-sample) involves reduction in amount and in particle size and thorough mixing to yield a portion representing the average composition of the bulk sample from which a several gram portion is removed for chemical analysis. For mineral determinations, initial air or oven drying of materials such as those of plant origin followed by grinding can precede bulk reduction. Dry samples such as ground solids, powders, grains and pulses can be reduced to representative subsamples by quartering or mechanical reduction. Pulverisation of solid and semi-solid foodstuffs can be effected by use of various mortars, food choppers, blenders, mixers and mills. Lyophilization (freeze drying) of bulk samples such as meat will facilitate pulverization, subsampling and storage. Freeze drying is accomplished by freezing the sample and evaporating off the ice under vacuum. As analyte levels may vary greatly among the different components and size fractions of the sample, steps must be taken to include, in the analytical sample, all components, dust, etc., to ensure representative sampling. Liquids can be mixed by repeated slow inversion of the container or by stirring prior to sub-sampling. Care must be taken to prevent stratification in liquids and finely divided powders. Storage should be in air-tight containers at room or low temperature as required.

Due regard must be paid to the question of contamination of bulk and analytical samples, especially when dealing with trace analysis. Soil splash, road dust and salt spray must be removed from vegetation surfaces by rinsing with water; in the summer, perspiration contamination during collection can be avoided by the use of polyethylene gloves. The grinding of hard materials may result in some contamination from the grinders, e.g. by Fe, Mn, Mo, Co, Cr, Ni from steel, and Al, B, Ca, K and Na from ceramic construction materials. Metal-to-metal contact in food choppers can be a source of contamination. It is conceivable that contamination of liquid and semi-solid food may occur via attack of the sample on the metal blades and container. Storage may also result in positive or negative contamination from the storage container, and pseudo-contamination due to sample segregation, if attention is not paid to container suitability and proper re-mixing prior to analysis. It may be necessary to monitor contamination and gauge its magnitude by comparing analyte levels in the prepared sample with those in a stringently prepared separate portion of the bulk sample, (less contamination) or with levels determined on a laboratory sample subjected to more extensive processing according to the prescribed procedure (more contamination). This area of food analysis remains incompletely explored.

Details are presented below for the preparation for elemental determinations, of analytical samples of some of the more common foodstuffs. It is assumed that bulk samples have been suitably collected; reduction to suitable size is to be by appropriate methods. Although in general these procedures are based on those advocated by official organizations for subsampling and storing samples, their suitability for all elements, particularly at low trace levels, has not necessarily been demonstrated. As some elements in certain food matrices may occur in volatile organoelemental forms, they may be lost during common drying procedures. It is in fact sometimes recommended, when analysing biological reference materials, to conduct determinations on the undried material (as supplied) and adjust the results to a dry sample basis using the moisture content measured on a separate portion of sample. This procedure is advocated here. Wherever results are to be expressed on a fresh-weight basis, and preliminary drying is indicated, the sample weight prior to drying must be recorded. For additional details on analytical sample preparation, consult refs. 7s, 8g, 21, 24, 30i, 37a, 38d, 39d, 125b, 128b. There is some discussion in the literature of errors in elemental determinations resulting from inhomogeneity [129—131], drying [132—136], and storage [137]. As results may often have to be reported on a moisture-free or dry-matter basis, and occasionally on an ash basis, a number of references are listed dealing with determinations in foodstuffs of moisture and solids [7b, 8b, 22b, 23b, 30b, 31a, 37b, 38b, 39e, 123b, 124b, 125c, 128c] and ash [7c, 8c, 21, 22c, 23c, 30b, 31b, 37b, 38c, 39b, 123c, 128d].

References pp. 201—210

1. *Recommended procedures for preparation of analytical samples*

(i) *Cereal products*

Deliver bulk wheat flour samples to air-tight containers. Prior to opening containers for analysis, thoroughly mix contents by inverting and rolling the container. For samples of wheat, rye, oats, buckwheat, corn, barley and rice and their products, grind the sample to pass a 1 mm mesh sieve and mix. For samples too moist to grind readily, dry at ca. 100°C to remove excess moisture before grinding. When removing a sample for analysis avoid temperature and humidity extremes. For all types of bread not containing fruit, cut into slices ca. 2 mm thick, dry at warm ambient temperature until sufficiently brittle, grind to pass a 1 mm mesh sieve, mix well and store in air-tight containers. Use a food chopper instead of a grinder for processing raisin bread. Mill-grind macaroni and similar products to produce ca. 500 g of material passing through a 1 mm mesh sieve and store in air-tight containers.

(ii) *Dairy products*

Milk and cream. Immediately before withdrawing portions for analysis, bring the sample to ca. 20°C and mix thoroughly by agitation, taking care not to cause frothing, until a uniform emulsion forms. If the cream does not disperse, warm the sample slowly in a water bath to 30—40°C and mix until homogeneous.

Unsweetened evaporated milk. Heat the unopened can in a 40—60°C water bath for 2 h, shaking vigorously every 15 min. Remove from bath, allow to cool to room temperature, open, and thoroughly stir the contents.

Sweetened condensed milk. Heat the unopened can in a water bath at 30—40°C, open, remove entire contents and stir thoroughly until sample is homogeneous.

Dried milk and nonfat dry milk. Transfer the sample to a dry, air-tight container with a capacity of ca. twice the volume of the sample and mix by shaking and inverting. Sift the sample through a 0.85 mm mesh sieve if necessary to break up any lumps and return it to the air-tight container.

Butter and margarine. Soften the sample in a sampling container by warming in a water bath at as low a temperature as practicable, but not exceeding 39°C. Shake at frequent intervals during softening procedure, remove from bath, shake vigorously until the sample cools to a creamy consistency and promptly remove the sample for analysis.

Cheese. Remove surface layer not usually consumed. Cut the sample into strips, pass several times through a food chopper and mix. Blend cream cheese and similar products in a high-speed blender for ca. 2—5 min until homogeneous. Alternatively mix well by intensive kneading. Store all samples in air-tight containers.

(iii) *Eggs and egg products*

Keep ca. 500 g of representative samples of liquid and frozen eggs frozen in air-tight containers. Warm in a 50°C bath and mix well before removing sample for analysis. For dried eggs, take ca. 500 g of a representative sample and store in an air-tight container; pass several times through a flour sifter to break up lumps before removing a sample for analysis.

(iv) *Meat and meat products*

Meats. Render uniform, a representative sample free of bone, by passing at least twice through a food chopper with plate openings, not larger than 4 mm, mixing thoroughly after each grinding. Remove casings from sausages before grinding. If not immediately required for analysis, dry either in vacuo at a temperature below 60°C or by evaporating on a steam bath with alcohol. Extract fat from the dried sample with petroleum ether (b.p. less than 60°C). Evaporate the ether and store the fat in a cool place for analysis if required.

Meat extracts. Completely remove liquid and semiliquid preparations including sediment from the container, and mix thoroughly, warming if necessary. Grind in a mortar, a representative quantity of products in solid form.

(v) *Fish and marine products*

Fresh and frozen fish. Cut representative samples from fish with or without heads, skin and bones as dictated by the nature of the test. Grind the sample several times using a meat chopper, with 1.5—3.0 mm holes, each time removing unground material from chopper and mixing with the ground material. Alternatively, use a blender for soft fish. For fish packed in brine, drain the brine, rinse off adhering salt crystals with saturated NaCl solution and drain again for 2 min prior to proceeding with grinding. Let frozen fish thaw at room temperature before processing.

Dried fish. Cut samples into small pieces, mix, reduce by quartering to ca. 100 g, and grind as finely as possible.

Canned products. Blend solid and liquid contents of a can in a blender until homogeneous or pass several times through a meat chopper. For large cans, drain the meat, collect the liquid, determine the weight of the meat and volume of liquid, recombine each in proportionate amounts and blend until homogeneous.

Shellfish. Before opening, wash exterior of the shell with a brush and potable water to remove all loose silt and dirt, and drain. Separate the edible portion from shells of at least 10 specimens, place in a suitable tared container and determine the wet weight. Blend the sample (solid and liquid) using a blender with stainless steel blades and Teflon gaskets. After removing edible portions from shell oysters, shell clams and scallops, transfer the shellfish meats to a skimmer, pick out pieces of shell and drain for 2 min before homogenizing meat and liquid.

References pp. 201—210

(vi) *Processed vegetable products*

Dried vegetables. Grind the sample to pass a ca. 0.5 mm mesh sieve and store in an air-tight container.

Canned products. For products containing solid and liquid components thoroughly grind the entire contents of a can in a food chopper and mix. If only the solid portion is required for analysis, grind drained vegetables. Add, if necessary, an equal weight of water for proper operation of the blender. Transfer to an air-tight container. If the sample is to be kept for some time before analysis, dry and store in an air-tight container. For comminuted products shake the unopened container to incorporate any sediment, transfer contents to a large receptacle and mix thoroughly by stirring. Store in an air-tight container, drying first, if the sample is to be kept for some time prior to analysis.

(vii) *Fruits and fruit products*

Pulp fresh fruits, dried fruits and preserves using a food chopper or suitable mixer or by grinding in a large mortar, and mixing thoroughly. For jellies and syrups, mix thoroughly to ensure homogeneity. For canned fruits, empty the contents on a 2 mm mesh sieve and leave to drain for two min. Invert all fruits having cups or cavities if they fall on the sieve with cups or cavities up, then proceed as for canned vegetable products.

(viii) *Fats and oils*

Melt solid fats and filter, using a heated funnel; filter oils that are not clear and make determinations on these homogeneous samples. Store in a cool place protected from light and air.

(ix) *Nuts and nut products*

Separate meats from shells, grind twice at least 250 g in a food chopper equipped with 3 mm plate openings or other suitable devices that give a smooth homogeneous paste without loss of oil. Mix well and store in an air-tight container. For butters and pastes, mix, warming semi-solid products.

(x) *Sugars and sugar products*

Sugars and syrups. Grind solids, if necessary, and mix thoroughly. With semi-solids, dissolve 50 g in a minimum volume of water, wash into a 250 ml volumetric flask, dilute to volume and mix. Alternatively, weigh and dilute 50 g of sample with water to 100 g. Mix uniformly by shaking to disperse any solid material before taking aliquots for analysis. With liquids, mix thoroughly. Dissolve any sugar crystals by gently heating, taking care to avoid evaporative loss of water. Calculate all analytical results to the original sample basis.

Molasses, molasses products and confectionery. Pass dry molasses samples (after reaching room temperature) containing lumps through a ca. 2 mm mesh sieve, crush the residue with a pestle in a dry mortar, and add to the portion passing through the sieve, mix thoroughly and store in an air-tight container. Process liquids as mentioned above. Grind confectionery samples, mix thoroughly and store in an air-tight container.

Honey. With liquid honey, mix by shaking and remove a portion for analysis. With granulated products, warm the closed container in 60—65°C water bath until the contents are liquefied, shaking occasionally. Mix thoroughly, cool, and remove a sample for analysis. Foreign material may be removed by straining warm (40°C) sample through cheesecloth in a heated funnel. With comb honey samples, separate the honey from the comb by straining through a 0.4 mm mesh sieve, straining through cheesecloth as outlined above, if comb or wax pass through the sieve.

(xi) *Beverages*

Remove CO_2 from beer and wine by transferring the sample to a large flask and shaking, or by pouring back and forth between large containers. Remove, by filtration, after removal of CO_2 and mixing, abnormal suspended matter or sediment present in beer, wine, fruit juices and other beverages marketed in the clear condition. Grind coffee and tea specimens to pass a 0.6 mm mesh sieve and store in air-tight bottles.

(xii) *Spices and other condiments*

Grind spices to pass through a 1 mm mesh sieve and mix. Immediately prior to analysis, thoroughly mix the material and remove a portion of not less than 2 g for analysis to insure uniformity. For prepared mustard, transfer the entire contents of the container to a vessel, stir well, store in an air-tight container and stir before removing a portion for analysis. Vinegars need only to be mixed and filtered.

For semi-solid and emulsified dressings, transfer a sample to a larger vessel, mix for ca. 2 min. until homogeneous and without delay, remove a portion for analysis. Blend separable dressings with 0.20 g of egg albumen powder as emulsifier per 100 g of sample in a high-speed blender. Mix thoroughly immediately prior to removing an aliquot for analysis. Correct analytical results for added emulsifier (include appropriate quantity of emulsifier in the blank determination).

(xiii) *Plants*

Remove all foreign matter such as adhering soil and dust by rinsing with distilled water but avoid prolonged rinsing to minimize leaching of soluble mineral constituents. Dry at room temperature or in an oven at 35—40°C until the sample is sufficiently dry to be ground. Grind the sample to pass a ca. 0.5 mm mesh sieve and store in an air-tight container. Avoid contamination by dust during drying, and from grinders and sieves when determining elements such as Ca, Al, Fe and Cr abundant in dust and equipment.

References pp. 201—210

(xiv) *Animal feeds*

Reduce an amount of representative material to a laboratory-sized sample by the technique of quartering as follows, first breaking or cutting into smaller pieces material originally present in long form. Mix the sample on a clean non-absorbent surface by adding to the top of the heap from the bottom thus minimizing size and density bias; divide the heap into quarters, reject two diagonally opposite sections, thoroughly mix the two remaining quarters and repeat the procedure until a sample of convenient size, ca. 200—300 g, remains. Dry as mentioned above for plants or follow other instructions specific to the material and determination to be made (e.g., ref. 21). Grind the sample to pass a 1 mm or smaller mesh sieve or otherwise reduce to as small particle size as possible, mix well and store in an air-tight container.

Notes

Although the foregoing procedures are based on those tested and advocated by official organizations for subsampling and storing samples, their suitability for all elements, particularly those at very low levels, has not been demonstrated. In particular, maceration, grinding and filtration may result in contamination, and these and other usual sample processing steps must be carefully evaluated, altered and avoided if necessary when dealing with trace analysis. Refer to section II.B for more discussion and to the references listed below for more details.

Selected references. These procedures were adapted essentially from AOAC Official Methods of Analysis [7s], The Society for Analytical Chemistry, Official, Standardised and Recommended Methods of Analysis [8g], UK Ministry of Agriculture Technical Bulletin 27 [21], Subcommittee on Procedures of the Chemistry Task Force of the (US) National Shellfish Sanitation Program [24] and Joint FAO/WHO Food Standards Programme, Codex Alimentarius Commission [125b]. Other pertinent references are: 30i, 37a, 38d, 39d and 128b.

III. SAMPLE TREATMENT

Prior to quantitation of analyte by atomic spectrometry it is usually necessary to destroy the organic matrix and bring the element into solution. Most of the multitude of decomposition procedures reported fall into one of two classes, wet digestion and dry ashing. Often many variants of each procedure provide adequate results with a variety of analytes and matrices. Several commonly used procedures of general applicability are described below; specific details are found in the sections dealing with the determination of individual elements. The reader is referred to several good sources of information on sample decomposition for fuller details and discussion [8h, 34, 138—140]. The procedures described below result in essentially total decomposition of the sample with the exception of complete attack on silicate and refractory organic compounds [141]. Many useful extractive, partial digestion procedures reported in the literature are not treated in this section.

A. Recommended procedures of sample treatment

1. Digestion with nitric and perchloric acids

Exercise caution with perchloric acid [8h]. Transfer to a 200 ml borosilicate Kjeldahl flask an accurately weighed amount of sample containing not more than 2 g of dry matter, add 25 ml of HNO_3 and boil gently for 30 min. Cool, add 15 ml of $HClO_4$ (3) and boil gently (about 1 h) until the solution is colorless or nearly so and white fumes of $HClO_4$ are evolved. *Do not allow contents to go dry.* Quantitatively transfer the contents, by rinsing out with water, into a suitable volumetric flask and make to volume.

Notes

(1) This procedure, adapted from ref. 8h, is applicable to a wide variety of biological materials, will result in the destruction of protein and carbohydrate, and will give a solution suitable for the determination of most of the mineral elements. Although small amounts of fat are tolerated, the method is not recommended for samples with a very high fat content. Due to the possibility of violent reaction with some highly reactive materials, this decomposition procedure should be first tested on small quantities of materials of unknown composition with due regard to possible hazards. Abide by safety precautions when working with perchloric acid [8h] (see also Chapter 3 of this book).

(2) With such large volumes of acids, charring of sample will not usually occur. Digestions should not, however, be left unattended and more HNO_3 in 1 ml portions should be added whenever darkening of the digestion mixture occurs. This procedure has the disadvantage of requiring a higher proportion of acids to sample than may be desired. The ratio of acids to sample may be reduced with attention to appropriate operational information. Close operator attention is required; charring of the sample, indicating the onset of potentially dangerous conditions must be rectified by the addition of HNO_3 as mentioned above. The references cited above and original literature dealing with the materials and elements of interest should be consulted for details of operation and safety.

2. Digestion with nitric, perchloric and sulfuric acids

Transfer to a 100 ml Kjeldahl flask 2 g of an accurately weighed sample and add 5 ml of HNO_3. After vigorous initial reaction subsides, heat gently until further reaction ceases, cool, slowly add 8 ml of H_2SO_4 and heat until the liquid darkens appreciably. Add gradually a 1—2 ml portion of HNO_3, and heat until darkening occurs again but do not heat so strongly as to cause excessive charring. Repeat the addition of HNO_3 and heating until the solution no longer darkens on heating for 5—10 min and is only pale yellow. Add 1 ml of $HClO_4$ and 2 ml of HNO_3 and heat for 15 min. Add a second 1 ml portion of $HClO_4$, heat for several minutes, cool, and add 10 ml of water. Gently boil the generally colorless solution to the evolution of white fumes of $HClO_4$, cool, add 5 ml of water and again boil to white fumes. Cool the solution, add 5 ml of water, quantitatively transfer to a suitable volumetric flask and make to volume.

Notes

(1) This procedure has been adapted from [8h]. For more reactive samples, begin decomposition with 10 ml of (1 + 2) HNO_3 and 10 ml of H_2SO_4 following the above procedure, prior to continuation with HNO_3—$HClO_4$ treatment.

(2) For samples which do not react vigorously, decomposition may be commenced with a mixture of 8 ml of H_2SO_4 + 10 ml of HNO_3, heating cautiously until reaction subsides, and boiling vigorously until the solution begins to darken before proceeding with the HNO_3—$HClO_4$ treatment.

3. *Dry ashing*

Accurately weigh 5—10 g of sample into a suitable silica or platinum crucible and spread thinly over the bottom. Add ashing aid if required. Dry and char the sample using an infrared lamp, hot plate or burner, taking care not to cause ignition. Place into a cold muffle furnace, raise the temperature slowly to 450—500°C and heat overnight, again not allowing the contents to ignite. If unoxidized organic matter remains, moisten the residue with water or (1 + 2) HNO_3, evaporate to dryness on a water bath and heat again in the furnace for a further period. When a suitable ash has been obtained, cool, moisten with water, carefully add 10 ml of (1 + 1) HCl and evaporate to dryness on a water bath. Dissolve the residue in (1 + 9) HCl or another suitable solvent.

Notes

(1) This is an outline of a dry ashing procedure adapted from [139] which should be generally suitable for a variety of samples; variations of the procedure and insertion of more details may, however be necessary.

(2) Incorporation of ashing aids such as H_2SO_4, MgO or $Mg(NO_3)_2$ may be required to facilitate ashing when samples have a low ash content.

Selected references to Section III: these procedures were adapted from The Society for Analytical Chemistry, Official, Standardised and Recommended Methods of Analysis [8h] and Gorsuch [139].

IV. RECOMMENDED ANALYTICAL PROCEDURES

A. General analytical protocol

1. *Apparatus and reagents*

(i) *Apparatus*

(a) Atomic absorption spectrometer: Spectrometer capable of operating over the wavelength region required, equipped with non-atomic absorption correction capability and with required light sources such as hollow cathode or electrodeless discharge lamps, and laminar flow burners operating with oxidants air or nitrous oxide and fuel gases acetylene, propane and hydrogen and other gases as required.

(b) Electrothermal atomisation device: Accessory with, e.g. graphite furnace cell to be used with atomic spectrometer, capable of pyrolyzing sample matrix and atomising elements of interest.

(c) Vapour generating apparatus: For mercury, any suitable commerical or custom-built apparatus or as described in refs. 7n and 13: For arsenic and selenium any suitable commercial or custom-built device giving adequate precision (refer to ref. 236 for discussion).

(d) Teflon digestion vessel: A vessel manufactured from polytetrafluoroethylene (Teflon) with a Teflon screw-cap or liner, supported, if necessary by a stainless steel body. A number of different vessels are manufactured by various companies; one of adequate volume must be selected to ensure suitability for intended procedure.

(e) Digestion flask for mercury determination: 250 ml flat-bottom boiling flask with 24/40 standard taper glass joint.

(f) Water condenser: 12—18 mm i.d. × 300 mm Liebig condenser with 24/40 standard taper joint. Modify by making indentations in the glass between lower standard taper and the water jacket with a pointed carbon rod to hold 6 mm Raschig rings. Add 8—10 rings and cover with ca. 3 mm i.d. glass helices to a height of 90 mm.

(g) Digestion apparatus for mercury determination: A digestion flask, a reservoir for collecting distillate, a two way stopcock, and a condenser as described on p. 8 of ref. 8, except that the two-neck flask is replaced by a 200 ml Kjeldahl flask with a B24 socket.

(h) Kjeldahl flasks: Flasks for acid digestion of samples, constructed of borosilicate glass, with capacities of 100, 200, 300, 500 and 800 ml.

(i) Beakers: Borosilicate glass beakers with capacities 50—2000 ml and 250 ml Vycor beakers.

(j) Crucibles for ashing: Vycor, silica, platinum or porcelain crucibles of ca. 100 ml volume.

(k) Muffle furnace, hot plate, oven, digestion rack and infrared lamp: as required for ashing and digestion.

(l) Pipets and volumetric flasks: Transfer and volumetric pipets and

References pp. 201—210

volumetric flasks meeting established tolerances with a range of capacities for preparation and transfer of solutions.

(m) Extraction tubes for iron and nickel determination: Standard glass-stoppered 25 × 200 mm test tubes which have been constricted to ca. 8 mm i.d. midway up the tube, giving a lower bulb of ca. 30—32 ml volume.

(n) Separatory funnels: Funnels of 100 ml and 125 ml capacity, with Teflon stopcocks.

(o) Centrifuge tubes, graduated cylinders and weighing bottles: Glass-stoppered with volumes 15—50 ml for storage of sample solutions.

(p) pH meter: Preferably with a small combination electrode.

(q) Filter paper: Acid-washed ashless filter paper such as Whatman No. 541 or Schleicher and Schuell No. 589-1H for rapid filtration of solutions.

(r) Plastic bottles: Polyethylene, polypropylene or Teflon bottles of volumes 25 ml to 1000 ml for storage of standard and sample solutions.

(ii) *Reagents*

Analytical grade if available, checked for element of interest and purified if necessary.

(1) Water: Distilled or deionized water with suitably low levels of analytes of interest.

(2) Nitric acid: Concentrated, sp. gr. ca. 1.42, 69—71% m/m.

(3) Perchloric acid: Concentrated, sp. gr. ca. 1.66, 70—72% m/m.

(4) Sulfuric acid: Concentrated, sp. gr. ca. 1.83, 95—98% m/m.

(5) Hydrochloric acid: Concentrated, sp. gr. ca. 1.18, 37—38% m/m. All acids are to be diluted with water as required, and the concentrated acids (2), (3), (5) may be purified by sub-boiling distillation in quartz or polypropylene stills if required for the determination of cadmium and lead.

(6) Ammonium hydroxide: Concentrated, sp. gr. ca. 0.90, 28—30% ammonia m/m; high purity reagent for cadmium and lead determination may be prepared by saturating cooled water with ammonia gas and filtering through a 0.3 μm porosity membrane filter; dilute as required.

(7) Hydrogen peroxide: 50% m/v and other dilutions as required.

(8) Elements and inorganic compounds: Purity 99.9% minimum for the preparation of aqueous standard solutions as described in Section IV.A.2.

(9) Tris (1-phenyl-1,3-butanediono) chromium (III): NBS No. 1078 ca. 9.6% chromium, for preparation of organo-chromium standard solution.

(10) Tris (1-phenyl-1,3-butanediono) iron(III): NBS No. 1079, ca. 10.3% iron, for preparation of organo-iron standard solution.

(11) Nickel cyclohexanebutyrate: NBS No. 1065b, 13.89% Ni, for preparation of organo-nickel standard solution.

(12) Bis(1-phenyl-1,3-butanediono) copper(II): NBS No. 1080, ca. 16.5% copper, for preparation of organo-copper standard solution.

(13) pH indicator solutions-0.1%: Dissolve 0.1 g of methyl red sodium salt or methylviolet in 100 ml of water; rub in an agate mortar 0.1 g of

bromophenol blue or bromthymol blue with 1.5 ml or 1.6 ml, respectively of 0.1 N NaOH solution and dilute to 100 ml with water.

(14) Petroleum ether.

(15) Ethanol, 95%.

(16) Chelating-extracting solution for boron determination: Prepare 15% (v/v) 2-ethyl-1,3-hexanediol in methyl isobutyl ketone. If extraction is carried out without making organic phase to fixed final volume, prepare water-saturated chelating-extracting solution by gently shaking the above solution with half its volume of water for ca. 1 min, let the layers separate and discard the aqueous phase.

(17) Methyl isobutyl ketone (4-methylpentan-2-one; MIBK): For lead determination, store over 0.5% v/v HNO_3 to remove lead.

(18) Barium hydroxide saturated solution: Add 30 g of $Ba(OH)_2$ to 1 l of water and mix; filter before use.

(19) Xylene.

(20) 2-Ethylhexanoic acid.

(21) Magnesium nitrate solution for ashing: Dissolve 50 g of $Mg(NO_3)_2 \cdot 6H_2O$ in water and dilute to 100 ml.

(22) Ammonium pyrrolidine carbodithioate (APDC) 1% m/v: Dissolve 1.0 g of APDC in water and dilute to 100 ml; prepare fresh daily and filter through a membrane filter to clarify if necessary.

(23) Metal-free sodium citrate buffer solution for iron and nickel determination, 1.0 M: Dissolve 147 g of sodium citrate dihydrate plus 105 g of citric acid monohydrate in water and dilute to ca. 900 ml. Add 10 drops of bromphenol blue indicator solution (13) and adjust the pH to 4.7.

(24) Sodium citrate solution for cobalt determination, 40% m/v: Dissolve 400 g of sodium citrate dihydrate in water and dilute to 1 l.

(25) 3-Heptanone.

(26) 2-Nitroso-1-naphthol, 1.0% m/v for cobalt determination: Dissolve 1.0 g of 2-nitroso-1-naphthol in 100 ml of ethanol.

(27) Chloroform.

(28) Sodium hydroxide solution, various concentrations: Transfer the mass of NaOH pellets required (e.g. 40.0 g for 1 N NaOH) to a beaker, dissolve in water, let cool, transfer to a 1 l volumetric flask and dilute to volume.

(29) Isoamyl acetate.

(30) Purified (1 + 30) perchloric acid for copper determination: Add 10 ml of 5% APDC to 1 l of (1 + 30) $HClO_4$. Mix, then extract the solution serially three times with 50 ml of MIBK to remove copper from the aqueous phase.

(31) Potassium iodide solution, 10% m/v for arsenic determination: Prepare in water and store in the cold.

(32) Sodium borohydride: Solid pellets or preferably an aqueous solution, stabilised with NaOH, of concentration geared to the hydride generator.

References pp. 201—210

(33) 8-Quinolinol-Methyl isobutyl ketone solution, 0.5% for molybdenum determination: Dissolve 0.5 g of 8-quinolinol in 100 ml of MIBK; prepare fresh daily.

(34) Potassium iodide solution, 0.1 M for cadmium determination: Dissolve 1.66 g of KI in water and dilute to 100 ml; prepare fresh daily.

(35) Amberlite LA-2 solution for cadmium determination: Prepare 1% (v/v) solution of Amberlite LA-2 (N-lauryl(trialkylmethyl)amine) in MIBK.

(36) Stripping solution for cadmium and lead determination: Add to water 30 ml of HNO_3 and 160 ml of 50% m/v H_2O_2 and dilute to 1 l to give a solution ca. 3% m/m and 8% m/m with respect to HNO_3 and H_2O_2, respectively.

(37) Modification solution for cadmium and lead determination: Prepare a 10% solution of ammonium dihydrogen phosphate $NH_4H_2PO_4$, purify by extracting with chloroform and dithizone and remove excess chloroform in vacuo. Dilute 33 ml of this phosphate solution plus 3.8 ml of high purity NH_4OH (6) to 1 l to give a solution 0.33% m/m and 0.096% m/m with respect to ammonium dihydrogen phosphate and ammonium hydroxide respectively.

(38) Dilution solution for cadmium and lead determination: Add to water, followed by dilution to 1 l the following: 7.5 ml of HNO_3 (concentration in final solution 0.75% m/m), 25 ml of 10% m/m ammonium dihydrogen phosphate (0.25%), 2.9 ml of NH_4OH (6) (0.072%), 40 ml of 50% m/v H_2O_2 (2%), and 5.4 ml of MIBK (17) (0.43%).

(39) Potassium iodide solution for tin determination, ca. 5 M: Dissolve 83 g of KI in water and dilute to 100 ml; prepare fresh daily.

(40) Toluene.

(41) Potassium hydroxide for tin determination, 5 M: Dissolve 281 g of KOH in water, let cool, transfer to a 1 l volumetric flask and dilute to volume.

(42) Ascorbic acid solution for tin determination, 5% m/v: Prepare fresh weekly.

(43) Vanadium pentoxide for mercury determination.

(44) Diluting solution for mercury determination: To a 1000 ml flask add 300—500 ml of water, 58 ml of HNO_3 and 67 ml of H_2SO_4 and dilute to volume with water.

(45) Potassium permanganate, 6% m/v.

(46) Reducing solution: Mix 50 ml of H_2SO_4 with ca. 300 ml of H_2O. Cool to room temperature, dissolve 15 g of NaCl, 15 g of hydroxylamine sulfate and 25 g of $SnCl_2$ in the solution and dilute to 500 ml.

(47) Hydroxylammonium chloride—sodium chloride reducing solution for mercury determination: Mix 20 ml of 15% m/v sodium chloride solution with 12 ml of 21% m/v hydroxylammonium chloride solution and dilute to 100 ml.

(48) Tin(II) chloride reducing solution for mercury determination: Heat 21 g of granulated tin under reflux with 50 ml of water and 50 ml of HCl

until no more tin will dissolve. Store the solution over a piece of metallic tin.

(49) Sodium sulfite, 10% m/v.

2. *Standard solutions*

Use pipets and volumetric flasks meeting appropriate established tolerances (e.g. ref. 142). For best precision, use pipets and volumetric flasks of volumes not less than ca. 2 ml and 25 ml, respectively. If using automatic dilution apparatus or other equipment, ensure performance to be compatible with accuracy desired. Prepare aqueous stock solutions of individual elements as described below, from minimum 99.9% pure metals or compounds by dissolution in pure acids and distilled or deionized water. Store in acid-cleaned, air-tight polyethylene containers. Prepare intermediate and working solutions by diluting stock solutions with acid solution of composition identical to that to be found in sample solutions. Include ionization suppressing and releasing agents, and major sample matrix elements if required. Prepare daily, solutions in the 0—100 $\mu g\,ml^{-1}$ range. To prevent excessive propagation of error, prepare intermediate solutions from stock solutions with one dilution, and working solutions with one additional dilution from the intermediate solution. Prepare 1000 ml of stock solutions containing 1000 μg of the analyte per ml, as follows:

Al: Dissolve 1.0000 g Al wire in a minimum volume of 2 M HCl and dilute to volume.

As: Dissolve 1.3203 g of As_2O_3 (dried 1 h at 110°C and cooled in desiccator) in a minimum volume of 20% m/v NaOH, dilute to 200 ml, neutralize with HCl and dilute to volume.

B: Dissolve 5.7199 g of fresh crystals of H_3BO_3 in ca. 500 ml of H_2O and dilute to volume.

Ca: Suspend 2.4973 g $CaCO_3$ (dried at 140°C and cooled in desiccator) in 50 ml of H_2O, carefully dissolve by adding dropwise a minimum volume (ca. 10 ml) of HCl and dilute to volume.

Cd: Dissolve 1.0000 g Cd in a minimum volume of 1 + 1 HCl and dilute to volume.

Co: Dissolve 1.0000 g Co in 10 ml of 2 M HCl and dilute to volume.

Cr: Dissolve 2.8289 g $K_2Cr_2O_7$ (dried for 2 h at 110°C and cooled in desiccator) in water and dilute to volume.

Cu: Dissolve 1.0000 g Cu in a minimum volume of HNO_3, add 5 ml of HCl, evaporate almost to dryness and dilute to volume with 0.1 M HCl.

Fe: Dissolve 1.0000 g Fe in 30 ml of 6 M HCl and dilute to volume.

Hg: Dissolve 1.0000 g Hg in 10 ml of 5 M HNO_3 and dilute to volume.

K: Dry KCl at 140°C, cool in a desiccator, dissolve 1.9067 g in H_2O and dilute to volume.

Mg: Dissolve 1.0000 g Mg in 50 ml of 5 M HCl and dilute to volume.

References pp. 201—210

Mn: Dissolve 1.0000 g Mn in 10 ml of HCl + 1 ml of HNO_3 and dilute to volume.

Mo: Dissolve 2.0425 g $(NH_4)_2MoO_4$ in water and dilute to volume.

Na: Dry NaCl at 140°, cool in a desiccator, dissolve 2.5421 g in H_2O and dilute to volume.

Ni: Dissolve 1.0000 g Ni in 20 ml of HNO_3 and dilute to volume.

Pb: Dissolve 1.0000 g Pb in 10 ml of HNO_3 and dilute to volume.

Se: Dissolve 1.0000 g Se in 5 ml of HNO_3 and dilute to volume.

Sn: Dissolve 1.0000 g Sn in 15 ml of warm HCl and dilute to volume.

V: Dissolve 2.2963 g NH_4VO_3 (dried at 110° and cooled in desiccator) in 100 ml of (1 + 10) HNO_3 and dilute to volume.

Zn: Dissolve 1.0000 g Zn in 20 ml of (1 + 1) HCl and dilute to volume.

Notes

(1) The requirement for calibrated laboratory-ware for containing and dispensing solutions extends to all facets of the analytical procedure. It rests with the analyst to ascertain accuracy suitable to the application, of all devices used, including micropipets used in conjunction with electrothermal atomizers. The temptation to use other than appropriately calibrated containers, and chemical reagents of suitably high purity, and not to follow good analytical procedures as practised by those with only peripheral contact with analytical chemistry, should be avoided.

(2) Ingot, rod, sheet, wire or other bulk forms of metals are preferred; powdered metals should be avoided as they may contain occluded gases. If necessary, metals should be etched, prior to weighing, in the type of acid used for dissolution, to remove surface oxide film. Gentle heating may be required if dissolution is slow.

(3) It has been reported that a crystal of thymol in the stock solution will inhibit bacterial growth [63b]; this preservation technique is recommended only when standard solutions are prepared free of acid. Solution stability is a function of nature of analyte and matrix and their concentrations, acid concentration, type of storage container, storage conditions, etc. At extremely low concentrations, contamination from solvents, water and containers must be considered.

Selected references: 7, 63b, 142—145.

3. Reference materials

For ultimate testing of procedure accuracy, and as part of a comprehensive data quality assurance program, analyse a reference material, compositionally similar to the sample, and certified with regard to the concentration of the analyte of interest. Take the usual weight of material (not less than that specified in the certificate provided by the issuing agency, and following specified drying, storage and other instructions) through the entire analytical procedure, preferably simultaneously with samples, and compare measured analyte level with certified value. Biological standard reference materials currently available from the US National Bureau of Standards, of use to the food analyst are listed in Table 2. Table 3 gives, for several reference materials, certified concentrations of elements discussed in this chapter.

Notes

(1) Other useful reference materials have been prepared by the International Atomic Energy Agency, Vienna [147, 148], and Bowen [149—151].

(2) In the absence of suitable reference materials, the procedure should be tested using different sample weights and also measuring recoveries of element added at the beginning of the procedure. It must be remembered, however, that these criteria although necessary, are not sufficient, for the complete demonstration of the validity of the analytical procedure. The application of an independent (different in all respects of sample treatment and analyte quantitation) analytical method to a homogeneous practice sample would provide very useful confirmation of method reliability.

Selected references: 146—152.

4. Spectrometer operating parameters

A listing, with respect to flame spectrometry, of some recommended instrumental operating parameters and spectral characteristics of elements of interest to the food analyst is presented in Table 4. As many commercial models of atomic absorption spectrometers are available, with different characteristics and operating requirements, these parameters must be considered as guides. The operator must be familiar with settings and procedures related to his instrument, and must follow the manufacturer's instructions. These and other parameters such as light source current or wattage, and burner position should be optimized, usually with respect to signal (or signal to noise for critical applications) as outlined in chapters 2 and 3 of this volume. For several elements, alternate wavelengths are listed when a less sensitive line or more linear calibration graph are required.

A listing of instrumental operating parameters and spectral characteristics of several elements determinable by electrothermal spectrometry are given in Table 5. The listed pyrolysis temperatures (maximum for 30 s pyrolysis), optimum atomization temperatures, characteristic concentrations and detection limits are normally for aqueous solutions and are typical guides to those obtainable with commercial graphite atomisers. Correction for non-atomic absorption is usually mandatory with electrothermal devices. Use of continuum light source and simultaneous correction is preferred to sequential measurement using a non-absorbing line.

Notes

(1) The absorption lines of four elements, Al, B, Na and V are closely

References pp. 201—210

TABLE 2

U.S. NATIONAL BUREAU OF STANDARDS, BIOLOGICAL MATRIX STANDARD REFERENCE MATERIALS FOR CHEMICAL COMPOSITION

SRM 1566	Oyster Tissue
SRM 1567	Wheat Flour
SRM 1568	Rice Flour
SRM 1569	Brewers Yeast
SRM 1570	Spinach
SRM 1571	Orchard Leaves
SRM 1573	Tomato Leaves
SRM 1575	Pine Needles
SRM 1577	Bovine Liver

spaced pairs (doublet plus a third nearby line for vanadium), usually not resolvable with ordinary monochromators. Non-absorbing line wavelengths are included in the Tables should the need arise for making non-atomic corrections and a continuum emission light source is unavailable. Characteristic concentrations and detection limits listed represent good typical values expected with suitably optimized spectrometers and aqueous solutions.

(2) Usual flame techniques are often insufficiently detective* to measure the low levels of As, Se and Hg present in foodstuffs. Mercury is commonly determined via the flameless cold-vapour technique, whereas there is much current activity in respect of the measurement of As and Se via their conversion to hydrides with subsequent decomposition in cool argon—hydrogen-entrained air flames or electrically-heated cells. Table 4 contains information on these techniques.

(3) Elements such as B, Cd, Co, Cr, Mo, Ni, Pb and V are also often found in foodstuffs at concentrations too low to be amenable to direct flame absorption. Resort must be made to concentration via, e.g., solvent extraction prior to introduction into the flame (SEFAAS) or to the newer technique of EAAS. The elements Na, K, Mg, Ca and Hg, although amenable to determination by EAAS are not listed in Table 5 as the flame provides adequate detectivity for the four macroelements, whereas the cold-vapour technique is clearly the one of choice for mercury. A good number of other elements in Table 5 can usually be determined in foods by flame spectrometry; EAAS information for them is included, however, to cover the occasional sample with extremely low trace-element content.

---

* The terms detective and detectivity rather than the frequently (mis)used terms sensitive and sensitivity, are used to denote analyte detection capability of the method. For a given method, although usually higher sensitivity (slope of calibration curve) is synonymous with greater detectivity, (lower detection limit), this is not invariably so, and the latter is the more accurate term. In addition, for performance comparisons among different methodologies, sensitivities cannot usually be compared due to the typically different response units inherent in the methods; comparisons, however, of detectivities, if expressed as, e.g., mg of analyte kg^{-1} of sample, can be made.

TABLE 3

CERTIFIED CONCENTRATIONS (mg kg^{-1}) OF CONSTITUENT ELEMENTS IN U.S. NATIONAL BUREAU OF STANDARDS BIOLOGICAL MATRIX STANDARD REFERENCE MATERIALS[a]

Element	SRM 1567 Wheat Flour	SRM 1568 Rice Flour	SRM 1570 Spinach	SRM 1571 Orchard Leaves	SRM 1573 Tomato Leaves	SRM 1577 Bovine Liver
Al			870 ± 50			
As		0.41 ± 0.05	0.15 ± 0.05	10 ± 2[b]	0.27 ± 0.05	0.055 ± 0.005
B				33 ± 3		
Ca	190 ± 10	140 ± 20	13500 ± 300	20900 ± 300	30000 ± 300	124 ± 6
Cd	0.032 ± 0.007	0.029 ± 0.004		0.11 ± 0.02		0.27 ± 0.04
Co		0.02 ± 0.01				
Cr			4.6 ± 0.3		4.5 ± 0.5	0.088 ± 0.012
Cu	2.0 ± 0.3	2.2 ± 0.3	12 ± 2	12 ± 1	11 ± 1	268 ± 8
Fe	18.3 ± 1.0	8.7 ± 0.6	550 ± 20	300 ± 20	690 ± 25	
Hg	0.001 ± 0.0008	0.0060 ± 0.0007	0.030 ± 0.005	0.155 ± 0.015		0.016 ± 0.002
K	1360 ± 40	1120 ± 20	35600 ± 300	14700 ± 300	44600 ± 300	9700 ± 60
Mg				6200 ± 200		604 ± 9
Mn	8.5 ± 0.5	20.1 ± 0.4	165 ± 6	91 ± 4	238 ± 7	10.3 ± 1.0
Na	8.0 ± 1.5	6.0 ± 1.5		82 ± 6		2430 ± 130
Ni				1.3 ± 0.2		
Pb			1.2 ± 0.2	45 ± 3	6.3 ± 0.3	0.34 ± 0.08
Se	1.1 ± 0.2	0.4 ± 0.1		0.08 ± 0.01		1.1 ± 0.1
Zn	10.6 ± 1.0	19.4 ± 1.0	50 ± 2	25 ± 3	62 ± 6	130 ± 13

[a] Concentration ± uncertainty on a dry sample basis from NBS certificates [146]. The certificates also contain informational data on other elements.
[b] Latest NBS value; personal communication from J. P. Cali and T. W. Mears.

TABLE 4

INSTRUMENTAL OPERATING PARAMETERS AND SPECTRAL CHARACTERISTICS OF SELECTED ELEMENTS IN FLAME ATOMIC ABSORPTION SPECTROMETRY[a]

Element	Absorbing wavelength (nm)	Spectral bandpass (nm)	Flame	Characteristic concentration ($\mu g\,ml^{-1}$)	Detection limit ($\mu g\,ml^{-1}$)	Non-absorbing wavelength (nm)
Al	309.27 / 309.28	0.2	NA (FR)	1.0	0.04	307.0
	396.15	0.2	NA (FR)	1.0	0.1	—
As	193.70	0.7	ArH (hydride gen)	0.0001	0.0003	192.0
B	249.68 / 249.77	0.1	NA (FR)	10	2	—
Ca	422.67	0.2	NA (FR)	0.03	0.002	—
	239.86	0.1	NA (FR)	4	—	—
Cd	228.80	0.5	AA (FL)	0.02	0.001	226.5
Co	240.73	0.1	AA (FL)	0.1	0.005	238.49 / 240.0(Ta) / 240.56(W)
Cr	357.87	0.2	NA (S)	0.15	—	352.0
	359.35		NA	0.3	—	
Cu	324.75	0.2	AA (FL)	0.04	0.005	323.1
Fe	248.33	0.2	AA (FL)	0.1	0.005	248.73(Re) 249.68(B) 249.22(Cu) 249.89 249.47(Be)
Hg	253.65	0.2	Cold vapour	—	0.00004	249.22(Cu)
K	766.49	0.5	AA	0.02	0.005	—
	404.41	0.1	AA	4	—	—
Mg	285.21	0.5	AA (FL)	0.003	0.0003	283.69(Cd) 284.00(Sn)
Mn	279.48	0.1	AA (S)	0.05	0.003	280.20(Pb) 282.44(Cu)
Mo	313.26	0.2	NA (FR)	0.4	0.03	311.2

165

TABLE 4 (CONTINUED)

Na	589.00	0.2	AA (FL)	0.02	0.002	—
	330.24 ⎫	0.2	AA (FL)	3	—	—
	330.30 ⎭					
Ni	232.00	0.2	AA (FL)	0.1	0.005	231.40 233.00
						231.72 233.53(Ba)
Pb	341.48	0.2	AA (FL)	0.5	—	280.1
	283.31	0.2	AA (FL)	0.3	0.01	283.69(Cd)
Se	217.00	1.0	AA (FL)	0.2	0.02	220.4
	196.03	0.5	ArH (hydride gen)	0.0008	0.0003	198.1
Sn	235.48	0.2	NA (FR)	2	—	—
	286.33	0.2	NA (FR)	4	0.5	283.9
	224.61	0.1	AH	0.5	0.05	226.50(Cd)
V	318.34 ⎫	0.2	NA (FR)	0.4	0.04	—
	318.40 ⎬					
	318.54 ⎭					
Zn	213.86	0.2	AA (FL)	0.01	0.002	212.5

[a] Recommended parameters and good typical spectral characteristics of elements in aqueous solution, adapted from refs. 43, 50, 51, 53, 63c, 153—159. Ionization of Al, Ca, Cr, K, Na, Sn and V has been suppressed by the presence of a high concentration of a more easily ionizable metal.

Abbreviations: AA, air—acetylene; AH, air—hydrogen; ArH, argon—hydrogen-entrained air; NA, nitrous oxide—acetylene; FR, fuel-rich (reducing); FL, fuel-lean (oxidising); S, "stoichiometric" or "normal stoichiometry". Characteristic concentration, usually denoted sensitivity in AAS literature, refers to the concentration of the element in aqueous solution required to produce 1% absorption (0.0044 absorbance). Detection limit usually refers to the concentration of the element producing an absorption equal to twice the standard deviation of a blank. Non-absorbing lines listed originate from the hollow-cathode lamp used for analyte determination unless otherwise indicated. Additional information is contained in notes to recommended procedures for the individual elements.

References pp. 201—210

TABLE 5

INSTRUMENTAL OPERATING PARAMETERS AND SPECTRAL CHARACTERISTICS OF SELECTED ELEMENTS IN ELECTROTHERMAL ATOMISATION ATOMIC ABSORPTION SPECTROMETRY[a]

Element	Absorbing wavelength (nm)	Pyrolysis temp. (°C)	Atomisation temp. (°C)	Characteristic concentration (pg)	Detection limit (pg)	Non-absorbing wavelength (nm)	Flame/electrothermal ratio Characteristic concentration	Flame/electrothermal ratio Detection limit
Al	309.27 309.28	1400	2700	50	5	307.0	14000	4000
As	193.70	600	2400	25	20	192.0	4000	12500
Cd	228.80	350	1900	1	0.1	226.5	30000	10000
Co	240.73	1100	2600	40	5	239.3	3800	5000
Cr	357.87	1200	2700	20	10	242.17(Sn) 352.0	5000	300
Cu	324.75	800	2600	30	2	323.1	3300	1000
Fe	248.33	1200	2500	25	5	249.22(Cu)	4000	2000
Mn	279.48	1100	2600	2	0.2	280.20(Pb) 282.44(Cu)	25000	10000
Mo	313.26	1900	2700	20	5	311.2	20000	40000
Ni	232.00	1000	2700	100	20	231.4[b] 231.7[b]	1000	500
Pb	283.31	600	2100	20	2	280.1 283.69(Cd)	25000	10000
Se	196.03	700	2500	200	100	198.1	5000	2500
Sn	224.61	1000	2500	100	100	226.50(Cd)	10000	1000
V	318.34 318.40	1600	2700	200	100	—	5000	200
Zn	213.86	500	2000	1	0.05	212.5	20000	40000

[a] Typical operating parameters and spectral characteristics with graphite furnaces are based on Fuller [53]; wavelengths are taken from Meggers et al. [159]. Characteristic concentration and detection limit (defined in Table 4) are typical absolute values realisable with aqueous solutions. For calculation of flame/electrothermal ratios, characteristic concentrations and detection limits obtainable with flame atomic absorption are based on aspiration of a 1 ml sample volume. Additional information is in section IV.A.4.

[b] The 231.4 nm and 231.7 nm non-absorbing lines recommended by Fuller [53] and Varian [154] respectively, may refer to the Ni atom lines at 231.35 nm and 231.72 nm.

Selected references: 43, 50, 51, 53, 63c, 153—159.

5. *Atomic absorption spectrometry*

(i) *Aqueous solution-flame*

Follow manufacturer's operating instructions, and own procedural modifications with custom-made equipment. Select and optimize parameters such as wavelength, flame gases and flame composition, burner position, light source current or wattage, slit widths, and solution aspiration rate (refer to Table 4 and chapter 2 of this volume) using a standard solution where required. Use scale expansion as required, commensurate with acceptable noise levels. Prepare five standard solutions to cover the anticipated concentration range of the element, including zero concentration. Aspirate and read absorbance of standard solutions alternately with 5—10 or more sample solutions (depending on spectrometer stability), flushing burner with water between aspirations of both standard and sample solutions. Automatic zero control may be used to re-establish zero absorbance for water, during each aspiration of water. Do not adjust instrument to zero absorbance while aspirating blank standard solution, but record actual reading.

(ii) *Solvent extraction—flame*

Refer to Section IV.A.5.(i). Select and optimize operating parameters using standard solution of the element prepared in the organic solvent. Prepare, together with samples, a series of five standard solutions in the organic solvent by taking aliquots of aqueous standard solutions through the entire complexation—extraction procedure. Aspirate and read standard solutions alternately with 5—10 or more sample solutions, aspirating blank water-saturated solvent between aspirations of standard and sample solutions.

(iii) *Electrothermal atomization*

Refer to Section IV.A.5.(i). Position the electrothermal atomization device in the light path of the spectrometer; select and optimize operating parameters. Maximize analytical signal in comparison to light emission from the atomizer by operating the photomultiplier at a low gain setting; alternatively, reduce light emission by reducing temperature of atomization if possible, or by installing a mask along optical path of the spectrometer. Calibrate response by injecting fixed volume of standard solutions of different concentrations covering range of interest, interspersing standards with samples. Due to the generally inferior precision of electrothermal techniques, perform all determinations in replicate, reject outlying observations, and use mean absorbances.

(iv) *Other discrete sampling techniques*

General procedures for other discrete sampling techniques such as sampling boat, cold-vapour mercury technique, and hydride methods for As and Se are similar to that mentioned above. Optimization under actual operating conditions and use of fixed volumes of standard solutions of varying concentrations is preferred.

Notes

(1) The dependence of spectrometer performance on instrument type has been mentioned [160]. It is not possible, therefore, to rely exclusively on published parameters; the analyst must conduct fairly comprehensive studies with his own instrument in order to delineate optimum operating conditions.

(2) Precision of delivery of the small volumes required and consistently proper positioning of the droplet in the furnace have an important bearing on precision in EAAS. The use of automatic sampling accessories is advantageous in minimizing these two problems, in addition to relieving operator tedium associated with the multitude of determinations required. Calibration curves in EAAS and with other discrete sampling techniques are often presented as plots of absorbance vs. absolute quantity of element. There are three possible ways of preparing the calibration graph, namely (a) fixed volume of standard solutions of varying concentrations, (b) varying volumes of solution of fixed concentration, (c) (for EAAS) multiple additions of fixed-concentration solution with drying after each addition. Slightly different calibration curves are to be expected; the first of these methods gives superior precision and accuracy. The idea of using solid samples has not yet been reliably put into practice. Either absorption peak height or area can be used; it has been observed [53] that the former appears to be superior for standard operation. Information recorded in Table 5 pertains to furnace cells made from graphite, presently the most commonly-used construction material for electrothermal atomizers.

Selected references: 7, 51, 53, 63b, 143, 154, 160.

6. *Reagent blanks*

Note the existence of two different reagent blanks in the analytical procedure — a standard solution reagent blank and a procedural reagent blank. Prepare a standard solution reagent blank by going through the standard solution preparation procedure (including solvent extraction) omitting the analyte. Use this blank to correct the zero-added analyte intercept of the calibration curve. Carry out a procedural reagent blank test by the entire procedure using the exact amounts of reagents (including water) used in the analysis, omitting only the sample. Use the procedural blank to correct calculated sample concentrations for positive and negative contamination from reagents and procedural manipulations. Do not zero-out blanks in the spectrometry step but quantitate both blanks for information on blank magnitudes and variability, and for estimation of detection limits.

Notes

(1) The frequency of blank determination depends on blank magnitude and variability in relation to the level of analyte in the sample and analytical precision desired. In standard applications, one or two procedural blanks are processed with a set of 10 or more samples. Special cases concerned with reliable analyses at unavoidably high blank levels, require as many blanks as samples. It must be remembered that conceptually and in practice, the procedural blank is different from the sample in that it contains no sample matrix. Physical and chemical behaviour of the blank, therefore, during sample treatment and subsequent operations will not simulate exactly the behaviour of the sample, with the consequence that the blank value is only an estimate of adventitious contamination and losses occurring during processing of actual samples.

7. *Interferences*

Minimize spectral, physical and chemical interferences by employing operating parameters and general instructions suggested in this section and specific instructions recommended for the different elements presented in Section IV.B. Use a reference material, or in its absence resort to tests with different aliquots of sample digest and measurement of recoveries of element added to digest, to get an indication of interference problems. Check for non-atomic absorption using simultaneous measurement with a continuum light source or adjacent wavelength and apply corrections if necessary.

Notes

(1) For highly accurate work, a thorough investigation of all possible interferences and means for their elimination or minimization is essential. The degree of interference is dependent on the spectrometer. For accurate work the analyst must use published operating conditions as guides for comprehensive studies with his own instrument in order to delineate optimum operating conditions. Various aspects of the vast question of interference in atomic spectrometry are discussed in the selected references listed below.

Selected references: 51, 53, 63a, b, 156, 160—163.

8. *Calibration and calculations*

Plot absorbance readings of standard solutions, corrected for reagent blank, as obtained in Section IV.A.5 against concentration of analyte when using continuous aspiration, or against absolute amount when using discrete sampling. Draw or calculate the regression straight line or quadratic curve through the points. Read off the calibration curve, concentrations or amounts of analyte in procedural reagent blank and sample, correct sample solution value for procedural blank and calculate analyte content (mg kg^{-1}) in the original sample. In special applications use the technique of bracketing the sample with closely spaced standards.

Selected references: 63b, 164.

B. Determination of specific elements

1. *Sodium and potassium*

(i) *Summary of procedure*

Sodium and potassium are extracted from solid samples by wet digestion or dry ashing followed by acid dissolution of residue, and are determined by flame AAS using an air—acetylene flame. Liquid samples are aspirated after dilution.

(ii) *Preparation of sample solution*

Marine products. Weigh 1 g of sample into a 50 ml Pyrex beaker, add 5 ml of HNO_3 and allow to digest overnight at ambient temperature. Heat on a hot plate set at low temperature until the sample dissolves and a clear red solution results, then take to dryness. Repeat the digestion twice more, finally add 2 ml of HNO_3 and dissolve the solids by warming. Filter the solution through nitric acid-cleaned glass wool into a 100 ml volumetric flask, rinsing out the beaker with hot distilled or deionized water, cool and dilute to volume. For determination of the elements using the most sensitive wavelengths, dilute the sample solution with water by a factor of 25 and 5 for sodium and potassium, respectively. For samples with high oil content, defat the dry sample (ca. 2 h at 110°C) by extracting on a steam bath with successive 10 ml portions of petroleum ether before proceeding with the digestion. Alternatively weigh 5 g of sample into a nitric acid-cleaned Vycor, silica or platinum crucible and char by heating under an infrared lamp or on a hot plate. Place in a cold muffle furnace, slowly bring the temperature to 500°C and ash for ca. 2 h. Dissolve the residue in 15 ml of (1 + 4) HNO_3, filter through acid-washed filter paper into a 100 ml volumetric flask and dilute to volume. For determinations using the most sensitive wavelengths, dilute the sample solution with water by factors of 100 and 25 for sodium and potassium, respectively.

Fruits and fruit products. Weigh 300 g of jelly, syrup, fresh or dried fruit or preserve into a 2 l beaker, add ca. 800 ml of water and extract by gently boiling for 1 h, replacing water lost by evaporation. Filter into a 2 l volumetric flask, cool, dilute to volume, and dilute again if necessary to bring the analyte concentration to a level suitable for FAAS.

Beverages. For wine samples dilute with water by a factor of 50—200; dilute mineral water samples as required. Dilute liquors with 50% ethanol by a factor required to bring absorbance onto the spectrometer scale. Prepare standards in identical diluent.

Plant materials. Transfer an accurately weighed 1-g sample into a silica crucible, ash for 2 h at 500°C and let cool. Wet the ash with 10 drops of water and carefully add 3—4 ml of (1 + 1) HNO_3. Evaporate excess HNO_3 on a 100—120°C hot plate, return the crucible to the furnace and ash for 1 h at 500°C. Cool, dissolve the ash in 10 ml of (1 + 1) HCl, transfer the solution to a 50 ml volumetric flask and dilute to volume. Alternatively, weigh 1 g of sample into a 150 ml Pyrex beaker, add 10 ml of HNO_3 and allow to soak thoroughly. Add 3 ml of $HClO_4$ and heat on a hot plate, slowly at first, until frothing ceases. Heat until the HNO_3 is almost evaporated. Should charring occur, cool, add 10 ml of HNO_3 and continue heating. Heat to fumes of $HClO_4$, cool, add 10 ml to (1 + 1) HCl, transfer the solution to a 50 ml volumetric flask and dilute to volume. In both the dry ashing and wet digestion procedures allow the silica to settle and use the supernatant for determination.

(iii) *Determination*

Aspirate solutions into an air—acetylene flame (fuel-lean for sodium measurement) and measure absorbances at 589.00 nm and 766.49 nm for sodium and potassium, respectively. Use less sensitive wavelengths of 330.24/330.30 nm and 404.41 nm for sodium and potassium, respectively, when more concentrated (by a factor of about 200) solutions are used.

Notes

(1) The spectral simplicity of sodium and potassium light sources permits the use of wide spectral bandpasses if required at the commonly used long wavelengths. For highest analytical sensitivity with Na, however, the most sensitive Na 589.00 nm line may be isolated from the slightly less sensitive 589.59 nm line using a 0.2 nm spectral bandpass; a wider spectral bandpass allowing both lines may be employed for less critical work. In the presence of high concentrations of calcium, band systems arising from CaO and CaOH molecules may interfere and impair the signal/noise ratio of the sodium absorption.

(2) Although ionization of sodium is negligible and potassium small in an air—propane flame, some ionization is experienced in the recommended hotter air—acetylene flame. Ionization should be suppressed by the incorporation of excess potassium or cesium (for sodium determinations) or excess sodium or cesium (for potassium determinations), at concentrations of 1000 μg ml^{-1} or greater, in the form of chlorides or nitrates, in both sample and standard solutions. Cesium is the more effective but more expensive ionization suppressant. Extent of ionization is inversely related to analyte concentration with errors due to incomplete suppression thus being greater at low concentrations. As it is difficult to obtain alkali metal salts free from traces of other alkali metals, possible contamination must be considered, especially at low analyte levels. Use of a branched capillary for introduction of ionization buffer has been advocated for flame spectrometry to

facilitate processing of samples and to conserve expensive reagents [165, 166].

(3) Few elements interfere with the determination of sodium and potassium. Important interferences via ionization suppression from other alkali metals present in the sample are minimized as mentioned above. Interferences from high mineral acid concentrations on sodium and potassium absorption may be compensated for by matching sample and standard solutions with respect to acid type and concentration. For samples containing very high concentrations of sodium or potassium, the burner may be angled to reduce the need for excessive dilution. Alternatively the less sensitive (by a factor of about 150 for sodium and 200 for potassium) 330.24/330.30 nm (Na) and 404.41 nm (K) absorption lines may be employed.

(4) Solutions for sodium determinations should be stored in containers made from plastics such as polyethylene. Potassium will precipitate as $KClO_4$ from concentrated $HClO_4$ solutions but will readily dissolve upon high dilution of the solution with water. Results for sodium measurements on silicon-containing samples taken through the common wet-digestion treatment may be low due to retention of the analyte in the insoluble silica. Accurate measurements require the dissolution of silica with hydrofluoric acid following the destruction of the organic matrix [167].

(5) The procedure for wines outlined by Ecrement [36] involves close matching of standard and sample matrices by dilutions with a synthetic solution made up of several organic compounds at concentrations close to those in a typical wine. The composition of the synthetic solution is: 100 ml of ethanol, 7 g of citric acid, 3 g of sugar, 2 g of glycerol, 3.8 g of tartaric acid, 0.15 ml of orthophosphoric acid made to 1 l with water. Samples are diluted 10- and 200-fold for the direct determination of sodium and potassium respectively.

Selected references: The procedure was adapted essentially from: AOAC Official Methods of Analysis [7h, 7u], J. Assoc. Off. Anal. Chem. [9] and the Subcommittee on Procedures of the Chemistry Task Force of the (US) National Shellfish Sanitation Program [24]. Other pertinent references are: 7r, 21, 34, 36, 37e, f, 41, 50, 51, 63c, 139, 154, 155, 159, 165, 166, 168.

2. Magnesium and calcium

(i) *Summary of procedure*

The sample is brought into solution by wet digestion or by dry ashing followed by acid dissolution. The concentrations of magnesium and calcium are determined by flame AAS using an air—acetylene and nitrous oxide—acetylene flame for magnesium and calcium, respectively.

(ii) *Preparation of sample solution*

In general, digest sample with HNO_3—$HClO_4$ (Section III.A.1), transfer the solution to a volumetric flask of suitable size and make up to volume. For plant materials follow either the dry ashing or wet digestion procedure described in Section IV.B.1(ii).

(iii) *Determination*

Aspirate solutions into a fuel-lean air—acetylene flame to measure magnesium at 285.21 nm and into a fuel-rich nitrous oxide—acetylene flame to determine calcium at 422.67 nm.

Notes

(1) No spectral line interferences have been reported for the absorption lines of magnesium and calcium; a bandpass up to 2 nm is permissible for absorption measurements using the magnesium 285.21 line. Magnesium sensitivity may be strongly dependent on hollow-cathode lamp current (e.g. Varian-Techtron lamp); a high lamp current is required for maximum stability when measuring calcium. The less sensitive (down by a factor of ca. 120 relative to 422.67 nm) 239.86 calcium line may be used if required. A less sensitive (ca. 25-fold) magnesium line variously reported at 202.5, 202.50 and 202.58 nm [41, 51, 153—155] is not recorded in the tables of Meggers et al. [159].

(2) Substantial ionization of calcium and the small ionization of magnesium in the nitrous oxide—acetylene flame are overcome by addition of 2000—5000 μg ml^{-1} of potassium to sample and standard solutions. In the determination of magnesium and calcium, chemical interferences from Al, Si, P and other elements observed in the air—acetylene flame, may be overcome by the addition of a known excess of a releasing agent such as lanthanum (ca. 10000 μg ml^{-1}) or strontium (ca. 1000—5000 μg ml^{-1}). Most common interferences are also eliminated in the nitrous oxide—acetylene flame, although both this hotter flame and lanthanum are required to reduce interference by titanium.

(3) Both elements can often be satisfactorily estimated without the need for complete digestion of sample, but only dilution of liquid samples, or simple treatments. Thus, for analysis of wines, Ecrement [36] has described a dilution procedure (1/100 for magnesium and 1/20 for calcium) with a synthetic solution (refer to note 5, Section IV.A.1) prior to determination by FAAS. Magnesium in beer has been determined after dilution by a factor of 50—100 [169]; magnesium and calcium were determined in milk following casein removal by trichloroacetic acid precipitation [170] and both elements were determined in orange and pineapple juice after suitable dilution with dilute HCl and centrifugation [171]. Calcium, but not magnesium, could be well extracted from a fine suspension of apple tissue, by boiling with HCl [172]. Magnesium and calcium have been determined in orange juice following simple hydrolysis with HNO$_3$ [173, 174]. A number of investigations of the efficacies of different wet-digestion and dry-ashing procedures have been reported [175—179].

References pp. 201—210

Selected references: 7d, 7q, 9, 21, 34, 36, 37g, 41, 50, 51, 63c, 139, 153–155, 159, 169–179.

3. Boron

(i) Summary of procedure

The sample is subjected to pressure digestion with HNO_3, or dry ashing, boron in the digest is chelated with 2-ethyl-1,3-hexanediol, extracted with MIBK and its concentration determined by FAAS using a nitrous oxide–acetylene flame.

(ii) Preparation of sample solution

Exercise caution when digesting organic materials in a pressure digestion vessel — Some advise against it. Transfer 1 g of an accurately weighed food sample into a Teflon pressure digestion vessel, of suitable volume, add 5.0 ml of HNO_3 and close tightly. Digest the sample by heating vessel at 150°C for 1 h in an oven. Remove from heat, allow to cool, and transfer contents with the aid of 10 ml of water to a 125 ml separatory funnel marked at 25 ml. Make the solution basic to litmus paper by slowly adding NH_4OH (ca. 5 ml is required). Acidify by the dropwise addition of (1 + 1) H_2SO_4. Add 10 ml of chelating–extracting solution consisting of 15% v/v 2-ethyl-1,3-hexanediol in methyl isobutyl ketone (MIBK) (16) and shake for 1 min. Let the phases separate and decant the organic layer into a 25 ml volumetric flask. Repeat the chelation–extraction with another 10 and 5 ml of chelating–extracting solution, combining organic phases into the volumetric flask. Dilute to volume with chelating–extracting solution and mix.

For plant material, weigh 10 g of sample into a platinum or porcelain crucible and moisten with ca. 20 ml of saturated $Ba(OH)_2$ solution. Dry 1 h at 150°C and ash for 10 h in a muffle furnace at 600°C. Remove from the furnace, add 10 ml of 5 N HCl and triturate with a polyethylene rod. Evaporate at 150°C to less than 6 ml, add 3 ml of extracting solution and stir vigorously. Decant into a 15 ml centrifuge tube and let the phases separate.

(iii) Determination

Aspirate organic solutions into a nitrous oxide–acetylene flame previously optimised on boron standards in an identical organic solvent matrix and measure boron absorbance at 249.68/249.77 nm. Prepare boron standards by chelating and extracting aqueous boron solutions in the same manner as sample digests. Flush the burner with extracting solution after aspiration of each standard and sample solution.

Notes

(1) The two lines at 249.68 nm and 249.77 nm separated by 0.09 nm may be difficult to resolve; a narrow 0.1 nm or less spectral bandpass is required

to isolate the more sensitive (by at least a factor of 2) 249.77 nm line. Careful adjustment of flame stoichiometry and burner height is necessary for best sensitivity.

(2) Little interference from other elements has been observed; 10000 µg ml^{-1} of Fe, Co or Ni is without effect. Varian [154], however, reports interference from high levels of sodium which could be minimized by adjusting the flame to normal stoichiometry (red cone 0.5—1.0 cm) with the consequent loss of sensitivity.

(3) The procedure for foods described above [180] was developed specifically for boron in caviar using flame emission spectroscopy; it has been adopted by the AOAC as official first action [180]. A similar extraction procedure following digestion with HNO_3—H_2SO_4—H_2O_2, has also been reported [7i, 181, 182] for determining boron, added as boric acid at high levels up to ca. 2000 mg boron kg^{-1} to foodstuffs such as dried fruits and caviar; this procedure, initially adopted by the AOAC, as official first action [7i], was later [12] declared surplus. The procedure for plants [183] was reported applicable to samples containing the element at levels as low as 3 mg kg^{-1}. A recent development [184] involves the conversion of boron to volatile methyl borate and sweeping the vapour into a nitrous oxide—acetylene flame for quantitation by FAAS.

Selected references: The procedure was adapted essentially from J. Assoc. Off. Anal. Chem. [180] and Melton et al. [183]. Other pertinent references are: 7i, 12, 21, 63c, 154, 159, 181, 182, 184.

4. Aluminium

(i) *Summary of procedure*
The sample is brought into solution by wet digestion and aluminium is quantitated by flame AAS using a nitrous oxide—acetylene flame.

(ii) *Preparation of sample solution*
Digest the sample by one of the acid wet-digestion procedures described above (Sections III.A.1, III.A.2), transfer the solution to a volumetric flask of suitable small size and make up to volume.

(iii) *Determination*
Aspirate solutions into a fuel-rich nitrous oxide—acetylene flame and measure Al absorbance at 309.27/309.28 nm.

Notes
(1) With the 309.27/309.28 doublet, sensitivity depends on spectral bandpass; a narrow 0.2 nm bandpass is recommended to minimise intense emission of the flame. Absorbance depends critically on flame stoichiometry and observation height. S/N can be improved by increasing lamp current and optimizing fuel flow.

References pp. 201—210

(2) Enhancement of absorbance by elements such as Cr, Mn, Fe, Co, Ni and Ti (and ionization of aluminium) can usually be controlled by selection of suitable flame conditions and incorporation of 2000 μg K ml^{-1} in both sample and standard solutions. The slight depressive effects of Ca, Si, perchloric and hydrochloric acids can be minimized by matching sample and standard solutions in respect of the major matrix elements and acids.

(3) The low level of aluminium in some samples may require concentration by chelation—solvent extraction prior to introduction into the flame. A procedure described for wines [36] involves digestion with HNO_3—H_2SO_4, extraction with 8-hydroxyquinoline in MIBK and determination with a nitrous oxide—acetylene flame. Aluminium has also been extracted with 2,4-pentanedione [185]. EAAS has been applied to measuring aluminium in beer [186]. Gorsuch [139] proposes wet oxidation to be preferred and mentions that adverse comments reported in respect of sample destruction have usually applied to dry ashing.

Selected references: 7e, 36, 37h, 41, 50, 51, 63c, 139, 154, 159, 185, 186.

5. Vanadium

(i) Summary of procedure

The sample is brought into solution by wet digestion or dry ashing—acid dissolution, and vanadium is determined by FAAS using a nitrous oxide—acetylene flame, or by EAAS.

(ii) Preparation of sample solution

Digest the sample by one of the wet digestion procedures described above (Sections III.A.1—III.A.2), transfer solution to a small volumetric flask and dilute to volume. Alternatively, dry ash according to Section III.A.3 and the following specific instructions. Transfer 5—20 g of sample into a platinum crucible, char or evaporate to dryness on a hot plate and ash at 450°C for 16 h in a muffle furnace. Remove, add 2 ml of HNO_3, and evaporate to dryness on a hot plate. Re-ash the residue for 2 h at 450°C, dissolve in 2 ml of 4 N HNO_3, warm on a hot plate, transfer to a 5 ml volumetric flask and dilute to volume.

(iii) Determination

Aspirate solutions into a fuel rich nitrous oxide—acetylene flame and measure vanadium absorbance at 318.34/318.40/318.54 nm using a spectral bandpass of 0.2 nm to pass all three emission lines. For samples with low vanadium contents, use the HNO_3 solution resulting from dry ashing and EAAS as follows: Introduce an appropriate small volume of solution into an electrothermal atomizer, purge with Ar and follow previously established drying, pyrolysis and atomization program (refer to Table 5 and ref. 115 for guide). Employ suitable non-atomic absorption correction.

Notes

(1) The doublet at 318.34/318.40 nm can be resolved from the adjacent 318.54 nm line with a 0.03 nm spectral bandpass. As all three lines have similar absorption sensitivity, transmission of the three lines using a 0.2 nm spectral bandpass is recommended for increased detectivity and precision. Optimum sensitivity is obtained with a slightly fuel-rich nitrous oxide—acetylene flame; ionization should be suppressed with $1000\,\mu g\,K\,ml^{-1}$. Variable enhancement of absorption by Al, Fe, Cr and other elements is removed by addition of $2000\,\mu g\,Al\,ml^{-1}$.

(2) Little has been published on the determination of vanadium in foodstuffs or other biological materials. The low concentrations commonly found in food materials necessitate resort to techniques more detective than FAAS.

Selected references: The dry ash—EAAS procedure was adapted essentially from Myron et al. [115], adapted with permission from J. Agric. Food Chem., 25 (1977) 297—300, copyright by the American Chemical Society. Other pertinent references are: 50, 51, 63c, 139, 154, 159.

6. Chromium

(i) *Summary of procedure*

The sample is brought into solution by wet digestion and chromium is determined by FAAS using a nitrous oxide—acetylene flame. Vegetable oils are dissolved in MIBK prior to FAAS determination.

(ii) *Preparation of sample solution*

Digest the sample with HNO_3—$HClO_4$ according to the procedure outlined in Section III.A.1, transfer the solution to a small volumetric flask and dilute to volume. With vegetable oils, accurately weigh 5 g into a 25 ml volumetric flask, dilute to volume with MIBK and mix.

(iii) *Determination*

Aspirate standards and solutions resulting from HNO_3—$HClO_4$ digestion into a stoichiometric or fuel-lean nitrous oxide—acetylene flame and measure chromium absorbance at 357.87 nm. For vegetable oils dissolved in MIBK use the tentative AOCS procedure [23d] as follows: Prepare organo-chromium stock solution containing $100\,\mu g\,Cr\,g^{-1}$ by dissolving with warming 0.0516 g of dried tris(1-phenyl-1,3-butanediono)chromium(III) (NBS No. 1078a, 9.7% Cr) in 3 ml of xylene and 3 ml of 2-ethylhexanoic acid and making to 50 ml with a base oil (vegetable oil containing no more than 0.02 mg of analyte kg^{-1}). Prepare working standard solutions by diluting stock solution with base oil—MIBK to give the required concentration of chromium in the oil—MIBK matrix similar to that of sample solution. Aspirate solutions into an air—acetylene flame optimised for these organic solutions and measure chromium absorbance at 357.87 nm.

References pp. 201—210

Notes

(1) A narrow spectral bandpass of ca. 0.2 nm is required with the 357.87 chromium line to eliminate nearby 357.66 nm and 358.23 nm argon lines when an argon-filled light source is used. In an air—acetylene flame, sensitivity and chemical interferences are critically dependent on flame stoichiometry and observation height.

(2) A recent detailed investigation [156] of interferences on Cr(VI) determination, in air—acetylene and nitrous oxide—acetylene flames, caused by a variety of acids, masking agents, model and real sample solutions, and flame parameters, has demonstrated the superiority of a nitrous oxide—acetylene flame in respect of interference and precision. Chromium ionization in the nitrous oxide—acetylene flame must be suppressed by 1000—2000 μg K ml^{-1}. The 359.35 Cr line has been recommended [63c]. It has been suggested [187] that in HClO$_4$, chromium in the sample and standard solutions be in the same oxidation state. It has also been reported [144] that in HNO$_3$ medium, both trivalent and hexavalent chromium give equal responses in a nitrous oxide—acetylene flame.

(3) The volatility of chromium during wet digestion or dry ashing has been discussed in the literature. Hartford [188] states that at microgram quantities, chromium is readily lost as volatile chromyl compounds; in fact, the element can be purposely volatilized as chromyl chloride, with HCl or NaCl. Cary and Allaway [189] observed the requirement for AgNO$_3$ in the ternary digestion mixture of HNO$_3$—HClO$_4$—H$_2$SO$_4$ to prevent losses of chromium whereas Jones et al. [190] reported good recoveries without AgNO$_3$. As some difficulties have been reported with dry ashing [191], probably wet digestion is to be preferred. An interesting study by Wolf and Greene [192] of effects of different sample treatments and different samples, demonstrated extremely wide variations in recoveries of endogenous chromium; publication of the solution to this problem is still pending. Digestion with H$_2$SO$_4$—H$_2$O$_2$ [193] is also being pursued [194].

(4) The procedures described above using FAAS are suitable for samples with chromium content $\gtrsim$ 2 mg kg^{-1}. To determine the very low (μg kg^{-1}) natural levels, resort must be made to techniques involving chelation—solvent extraction [189, 195] and/or EAAS [196—198]. No procedure for μg kg^{-1} levels of chromium in biological tissues has yet reached official status.

Selected references: The procedure for vegetable oils was adapted from Official and Tentative Methods of the American Oil Chemists' Society [23d]. Other pertinent references are: 7q, 23d, 37i, 41, 50, 51, 63c, 139, 144, 154, 156, 159, 187—201.

7. Manganese

(i) *Summary of procedure*

Samples are wet digested or dry ashed and manganese is determined by FAAS using an air—acetylene flame.

(ii) *Preparation of sample solution*

In general, digest the sample with HNO_3—$HClO_4$ (Section III.A.1), transfer the solution to a small suitable volumetric flask and dilute to volume. Alternatively, dry ash the sample according to instructions in Section III.A.3, transfer the solution to a volumetric flask and dilute to volume. For plant materials follow either the dry-ashing or wet-digestion procedure described in Section IV.B.1(ii).

(iii) *Determination*

Aspirate solutions into a stoichiometric air—acetylene flame and measure manganese absorbance at 279.48 nm.

Notes

(1) A spectral bandpass of 0.2 nm or less is necessary to isolate the manganese 279.48 nm line from the strongly absorbing 279.83 nm line for maximum sensitivity and calibration linearity. Air—acetylene flames with fuel-lean, normal, and fuel-rich stoichiometry have been suggested. The FAAS determination of manganese is virtually interference-free. Excessive concentrations of iron enhance manganese absorption; the depressive effect of silicon can be overcome by addition of $CaCl_2$ to the solutions.

(2) Liquid samples such as fruit juices and wines may not have to be digested. Fruit juices have been centrifuged with dilute acid to give clear solutions for analysis [171]. Determination on wines is by direct aspiration after filtration if required, comparing absorbances to those of standards prepared in a synthetic solution simulating the wine matrix as outlined in note 5 to Section IV.B.1 [36].

(3) Several publications have reported on comparison studies of wet digestion and dry-ashing of biological materials for manganese determinations [139, 175, 179]. Although some differences may occur between the two, and manganese is lost at high ashing temperatures, when carefully applied, both procedures should generally perform well.

(4) The FAAS procedure described above, is suitable for the majority of food-analysis applications. In some instances, however, low manganese contents will dictate the use of SEFAAS or EAAS techniques [202—205]. An interesting multielement (Cd, Cu, Co, Mn, Ni, Pb and Zn) scheme using chelating ion exchange has been outlined by Baetz and Kenner [206].

Selected references: 7d, 7g, 7t, 9, 21, 34, 36, 37j, 41, 50, 51, 63c, 139, 154, 159, 171, 175, 179, 193, 200—207.

8. *Iron*

(i) *Summary of procedure*

After wet digestion or dry ashing, iron in samples is determined by FAAS using an air—acetylene flame.

(ii) *Preparation of sample solution*

For a variety of cereal products, weigh 10 g into a platinum, Vycor or silica crucible and ash in a furnace at ca. 550°C until a light grey ash results. To facilitate removal of the last traces of carbon, moisten the ash with 0.5–1.0 ml of $Mg(NO_3)_2$ solution (21) or with HNO_3. Do not add these ashing aids to products containing NaCl in a platinum crucible. Dry, carefully ignite in a furnace and cool. Add 5 ml of HCl and evaporate on a steam bath. Dissolve the residue in 2.0 ml of HCl, cover with a watch glass, and heat for 5 min on a steam bath. Transfer via a filter quantitatively into a 100 ml volumetric flask and dilute to volume. Alternatively, transfer 10 g of sample to an 800 ml Kjeldahl flask, separately add 20 ml of water, 5 ml of H_2SO_4 and 25 ml of HNO_3, mixing after the addition of each reagent. Allow to stand for several minutes, then heat gently at intervals until heavy evolution of NO_2 stops. Continue heating gently until charring begins and cautiously add several ml of HNO_3 at intervals until SO_3 is evolved and a colorless or very pale yellow liquid results. Cool, add 50 ml of water and one Pyrex glass bead, heat to SO_3 fumes, cool, add 25 ml of water, transfer via a filter to a 100 ml volumetric flask and dilute to volume.

For vegetable oils containing at least 2 mg Fe kg^{-1} accurately weigh 5 g into a 25 ml volumetric flask, dilute to volume with MIBK and mix.

For wines, dilute 20 ml of sample, with 88 ml of 95% ethanol and water, to 200 ml in a volumetric flask. Analyse distilled liquor without preparation; prepare standard solutions of Fe in 43% v/v ethanol.

With beer, pipet 25 ml into a 100 ml stoppered flask. Equilibrate flask in water bath at 20°C for 30 min, add 2.0 ml of 1% APDC solution (22), mix and add 10.0 ml of MIBK. Shake vigorously for 5 min and centrifuge to separate layers. Equilibrate at 20°C for 10 min.

Process plant samples by dry ashing according to the procedure in Section IV.B.1(ii), reducing by heating the volume of (1 + 1) HCl to ca. 5 ml so that when diluted to 50 ml, the final solution will be ca. 0.5 M with respect to HCl. Pipet 20 ml or less of the digest into an extraction tube (m) and add 5.0 ml of 1.0 M metal-free citrate buffer adjusted to pH 4.7 (bromphenol blue) (23). Add dropwise (1 + 1) NH_4OH to the colour change, plus three drops in excess. Add 2 to 10 ml of 3-heptanone followed by water to bring the top of the solution near the constriction. Add 2.0 ml of 1% APDC (22) and additional water to bring the solvent interface to the centre of the constriction. Stopper and shake for 1 min.

(iii) *Determination*

Aspirate solutions into a fuel-lean air—acetylene flame and measure iron absorbance at 248.33 nm. For procedures involving chelation—extraction, treat standard solutions in the same manner as samples. Aspirate organic solutions into a flame previously optimized on the organic solvent. Flush burner with extracting solution after aspiration of each standard and sample solution. Carry out determinations on beer and plant sample extracts within

30 min and 2 h, of extraction respectively. For vegetable oils in MIBK, follow the procedure described in Section IV.B.6(iii). Prepare organo-iron stock solution from 0.0486 g of dry tris(1-phenyl-1,3-butanediono)iron(III) (NBS No. 1079a, 10.30% Fe, dried 1 h at 100°C).

Notes
(1) Even the narrowest spectral bandpass obtainable with most spectrometers will not isolate the most sensitive 248.33 nm iron line from the non-absorbing iron line at 248.42 nm and other nearby lines. Iron calibration curves, therefore, will usually be non-linear. The 302.05/302.06 nm iron line has been reported to have a better signal/noise ratio [194].

(2) Although only slightly ionized in an air—acetylene flame, iron is substantially ionized in a nitrous oxide—acetylene flame and 1000 μg K ml^{-1} is used to overcome this problem. Non-atomic light absorption and scattering at the Fe 248.33 line should be checked and compensated for using either a continuum light source or one of the following lines: Fe 249.89, Re 248.73, Cu 249.22, Be 249.47 and B 249.68 nm. Chemical interferences are minimised in a fuel-lean air—acetylene flame with only silicon interfering significantly. Silicon interference can be suppressed by the addition of $CaCl_2$; signal suppression by citric acid is overcome by the addition of phosphoric acid.

(3) Digestion of liquid samples may not be necessary. Analyses of a distilled liquor, beer and wines have been reported [25, 36, 169, 208]. Determination of wines by direct aspiration after filtration if required, comparing absorbances to those of standards prepared in a synthetic solution simulating the wine matrix has been described by Ecrement [36] and is outlined in note 5 to Section IV.B.1. Iron in orange juice has been determined following hydrolysis with HNO_3 [174].

(4) Wet digestion and dry ashing of samples for Fe estimation have been discussed in a number of publications [139, 175, 176, 178, 179, 191, 202, 209]. Although some difficulties have been reported with both approaches, wet digestion may be preferred.

(5) The direct aspiration-FAAS procedure described above, is suitable for the majority of food samples as iron usually occurs at relatively high concentrations. In the few instances of low iron levels, techniques of SEFAAS or EAAS will have to be resorted to. The advantage of the lower solubility in water of 3-heptanone has been discussed [210]. Electrothermal atomization has been applied to the measurement of iron in fats and oils [205].

Selected references: The procedures were adapted essentially from: AOAC Official Methods of Analysis [7g, 7v, 7w], Official and Tentative Methods of the American Oil Chemists' Society [23d], The Institute of Brewing Analysis Committee [25] and Chaney [211]. Other pertinent references are: 7d, 7q, 7t, 34, 36, 37k, 41, 50, 51, 63, 139, 154, 159, 169, 174—176, 178, 179, 191, 202, 205, 209, 212.

9. Cobalt

(i) Summary of procedure

Samples are wet digested, cobalt is chelated with 2-nitroso-1-naphthol, solvent-extracted and determined either by FAAS using an air—acetylene flame, or by EAAS.

(ii) Preparation of sample solution

For SEFAAS determination on plant materials weigh 2 g of sample into a 100 ml Kjeldahl flask and digest with 30 ml of HNO_3 + 5 ml of $HClO_4$. Continue heating for 1 h past the appearance of white fumes of $HClO_4$ to destroy refractory organic matter and to dehydrate the silica. Dilute to ca. 40 ml, heat to dissolve any precipitate of $KClO_4$ and let cool. Add 10 ml of 40% m/v sodium citrate solution (24) and one drop each of bromthymol blue and methyl red indicator solutions (13). Adjust pH to within the range 5.3—5.7 (orange to golden orange colour) with 10 M NH_4OH correcting any overshoot with ca. 6 M HCl. Add 5 ml of 6% v/v H_2O_2 to reduce Fe^{3+} to Fe^{2+} and decant the solution from the precipitate into a 100 ml separatory funnel. Add 1 ml of 1% m/v ethanolic solution of 2-nitroso-1-naphthol (26), mix and allow to stand for 1 h. Serially extract the cobalt chelate with 10, 5 and 5 ml portions of $CHCl_3$ by shaking each time for 1 min and collect the pooled extracts in a 50 ml glass-stoppered weighing bottle. Discard the aqueous phase by pouring out of the funnel and rinse the funnel with water taking care that no aqueous phase enters the stopcock or stem of the funnel. Return the organic phase to the separatory funnel rinsing in with 5 ml of $CHCl_3$. Add 10 ml of 2 N NaOH (28) to the extract and shake for 15 s to remove excess 2-nitroso-1-naphthol. Return the organic phase to the weighing bottle and evaporate to dryness. Let cool, add 1 ml of MIBK, stopper, and carefully dissolve the residue.

For EAAS determination on plant materials, digest 0.5 g of sample with 15 ml of HNO_3 + 2.5 ml of $HClO_4$ and reduce the volume of $HClO_4$ to ca. 0.5 ml. Let cool, rinse the wall of the flask with 5 ml of water and warm the solution to dissolve any $KClO_4$ precipitate. When the digests have cooled, add 1 ml of 40% m/v sodium citrate solution (26), one drop each of bromthymol blue and methyl red indicator solutions (13) and adjust pH as described above with concentrated NH_4OH. Correct any overshoot with 3 M HCl or 2 M NH_4OH. Add 1 ml of 30% H_2O_2 followed by 0.3 ml of 1% m/v ethanolic solution of 2-nitroso-1-naphthol (26) and transfer to a 15 ml glass-stoppered graduated cylinder. Rinse the flask with two portions of 1.5 ml of water and mix the solution. Make the final volume of solution in the cylinder to 11—12 ml, add 1 ml of isoamyl acetate and shake vigorously for 30 s.

(iii) Determination

For FAAS determination, aspirate MIBK solution into an air—acetylene flame previously optimized on cobalt standards in an identical matrix and measure cobalt absorbance at 240.73 nm. Flush the burner with MIBK after aspiration of each standard and sample solution. For EAAS determination, introduce an appropriate small volume of isoamyl acetate solution into an electrothermal atomizer, purge with N_2 and follow a previously established drying, pyrolysis and atomization program (refer to Table 5 and ref. 214 for guide). For both types of determinations, prepare cobalt standard solutions by chelating and extracting aqueous cobalt standard solutions in the same manner as sample digests, beginning with the addition of the sodium citrate solution.

Notes

(1) For optimum sensitivity and linear calibration, it is necessary to resolve, using a narrow 0.1 nm or less spectral bandpass, the 240.73 nm line from non-absorbing cobalt lines at 240.77 nm and 240.88 nm. As most AA spectrometers cannot perform this well, cobalt calibration curves will be non-linear. Sensitivity is highest in a fuel-lean air—acetylene flame and quite invariable with burner height. Non-specific light losses at the cobalt wavelength arising from light scattering and molecular absorption should be checked-for and corrected using a continuum light source or non-absorbing wavelengths listed in Table 4.

(2) Of the few interferences reported in the air—acetylene flame, the enhancement caused by high concentrations of iron (e.g. 10000 μg ml^{-1}) in perchloric acid is of greatest interest to the food analyst. Reported effects of acids will be minimized when following the usual practice of acid matching with sample and standard solutions. Chemical interferences can be almost completely eliminated in the nitrous oxide—acetylene flame, and by extracting cobalt from the sample matrix as described above.

(3) The typically low levels of cobalt in foodstuffs dictate the need for some form of preconcentration prior to FAAS, or resort to the more detective technique of EAAS for all but the infrequent high-level samples. Simmons [213, 214] used 2-nitroso-1-naphthol for chelation whereas the 1-nitroso-2-naphthol isomer was used by Hageman and co-workers [215]. Excess chelating agent is removed in the FAAS procedure to prevent nebuliser blockage. Other publications also describe FAAS [216, 217] and EAAS procedures [201] for animal feeds and fish tissue. Although wet digestion sample treatment procedures may be suitable, Gorsuch [139] favours dry ashing for excellent recoveries and minimum reagent-blank levels.

Selected references: The procedures were adapted essentially from Simmons [213, 214], adapted with permission from Anal. Chem., 45 (1973) 1947—1949; ibid., 47 (1975) 2015—2018; copyright by the American Chemical Society. Other pertinent references are: 41, 50, 51, 63c, 139, 154, 159, 201, 215—217.

10. Nickel

(i) Summary of procedure

The sample is brought into solution by wet digestion or dry ashing. Nickel is concentrated if required by chelation—solvent extraction, and is determined by FAAS using an air—acetylene flame. Vegetable oils are dissolved in MIBK prior to FAAS measurement.

(ii) Preparation of sample solution

Subject the sample to one of the wet-digestion or dry-ashing procedures described in Sections III.A.1—III.A.3, transfer the solution to an appropriate volumetric flask and dilute to volume. Use this solution for direct FAAS determination when the nickel content of the product (e.g. tea) is sufficiently high. For the more typical case of low nickel levels, follow for plant materials the dry ashing APDC-3-heptanone extraction procedure outlined in Section IV.B.8(ii) using an agitation time of 20 s after the addition of APDC solution. With vegetable oils, accurately weigh 5 g into a 25 ml volumetric flask, dilute to volume with MIBK and mix.

(iii) Determination

Aspirate solutions into a fuel-lean air—acetylene flame and measure nickel absorbance at 232.00 nm. Aspirate organic solvent solutions into the same flame previously optimized on nickel standards in an identical matrix. Flush the burner with 3-heptanone after aspiration of each standard and sample solution. For vegetable oils in MIBK, follow the tentative AOCS procedure [23d] as described in Section IV.B.6(iii). Prepare organo-nickel stock solution from 0.0360 g of Ni cyclohexane butyrate (NBS No. 1065b, 13.89% Ni, desiccated for 48 h over P_2O_5) according to the procedure in the same section. Aspirate sample and standard solutions into an optimized air—acetylene flame.

Notes

(1) The rich nickel spectrum in the vicinity of the most sensitive atomic absorption Ni line, 232.00 nm, leads to pronounced curvature of the calibration graph. For highest sensitivity and somewhat reduced calibration curvature, a 0.2 nm or less spectral bandpass must be employed to attempt to isolate the 232.00 nm line from adjacent non-absorbing Ni lines at 231.72 nm and 232.14 nm.

(2) A fuel-lean air—acetylene flame provides for relatively interference-free operation. When using the 232.00 nm line, light scattering and molecular absorption must be considered and corrected with a hydrogen or deuterium continuum source or one of the following non-absorbing lines: Ni 231.40, Ni 233.00 or Ba 233.53 nm. It must be ascertained that nickel is not present in the cathode of the continuum light source. The four-fold less sensitive

absorption line at 341.48 nm may be preferred as it provides more linear calibration and reduced non-atomic absorption.

(3) In the analysis of teas, the AOAC [7k] recommends matrix matching by preparing standard solutions to contain Na, K, Mg, Ca and Al at the levels found in sample digests, in addition to digestion acids. The AOCS procedure [23d] is intended for oil samples containing at least 2 mg Ni kg^{-1}. Other published procedures for oils involve FAAS after extraction with HNO_3 [218], and EAAS [201, 205].

Selected references: The procedures were adapted essentially from Official and Tentative Methods of the American Oil Chemists' Society [23d] and Chaney [211]. Other pertinent references are: 7k, 37l, 63c, 139, 154, 159, 201, 205, 218.

11. Copper

(i) Summary of procedure

The sample is brought into solution by wet digestion or dry ashing and copper is determined by FAAS using an air—acetylene flame. Vegetable oils are dissolved in MIBK, whereas other liquid samples require no or minimal treatment. When necessary, copper is concentrated by chelation—solvent extraction prior to measurement by FAAS or EAAS.

(ii) Preparation of sample solution

In general, destroy organic matter using one of the wet digestion or dry ashing procedures described in Sections III.A.1—III.A.3, and transfer the solution to a volumetric flask of suitable small volume. For dry ashing use platinum or silica crucibles; avoid porcelain as copper may be extracted from the glaze. If an ashing aid is required, use $Mg(NO_3)_2$. Dissolve the residue in (1 + 1) HCl or (2 + 1 + 3) $HCl—HNO_3—H_2O$, provided that the latter mixture is not used in a platinum vessel.

Alternatively for solid foodstuffs in general, proceed according to the following wet-digestion procedure. Accurately weigh a sample containing not more than 20 g solids, according to expected Cu content, and transfer to an 800 ml Kjeldahl flask. If sample contains less than 75% water, add water to obtain this dilution. Add an initial volume of HNO_3 to equal ca. twice the dry sample weight and as many ml H_2SO_4 as g of dry sample but not less than 5 ml. Warm slightly and carefully to avoid excessive foaming. When reaction has quieted, heat cautiously and maintain oxidizing condition by cautiously adding small amounts of HNO_3 whenever the mixture darkens. Digest until organic matter is destroyed and SO_3 fumes are evolved and the solution is colorless or light straw colored. Cool, add 25 ml of water and remove nitrosylsulfuric acid by heating to fumes of SO_3. Repeat addition of water and fuming, cool, transfer to 100 ml volumetric flask and make to volume. When the sample contains a large amount of fat, make a partial digestion with HNO_3 until only the fat remains undissolved. Cool, filter free of solid fat, wash residue with water, add H_2SO_4 to the filtrate and proceed as above.

For vegetable oils containing at least 2 mg Cu kg^{-1}, accurately weigh 5 g into a 25 ml volumetric flask, dilute to volume with MIBK and mix. Wines, distilled liquors, beer and other beverages may be analyzed directly, filtering through a membrane filter if necessary to remove suspended matter, and diluting if required to lower dissolved solids content. For plant material follow the dry ashing or wet digestion procedure described in Section IV.B.1(ii).

In instances of low copper levels, treat the sample digest resulting from any of the preceding ashing or digestion procedures according to the following procedure of chelation—extraction and FAAS. Ensure that the acidity of the digest is not greater than 5 N when chelation—extraction is performed. Pipet a suitable aliquot into a narrow neck flask and dilute to the base of the neck. Add 2 ml of 1% aqueous solution of APDC (22), mix and allow to stand for 2 min. Add 4 ml of MIBK or 3-heptanone, shake for 1 min and let the phases separate. Use the organic layer without delay for FAAS measurement.

Alternatively, for small quantities of plant material, use the following procedure in conjunction with EAAS. Weigh 0.01—0.05 g of very finely divided sample into a 15 × 150 mm Pyrex test tube and add 1 ml of (1 + 8) $HClO_4$/HNO_3. For larger quantities of sample use larger volumes of acids and ensure that sufficient HNO_3 is present for safe digestion. Heat to fumes of $HClO_4$, dissolve and transfer the residue to a 10 ml glass-stoppered cylinder using 9 ml of purified (1 + 30) $HClO_4$ (30). Add 1 ml of 1% aqueous APDC (22), mix, add 1 ml of MIBK and shake vigorously for 15 s. Allow the solution to stand for 1 h to permit phase separation prior to EAAS measurement.

(iii) *Determination*

Aspirate solutions into a fuel-lean air—acetylene flame and measure copper absorbance at 324.75 nm. For vegetable oils in MIBK, follow the procedure describes in Section IV.B.6(iii). Prepare organo-copper stock solution from 0.0303 g of bis(1-phenyl-1,3-butanediono)Cu(II) (NBS No. 1080, 16.5% Cu, dried for 30 min at 110°C) according to the procedure. For distilled liquor and beer, prepare standard solutions in 50% and 3% v/v ethanol, respectively. For FAAS measurements on organic solutions, aspirate solutions into an air—acetylene flame previously optimized on copper standards in an identical matrix. Flush burner with appropriate solvent after aspiration of each standard and sample solution. For EAAS determination, introduce an appropriate small volume of MIBK extract into an electrothermal atomizer and follow a previously established drying, pyrolysis and atomization program (refer to Table 5 and ref. 224 for guide). For both chelation—solvent extraction procedures, prepare Cu standard solutions by chelating and extracting aqueous standard solutions in the same manner as sample digests.

Notes

(1) The spectral interference on the Cu 324.75 nm line caused by light absorption by europium at 324.75 nm [155] need not be considered by the food analyst. Apart from effects of very high concentrations of iron and zinc, chemical interferences are not a problem.

(2) In the analysis of teas, the AOAC [7k] recommends matching the matrices of standard solutions to those of samples by the incorporation in standard solutions of Na, K, Mg, Ca and Al, at levels found in sample digests, in addition to digestion acids. The procedure of Ecrement [36] for direct FAAS determination of copper in wines also involves matrix matching as outlined in note 5 to Section IV.B.1.

(3) Investigations of wet digestion and dry ashing procedures have been reported by several workers [139, 176, 178, 179, 202, 209, 219]. Wet digestion is reported by Gorsuch [139] to have been successful with a range of acid and oxidant combinations, whereas the situation with dry ashing is less clear-cut. Complete digestion may not always be necessary; copper has been determined in orange juice and butteroil following HNO_3 extraction [174, 220], and in edible fats and oils after extraction with acid—EDTA solutions [221]. Materials such as oils, fats and fatty foods which are difficult or impossible to wet-oxidize, may be simply dissolved in organic solvents and solutions aspirated into the flame [8d, 23d, 222].

(4) With the majority of food products, copper can usually be determined by FAAS. In the few instances of low copper levels, resort has commonly been made to chelation—solvent extraction prior to FAAS measurement [8d, 34, 37m, 217]; recent reports on the application of EAAS to analysis of foodstuffs and biological materials have appeared [193, 201, 204, 205, 223, 224]. The importance of non-atomic absorption in the determination of copper in plant digests by FAAS has been studied by Simmons [225]. Correction was necessary to arrive at good results, with a continuum light source providing more accurate data than use of a nearby non-absorbing line.

Selected references: The procedures were adapted essentially from AOAC Official Methods of Analysis [7x], the Society for Analytical Chemistry, Official, Standardized and Recommended Methods of Analysis [8d], American Oil Chemists Society [23d], Allen [37m] and Simmons and Loneragan [229]; the latter was adapted by permission from Anal. Chem., 47 (1975) 566—568, copyright by the American Chemical Society.

Other pertinent references are: 7d, 7f, 7k, 7q, 7t, 9, 21, 25, 34, 36, 41, 50, 51, 63c, 139, 154, 155, 159, 174, 176, 178, 179, 193, 201, 202, 204, 205, 208, 209, 212, 217, 219—223, 225, 226.

12. *Zinc*

(i) *Summary of procedure*

The sample is brought into solution by wet digestion or dry ashing, and zinc is determined by FAAS using an air—acetylene flame. Liquid samples require no or minimal treatment.

(ii) *Preparation of sample solution*

In general, destroy organic matter using one of the wet digestion or dry ashing procedures described in Sections III.A.1—III.A.3, and transfer the solution to a volumetric flask of suitable volume. For dry ashing use platinum or silica crucibles; avoid porcelain as zinc may be extracted from the glaze. Addition of 0.2 g of $CaCO_3$ to the sample before ashing facilitates ashing and dissolution of residue. Dissolve the residue in $(1+1)$ HCl or $(2+1+3)$ HCl—HNO_3—H_2O, provided that the latter mixture is not used in a platinum vessel.

Alternatively for foodstuffs in general, proceed according to one of the following wet-digestion or dry-ashing procedures. Where glassware is called for, use only borosilicate glassware thoroughly cleaned with hot HNO_3. Clean platinum crucibles by fusion with $KHSO_4$ followed by leaching with 10% HCl. Wet digestion: Accurately weigh into a 300 ml or 500 ml Kjeldahl flask, up to 10 g of sample estimated to contain 25—100 µg Zn. Evaporate liquid samples to small volume before digesting. Add ca. 5 ml of HNO_3 and cautiously heat until vigorous reaction subsides. Add 2.0 ml of H_2SO_4 and continue heating, maintaining oxidizing conditions by adding small quantities of HNO_3 until the solution is colorless and fumes of SO_3 are evolved. Cool, dilute with 20 ml of water, transfer quantitatively to a 100 ml volumetric flask and dilute to volume. Dilute further if required, with $(1+49)$ H_2SO_4. Dry ashing: Accurately weigh, into a platinum crucible, sufficient sample to contain 25—100 µg of zinc. Char under an IR lamp and ash until carbon-free in furnace with temperature slowly raised to not over 525°C. Dissolve the ash in a minimum volume of $(1+1)$ HCl, add 20 ml of water and evaporate to near-dryness on a steam bath. Add 20 ml of 0.1 N HCl and heat for 5 min. Transfer quantitatively with 0.1 N HCl into a 100 ml volumetric flask and dilute to volume with 0.1 N HCl. Dilute further if required, with 0.1 N HCl.

Analyze beer, wines, other beverages and liquid samples with no or minimal treatment. Filter through a membrane filter if necessary to remove suspended matter and dilute if required to reduce dissolved solids concentration and to bring zinc concentration to scale of spectrometer. For plant material follow the dry-ashing or wet-digestion procedure described in Section IV.B.1(ii).

(iii) *Determination*

Aspirate solutions into a fuel-lean air—acetylene flame and measure zinc absorbance at 213.86 nm.

Notes

(1) With the sensitive 213.86 nm absorption line, spectral bandpass is not

critical unless the light source contains a brass cathode. In that instance, a bandpass of not greater than 2 nm is required to avoid transmission of Cu lines at 216.51 and 217.89 nm. The recommended fuel-lean air—acetylene flame absorbs at the zinc line; constant flame conditions must be maintained for precise results. Non-atomic absorption by the flame and dissolved solids must be checked and suitably corrected. Atomic absorption spectrometric determinations of zinc are otherwise relatively interference-free with only silicon and phosphate exhibiting some depression, and copper enhancing zinc absorption.

(2) The normally moderate to high zinc contents in foodstuffs coupled with excellent AAS detectivity for this element, indicate FAAS to be the technique of choice for virtually all food analyses. Should in the odd instance, greater detectivity be required, resort can be made to chelation—extraction with APDC—MIBK as outlined in Section IV.B.11(ii), or EAAS [227]. The ubiquitous nature of zinc can lead to problems with contamination from reagents, glassware, plastic ware and other apparatus when low levels are determined.

(3) Errors can result from differences in pH among sample and standard solutions; zinc absorption has been reported constant in the pH range of about 2—5 [63c] or below pH 3.6 [228]. In the procedure of Ecrement [36] for FAAS determination of zinc in wines, matrix interferences are minimised by approximately matching the matrix of standard solutions to samples using a synthetic diluent.

(4) Investigations of wet digestion and dry ashing procedures have been reported by several workers [139, 175, 176, 178, 179, 191, 202, 207, 209, 229]. Gorsuch [139] states that whereas wet-digestion methods seem to give uniformly good recoveries, dry-ashing methods have generated an appreciable amount of controversy. The presence of large amounts of chlorine in any chemical form may cause losses of zinc by volatilization or retention. Apart from this consideration, however, dry ashing at ca. 500°C should be satisfactory for the majority of materials. Although porcelain crucibles are not advocated for dry ashing by one official organization [8d], they are used in another official procedure [9]. Complete digestion of sample may not always be necessary; zinc has been determined in fruit juices following acid extraction [171, 174], and in edible fats and oils after extraction with acid EDTA solutions [221].

Selected references: The procedures were adapted essentially from AOAC Official Methods of Analysis [7o] and the Society for Analytical Chemistry, Official, Standardized and Recommended Methods of Analysis [8d]. Other pertinent references are: 7d, 7g, 7t, 9, 21, 36, 37n, 41, 50, 51, 63c, 139, 154, 159, 171, 174—176, 178, 179, 191, 202, 207, 209, 221, 226—230.

13. *Arsenic and selenium*

(i) *Summary of procedure*

Sample organic matter is destroyed by wet digestion, arsenic and selenium are converted to gaseous AsH_3 and H_2Se, respectively, through reduction by sodium borohydride in a hydride generator. The volatile hydrides are transported into an argon—hydrogen-entrained air flame in which absorbances of arsenic and selenium are measured at 193.70 nm and 196.03 nm, respectively.

(ii) *Preparation of sample solution*

Accurately weigh appropriate amount of samples (e.g. 5—10 g of products containing > 50% water and low in fat, starch or sugar, such as meat, fish, fruits and vegetables; 3—5 g of products containing 10—50% water; 1—3 g of products containing less than 10% water such as cereal products and dried foods; 1—2 g of products high in fat or sugar such as cheese, butteroil, syrups and jams) and transfer with three glass beads into a 100 ml Kjeldahl flask. Add 30 ml of (5 + 1) HNO_3—$HClO_4$ and let the sample digest overnight at room temperature or heat continuously until foaming subsides and the sample dissolves. Gradually increase the temperature to achieve steady vigorous boiling, reduce volume by one half, cool, and add 10 ml of (1 + 1) HNO_3—H_2SO_4. Continue heating through $HClO_4$ oxidation characterized by vigorous surface reaction and evolution of white fumes. If the yellow digest begins to darken, avoid charring by the cautious addition of 1 ml portions of HNO_3. Continue heating until the solution becomes clear and colourless and dense white fumes of SO_3 appear. Heat 5 min past this stage, cool, transfer to a 100 ml volumetric flask containing 30 ml HCl (or a volume to give final concentration appropriate to the specific hydride generator used) and 20 ml of water, and dilute to volume.

(iii) *Determination*

Transfer a 20 ml aliquot of digest (or volume required by the hydride generator used) to a hydride generator affording suitable precision, for As determination, add 2.0 ml of 10% solution of KI (31), let stand 2 min, and generate arsine and hydrogen selenide with $NaBH_4$ (32), following procedures specific to the particular generator used, for this and all relevant operations. Transport the gaseous hydrides to an argon—hydrogen-entrained air flame and measure absorbances of arsenic and selenium at 193.70 nm and 196.03 nm, respectively.

Notes

(1) Flame gases, especially in air—hydrocarbon flames, absorb light strongly in the UV regions where arsenic and selenium absorption lines occur. The commonly used arsenic line at 193.70 nm is less strongly absorbed than the one at 188.99 nm and is more sensitive than the line at 197.20 nm. Superior sensitivities and detection limits for both elements are obtained in the more transparent argon—hydrogen-entrained air flames. Absorbance and signal/noise, particularly for arsenic, depends on hollow-cathode lamp current; hollow-cathode lamp current and electrodeless discharge lamp wattage should be optimised for best performance. The

occurrence of more severe chemical inteferences in the cooler argon—hydrogen-entrained air flame is reduced by the necessary resort to the more detective technique of hydride generation when determining these elements in foods by atomic spectrometry [refs. 158 (and references therein), 231—234]. Reduction of arsenic and selenium to volatile AsH_3 and H_2Se, respectively, with their subsequent release from solution and transport to the flame leaves behind many potential interferants and substantially increases detectivity. Interferences in the flame from other hydride-forming elements and interferences of elements with the hydride reaction may, however, occur, and should be checked. Non-atomic absorption arising from absorbance changes of the flame due to alteration of gas composition during hydride transport should be checked. Stability of a "soft", low gas flow, more sensitive flame may be enhanced by use of custom-made burner shielding.

(2) Hydrides may be decomposed by atomization techniques other than the flame; techniques reported in the literature include helium or argon plasmas, tube furnaces (electrically heated, with or without use of a hydrogen diffusion flame, externally flame-heated, tube-confined hydrogen—oxygen or air flames) and a graphite furnace of the type used for EAAS. Many other variants of generation, storage and transport into the atomiser have also been reported. Information regarding the performance of the hydride-AAS technique for arsenic and selenium amassed from a recent large scale AOAC collaborative study [235, 236], indicates somewhat high instrument-related imprecision and suggests the need for caution in applying the technique. Several reports dealing with the applications of EAAS have appeared [157, 237—242].

(3) Ashing of organic samples for the estimation of these two volatile elements have been extensively studied [139, 243—248]. Wet-digestion procedures are satisfactory. It has been demonstrated that for selenium, oxidizing conditions must be maintained throughout the digestion; incorporation of $HClO_4$ into the digestion mixture and avoiding sample charring leads to complete oxidation and quantitative recovery. Dry ashing has generally been less favoured due to losses of the elements by volatilization; use of magnesium nitrate as an ashing aid, however, appears to yield favorable recoveries.

Selected references: The procedure was adapted essentially from Ihnat and Miller [158]. Other pertinent references are: 7y, 34, 50, 51, 63c, 139, 153, 154, 157, 159, 231—251.

14. *Molybdenum*

(i) *Summary of procedure*

The sample is wet digested, molybdenum is concentrated by chelation—solvent extraction, and determined by FAAS using a nitrous oxide—acetylene flame.

(ii) *Preparation of sample solution*

Digest several g of air-dried plant sample with HNO_3–$HClO_4$–H_2SO_4 and remove all of the $HClO_4$ by boiling. Dilute the H_2SO_4 digest with water to ca. 20 ml, and filter into a 50 ml centrifuge tube. Adjust pH with ammonium hydroxide to 1.3 using a pH meter, add 5 ml of 0.5% 8-quinolinol (oxine) solution in MIBK (33), shake for 25 min, and let the phases separate.

(iii) *Determination*

Aspirate the MIBK solution into an optimized fuel-rich nitrous oxide–acetylene flame and measure molybdenum absorbance at 313.26 nm. Prepare standard solutions in the organic solvent by taking aqueous standards through the chelation–extraction procedure.

Notes

(1) A strongly reducing fuel-rich nitrous oxide–acetylene flame is superior to other flames for sensitivity and freedom from interferences. Optimisation of burner height is important as absorption signal is fairly dependent on observation height. In aqueous systems interference from calcium has been controlled by the addition of aluminium or Na_2SO_4. Reduced sensitivity has been reported in the presence of acetone vapour from depleted acetylene cylinders.

(2) With the procedure described above it has been observed [252] that noise was decreased and flame stability and measurement precision were increased if water was sprayed into the flame for several seconds after each measurement on MIBK solutions. The authors also reported excessive foaming when digesting 2.5 g portions of plant samples in 250 ml Kjeldahl flasks and a mixture of 5 ml $HClO_4$ + 1 ml H_2SO_4 and, in total, 50 ml of HNO_3. They resorted to using special double flasks and successfully digested 5 g of plant material with the combination 3 ml $HClO_4$ + 1 ml H_2SO_4 + 30 ml HNO_3. The detection limit for a 5 g sample was reported to be 0.025 µg Mo g^{-1}. This would permit application to that population of samples containing moderate to high natural levels of the element. More detective AAS approaches are required for the bulk of foods. Other extraction studies reported deal with the systems oxine–$CHCl_3$–methyl isoamyl ketone [253], APDC–MIBK [34] and thiocyanate–$CHCl_3$–long chain alkylamines [254, 255].

Selected references: The procedure was adapted essentially from Stupar et al. [252]. Other pertinent references are: 34, 41, 50, 51, 63c, 154, 253–255.

15. Cadmium

(i) *Summary of procedure*

The sample is digested with HNO_3–H_2SO_4–H_2O_2, cadmium is converted to the iodocadmate ion with KI, is extracted with a solution of a liquid ion-exchange resin in MIBK and determined by FAAS in an air—acetylene flame. Alternatively, the sample is digested with HNO_3—$HClO_4$, cadmium is chelated and extracted with APDC—MIBK, back-extracted into an aqueous solution of HNO_3 and H_2O_2 and determined by EAAS.

(ii) *Preparation of sample solution*

(a) Transfer to a 100 ml Kjeldahl flask 2 g or more of an accurately weighed food sample together with several glass beads and add 10 ml of H_2SO_4. To the cold mixture add 50% H_2O_2 dropwise until the reaction slows down or until the solution becomes colourless. If the reaction does not begin within two min of the addition of H_2O_2, initiate by gentle warming. Heat the solution to fumes of H_2SO_4 adding more H_2O_2 as necessary until a colourless solution is obtained. Observe all necessary precautions to avoid mechanical loss of solution. When the oxidation is complete, cool the digest, carefully add 20 ml of water and again cool the digest. Transfer the solution to a 100 ml separatory funnel with a minimum amount of water. Add 5 ml of 0.1 M KI solution (34), dilute to ca. 50 ml with water and mix well. Add with a pipette 10 ml of Amberlite LA-2/MIBK solution (35) and shake vigorously for 20 s. Allow the phases to separate, run-off and discard the aqueous phase and filter the organic phase through a dry Whatman No. 541 or equivalent filter paper to remove suspended droplets of aqueous phase collecting the organic phase in a small glass-stopped vessel.

(b) Alternatively, accurately weigh 1—15 g (dry mass) of food sample into a 250 ml Vycor beaker covered with a non-ribbed watch glass and dry overnight at 120°C if necessary. Add HNO_3 according to the formula: ml HNO_3 = 10 + (5 × dry mass of sample in g) and allow digestion at room temperature for 2—12 h. Boil the solution until the volume is decreased to about one half of the original volume of HNO_3, and add $HClO_4$ according to the formula: ml $HClO_4$ = 20 + (20 × mass of fat or oil present in g) + 7 × dry mass of sample in g. Heat again, adding 2 ml portions of HNO_3 in the unlikely event that evolution of gases becomes excessive or the froth above the solution darkens, and reduce the volume to 4—8 ml. Rinse the watch glass and walls of the beaker with ca. 40 ml of water, heat the solution to ca. 90°C and transfer to a 125 ml separatory funnel, rinsing the beaker with 10—15 ml of water. Adjust the acidity with NH_4OH solution to a blue-green hue of the methyl violet indicator (one drop of 0.1% aqueous solution (13)). Carry out the adjustment rapidly as the indicator colour is stable for only a few minutes. Use a pH meter to adjust pH to 1.6 in instances where large amounts of metal interfere with the indicator. Add 5 ml of 1% APDC solution (22) while swirling for 10 s. Immediately after, add 70 ml of MIBK, stopper and shake for 60 s. Unstopper, let the phases separate for 20 s and drain and discard the aqueous phase, leaving behind ca. 1 ml. Pipette in stripping solution (36) (5 ml when Cd < 0.2 µg; 10 ml when 0.2 µg < Cd < 0.4 µg, and 20 ml when Cd > 0.4 µg), stopper the funnel and rinse the stem with water. Shake gently for 5 min, invert to avoid seepage, stand overnight, shake a further 5 min and return to an upright position. After 1 h drain the aqueous phase into a beaker, discarding the first 1—2 ml, and pipette 2.00 ml into a 25 ml polypropylene bottle containing 6.00 ml of modification solution (37). In both sample preparation schemes, carry a procedural reagent blank through identical respective operations.

(iii) *Determination*

(a) Aspirate the MIBK solutions into an air—acetylene flame with air flow adjusted to give a flame just short of luminous while aspirating MIBK and measure cadmium absorbance at 228.80 nm. Prepare standard solutions in an identical organic matrix by taking inorganic cadmium standard solutions through the same extraction procedures as for the samples. Flush the burner with MIBK after aspiration of standard and sample solutions. Exercise care when changing solutions as with some burner systems, such as the three slot Boling burner, the removal of the ketone renders the flame weak and there is a tendency to flash back.

(b) For EAAS determination, introduce an appropriate small volume of solution into an electrothermal atomizer and follow a previously established drying, pyrolysis and atomization program (refer to Table 5 and ref. 256 for guidance). Prepare standards in an aqueous dilution solution (38) of composition identical to the bulk matrix of the sample solutions; there is no need to process standards through the complete chelation/extraction/back extraction procedure. When absorbances of sample solutions exceed those of ca. 6 ng ml^{-1} cadmium standards, reduce by diluting with dilution solution.

Notes

(1) Absorbance is highly dependent on lamp current and flame stoichiometry. Hollow-cathode lamps must be operated at fairly low currents to prevent self-absorption; good lamp stability, however, permits use of high instrumental scale expansion. The spectral bandpass used is not critical. For aqueous solutions very few chemical interferences have been reported in the air—acetylene flame; depression of Cd absorbance is caused by large amounts of silicon.

(2) The occurrence in foodstuffs of cadmium at typically low levels, often necessitates some concentration of the element or resort to more detective EAAS. However, in instances of higher contents at the fractional mg kg^{-1} and greater levels, the high AAS detectivity of cadmium permits its determination by direct aspiration-FAAS as described in the literature [8d, 226, 257, 258]. Baetz and Kenner [206] have applied their chelating ion-exchange technique to effect separation and concentration prior to FAAS measurement. APDC is a common chelating agent for cadmium, and a number of papers report on its use together with MIBK or n-butyl acetate as extracting solvents [7q, 226, 256, 258—260]. Procedures using extraction with Aliquat 336 (tricaprylylmonomethylammonium chloride) or Amberlite LA-2/MIBK, and FAAS have been reported [17, 261]. Several publications

have appeared describing the application of EAAS to cadmium determination [193, 197, 201, 227, 256, 257, 259, 262—265], with chelation—solvent extraction preceding the measurement step in some cases [197, 256, 265]. Both wet-digestion and dry-ashing procedures have been applied to biological materials with wet digestion being more reliable [139, 178, 191, 266, 267].

Selected references: Procedures (a) and (b) were adapted essentially from Analytical Methods Committee [8h, 17] and Dabeka [256], respectively; the latter adapted with permission from Anal. Chem., 51 (1979) 902—907, copyright by the American Chemical Society. Other pertinent references are: 7j, 7q, 8d, 37o, 41, 50, 51, 63c, 139, 154, 178, 191, 193, 197, 201, 206, 226, 227, 257—267.

16. Tin

(i) Summary of procedure

The sample is wet digested, if necessary, tin is extracted as $Sn(IV)I_4$ into toluene and back-extracted as stannate with aqueous KOH to reduce interferences in the flame, and is determined by FAAS using a nitrous oxide—acetylene or air—hydrogen flame.

(ii) Preparation of sample solution

Digest an accurately weighed appropriate amount of food sample according to the H_2SO_4—H_2O_2 wet-digestion procedure described in Section IV.B.15(ii)(a). Transfer the digest to a graduated cylinder and add H_2SO_4 to give a volume one quarter that of a suitable calibrated flask. Transfer the solution to the flask and rinse both digestion flask and cylinder with water, combining washings into the calibrated flask. Carry out the following operations rapidly. Cool the solution, make to volume with water, pipet an aliquot containing 0 to 400 µg Sn into a separatory funnel and add 2.5 ml of 5 M KI solution (39) to 25 ml of diluted digest. Mix, add 10 ml of toluene, shake vigorously for 2 min, allow phases to separate and discard the aqueous layer. Wash the toluene layer, without shaking, with 5 ml of a mixture containing the same proportions of KI solution and 9 N H_2SO_4 as the diluted digest (1 volume of KI solution to 10 volumes of the acid). Discard the aqueous phase; toluene layer should be pink with extracted iodine. Repeat the washing, rinsing the inner surface of the separatory funnel, and discard the wash solution. Add 5.0 ml of water, and 0.50 ml of 5 M KOH (41) and shake the funnel for 30 s. Should the toluene layer not be colorless, add 5 M KOH drop by drop from a graduated pipette until the pink colour disappears, then 2 drops in excess. Note the total volume of KOH added and shake the funnel for 30 s. Drain the KOH phase quantitatively into a 20 ml calibrated flask, add 5.0 ml of 6 M HCl and mix. If more than 0.5 ml of KOH solution has been used for the extraction, add an equal volume of 6 M HCl in addition to the 5.0 ml. Remove the yellow iodine with 0.5 or 1 ml of 5% m/v ascorbic acid solution (42) and dilute to volume with water.

Alternatively, digest an accurately weighed amount of sample according to any of the two wet-digestion procedures described in Sections III.A.1– III.A.2, transfer to a volumetric flask of suitable volume and dilute to volume. Use this solution directly for FAAS.

(iii) *Determination*

Aspirate solutions resulting from the first procedure, in which Sn is extracted from potential interfering elements, into an air—hydrogen flame and measure tin absorbance at 224.61 nm. Check and correct for non-atomic absorption. For solutions from the second procedure, from which tin is not isolated from interferants, aspirate solutions into a fuel-rich nitrous oxide—acetylene flame and measure tin absorbance at 235.48 nm or 286.33 nm.

Notes

(1) The fuel-rich nitrous oxide—acetylene flame provides for interference-free operation. Somewhat greater (ca. two-fold) sensitivity is observed at the 235.48 nm line over absorption at the 286.33 nm line; a narrow 0.2 nm (or less) spectral bandpass must be used to avoid spectral interferences at the latter line. Measurements should be conducted at 286.33 nm when non-atomic absorption is encountered at 235.48 nm. Ionization in the nitrous oxide—acetylene flame should be suppressed by the addition of 1000 μg ml^{-1} of potassium or cesium.

(2) For the determination of tin in canned foods, H_2SO_4–H_2O_2 is a suitable digestion medium. HNO_3 together with H_2SO_4 is preferred for complete digestion of samples containing organotin to prevent loss of analyte [268]. When using 50% H_2O_2, care must be taken to ensure that the reagent does not contain sodium stannate as a stabilizer. If antimony is present in the sample, the concentration of iodide in the solution is increased to 1.5 M to prevent co-extraction of antimony. Although both wet digestion and dry ashing can be suitable [139, 269], wet digestion is preferred. A procedure for dibutyltin dilaurate in feeds involves solvent extraction of the compound prior to FAAS [15].

(3) Increased detectivity obtainable via conversion of tin to SnH_4 prior to absorption measurements [270], and by EAAS [271, 272] are of interest but further developments in these techniques must be awaited before their recommendation for general usage.

Selected references: The extractive procedure was adapted from Engberg [268]. Other pertinent references are: 7z, 8d, 15, 41, 50, 51, 63c, 139, 154, 159, 269—272.

17. *Mercury*

(i) *Summary of procedure*

Organic matter in the sample is destroyed by wet digestion with

H_2SO_4—HNO_3—V_2O_5, with HNO_3 under pressure in a Teflon vessel, or with H_2SO_4—HNO_3—H_2O_2; the digest is reduced with hydroxylamine sulfate—$SnCl_2$ or hydroxylammonium chloride—$SnCl_2$, elemental mercury vapour is removed by aeration, and transported to a cell with quartz end windows situated in an AA spectrometer where mercury is determined by AAS.

(ii) *Preparation of sample solution*

(a) Accurately weigh 5 g (fresh weight) of fish sample into a digestion flask (e) and if necessary, rinse the neck with not more than 5 ml of water. Add ca. twenty 6—8 mesh boiling stones, 10—20 mg V_2O_5 and 20 ml of (1 + 1) H_2SO_4—HNO_3. Quickly connect flask to condenser (f) through which cold water circulates, and swirl to mix contents. Heat to produce a low initial boil (ca. 6 min) and complete the digestion with a strong boil (ca. 10 min), swirling flask intermittently during the digestion. No solid material should be apparent after ca. 4 min except for globules of fat. Remove the flask from heat and wash the condenser with 15 ml of water. Add 2 drops of 30% H_2O_2 through the condenser and wash into the flask with 15 ml of water. Cool to room temperature, disconnect the flask, rinse the ground joint with water and quantitatively transfer, with rinsing, to a 100 ml volumetric flask, diluting to volume with rinse water. Ignore solidified fat.

(b) *Exercise caution when digesting organic materials in a pressure digestion vessel — some advise against it.* Accurately weigh 1 ± 0.1 g of fish sample into a Teflon pressure digestion vessel (d) (this procedure is specified for a 23 ml vessel available from Uni-Seal Decomposition Vessels, P.O. Box 9463, Haifa, Israel; do not change sample weight or acid volume substantially as excessive pressure may damage the vessel), add 5.0 ml of HNO_3 and close the vessel with a screw cap. Place in a preheated oven at 150°C for 30—60 min or until the solution is clear. Remove the vessel, allow to cool to room temperature, and transfer the digest with the aid of 95 ml of diluting solution (44) to a flask suitable for mercury generation.

(c) Weigh 2.5 g of fish sample into a 200 ml Kjeldahl flask having a B24 socket (g), add 9 ml of H_2SO_4 and attach the upper part of the digestion apparatus to the flask. Heat on a heating mantle, and swirl vigorously until a homogeneous tarry fluid is obtained, cool the flask in ice and add 2 ml of 50% H_2O_2 through the top of the condenser. Open tap A to slowly introduce the H_2O_2 into the mixture, remove the flask from the ice-bath and swirl it slowly until the reaction begins. As the reaction slows down, apply heat. Close tap A, add 2 ml of HNO_3 through the top of the condenser, and allow the acid to run, through tap A, slowly into the flask while the contents are still hot. After 2 min close tap A, heat the flask until fumes are evolved, and run off the condensate that has collected in B into a beaker. Add, through the top of the condenser, 1 ml of 50% H_2O_2 and 1 ml of HNO_3 to the flask, close tap A and again heat until fumes are evolved; transfer the condensate from B into the same beaker as before. Repeat this operation with 0.5 ml portions of both H_2O_2 and HNO_3 until the digest is a pale straw color. Return the cool condensate from the beaker to the flask through the condenser. Cool the contents of the flask and add 6% m/v $KMnO_4$ solution until a permanent pink color is produced. Transfer the digest to a 50 ml calibrated flask, rinse the reflux system and Kjeldahl flask with water into the calibration flask and dilute to volume. Set aside for 24 h before proceeding with the AAS determination.

(iii) *Determination*

(a) Pipet 25 ml of solution from digestion (a) into the original digestion flask (e) and add 75 ml of diluting solution (44); alternatively, use entire digest from procedure (b). Follow operating instructions pertaining to the specific apparatus used [7n, 13] generate mercury with 20 ml of reducing solution (46), aerate for ca. 1 min or as required.

(b) Alternatively, pipette 10 ml of the solution resulting from digestion procedure (c) into the aeration test tube of the vapour generation apparatus [19]. Add water to bring the volume to 13 ml, then add 2.0 ml of reducing solution (47) and 0.20 ml of reducing solution (49). Mix well and aerate according to the pertinent operating instructions [19].

Transport the mercury vapour in the air stream to the absorption cell and measure the absorbance of the Hg vapour at 253.65 nm.

Notes

(1) Although the Hg 184.96 nm line provides greater sensitivity than the 253.65 nm line, the former is in a wavelength region where the atmosphere and most flames absorb strongly. Elemental mercury vapour is monoatomic and no atomization device is necessary once Hg has been released in this state from the sample. The greatly enhanced detectivities of cold vapour AA techniques have led to their widespread application to determination of Hg in biological materials. Broadband molecular absorption in the 254 nm region by organic vapours must be checked and corrected for using a continuum light source. Interference of inorganic ions which precipitate mercury from solution or are reduced by the stannous chloride reductant can be reduced or checked using standard solutions of composition similar to that of samples.

(2) References 7n, 13 and 19 should be consulted for information on the construction of the vapour generation apparatuses, and operational and other details. A blank determination using a similar sample with less than 0.05 mg Hg kg^{-1} should be made. Should a blank mercury value greater than 0.05 mg be obtained, reagents should be examined for mercury and those with high levels should be replaced. The volatility of mercury and its compounds necessitates care in the decomposition of organic materials. Dry-ashing techniques are not generally applicable; wet-digestion procedures continue to be the most commonly used and the most successful [139]. Of the vast literature on mercury determination, the reader is referred to the following reviews [58, 273, 274], and methodology reports [28, 275—278].

Selected references: The above digestion procedures (a) and (b) and determination procedure (a) are from work of the AOAC [7n, 13]; digestion procedure (c) and determination procedure (b) result from studies of a combined subcommittee of the Analytical Methods Committee of the UK Chemical Society, AOAC and the UK Ministry of Agriculture, Fisheries and Food [19]. Other pertinent references are: 8d, 14, 50, 51, 58, 63c, 139, 154, 159, 273—278.

18. *Lead*

(i) *Summary of procedure*

The sample is digested with HNO_3—$HClO_4$ or H_2SO_4—H_2O_2; lead is chelated and extracted with APDC—MIBK and is determined by aspiration of the MIBK phase into an air—acetylene flame, or is back extracted from the organic phase into an aqueous solution of HNO_3 and H_2O_2 followed by EAAS determination.

(ii) *Preparation of sample solution*

(a) Digest an appropriate mass of sample with H_2SO_4—H_2O_2 according to the procedure in Section IV.B.15(ii)(a) to the point prior to addition of water. Dilute the cold solution with 10 ml of water, add 2 ml of 10% sodium sulfite and boil to fumes of SO_3. Dilute the digest to 50 ml with water, cool and transfer to a 125 ml separatory funnel. Add 2 ml of 1% APDC solution (22), shake and let stand for 5 min. Accurately add 10 ml of MIBK and shake the mixture vigorously for 1 min. Allow the layers to separate, discard the aqueous layer and filter the organic layer through a small dry filter paper into a suitable glass-stoppered flask.

(b) Take an appropriate mass of sample through the HNO_3—$HClO_4$ digestion, and chelation—extraction, back extraction procedures described in Section IV.B.15(ii)(b).

(iii) *Determination*

(a) Aspirate MIBK solutions into an optimised air—acetylene flame and measure lead absorbance at either 283.31 or 217.00 nm, whichever wavelength gives the greater signal-to-noise ratio. Prepare standard solutions by taking aliquots of aqueous lead standard solutions with 5 ml of H_2SO_4 diluted to 50 ml through the same chelation—extraction procedure. Flush the burner with MIBK after aspiration of standard and sample solutions. Exercise care when changing solutions as with some burner systems, such as the three-slot Boling burner, the removal of the ketone renders the flame weak and there is a tendency to flash-back.

(b) For EAAS determination, use the solution resulting from Section IV.B.18(ii)(b) and proceed according to instructions in Section IV.B.15(iii)(b). When absorbances of sample solutions exceed those of ca. 60 ng ml^{-1} lead standards, reduce by diluting with dilution solution (38).

References pp. 201—210

Notes

(1) Although the 217.00 nm line gives three-fold greater sensitivity than the 283.31 nm line, the latter is preferred due to the possible spectral interference near 217 nm from argon filler gas, and the increased non-atomic absorption by the flame and other species at the lower wavelength. With aqueous solutions, no significant chemical interferences occur in the fuel-lean air—acetylene flame. Effects of anions which give precipitates with lead can usually be reduced by making the solutions 0.1 M with respect to EDTA.

(2) Careful cleaning of apparatus used and checking of reagents for excessive lead contamination is necessary when dealing with low levels of the element; procedures have been published [7l, 7m, 7aa].

(3) A routine method for screening beer samples with lead at levels 0.2 mg kg^{-1} and higher [26] based on the work of Roschnik [279], involves chelation of the original sample with diethylammonium diethyldithiocarbamate, extraction into xylene followed by determination by FAAS. Other reports on methods utilizing SEFAAS have been published [169, 260]. The new technique of EAAS has been the subject of several studies [197, 227, 256, 280—282]. Two of these methods involve chelation—solvent extraction and back extraction into an aqueous phase prior to measurement by EAAS [197, 256]. It appears that wet digestion rather than dry ashing is preferred for destruction of organic matter [139, 178, 267].

Selected references: The procedures (a) and (b) have been adapted essentially from Analytical Methods Committee [18] and Dabeka [256], respectively. Other pertinent references are: 7l, 7m, 7aa, 26, 36, 37p, 41, 50, 51, 63c, 139, 154, 159, 169, 178, 197, 227, 260, 267, 279—282.

ACKNOWLEDGEMENTS

The author thanks the following organizations and authors for permission to use published material: Association of Official Analytical Chemists (Official Methods of Analysis; J. Assoc. Off. Anal. Chem.), American Chemical Society (Anal. Chem., J. Agric. Food Chem.), The Chemical Society and N. W. Hanson (Official, Standardised and Recommended Methods of Analysis), The Chemical Society and P. W. Shallis and A. Engberg (Analyst), The Institute of Brewing and P. A. Martin and the joint editor of the J. Inst. Brew. (J. Inst. Brew.), American Oil Chemists' Society (Official and Tentative Methods of the American Oil Chemists' Society), Pergamon Press Ltd. (T. T. Gorsuch, The Destruction of Organic Matter), J. D. Sauerländer's Verlag (Landwirtsch. Forsch.), and R. L. Chaney (R. L. Chaney, The Effect of Nickel on Iron Metabolism by Soybean, Ph.D. Thesis, Purdue Univ., West Lafayette, IN). Appreciation is expressed to T. C. Rains, Center for Analytical Chemistry, National Bureau of Standards, Washington, D.C., for critical review of the manuscript.

Contribution No. 1160 from the Chemistry and Biology Research Institute.

REFERENCES

1. A. Walsh, Spectrochim. Acta, 7 (1955) 108—117.
2. C. T. J. Alkemade and J. M. W. Milatz, Appl. Sci. Res. Sect. B, 4 (1955) 289—299.
3. C. T. J. Alkemade and J. M. W. Milatz, J. Opt. Soc. Am., 45 (1955) 583—584.
4. K. G. Sloman, A. K. Foltz and J. A. Yeransian, Anal. Chem., 41 (1969) 63R—89R.
5. A. J. MacLeod, Instrumental Methods of Food Analysis, Elek Science, London, 1973, (a) p. 652, (b) Ch. 6.
6. B. Welz, Atomic Absorption Spectroscopy, Verlag Chemie, New York, 1976, (English translation of 1975 German edn.), (a) vii, (b) 194—198.
7. W. Horwitz (Ed.), Official Methods of Analysis, 12th edn., Association of Official Analytical Chemists, Washington, D.C., 1975, (a) sects. 3.001, 7.001, 10.078, 10.111, 10.132, 10.165—10.167, 10.179, 14.001, 14.124, 16.001—16.019, 16.172, 16.185, 16.215, 17.001, 22.001, 31.065—31.066, (b) sects. 3.003, 7.003—7.009, 10.088—10.090, 10.136, 10.181, 13.002, 14.002—14.004, 14.076, 14.083—14.084, 14.113, 14.122, 14.125—14.126, 15.010—15.011, 15.035, 16.032—16.034, 16.165, 16.174, 16.187, 16.217—16.222, 16.252, 17.006—17.007, 18.019—18.020, 22.013—22.019, 23.002, 23.009, 24.002—24.003, 24.058, 27.005, 28.002, 30.005, 30.034, 30.045, 31.005—31.008, 31.068, 31.111—31.112, 31.150—31.151, 31.192—31.196, 32.036—32.042, 32.046—32.048, (c) sects. 3.004, 7.010, 8.025, 9.023, 9.113, 10.035, 10.130, 10.150, 11.017, 12.009, 13.003, 13.029, 14.006—14.008, 14.059, 14.077, 14.095, 14.127—14.128, 15.013, 15.037, 16.035, 16.135, 16.152, 16.178, 16.190, 16.223, 18.021, 22.025, 23.003, 23.010, 23.019, 24.006, 24.059, 27.009, 30.006—30.008, 30.065—30.066, 31.012—31.015, 31.073, 31.092—31.093, 31.113—31.114, 31.152—31.153, 31.200—31.201, 32.012, (d) sects. 7.077—7.082, (e) sects. 8.021—8.024, (f) sects. 9.029—9.031, 10.019, (g) sects. 9.036—9.038, (h) sects. 18.033—18.038, (i) sects. 20.042—20.045, (j) sects. 25.026—25.030, (k) sects. 25.041—25.045, (l) sects. 25.060—25.070, (m) sects. 25.077—25.082, (n) sects. 25.103—25.107, (o) sects. 25.143—25.147, (p) sects. 32.028—32.033, (q) sects. 33.089—33.094, (r) sects. 33.098—33.105, (s) sects. 3.002, 7.002, 8.001, 10.001, 10.079, 10.112, 10.135, 10.155, 10.168, 10.180, 13.001, 14.001, 14.057, 14.082, 14.124, 15.009, 15.034, 16.020, 16.127, 16.128, 16.149, 16.163, 16.173, 16.186, 16.216, 16.251, 17.001, 18.011, 22.008, 23.001, 24.001, 24.057, 27.004, 28.001, 30.001, 31.001, 31.067, 31.090, 31.108, 31.147, 31.191, 32.003, 32.036, 33.092, (t) sects. 2.096—2.100; (u) 3.016—3.019, 9.024—9.027, 11.022—11.025, 22.033—22.034, (v) 11.020, (w) sect. 14.013, (x) sects. 25.008, 25.038, (y) 25.008, 25.120, 41.010, (z) sects. 25.008, 25.127, (aa) sect. 25.059.
8. N. W. Hanson (Ed.), Official, Standardised and Recommended Methods of Analysis, 2nd edn., The Society for Analytical Chemistry, London, 1973, (a) pp. 115—116, (b) pp. 153—154, (c) pp. 154—155, (d) pp. 3—58, (e) pp. 494—495, (f) pp. 503—858, (g) pp. 107, 116, 128, 131—132, 152—153, 260—261, (h) pp. 3—23.
9. J. Assoc. Off. Anal. Chem., 58 (1975) 383—384.
10. J. Assoc. Off. Anal. Chem., 59 (1976) 470.
11. J. Assoc. Off. Anal. Chem., 60 (1977) 465.
12. J. Assoc. Off. Anal. Chem., 60 (1977) 469.
13. J. Assoc. Off. Anal. Chem., 60 (1977) 470—471.
14. J. Assoc. Off. Anal. Chem., 60 (1977) 474—476.
15. J. Assoc. Off. Anal. Chem., 60 (1977) 486.
16. J. Assoc. Off. Anal. Chem., 61 (1978) 462—463.
17. Analytical Methods Committee, Analyst, 100 (1975) 761—763.
18. Analytical Methods Committee, Analyst, 100 (1975) 899—902.
19. Analytical Methods Committee, Analyst, 102 (1977) 769—776.
20. Analytical Methods Committee, Analyst, 103 (1978) 643—647.

21 Ministry of Agriculture, Fisheries and Food, Technical Bulletin 27, The Analysis of Agricultural Materials, A Manual of the Analytical Methods used by the Agricultural Development and Advisory Service, Her Majesty's Stationery Office, London, 1973.
22 Approved Methods Committee, W. C. Schaefer (Chairman), Approved Methods of the American Association of Cereal Chemists, American Association of Cereal Chemists Inc., St. Paul, MN, Revised Oct. 1976, (a) Methods 62-05 to 62-80, 64-40 to 64-71, (b) Methods 44-01 to 44-60, (c) Methods 08-01 to 08-18, (d) Method 40-42, (e) Method 40-70.
23 W. E. Link (Ed.), Official and Tentative Methods of the American Oil Chemists' Society, 3rd ed., American Oil Chemists' Society, Campaign, IL, 1977, (a) Methods Aa 1-38, Ab 1-49, Ac 1-45, Ad 1-48, Af 1-54, Ba 1-38, Bb 1-38, Bc 1-50, C1-47, (b) Methods Ca 2a-45, Ca 2b-38, Ca 2c-25, Ca 2d-25, Ca 2e-55, Ba 2-38, Bd 2-52, Bc 2-49, Ae 2-52, Aa 3-38, Af 2-54, Ab 2-49, Ac 2-41, Ad 2-52, (c) Methods Ba 5a-68, Ba5-49, Bc 5-49, Ca 11-55, (d) Tentative Method Ca 15-75.
24 Subcommittee on Procedures of the Chemistry Task Force of the National Shellfish Sanitation Program, Collection, Preparation and Analysis of Trace Metals in Shellfish, USDHEW, PHS, FDA, Publ. No (FDA) 76-2006, 1975.
25 The Institute of Brewing Analysis Committee (P. A. Martin, Chairman), J. Inst. Brew., 79 (1973) 289—293.
26 The Institute of Brewing Analysis Committee (J. Weiner, Chairman), J. Inst. Brew., 83 (1977) 82—84.
27 The Institute of Brewing Analysis Committee (J. Weiner, Chairman), J. Inst. Brew., 80 (1974) 486—488.
28 Mercury Analysis Working Party of the Bureau International Technique du Chlore, Anal. Chim. Acta, 84 (1976) 231—257.
29 International Association of Cereal Chemistry, Reports of the ICC, Working and Discussion Meetings, Vienna, 12—14 May, 1976.
30 R. Lees, Food Analysis: Analytical and Quality Control Methods for the Food Manufacturer and Buyer, 3rd edn., Leonard Hill Books, London, 1975. (a) p. 191, (b) Methods M7a-b, M9a-k, (c) Method A17, (d) Method C1, (e) Method C3a-b, (f) Method C17a-b, (g) Method M1, (h) Methods L3, M5, Z1, (i) pp. 45—46.
31 D. Pearson, The Chemical Analysis of Foods, 7th edn., Churchill Livingstone, Edinburgh, 1976, (a) pp. 6—7, (b) pp. 7—9, (c) pp. 20, 79—80, 83, 88—89.
32 F. L. Hart and H. J. Fisher, Modern Food Analysis, Springer-Verlag, New York, 1971, Method 6-24 pp. 122—123, pp. 459—460.
33 M. Pinta (Ed.), translated by K. M. Greenland and F. Lawson, Atomic Absorption Spectrometry, Halstead Press, New York, 1975.
34 J. Laporte and G. Kovacsik, with J. Bellanger, in M. Pinta (Ed.), translated by K. M. Greenland and F. Lawson, Atomic Absorption Spectrometry, Halstead Press, New York, 1975, Ch. 9.
35 M. Pinta, in M. Pinta (Ed.), translated by K. M. Greenland and F. Lawson, Atomic Absorption Spectrometry, Halstead Press, New York, 1975, Ch. 16.1.
36 F. Ecrement in M. Pinta (Ed.), translated by K. M. Greenland and F. Lawson, Atomic Absorption Spectrometry, Halstead Press, New York, 1975, Ch. 16.2.
37 S. E. Allen, H. M. Grimshaw, J. A. Parkinson and C. Quarmby, Chemical Analysis of Ecological Materials, Blackwell Scientific Publications, Oxford, 1974, (a) pp. 71—80; (b) pp. 80—81; (c) pp. 81—82; (d) pp. 121—236, 305—374, 388—393; (e) pp. 83—92; (f) pp. 214—216, 220—221; (g) pp. 133—136, 174—176; (h) pp. 124—125; (i) p. 316; (j) pp. 178—179; (k) pp. 167—168; (l) pp. 328—329; (m) pp. 161—163; (n) pp. 231—232; (o) p. 313; (p) pp. 319—320.
38 M. A. Joslyn, in M. A. Joslyn (Ed.), Methods in Food Analysis, Physical, Chemical, and Instrumental Methods of Analysis, 2nd edn., Academic Press, New York, 1970, (a) Ch. 2, (b) Ch. 4, (c) Ch. 5, (d) Ch. 3.
39 Y. Pomeranz and C. E. Meloan, Food Analysis: Theory and Practice, revised edn.,

Avi Publishing Co. Inc., Westport, CT, 1978, (a) Ch. 10, (b) Ch. 34, (c) Ch. 2, (d) Ch. 3, (e) Ch. 33.
40 J. W. Robinson, Atomic Absorption Spectroscopy, Marcel Dekker, New York, 1966, Ch. 4.
41 W. T. Elwell and J. A. F. Gidley, Atomic Absorption Spectrophotometry, 2nd revised edn., Pergamon Press, Oxford, 1966.
42 W. J. Price, Spectrochemical Analysis by Atomic Absorption, Heyden, London, 1979, pp. 248—256.
43 W. Slavin, Atomic Absorption Spectroscopy, Interscience Publishers, New York, 1968, Ch. 5.
44 J. Ramirez-Muñoz, Atomic Absorption Spectroscopy and Analysis by Atomic Absorption Flame Photometry, Elsevier, Amsterdam, 1968, Ch. 19.
45 I. Rubeška and B. Moldan, Atomic Absorption Spectrophotometry, Iliffe, London, 1969.
46 R. Mavrodineanu in R. Mavrodineanu (Ed.), Analytical Flame Spectroscopy — Selected Topics, Springer-Verlag, New York, 1970, Ch. 13.
47 G. D. Christian and F. J. Feldman, Atomic Absorption Spectroscopy — Applications in Agriculture, Biology and Medicine, Wiley-Interscience, New York, 1970, pp. 181—445.
48 R. J. Reynolds and K. Aldous, Atomic Absorption Spectroscopy — A Practical Guide, C. Griffin and Co., London, 1970, Ch. 5.
49 J. Dvořak, I. Rubeška and Z. Řezač, Flame Photometry-Laboratory Practice, English translation edited by R. E. Hester, CRC Press, Chemical Rubber Co., Cleveland, 1971, Appendix 3.
50 W. J. Price, Analytical Atomic Absorption Spectrometry, Heyden and Sons, London, 2nd printing (with corrections), 1974, pp. 154—161.
51 G. F. Kirkbright and M. Sargent, Atomic Absorption and Fluorescence Spectroscopy, Academic Press, London, 1974.
52 S. R. Koirtyohann and E. E. Pickett, in J. A. Dean and T. C. Rains (Eds.), Flame Emission and Atomic Absorption Spectrometry, Vol. 3, Elements and Matrices, Marcel Dekker, New York, 1975, (a) Ch. 17, (b) Ch. 15.
53 C. W. Fuller, Electrothermal Atomisation for Atomic Absorption Spectrometry, The Chemical Society, London, 1977, (a) pp. 96—99.
54 M. Varju, Elelmiszervizsgalati kozlemenyek 17 (1-2) (1971) 64—70. Chem. Abstr., 76 (1972) 71059f.
55 C. C. Saarloos, Chem. Tech. (Amsterdam), 27 (8) (1972) 205—209. Chem. Abstr., 77 (1972) 70970e.
56 J. Morre, Lait, 54 (1974) 139—152.
57 K. Sakai, Shokuhin Kogyo, 18 (1975) 71—78. Chem. Abstr., 83 (1975) 204864h.
58 P. L. Schuller and H. Egan, Cadmium, Lead, Mercury and Methylmercury Compounds, A Review of Methods of Trace Analysis and Sampling with Special Reference to Food, FAO, Rome, 1976.
59 I. Teper, Krmivarstvi Sluzby, 13 (3) (1977) 63. Chem. Abstr., 87 (1977) 100691r.
60 N. T. Crosby, Analyst, 102 (1977) 225—268.
61 H. Seiler, Mitt. Geb Lebensmitt-u-Hyg., 63 (1972) 180—187.
62 R. Mavrodineanu (Ed.), Analytical Flame Spectroscopy-Selected Topics, Springer-Verlag, New York, 1970.
63 J. A. Dean and T. C. Rains (Eds.), Flame Emission and Atomic Absorption Spectrometry, (a) Vol. 1, Theory, 1969, (b) Vol. 2, Components and Techniques, 1971, (c) Vol. 3, Elements and Matrices, 1975, Marcel Dekker, New York.
64 E. J. Underwood, Trace Elements in Human and Animal Nutrition, 4th edn., Academic Press, New York, 1977, (a) p. 425, (b) p. 6.
65 B. K. Watt and A. L. Merrill, Composition of Foods, Agriculture Handbook No. 8,

United States Dept. of Agriculture, Wash., D.C. Revised Dec. 1963, Reprinted Oct. 1975.
66 L. P. Posati and M. L. Orr, Composition of Foods, Dairy and Egg Products, Raw, Processed, Prepared, Agriculture Handbook No. 8-1, United State Dept. of Agriculture, Washington, D.C. Revised Nov. 1976.
67 A. C. Marsh, M. K. Moss and E. W. Murphy, Composition of Foods, Spices and Herbs, Raw, Processed, Prepared, Agriculture Handbook No. 8-2, United States Dept. of Agriculture, Washington, D.C. Revised 1977.
68 D. Schlettwein-Gsell and S. Mommsen-Straub, Int. J. Vitamin Nutr. Res., Beiheft, 13 (1973) 9—22, 13 (1973) 23—33.
69 D. Schlettwein-Gsell and S. Mommsen-Straub, Int. J. Vitamin Nutr. Res., 41 (1971) 116—125, 41 (1971) 268—285, 41 (1971) 429—437, 41 (1971) 554—582, 42 (1972) 324—352, 42 (1972) 607—617, 43 (1973) 93—109, 43 (1973) 110—119, 43 (1973) 242—250, 43 (1973) 251—263.
70 Great Britain Ministry of Agriculture, Fisheries and Food, Survey of Lead in Food, Her Majesty's Stationery Office, London, 1972.
71 Great Britain Ministry of Agriculture, Fisheries and Food, Survey of Cadmium in Food, Her Majesty's Stationery Office, London, 1973.
72 Great Britain Ministry of Agriculture, Fisheries and Food, Survey of Mercury in Food; A Supplementary Report, Her Majesty's Stationery Office, London, 1973.
73 D. A. Ratkowsky, T. G. Dix and K. C. Wilson, Aust. J. Mar. Freshwat. Res., 26 (1975) 223—231.
74 D. C. Kirkpatrick and D. E. Coffin, Can. Inst. Food Sci. Technol. J., 7 (1974) 56—58.
75 J. Jaffray and J. M. DeMan, Can. Inst. Food Sci. Technol. J., 7 (1974) 159—161.
76 V. Jiranek and R. Bludovsky, Collect. Czech. Chem. Commun., 41 (1976) 2690—2695.
77 T. Szprengier, Bull. Vet. Inst. Pulawy, 19 (1975) 99—103.
78 M. I. Gomez and P. Markakis, J. Food Sci., 39 (1974) 673—675.
79 E. T. Hall, J. Assoc. Off. Anal. Chem., 57 (1974) 1068—1073.
80 E. G. Zook, J. J. Powell, B. M. Hackley, J. A. Emerson, J. R. Brooker and G. M. Knobl, Jr., J. Agric. Food Chem., 24 (1976) 47—53.
81 S. D. Dassani, B. E. McClellan and M. Gordon, J. Agric. Food Chem., 23 (1974) 671—674.
82 A. G. Hugunin and R. L. Bradley, Jr., J. Milk Food Technol., 38 (1975) 354—368.
83 R. Schelenz and J. F. Diehl, Z. Lebensm. Unters.-Forsch., 151 (1973) 369—375.
84 R. E. Simpson, W. Horwitz and C. A. Roy, Pestic. Monit. J., 7 (3/4) (1974) 127—138.
85 H. -G. Essing, K. -H. Schaller, D. Szadkowski and G. Lehnert, Arch. Hyg. Bakteriol., 153 (1969) 490—494.
86 T. Kjellstrom, B. Lind, L. Linnman and C. -G. Elinder, Arch. Environ. Health, 31 (1975) 321—328.
87 J. C. Bruhn and A. A. Franke, J. Dairy Sci., 59 (1976) 1711—1717.
88 E. A. Childs and J. N. Gaffke, J. Food Sci., 39 (1974) 853—854.
89 R. J. Lovett, W. H. Gutenmann, I. S. Pakkala, W. D. Youngs, D. J. Lisk, G. E. Burdick and E. J. Harris, J. Fish. Res. Board Can., 29 (1972) 1283—1290.
90 L. Friberg, M. Piscator and G. Nordberg, Cadmium in the Environment, CRC Press, Cleveland, 1971, p. 23.
91 H. Jonsson, Z. Lebensm. Unters.-Forsch., 160 (1976) 1—10.
92 G. Lehnert, G. Stadelmann, K. -H. Schaller and D. Szadkowski, Arch. Hyg. Bakteriol., 153 (1969) 403—412.
93 H. L. Huffman, Jr. and J. A. Caruso, J. Agric. Food Chem., 22 (1974) 824—827.
94 J. F. Reith, J. Engelsma and M. van Ditmarsch, Z. Lebensm. Unters.-Forsch., 156 (1974) 271—278.

95 B. Boppel, Z. Lebensm. Unters.-Forsch., 153 (1973) 345—347.
96 J. W. White, Jr., Adv. Food Res., 24 (1978) 287—374.
97 F. Bermejo-Martinez, C. Baluja-Santos and J. A. Ravina-Pereiro, Analusis, 3 (1975) 157—163.
98 S. J. Yeh, P. Y. Chen, C. N. Ke, S. T. Hsu and S. Tanaka, Anal. Chim. Acta, 87 (1976) 119—124.
99 W. R. Penrose, CRC Crit. Rev. Environ. Control, 4 (1974) 465—482.
100 E. G. Zook, F. E. Greene and E. R. Morris, Cereal Chem., 47 (1970) 720—731.
101 M. Moll, Brauwissenschaft, 30 (1977) 347.
102 J. C. Meranger, Bull. Environ. Contam. Toxicol., 5 (1970) 271—275.
103 W. J. Price and J. T. H. Roos, J. Sci. Food Agric., 20 (1969) 437—439.
104 K. Lorenz and R. Loewe, J. Agric. Food Chem., 25 (1977) 806—809.
105 C. R. Meiners, N. L. Derise, H. C. Lau, M. G. Crews, S. J. Ritchey and E. W. Murphy, J. Agric. Food Chem., 24 (1976) 1126—1130.
106 N. J. Daghir and N. N. Hariri, J. Agr. Food Chem., 25 (1977) 1009—1010.
107 J. F. Uthe and E. G. Bligh, J. Fish. Res. Board Can., 28 (1971) 786—788.
108 E. Amakawa, K. Ohnishi, N. Taguchi and H. Seki, Jpn. Anal., 27 (1978) 81—84.
109 L. T. Black, J. Am. Oil Chem. Soc., 52 (1975) 88—91.
110 R. D. Wauchope, J. Agric. Food Chem., 26 (1978) 226—228.
111 D. Arthur, Can. Inst. Food Sci. Technol. J., 5 (1972) 165—169.
112 M. A. Amer and G. J. Brisson, Can. Inst. Food Sci. Technol. J., 6 (1973) 184—187.
113 T. R. Shearer and D. M. Hadjimarkos, Arch. Environ. Health, 30 (1975) 230—233.
114 S. D. Senter, J. Food Sci., 41 (1976) 963—964.
115 D. R. Myron, S. H. Givand and F. H. Nielsen, J. Agric. Food Chem., 25 (1977) 297—300.
116 D. C. Smith, Pestic. Sci., 2 (1971) 92—95.
117 D. C. Smith, E. Sandi and R. Leduc, Pestic. Sci., 3 (1972) 207—210.
118 P. E. Corneliussen, Pestic. Monit. J., 2 (1969) 140—152.
119 R. Bradicich, N. E. Foster, F. E. Hons, M. T. Jeffus and C. T. Kenner, Pestic. Monit. J., 3 (1969) 139—141.
120 A. A. Benedetti-Pichler, Essentials of Quantitative Analysis, The Ronald Press, New York, 1956. Chs. 17, 18, 19.
121 W. W. Walton and J. I. Hoffman, in I. M. Kolthoff and P. J. Elving (Eds.), Treatise on Analytical Chemistry, The Interscience Encyclopedia Inc., New York, 1959, Part 1, Vol. 1, Ch. 4.
122 C. A. Bicking, in I. M. Kolthoff and P. J. Elving (Eds.), Treatise on Analytical Chemistry, 2nd edn., Part I, Theory and Practice, Vol. 1, John Wiley, New York, 1978, Ch. 6.
123 International Association for Cereal Chemistry, ICC Standards, Vienna, (a) ICC Standard No. 101, approved 1960, ICC Standard No. 120, 1972, (b) ICC Standard No. 109, approved 1960, Draft ICC Standard No. 109/1, ICC Standard No. 110, approved 1960, Draft ICC Standard No. 110/1, (c) ICC Standard No. 104, approved 1960.
124 International Dairy Federation, Brussels, (a) International Standard FIL-IDF 50: 1969, (b) International Standards FIL-IDF 4: 1958, 10: 1960, 15: 1961, 21: 1962, 23: 1964, 26: 1964, 58: 1970, 70: 1972, (c) International Standard FIL-IDF 27: 1964.
125 FAO/WHO, Code of Principles Concerning Milk and Milk Products, International Standards and Standard Methods of Sampling and Analysis for Milk Products, 7th edn., Joint FAO/WHO Food Standards Programme, Codex Alimentarius Commission, CAC/M 1-1973, Rome, 1973, (a) pp. 59—68, (b) pp. 70, 75—76, 84, 89—90, 94, 101—102, 104, 111, 117, (c) pp. 101—102.
126 American Public Health Association, W. J. Hausler, Jr. (Ed.), Standard Methods for the Examination of Dairy Products, 13th edn., American Public Health Association, Washington, D.C., 1972, pp. 28—46.

127 British Standards Institution, London, BS 769: 1961, Methods for the Chemical Analysis of Butter, BS 770: 1963, Methods for the Chemical Analysis of Cheese, BS 2472: 1966, Methods for the Chemical Analysis of Ice Cream, BS 809: 1974, Methods of Sampling Milk and Milk Products.
128 International Organisation for Standardisation, Geneva, (a) ISO/R542-1967, Oilseeds-Sampling; ISO/R661-1968, Crude Vegetable Oils and Fats-Preparation of Contract Sample for Analysis; ISO/R707-1968; Milk and Milk Products-Sampling; ISO/R874-1968; Fresh Fruits and Vegetables-Sampling; ISO/R948-1969, Spices and Condiments-Sampling; ISO/R950-1969, Cereals-Sampling (as Grain); ISO/R951-1969, Pulses-Sampling; ISO/R966-1969, Seeds-Sampling and Methods of Test; ISO1839/1-1975, Tea-Sampling — Part I — Sampling from Large Containers; ISO1839/2-1976, Tea-Sampling — Part II — Sampling from Small Containers; ISO2170-1972, Cereals and Pulses-Sampling of Milled Products; ISO2292-1973, Cocoa Beans-Sampling; ISO3100/1-1975, Meat and Meat Products-Sampling, Part I: Taking Primary Samples. (b) ISO664-1977, Oilseeds — Reduction of Contract Samples to Analysis Samples; ISO1572-1975, Tea-Preparation of Ground Sample of Known Dry Matter Content; ISO2825-1974, Spices and Condiments — Preparation of Ground Sample for Analysis. (c) ISO/R662-1968, Crude Vegetable Oils and Fats — Determination of Moisture and Volatile Matter; ISO/R711-1968; Cereals and Cereal Products — Determination of Moisture Content (Basic Reference Method); ISO/R712-1968, Cereals and Cereal Products — Determination of Moisture Content (Routine Method); ISO/R771-1968, Oilseed Residues — Determination of Moisture and Volatile Matter; ISO/R934-1969, Animal Fats — Determination of Water (Entrainment-Distillation Method); ISO/R939-1969, Spices and Condiments — Determination of Moisture Content (Entrainment Method); ISO/R993-1969, Animal Fats — Determination of Moisture and Volatile Matter; ISO1422-1973, Meat and Meat Products — Determination of Moisture; ISO/R1446-1970, Green Coffee Beans — Determination of Moisture Content (Basic Reference Method); ISO/R1447-1970, Green Coffee Beans — Determination of Moisture Content (Routine Method); ISO1573-1975, Tea — Determination of Loss in Mass at 103°C; ISO2291-1972, Cocoa Beans — Determination of Moisture Content (Routine Method); ISO2920-1974, Whey Cheese — Determination of Dry Matter Content (Reference Method). (d) ISO/R728-1969, Spices and Condiments — Determination of Total Ash; ISO/R749-1968, Oilseed Residues — Determination of Total Ash; ISO/R763-1971, Fruit and Vegetable Products — Determination of Ash Insoluble in Hydrochloric Acid; ISO/R936-1969; Meat and Meat Products — Determination of Ash; ISO1575-1975, Tea — Determination of Total Ash; ISO2171-1972; Cereals, Pulses and Derived Products — Determination of Ash.
129 M. A. Perring, J. Sci. Food Agric., 25 (1974) 247—250.
130 N. D. Michie and E. J. Dixon, J. Sci. Food Agric., 28 (1977) 215—224.
131 J. W. Jones and K. W. Boyer, J. Assoc. Off. Anal. Chem., 62 (1979) 122—128.
132 P. Strohal, S. Lulic and O. Jelisavcic, Analyst, 94 (1969) 678—680.
133 P. D. LaFleur, Anal. Chem., 45 (1973) 1534—1536.
134 H. O. Fourie and M. Peisach, Radiochem. Radioanal. Lett., 26 (1976) 277—290.
135 D. Behne, P. Brätter, H. Gessner, G. Hube, W. Mertz and U. Rösick, Fresenius Z. Anal. Chem., 278 (1976) 269—272.
136 H. O. Fourie and M. Peisach, Analyst, 102 (1977) 193—200.
137 H. L. Huffman, Jr. and J. A. Caruso, Talanta, 22 (1975) 871—875.
138 E. C. Dunlop, in I. M. Kolthoff and P. J. Elving (Eds.), Treatise on Analytical Chemistry, Part I, Vol. 2, Interscience, New York, 1961, Ch. 25.
139 T. T. Gorsuch, The Destruction of Organic Matter, Pergamon, Toronto, 1970.
140 R. Bock, A Handbook of Decomposition Methods in Analytical Chemistry, John Wiley, New York, 1979 (English translation by I. L. Marr of 1972 German edn.).
141 G. D. Martinie and A. A. Schilt, Anal. Chem., 48 (1976) 70—74.

142 J. C. Hughes, Testing of Glass Volumetric Apparatus, National Bureau of Standards circular 602, Washington, D.C., 1958.
143 American Public Health Association, Standard Methods of the Examination of Water and Wastewater, 14th edn., American Public Health Association, Washington, D.C., 1976.
144 Scientific Committee on Problems of the Environment (SCOPE), Working Group on Methodology of Determination of Toxic Substances in the Environment, Environmental Pollutants Selected Analytical Methods (SCOPE 6), Ann Arbor Science Publishers, Ann Arbor, MI, 1975.
145 C. Duval, Inorganic Thermogravimetric Analysis, 2nd revised edn., Elsevier, Amsterdam, 1963.
146 US Department of Commerce, National Bureau of Standards Certificate of Analysis, Standard Reference Material 1567, Wheat Flour, Washington, D.C., 1978; Standard Reference Material 1568, Rice Flour; Standard Reference Material 1570, Spinach; Standard Reference Material 1571, Orchard Leaves; Standard Reference Material 1573, Tomato Leaves; Standard Reference Material 1577, Bovine Liver.
147 L. Gorski, J. Heinonen and O. Suschny, Final Report on the Intercomparison of Trace Multielement Analysis in Dried Animal Whole Blood (A-2), Calcinated Animal Bone (A-3/1), Powdered Milk (A-8), Wheat Flour (V-2/1), and Dried Potatoes (V-4), International Atomic Energy Agency, Vienna, No. IAEA/RL/25, 1974.
148 R. Fukai, B. Oregioni and D. Vas, Oceanologica Acta, 1 (1978) 391—396.
149 H. J. M. Bowen, Analyst, 92 (1967) 124—131.
150 H. J. M. Bowen, Adv. Act. Anal., 1 (1969) 101—113.
151 H. J. M. Bowen, J. Radioanal. Chem., 19 (1974) 215—226.
152 G. A. Uriano and C. C. Gravatt, Crit. Rev. Anal. Chem., 6 (1977) 361—411.
153 Varian Techtron, Hollow Cathode Lamp Data, Springvale, Australia, 1972.
154 Varian Techtron, Analytical Methods for Flame Spectroscopy, Springvale, Australia, 1972.
155 J. W. Robinson (Ed.), Handbook of Spectroscopy, Vol. 1, CRC Press, Cleveland, 1974.
156 M. Ihnat, Can. J. Spectrosc., 23 (1978) 112—125.
157 M. Ihnat, J. Assoc. Off. Anal. Chem., 59 (1976) 911—922.
158 M. Ihnat and H. J. Miller, J. Assoc. Off. Anal. Chem., 60 (1977) 813—825.
159 W. F. Meggers, C. H. Corliss and B. F. Scribner, Tables of Spectral Line Intensities, Part I, Arranged by Elements, Part II, Arranged by Wavelength, NBS Monograph 145, 2nd edn., National Bureau of Standards, Washington, D.C., 1975.
160 M. S. Cresser, Lab. Pract., 26 (1977) 171—173.
161 R. J. Lovett, D. L. Welch and M. L. Parsons, Appl. Spectrosc., 29 (1975) 470—477.
162 B. E. Buell, in E. L. Grove (Ed.), Applied Atomic Spectroscopy, Vol. 2, Plenum Press, New York, 1978, Ch. 2.
163 C. Hendrikx-Jongerius and L. De Galan, Anal. Chim. Acta, 87 (1976) 259—271.
164 G. H. Morrison, Pure Appl. Chem., 41 (1975) 395—403.
165 I. Rubeška, M. Mikšovsky and M. Huka, At. Absorpt. Newsl., 14 (1975) 28.
166 F. J. Szydlowski, At. Absorpt. Newsl., 17 (1978) 65—69.
167 T. C. Rains, personal communication.
168 W. Schuhknecht and H. Schinkel, Fresenius Z. Anal. Chem., 194 (1963) 161—183.
169 J. P. Weiner and L. Taylor, J. Inst. Brew., 75 (1969) 195—199.
170 I. B. Brooks, G. A. Luster and D. G. Easterly, At. Absorpt. Newsl., 9 (1970) 93—94.
171 J. T. H. Roos and W. J. Price, J. Sci. Food Agric., 21 (1970) 51—52.
172 M. A. Perring, J. Sci. Food Agric., 25 (1974) 237—245.
173 J. A. McHard, J. D. Winefordner and J. A. Attaway, J. Agric. Food Chem., 24 (1976) 41—45.
174 J. A. McHard, J. D. Winefordner and S.-V. Ting, J. Agric. Food Chem., 24 (1976) 950—953.

175 J. Kowalczuk, J. Assoc. Off. Anal. Chem., 53 (1970) 926—927.
176 W. J. Adrain, Analyst, 98 (1973) 213—216.
177 W. J. Adrain and M. L. Stevens, Analyst, 102 (1977) 446—452.
178 E. E. Menden, D. Brockman, H. Choudhury and H. G. Petering, Anal. Chem., 49 (1977) 1644—1645.
179 M. Prasad and M. Spiers, J. Agric. Food Chem., 26 (1978) 824—827.
180 J. Assoc. Off. Anal. Chem., 58 (1975) 392.
181 W. Holak, J. Assoc. Off. Anal. Chem., 54 (1971) 1138—1139.
182 W. Holak, J. Assoc. Off. Anal. Chem., 55 (1972) 890—891.
183 J. R. Melton, W. L. Hoover, P. A. Howard and J. L. Ayers, J. Assoc. Off. Anal. Chem., 53 (1970) 682—685.
184 R. R. Elton-Bott, Anal. Chim. Acta, 86 (1976) 281—284.
185 J. W. Robinson, P. F. Lott and A. J. Barnard, Jr., in H. A. Flaschka and A. J. Barnard, Jr. (Eds.), Chelates in Analytical Chemistry, Vol. 4, Marcel Dekker, New York, 1972, pp. 233—275.
186 T. R. M. Helin and J. C. Slaughter, J. Inst. Brew., 83 (1977) 15—16.
187 H. C. Green, Analyst, 100 (1975) 640—642.
188 W. H. Hartford, in I. M. Kolthoff and P. J. Elving (Ed.), Treatise on Analytical Chemistry, Part II, Vol. 8, Interscience, New York, 1963, Ch. Chromium.
189 E. E. Cary and W. H. Allaway, J. Agric. Food Chem., 19 (1971) 1159—1161.
190 G. B. Jones, R. A. Buckley and C. S. Chandler, Anal. Chim. Acta, 80 (1975) 389—392.
191 S. R. Koirtyohann and C. A. Hopkins, Analyst, 101 (1976) 870—875.
192 W. R. Wolf and F. E. Greene, in P. D. LaFleur (Ed.), Accuracy in Trace Analysis; Sampling, Sample Handling, Analysis, Vol. 1, NBS Special Publication 422, Washington, D.C., 1976, pp. 605—610.
193 P. Schramel, Anal. Chim. Acta, 67 (1973) 69—77.
194 W. R. Wolf, personal communication.
195 E. E. Cary and O. E. Olson, J. Assoc. Off. Anal. Chem., 58 (1975) 433—435.
196 J. J. Christensen, P. A. Hearty and R. M. Izatt, J. Agric. Food Chem., 24 (1976) 811—815.
197 I. Okuno, J. A. Whitehead and R. E. White, J. Assoc. Off. Anal. Chem., 61 (1978) 664—667.
198 W. Wolf, W. Mertz and R. Masironi, J. Agric. Food Chem., 22 (1974) 1037—1042.
199 D. Arthur, Can. Spectrosc., 15 (1970) 134—136.
200 W. J. Garcia, C. W. Blessin and G. E. Inglett, Cereal Chem., 51 (1974) 788—797.
201 S. Slavin, G. E. Peterson and P. C. Lindahl, At. Absorpt. Newsl., 14 (1975) 57—59.
202 A. S. Baker and R. L. Smith, J. Agric. Food Chem., 22 (1974) 103—107.
203 G. B. Belling and G. B. Jones, Anal. Chim. Acta, 80 (1975) 279—283.
204 J. Smeyers-Verbeke, G. Segebarth and D. L. Massart, At. Absorpt. Newsl., 14 (1975) 153—154.
205 J. T. Olejko, J. Am. Oil Chem. Soc., 53 (1976) 480—484.
206 R. A. Baetz and C. T. Kenner, J. Agric. Food Chem., 23 (1975) 41—45.
207 D. L. Smith and W. G. Schrenk, J. Assoc. Off. Anal. Chem., 55 (1972) 669—675.
208 M. E. Varju, At. Absorpt. Newsl., 11 (1972) 45.
209 R. A. Isaac and W. C. Johnson, J. Assoc. Off. Anal. Chem., 58 (1975) 436—440.
210 R. J. Everson and H. E. Parker, Anal. Chem., 46 (1974) 2040—2042.
211 R. L. Chaney, The Effect of Nickel on Iron Metabolism by Soybean, Ph.D. Thesis, Purdue University, West Lafayette, IN, 1970.
212 D. R. Boline and W. G. Schrenk, J. Assoc. Off. Anal. Chem., 60 (1977) 1170—1174.
213 W. J. Simmons, Anal. Chem., 45 (1973) 1947—1949.
214 W. J. Simmons, Anal. Chem., 47 (1975) 2015—2018.
215 L. Hageman, L. Torma and B. E. Ginther, J. Assoc. Off. Anal. Chem., 58 (1975) 990—994.

216 K. Julshamn and O. R. Braekkan, At. Absorpt. Newsl., 12 (1973) 139—141.
217 S. A. Popova, L. Bezur and E. Pungor, Fresenius Z. Anal. Chem., 271 (1974) 269—272.
218 W. J. Price, J. R. H. Roos and A. F. Clay, Analyst, 95 (1970) 760—762.
219 M. Heckman, J. Assoc. Off. Anal. Chem., 54 (1971) 666—668.
220 R. K. Roschnik, J. Dairy Sci., 55 (1972) 750—752.
221 R. A. Jacob and L. M. Klevay, Anal. Chem., 47 (1975) 741—743.
222 G. R. List, C. D. Evans and W. F. Kwolek, J. Am. Oil Chem. Soc., 48 (1971) 438—441.
223 L. Maurer, Z. Lebensm. Unters-Forsch., 156 (1974) 284—287.
224 W. J. Simmons and J. F. Loneragan, Anal. Chem., 47 (1975) 566—568.
225 W. J. Simmons, Anal. Chem., 50 (1976) 870—873.
226 S. G. Capar, J. Assoc. Off. Anal. Chem., 60 (1977) 1400—1407.
227 N. M. Morris, M. A. Clarke, V. W. Tripp and F. G. Carpenter, J. Agric. Food Chem., 24 (1976) 45—47.
228 W. W. Brachaczek, J. W. Butler and W. R. Pierson, Appl. Spectrosc., 28 (1974) 585—587.
229 J. G. van Raaphorst, A. W. van Weers and H. M. Haremaker, Analyst, 99 (1974) 523—527.
230 The Institute of Brewing Analysis Committee, (J. Weiner, Chairman), J. Inst. Brew., 80 (1974) 486—488.
231 R. D. Wauchope, At. Absorpt. Newsl., 15 (1976) 64—67.
232 O. E. Clinton, Analyst, 102 (1977) 187—192.
233 Subcommittee 6 (T. J. Kneip, Chairman), Health Lab. Sci., 14 (1977) 53—58.
234 J. Flanjak, J. Assoc. Off. Anal. Chem., 61 (1978) 1299—1303.
235 M. Ihnat and H. J. Miller, J. Assoc. Off. Anal. Chem., 60 (1977) 1414—1433.
236 M. Ihnat and B. K. Thompson, J. Assoc. Off. Anal. Chem., 63 (1980) 814—839.
237 H. Freeman, J. F. Uthe and B. Flemming, At. Absorpt. Newsl., 15 (1976) 49—50.
238 P. R. Walsh, J. L. Fasching and R. A. Duce, Anal. Chem., 48 (1976) 1014—1016.
239 G. T. C. Shum, H. C. Freeman and J. F. Uthe, J. Assoc. Off. Anal. Chem., 60 (1977) 1010—1014.
240 F. J. Szydlowski, At. Absorpt. Newsl., 16 (1977) 60—63.
241 J. E. Poldoski, At. Absorpt. Newsl., 16 (1977) 70—73.
242 A. J. Thompson and P. A. Thoresby, Analyst, 102 (1977) 9—16.
243 G. M. Beorge, L. J. Frahm and J. P. McDonnell, J. Assoc. Off. Anal. Chem., 56 (1973) 1304—1305.
244 P. J. LeBlanc and A. L. Jackson, J. Assoc. Off. Anal. Chem., 56 (1973) 383—386.
245 J. F. Uthe, H. C. Freeman, J. R. Johnston and P. Michalik, J. Assoc. Off. Anal. Chem., 57 (1974) 1363—1365.
246 M. Ihnat, J. Assoc. Off. Anal. Chem., 57 (1974) 368—372.
247 H. Woidich and W. Pfannhauser, Fresenius Z. Anal. Chem., 276 (1975) 61—66.
248 K. H. Tam and H. B. S. Conacher, J. Environ. Sci. Health B, 12 (1977) 213—227.
249 A. E. Smith, Analyst, 100 (1975) 300—306.
250 F. D. Pierce and H. R. Brown, Anal. Chem., 48 (1976) 693—695.
251 T. J. Kneip (Chairman), Health Lab. Sci., 14 (1977) 53—58.
252 J. Štupar, F. Dolinšek, M. Špenko and J. Furlan, Landwirtsch, Forsch., 27 (1974) 51—61.
253 S. R. Koirtyohann and M. Hamilton, J. Assoc. Off. Anal. Chem., 54 (1971) 787—789.
254 C. H. Kim, P. W. Alexander and L. E. Smythe, Talanta, 22 (1975) 739—744.
255 C. H. Kim, P. W. Alexander and L. E. Smythe, Talanta, 23 (1976) 229—233.
256 R. W. Dabeka, Anal. Chem., 51 (1979) 902—907.
257 D. R. Boline and W. G. Schrenk, Appl. Spectrosc., 30 (1976) 607—610.
258 C. Delage, N. Oudart and C. Guichard, Ann. Pharm. Fr., 34 (1976) 315—322.

259 T. Kjellström, B. Lind, L. Linnman and G. Nordberg, Environ. Res., 8 (1974) 92—106.
260 U. Anders and F. Hailer, Fresenius Z. Anal. Chem., 278 (1976) 203—206.
261 A. Dewitt, R. Duwijn, J. Smeyers-Verbeke and D. L. Massart, Bull. Soc. Chim. Belg., 84 (1975) 91—98.
262 L. Linnman, A. Anderson, K. O. Nilsson, B. Lind, T. Kjellström and L. Friberg, Arch. Environ. Health, 27 (1973) 45—47.
263 E. R. Blood and G. C. Grant, Anal. Chem., 47 (1975) 1438—1441.
264 K. -R. Sperling, At. Absorpt. Newsl., 14 (1975) 60—62.
265 A. M. Ure and M. C. Mitchell, Anal. Chim. Acta, 87 (1976) 283—290.
266 T. J. Ganje and A. L. Page, At. Absorpt. Newsl., 13 (1974) 131—134.
267 M. P. C. de Vries, K. G. Tiller and R. S. Beckwith, Commun. Soil Sci. Plant Anal., 6 (1975) 629—640.
268 A. Engberg, Analyst, 98 (1973) 137—145.
269 C. Wehrer, J. Thiersault and P. Langel, Ind. Aliment, Agric., 93 (1976) 1439—1446.
270 P. N. Vijan and C. Y. Chan, Anal. Chem., 48 (1976) 1788—1792.
271 H. L. Trachman, A. J. Tyberg and P. D. Branigan, Anal. Chem., 49 (1977) 1090—1093.
272 P. Hocquellet and N. Labeyrie, At. Absorpt. Newsl., 16 (1977) 124—127.
273 S. Chilov, Talanta, 22 (1975) 205—232.
274 A. M. Ure, Anal. Chim. Acta, 76 (1975) 1—26.
275 M. Malaiyandi and J. P. Barrette, J. Assoc. Off. Anal. Chem., 55 (1972) 951—959.
276 T. C. Rains and O. Menis, J. Assoc. Off. Anal. Chem., 55 (1972) 1339—1344.
277 R. K. Munns and D. C. Holland, J. Assoc. Off. Anal. Chem., 60 (1977) 833—837.
278 W. R. Simpson and G. Nickless, Analyst, 102 (1977) 86—94.
279 R. K. Roschnik, Analyst, 98 (1973) 596—604.
280 G. K. Pagenkopf, D. R. Neuman and R. Woodriff, Anal. Chem., 44 (1972) 2248—2250.
281 J. K. Kapur and T. S. West, Anal. Chim. Acta, 73 (1974) 180—184.
282 G. Velghe, M. Verloo and A. Cottenie, Z. Lebsenm. Unter-Forsch., 156 (1974) 77—80.

Chapter 4e

Applications of atomic absorption spectrometry in ferrous metallurgy

K. OHLS and D. SOMMER
Hoesch Hüttenwerke Aktiengesellschaft, Hörder Burgstr. 15—17, 4600 Dortmund (W. Germany)

I. INTRODUCTION

Atomic absorption spectrometry was introduced approximately 18 years ago into laboratories concerned with the analysis of materials connected with ferrous metallurgy. Thus, literature references for this area of analysis only cover this period [89, 103, 134, 135, 151, 154]. Atomic absorption apparatus has been improved significantly during the past decade. Analysts have made demands upon manufacturers which have led to the development of sophisticated electronics and improvements in optical systems [33, 34, 52, 53, 59, 148, 149].

Atomic absorption analysis made available to the routine laboratory an analytical technique which initially was intended to produce considerable simplification of procedures for the analysis of aqueous, acidic or basic solutions, and thereby contribute to a reduction in costs. Numerous reviews show the worldwide application of this technique [15, 40, 77, 126—129, 137]. Nevertheless, some 10 years passed before atomic absorption became part of the international standardisation of analytical methods. At present, there are many standard methods being developed on the basis of atomic absorption [35, 67]. Some, dealing with the determination of metals in lubricating oils, are already in use [35, 66], although the overwhelming majority, for example those dealing with the analysis of iron ores [67], are still being developed. The first indication of standardisation of atomic absorption methods for iron and steel analysis was seen in 1973 [8].

The industrial application of atomic absorption for routine analysis can be divided into four areas:

(1) The incoming inspection of all raw materials.
(2) Production testing (indirect).
(3) Final inspection of all products.
(4) Environmental analysis.

A characteristic of the analytical requirement at the incoming inspection stage is the accuracy of the results, because these are frequently exchanged between the vendor and the purchaser, and become the subject of financial negotiation. Raw materials include iron ore, ferrous alloys and raw iron, but also casting aids, refractories, as well as lubricating oils, fats and fuels

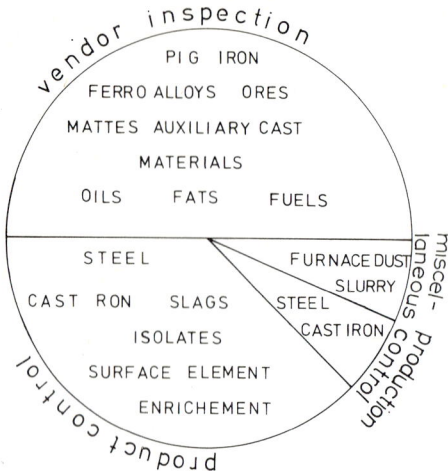

Fig. 1. Application of AAS.

(see Fig. 1). Trace concentrations in raw materials are also significant for environmental reasons. Analytical methods must be capable of working over concentration ranges between 10^{-5} and 10^2 weight percent. Atomic absorption analysis has made a significant contribution in these areas.

For rapid analysis during the production process atomic absorption is mainly of indirect value because, due to the sequential character of the technique, it cannot be used for complete steel or slag analysis in a two to three minute period. The analytical requirements for the testing of rapid continuous production processes are fulfilled by the techniques of emission and X-ray spectrometry. These techniques are characterised by great speed, high precision and simultaneous multi-element analysis. Accuracy must, however, be constantly checked with a variety of special calibration samples. This requires the determination of the true concentrations of the calibration samples with chemical methods of solution analysis, whose precision is often only equal to or, when compared with X-ray spectrometry, frequently poorer. Chemical analysis is, however, the basis of all comparisons, and must be repeated frequently for the determination of the true concentrations. Atomic absorption, with its relatively good precision, has greatly simplified the analytical control of numerous elements.

In slower production processes, atomic absorption can be used directly for production control of those processes for which rapid sample preparation techniques have been developed. One such example is the determination of magnesium concentration during the production of cast iron [70]. A further example is the determination of acid soluble aluminium in steel [147], where in large steel companies 100 samples or more per day may be required. (See also section II.A.)

Considerable significance is attached to atomic absorption in production testing. The products of ferrous industries are materials with specific qualities which must be guaranteed world-wide, often regulated by law. Since finished products cannot be tested by techniques requiring mechanical destruction, correlations have been established between material qualities and chemical structure. Thus, chemical analysis has become a significant aid in the testing of materials and their qualities. Because of the need for a guaranteed product quality, the chemical analysis must be highly accurate. Because of time limitations, and therefore cost, product analyses are also done using emission and X-ray spectrometry, but here the number of check measurements with standardised methods must be increased considerably. This is one of the main applications of atomic absorption in the laboratories of the iron and steel industry.

In addition to the metallic products there are also by-products such as fertilisers, slags and recycling materials. Here too the concentration ranges between 10^{-4} to 50 weight percent describe metallic materials and oxide products.

Another application of atomic absorption is in the determination of concentrations on steel surfaces after special sample preparation, and the analysis of steel residues (purity tests) after isolation and possible selective dissolution of the iron matrix [18, 124, 139] (Fig. 1). Atomic absorption is particularly useful for environmental analysis where dust samples can be analysed in a similar manner to steel residues; water and effluents are the main examples.

Table 1 summarises the most common materials in ferrous metallurgy, listed by element, which are analysed by atomic absorption spectrometry.

Because of the growing requirements for accurate analytical results, it is sensible to work with at least two different techniques. If the results of combined methods, such as photometry/atomic absorption, ICP spectrometry/atomic absorption, polarography/atomic absorption, etc., agree within predetermined limits in standard deviation, there is a great probability that the results are also accurate.

The future uses of atomic absorption also depend on the ability of analysts to produce rapid and simple techniques for sample preparation, and to compensate for the small number of interferences that occur because of matrix elements. In routine laboratories particularly, there is an urgent need for simple methods.

By means of the microprocessors built into recent instruments, operations such as calibration and checking of stability have been simplified. The analyst needs only to use universally accepted standard methods which can be easily applied to as many classes of materials as possible. Such methods will be described in the following sections for chemical analysis in the ferrous metallurgy industry.

References pp. 246—249

TABLE 1

ELEMENTS DETERMINED BY AAS ROUTINE ANALYSIS IN DIFFERENT MATERIALS

Elements	Materials
Al	Steel, pig iron, cast iron, ferro alloys, ores, slags, refractories, furnace dust, slurry, auxiliary cast materials, isolates, fats
B	Auxiliary cast materials ($>10\%$ B)
Ba	Ores, slags
Bi	Furnace dust, cast iron
Ca	Steel, pig iron, cast iron, ferro alloys, ores, slags, refractories, furnace dust, slurry, auxiliary cast materials, isolates, oils, fats
Cd	Furnace dust, slurry
Co	Steel, pig iron, cast iron, ores, slags, furnace dust
Cr	Steel, pig iron, cast iron, ferro alloys, ores, slags, refractories, furnace dust, slurry
Cu	Steel, pig iron, cast iron, ferro alloys, ores, slags, furnace dust, slurry
Fe	Slurry, isolates, oils
Li	Ores, slags, auxiliary cast materials, fats
Mg	Steel, pig iron, cast iron, ferro alloys, ores, slags, refractories, furnace dust, slurry, auxiliary cast materials, oils, fats
Mn	Steel, pig iron, cast iron, ferro alloys, ores, slags, refractories, furnace dust, slurry, auxiliary cast materials, isolates
Mo	Steel, pig iron, cast iron, ores, furnace dust, slurry, isolates
Na	Oils, fats (slags, refractories)
Ni	Steel, pig iron, cast iron, ores, slags, oils
Pb	Steel, pig iron, cast iron, ores, slags, furnace dust, slurry, petrol
Sb	Steel, pig iron, cast iron, ores, slurry, oils
Si	Steel, pig iron, cast iron, ores
Sn	Steel, pig iron, cast iron, ores, slurry, slags, furnace dust
Ti	Steel, pig iron, cast iron, ferro alloys, ores, slags, refractories, furnace dust, slurry, auxiliary cast materials
V	Steel, pig iron, cast iron, slags, furnace dust, slurry, isolates, oils
W	Steel
Zn	Steel, pig iron, cast iron, ores, slags, furnace dust, slurry, auxiliary cast materials, oils, fats

II. ANALYSIS OF IRON, STEEL AND ALLOYS

Many authors [15, 21, 25, 28, 29, 31, 81, 82, 106—108, 111, 114, 119, 123, 126—129, 131, 137, 141, 142, 147, 153] have reported studies on the analysis of common acid solutions of different steels where one to a maximum of six elements are determined sequentially from one sample weighing. Elements commonly analysed include Al, Co, Cr, Cu, Mg, Mn, Mo, Ni, Si, Ti, V and W. There has also been no lack of effort to determine elements such as As [54, 101, 142] that are difficult to analyse using atomic absorption. Efforts to produce standard methods have also been described [5, 56]. The closest to a universal method is the description of the determination of 12 elements sequentially from two different solution strengths [140].

For almost all types of steel there is a similar problem in the dissolution step. Three elements, Al, Si, and W, behave differently in the analysis of the total content and require special methods. These methods are simple extensions of the technique, which in the case of the determination of acid soluble Al or Si, can simply be omitted. When, during the dissolution, an oxide residue remains, which in general consists of SiO_2 or Al_2O_3, this must be brought into solution with a fusion, and added to the rest of the sample. If the W concentration is higher than 0.5%, the dissolving acid solution should also include phosphoric acid. Thus, it is possible to produce a universal method for the elements that are contained in steel and which influence its properties.

A. General solution methods

A general method for the determination of all elements of interest in steel samples must be simple enough for routine work and able to be set out in a flow diagram (Fig. 2).

The acid mixture is chosen such that the majority of steels can be dissolved in it. When a residue remains this must be fused using the well-known sodium tetraborate method. This sequential determination of 15 elements in one weighing obviously requires the previous setting up of 15 calibration curves in the appropriate concentration range (Fig. 2). This can be done, in principle, in two ways; either the calibration curves are entered using known data with the aid of a microprocessor or appropriate standard samples are used with every analysis.

It has been demonstrated that a calibration curve set up via direct entry through a microprocessor can be reliably recalibrated with one standard solution. This will be described in more detail in section V.

The general solution method does not include the elements Sn and W. Tungsten can be kept in solution with a different acid (H_3PO_4 addition) and can be measured in the concentration range 0.01 to 15 weight percent at 255.1 nm without difficulty. Tin determination in iron and unalloyed steel can be performed directly using the general solution method, but higher Sn concentrations require a fusion. Tin concentrations in the range 0.001 to 1 weight percent can be measured without problems at 224.0 nm.

Far more difficult, however, is the atomic absorption analysis of some other elements typically found in steel, for example Nb, As, Sb, Se, Te, or Bi. The determination of boron in steel in the range between 10^{-3} to 10^{-5} weight percent has been practically impossible until recently.

Future goals for the increased use of atomic absorption are the production of simple general methods and the further simplification of analytical techniques. One example is the determination of acid-soluble aluminium in steel, which is still significant during steel production. With the help of a con-

216

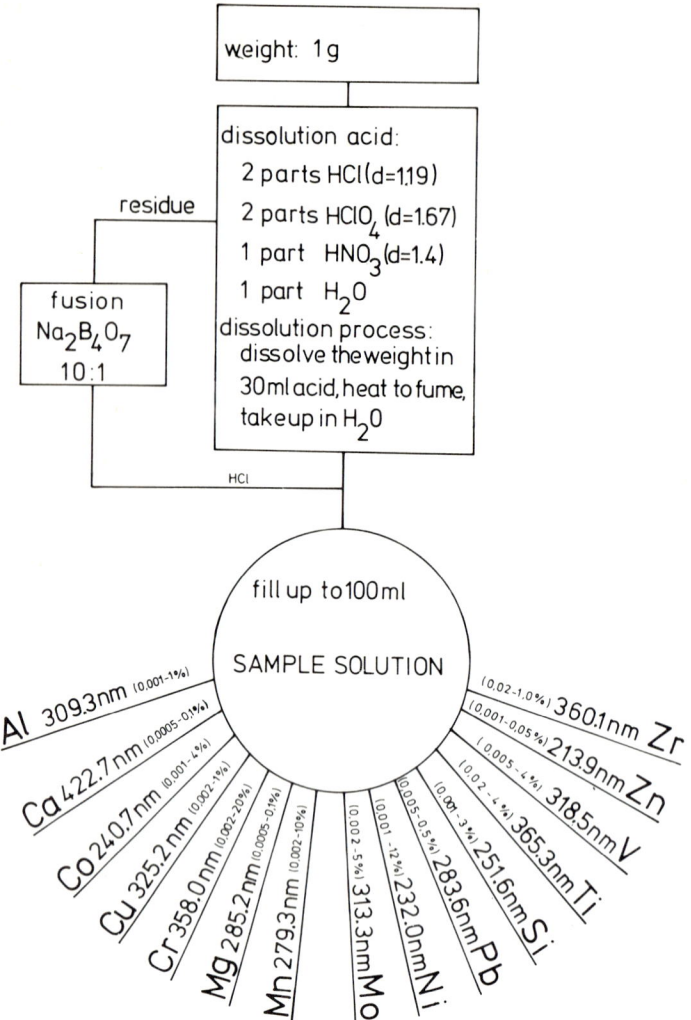

Fig. 2. General solution method.

ventionally set up acid solution (3 parts HCl plus 1 part HNO_3), the analyst wishes to determine the unoxidised, and therefore still reactive, concentration of Al in steel rapidly and accurately. It is assumed that Al_2O_3 will not be dissolved. A large quantity of such samples need to be determined daily. A large number of rapid determinations of nickel concentrations in steel alloys during the production may also be required, for example as a check on X-ray analysis, in order to measure the nickel concentration in the alloy as exactly as possible.

Using a double-channel spectrometer the normal nickel or aluminium determination by atomic absorption spectrometry can be considerably simplified as a method that does not require weighing the sample has been developed in our laboratory.

In various samples of a specific type of steel the ratio of the concentration of an element to the concentration of iron can be regarded as constant:

C_X/C_{Fe} = Const.

According to Beer's Law absorbance is proportional to concentration; therefore, for the steel sample A_X/A_{Fe} is approximately constant.

For low-alloy steels the total iron concentration can be regarded as being 99 weight percent, so that the concentration C_X can be computed from the absorbance A_X independently, if the absorbance of iron (A_{99}) is measured simultaneously in the second channel.

A_X = Const. $\times A_{99}$

Although the absolute values of A_X and A_{99} vary, dependent upon the amount of sample that happens to be used, the ratio always remains constant.

This requires a very insensitive iron line which produces a reproducible signal. The spectral line Fe 282.3 nm is suitable. The ratio of both values (Channel A : Channel B) is shown directly and also printed. The recorder signals show the constancy of these for iron and the dependence on concentration of those for Ni/Fe (Fig. 3). A similar ratio is true for Al, whose determination is thereby greatly simplified. A chip or chips of steel of about 1—2 g are placed into a beaker, dissolved with 20 ml aqua regia, filtered and analysed directly. Standard samples prove that no significant variations occur in results (Fig. 4). A standard sample containing 0.067% aluminium is used to show that the method without sample weighing is independent of the sample quantity and the dilution for a certain range (Table 2). This method can be applied to different elements and different concentration ratios in low-alloy steels. It could also be useful in cases where sample weighings are difficult to make, as for example on research vessels [94]. Atomic absorption is also applied to the analysis of accompanying elements in ferro alloys [113]. Methods have been published for ferro silicon, with the elements Al, Ca, Mn, Ti [32], ferro manganese [45] with silicon [46], and ferro chromium with Pb and Sn [105] and Bi [146].

In general, ferrous alloys are difficult to dissolve with acids so that a fusion, for example with sodium/potassium carbonate, is recommended. The resulting high salt concentration can produce difficulties (viscosity, nebuliser/burner system), as is discussed in Chapter 3. Once sample solutions are available, there is no difference in analysis from the methods for iron and steel.

B. Trace-element analysis

Trace-metal analysis in pure iron and low-alloy steels as well as their high-alloyed variants is still relatively undeveloped. Traditionally, this was

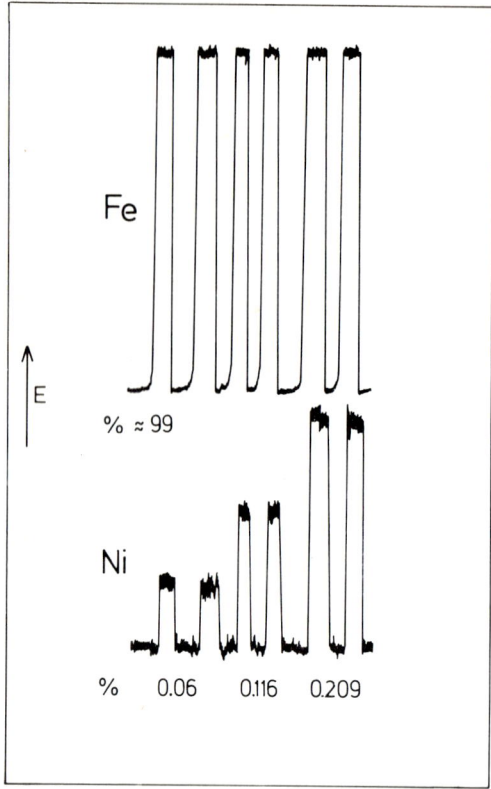

Fig. 3. Determination of Ni without sample weighing.

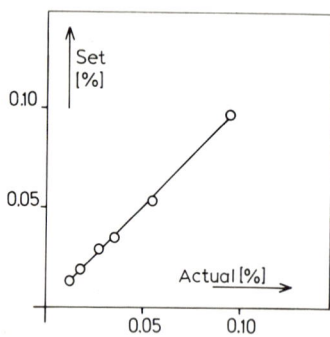

Fig. 4. Accuracy of the Al determination controlled by standard samples.

done by means of complicated concentration procedures, for example dissolution of the iron matrix by so-called isolation or chloride techniques, and microanalysis of the residues. It is possible to determine main components in these residues with atomic absorption spectrometry [18, 124, 139]. More interesting, however, are the methods which can be used to

TABLE 2

EFFECTS OF WEIGHT AND DILUTION RATIO. SET VALUE OF STANDARD SAMPLE: 0.067 MASS % Al

Weight (g)	Volume of solution (ml)	Actual value (%)
1.8	75	0.068
1.5	75	0.066
0.9	75	0.067
2.7	100	0.069
1.5	100	0.070
0.9	100	0.067
2.1	125	0.067
1.8	125	0.067
1.2	125	0.068
2.4	150	0.070
1.2	150	0.068
0.9	150	0.068

determine trace concentrations directly in the steel sample or trace concentrations in sample sizes of the order of micrograms.

The few articles currently available regarding trace analysis without preconcentration, use in general the graphite furnace technique [102, 120, 138] with sample sizes of the order of microliters, and deal with the elements Sb [47, 83], Pb and Bi [48—50], As, Sb, Bi, Sn, Cd, Pb [10, 57, 116] as well as Al, Cr, Sn [6, 62], Co, and Mg [104]. Alkaline earths can be determined directly with the flame method [122, 147]. Further techniques of atomic absorption by flame use concentration methods, for example for the determination of small concentrations of tin [17], Te [26], Co, Pb, and Bi [104], and W [106]. From the analytical viewpoint, it is only useful to remove the iron matrix. The extraction of the elements to be determined from the matrix always carries with it the danger of losses and therefore results showing concentrations that are too low.

1. *Solid sample technique*

The use of a two-channel spectrometer in combination with a furnace atomiser with a temperature program makes available an entirely new possibility for direct trace-element analysis.

The use of a rectangular graphite tube makes possible the placing of the solution in a sample boat (Fig. 5). The authors propose two changes: reduction of the size of the sample space, and use of a very pure graphite from emission spectroscopy (RW-O) instead of the relatively expensive pyrolytic

Fig. 5. Graphite tube oven with standard and modified boat.

graphite (Fig. 5). Instead of solutions, a single steel chip is introduced, heated to a minimum of 2600°C, and the resultant atomic vapor is analysed directly. When the sample is weighed (about 1 mg), at least two elements can be determined simultaneously with the two-channel instrument. Because there is no dilution, the same sensitivity produces a much lower detection limit without degrading the precision of the results. In particular, accuracy can be easily checked by standard materials. There are no contaminations due to acids or other chemicals and small amounts of ions are not adsorbed on the walls of vessels.

Since the weighing of such small sample quantities can influence the analytical error, it is desirable to avoid the need to weigh the sample. One channel is occupied by the reference absorption measurement, for example with pure iron or low-alloy steels an iron hollow-cathode lamp is used and the measurement takes place at 372 nm. In this case, only one further element can be determined. The carrying out of the method is simple (Fig. 6). Tests of this compact sample technique with standard materials, of which there are at present only a few containing certified trace elements, show satisfactory agreement (Fig. 7). As regards traces, the chips of standard materials are very homogeneous.

It has been found that with repeated rapid heating, for example at 2600°C, numerous elements can be determined from one chip. Repeated heating does, however, produce increasingly smaller but determinable peaks; at present it is not yet known how many elements can indeed be determined in one chip.

Figure 8 is an example of the reduction of peak heights with increasing numbers of heating periods in the determination of copper in pure iron.

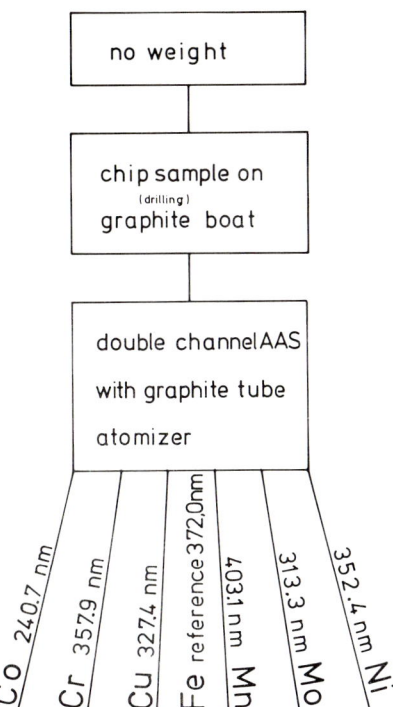

Fig. 6. Solid sample method.

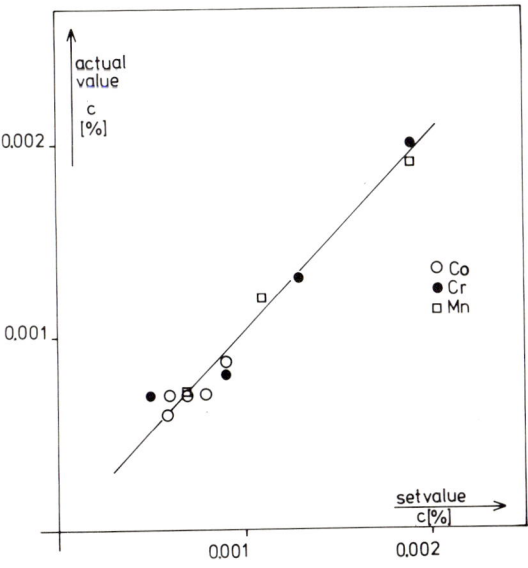

Fig. 7. Accuracy of the solid sample methods controlled by standard samples.

References pp. 246—249

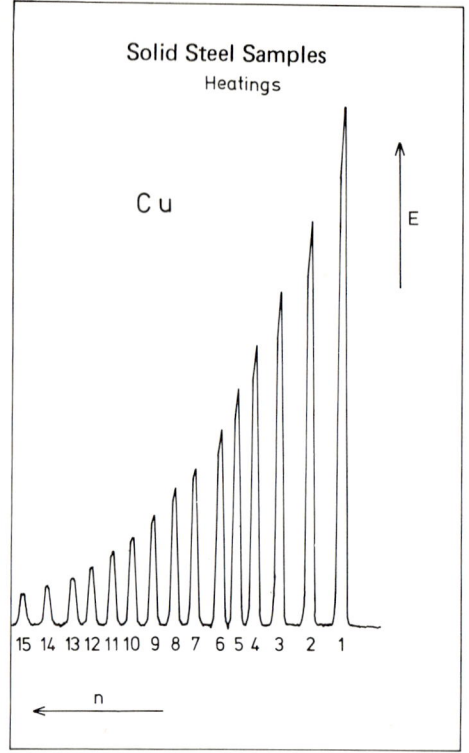

Fig. 8. Decrease of peak height by repeated heatings of the same sample.

The copper concentration is 0.003 weight percent. If the peak after the tenth heating is still measurable with sufficient precision, then at least 10 elements can be determined per chip.

When samples are less homogeneous, this method can be repeated frequently, since the technique is simple and the time requirement relatively small. The numerous single results permit statistical calculations, and in general lead to the correct mean value. The detection limits in the compact sample method have not yet been found exactly, because the appropriate samples are not available but could lie in the range 10^{-5}–10^{-7} weight percent.

2. Preconcentration

Flame atomic absorption can also be used for trace analysis when the iron matrix is extracted [39, 147]. When this extraction is combined with the Hoesch injection technique [3, 130], trace analysis can also be performed [31, 99, 142]. By using less than 100 µl of sample solution, the improvement compared to conventional techniques is at least a factor of 10 [13, 14].

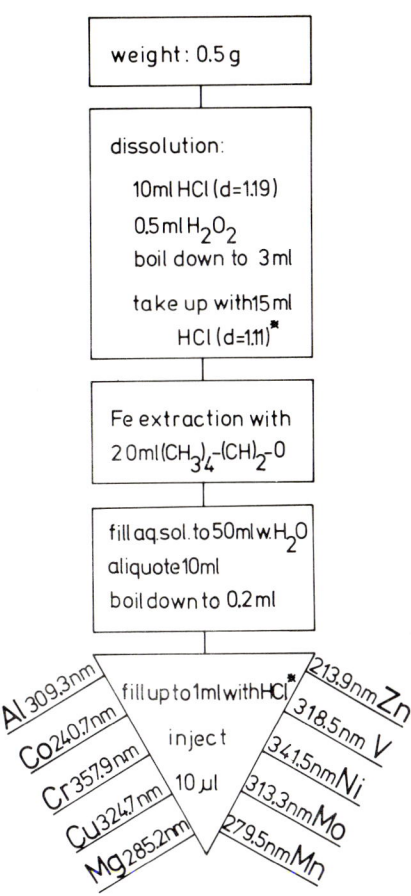

Fig. 9. General injection method with pre-concentration.

In general, a simple technique (Fig. 9), which can be used with the majority of iron alloys is again available.

Through the use of a two-channel spectrometer two elements can be determined simultaneously in each case. Reduction of the sample solution to $10\,\mu l$ degrades the reproducibility of the results only slightly; the RSD is in the range 5—10% for concentrations of 10^{-3} to 10^{-4} weight percent (Fig. 10). With an aspiration volume of $10\,\mu l$, a triple determination per element can be carried out, theoretically, for more than 50 elements in 1 ml.

Because standard samples contain only a limited number of elements whose concentration lies in the range of this method, we would determine, for example, 9 trace elements in 15 attempts, 3 times each, involving a consumption of $150\,\mu l$ of sample solution (Table 3). The pairs Cu 324.7 nm/Cr 357.9 n, Ni 314.5 nm/Mn 279.5 nm and Zn 213.9 nm/Co 240.7 nm were determined with the air/acetylene flame, and Al 309.3 nm/Mo 313.3 nm and Mg 285.2 nm were determined with the nitrous oxide/acetylene flame.

References pp. 246—249

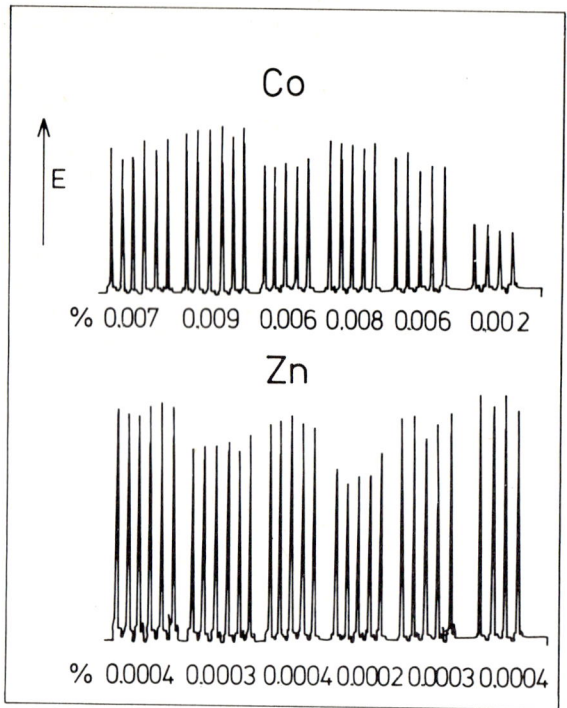

Fig. 10. Simultaneous determination of Co and Zn using the injection technique with 10 µl.

C. Interference of sample matrix

Interferences in atomic absorption occur with flame methods as well as electrothermal methods. Although a great number of articles have appeared that discuss interferences with flame methods [27, 41, 60, 84, 122, 136], only few articles have appeared regarding interferences with the furnace atomiser [117, 155]. With flame methods, the type of interference often depends upon the gas composition of the flame. Anionic interferences, that is, the appearance of chemical combinations [110] that cannot be broken up by the temperature of the air/acetylene flame (2400°C), can be prevented by the use of a hotter flame, e.g. nitrous oxide/acetylene (2800°C) although ionisation interferences are particularly significant with hot flames. These can be suppressed by the addition of easily ionised elements such as potassium or cesium (ionisation buffers) [85]. Additionally, physical interferences are observed which can be shown to be due to differing viscosities of standard and sample solutions (high salt acid concentrations) [100] and high salt concentrations after fusion). By the addition of increased dilution steps, viscosity differences can easily be removed.

In ferrous metallurgy interference effects have a very insignificant effect,

TABLE 3

CERTIFIED AND MEASURED VALUES BY THE MODIFIED INJECTION TECHNIQUE [in mass %]

	Mn	Ni	Cr	Mo	Cu	Co	Al	Zn	Mg	
Fe reductum	0.0020				0.0020	0.0002	0.0005	0.0004		Certified
	0.0009	0.0022	0.0015	0.0067	0.0012				0.0003	Measured
AKP 043-1	0.0007	0.0056	0.0052	0.0062	0.0020	0.0060	0.0014			Certified
	0.0001	0.0004	0.0009	0.0009	0.0002	0.0004	0.0001			SD
	0.0008	0.0055	—	0.0056	0.0020	0.0058	0.0014	0.0003	0.0004	Measured
AKP 044-1	0.0011	0.0879	0.0009	0.0070	0.0043	0.0078	(0.0001)[b]			Certified
	0.0002	0.0032	0.0001	0.0008	0.0002	0.0003				SD
	0.0012	—	0.0011	0.0078	0.0041	0.0083	0.0007	0.0003	0.0004	Measured
BCS 149-3[a]	0.019	0.004	0.001	0.001	0.001	0.007	(0.002)			Certified
	0.018	0.0036	0.0007	0.0012	0.0010	0.0067	0.0009	0.0004	0.0003	Measured
BCS 260-2[a]	0.013	0.0011	(0.001)	(0.002)	0.002	0.009	(0.0012)	(0.0003)	(0.0003)	Certified
	0.0137	—	0.0013	0.0016	0.0020	0.0088	0.0007	0.0003	0.0003	Measured
BCS 260-4[a]	0.002	0.003	0.002	0.002	0.003	0.006	(0.001)	0.0004	0.0003	Certified
	0.0016	0.0035	0.0018	0.0017	0.0029	0.0062	0.0009			Measured

[a] BCS standard have no certified SD.
[b] Values in parentheses are not exactly determined.

being theoretical in character. For the production of standard curves, due to the complex matrix, pure element solutions are never used. The correction of a single interference would not solve the problem. Further effects must be corrected, which leads to complicated and time consuming standardisations.

For the analysis of iron, steel, and ferro compounds, therefore, standards are always used whose chemical composition is very similar to the sample. These standards undergo the same dissolution and dilution procedure as the sample. In special cases they are fused as for the sample. Therefore, the standard solutions differ neither in their physical characteristics nor in their concentration of acid or total salt. Any effects that occur will be the same for sample and standard measurements. They are easily eliminated by including them in the calibration.

If in special cases no appropriate standards are available, model solutions must be prepared (section V.A). These are obtained by including in the model solution all the main components of the sample made up out of analytically pure solutions. In addition, the acids and fusion materials must be added in the same concentration as is apparent in the sample. Such model solutions can be regarded as adequate standards. Interferences are thereby eliminated.

When a graphite tube is used, different interference effects appear. These can be caused by the sample as well as by the specific instrument [24]. Interferences caused by the sample occur when the element is vaporised as a molecule, the chemical compound hinders atomisation, or when after vaporisation of the matrix the sample has poor contact with the graphite tube. Also, unspecific absorption can occur by simultaneous vaporisation of the matrix. Here, molecular absorption or light loss by scattering from solid particles are the most common effect.

Most of these effects can easily be removed by careful temperature programming — vaporisation of the liquid, separation of the matrix by slow ramping of the temperature and rapid atomisation. Unspecific molecular absorption can be removed by the use of a continuum source, in so far as broad absorption bands occur. In practice, pipetting of the sample causes the greatest error. When the sample is pipetted with a piston pipette, the solution drops do not always strike the same place in the graphite tube. As a result, the temperature conditions could be quite different for the calibration and measurement of the sample. Thus, the reproducibility of the method becomes distinctly worse. An improvement is achieved by the use of an automatic sampling system, which always places the same size of droplet at the same angle and the same position in the graphite tube.

While chemical and physical interference possibilities can be largely removed by making the standard solution and the sample solution essentially similar, unreproducible temperature settings frequently lead to differences which cannot be corrected. Though there has been no lack of effort to achieve high temperature constancy by the construction of special feedback systems, the lifetime of these systems is very limited when the very high temperatures required with ferrous metallurgy are used. The analytical solutions, which always have corrosive acids (perchloric acid, aqua regia), required for iron and steel make even pyrolytically coated graphite tubes usable only a few times. Even after a single use the inner wall of the graphite tube has been attacked so much that reproducible temperature conditions and constant heat transfer can no longer be taken for granted. In iron metallurgical solutions the acid content must be regarded as the actual matrix and is of course present in far higher concentrations than the iron content itself.

These matrix effects can be avoided by the solid-sample system described in section II.B.1. At temperatures around 2600°C the steel sample placed into the rectangular graphite tube melts without measurable vaporisation of larger particles. Effects from the iron matrix are not observed.

The interferences caused by the instrumentation include carbide formation, which, for example in the determination of vanadium, become apparent as a long tailing. The analytical results can be improved if not only the peak height but also, by area integration, the peak area, are measured. The determinations of B, Nb, Ta, W, and Zr are prevented completely by the formation of stable carbides.

Influence of high DC light emission from the graphite tube are difficult to remove. Though the DC and AC light emissions are electrically separable, the reflection of the DC light on the cathode of the hollow-cathode lamp produces pulsing of the light. This emission has the same frequency as the elements' specific hollow-cathode light and cannot be electrically separated.

D. Precision and accuracy

Reproducibility is a measure of the constancy of the data from the total procedure — dissolution of the sample, dilution, measurement, and readout [65]. It can be determined when the procedure is repeated several times with a single instrument. For the determination of the accuracy of analytical data a second independent analytical procedure is always required. In the analysis of concentrations between 0.001% and 20.0% checking of the accuracy is readily done with photometric or gravimetric procedures. It is difficult to check accuracy with concentrations below one microgram per milliliter. Contamination or losses on the walls of test tubes, borders of phases, or in the mixing chamber of the atomic absorption spectrometer can only be checked with difficulty. Also, contaminations from laboratory glassware, laboratory air, and the solvents used must be taken into account. Only after blank solutions are determined can very pure chemicals be added. The problem of the accuracy of analysis is most significant when the graphite tube, cold vapour apparatus or special concentration methods are used. When particularly low concentrations are measured, it is desirable not only to use two analytical systems, but also when possible different sample preparation

References pp. 246—249

methods. Thus, changes that occur even before the actual analysis can be noted.

The optics of all common atomic absorption spectrometers of similar price are much the same, but nebulisation systems are designed according to various principles (direct injection, counter flow). When normal flame methods are used, with constant aspiration rate of the sample solution, the reproducibility of the measurements depends mainly upon the nebuliser system. There are exceptions, namely smaller atomic absorption instruments which work according to the single beam principle, and where drifting and variability of the hollow-cathode lamp have a direct effect upon the result. Double-beam atomic absorption spectrometers compensate completely for changes in light intensity by the measurement of the light intensity in the reference beam.

As an example of the reproducibility of a modern instrument a steel sample containing 1.04 weight percent nickel is dissolved, (Fig. 2) and continually aspirated. Between two integration values (integration time one second) water is always aspirated to clean the mixing chamber.

The measured values (in absorbance) are: 0.313, 0.313, 0.314, 0.318, 0.315, 0.318, 0.315, 0.314, 0.317, 0.317

mean value: 0.316 wt. %; $S = \pm 0.002$ wt. %.

This SD is equivalent to an RSD of 0.65%. It includes all instrumental variations and can be regarded as characteristic for the concentration range between 0.0001 weight percent and 20 weight percent. If the same sample is analysed several times, the standard deviation is increased by the variations that occur because of sample preparation, to a level of about two relative percent.

If the total salt content of a solution is high or if only small amounts of sample are available the injection technique may be used. If a piston burette is used for the injection, individual errors cannot be avoided, and reproducibility becomes somewhat poorer than with continuous aspiration. Also, the uncertainty of the volume measurement with small injection volumes becomes noticeable in the reproducibility. In Table 4 the data and statistical values of a simultaneous copper and nickel determination are shown.

The RSD with relatively large injection volumes and pure element solutions is not significantly larger than with continuous aspiration.

Trace analysis including previous element concentration procedures as well as partial matrix removal (II.B.2) produces an increase in RSD with an injection volume of only 10 μl of 7.6 relative percent in a copper concentration of 0.0021 weight percent and to 16 relative percent with a manganese content of 0.00088 weight percent. Use of an automatic injector can improve the reproducibility of the sample volume considerably [14].

When the graphite furnace is used for extreme-trace analysis, relative standard deviations are in general above 10 relative percent. The solid sample analysis described in section II.B.1, using a steel chip and the rectangular cuvette, leads to an RSD of 13 relative percent in the determination of a

TABLE 4

REPRODUCIBILITY OF THE INJECTION TECHNIQUE. SIMULTANEOUS MEASUREMENT OF A SOLUTION CONTAINING $2\,\mu g\ Cu\ ml^{-1}$ AND $5\,\mu g\ Ni\ ml^{-1}$, INJECTED VOLUME: $100\,\mu l$

	Cu (E)	Ni (E)
	0.099	0.119
	0.101	0.122
	0.094	0.113
	0.095	0.115
	0.100	0.115
	0.097	0.118
	0.093	0.119
	0.100	0.118
	0.104	0.115
	0.099	0.121
$\bar{x}$	0.0982	0.1175
$s_{f=9}$	0.0034	0.0029
s_Γ	0.03	0.02

copper concentration of 0.00125 weight percent, and an RSD of 20 relative percent with a manganese concentration of 0.00071 weight percent, and lies in the range of trace analysis. The frequently mentioned assumption of poor reproducibility with solid-sample analyses in the graphite tube thus does not apply for steel chips.

III. ANALYSIS OF ORES, SLAGS AND OTHER OXIDES

Oxide materials are distinguished by the fact that the compound forms of the elements may be very different, depending on the many geological types. In general, various main components (greater than 10 weight percent) and trace concentrations (less than 0.01 weight percent) are present. The analysis must take this into account not only in the sample preparation (preparation of solutions) but also in the desired indication of results.

Basically, the literature provides two dissolution methods: sample preparation with sample weights of 0.2—1 g and large dilutions, or smaller sample weights with less dilution (III.B). The relatively large dilution, in general after a fusion [51], for the determination of main and lesser components, as for example in silicate analysis [2], the determination of Al, Ca, Mg, Mn and Si in slags [4], Si [55], Pb and Mn [143], and also Cd, Ca, Cu, Pb, Mg and Si in ores or iron sinter [97, 147] and Cr, Mg in refractories [93] is presently used in routine analysis.

References pp. 246—249

A. General solution methods

The problem is to prepare solutions suitable for the desired analytical results. Once a homogeneous analytical solution is available, the differences between the techniques used with dissolved metals are no longer great since the same nebuliser systems can be employed. For the determination of major components it is preferable to use less sensitive spectral lines of the same elements rather than use greater dilutions.

A significant difference in iron solutions can arise from the method of sample preparation; the salt concentration can be very high and clogging of the burner by crystallisation may be a problem.

1. *Direct dissolution*

It can be seen from the above that it is desirable, wherever possible, to dissolve the oxides directly in mixed acids. Small quantities of residue can be fused and added, as described for the analysis of steel. Dissolution in pressure bombs [12, 57] is a method used to keep the salt concentration within limits. This technique is usable for single determinations in particular cases (small quantities); it has not been extensively applied in routine analysis up to now because the use of larger quantities is impossible or too dangerous.

The analytical procedure is rather simple (Fig. 11).

2. *Universal fusion technique*

The great majority of oxide materials can be brought into solution using a fusion technique, mixtures of tetraborates being the most widely used fusing materials. For measuring sodium in solution $Li_2B_4O_7/LiBO_2$ must be used instead of $Na_2B_4O_7$. Sodium can be determined with the necessary precision only with great difficulty owing to contamination present in industrial laboratories. Where it is done, the nebuliser and burner systems must be clean and air pollution must be reduced.

With sodium tetraborate fusions, small additions of oxidising materials (Na_2O_2, $NaNO_3$) have been found successful for substances that are difficult to fuse.

The procedure for analytical methods employing fusion is again very simple (Fig. 12). There are only rarely difficulties involving burner clogging, if burner slots of 0.8 to 1 mm can be used. If it is necessary for technical reasons to work with narrower slots, or when great numbers of samples are to be analysed without cleaning the system, the use of the Hoesch injection technique [130] is recommended, for example with 100 μl injections, either manually or with the help of an automatic sampling system [14]. Different matrices lead to the consideration of different classes of material, for example based on iron (iron ores, drauses, dusts), calcium (lime, slags),

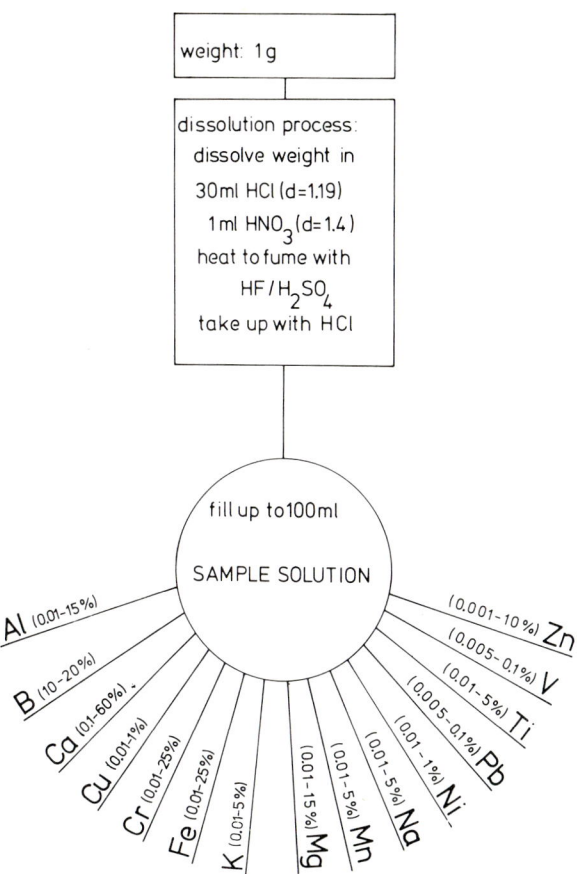

Fig. 11. General dissolution method for oxides.

aluminium (fire clay), magnesium (magnesite) or silicon (silicate rocks, sand, residues) as well as substances without a unique basis or with various major constituents (slimes, blast furnace slags). This is not really necesary from the analytical point of view in which there are only differences in the sample preparation, the dilution or the choice of spectral lines, according to differing sensitivity.

Through calibration and re-calibration techniques possible interelement effects or influences from the matrix can be largely eliminated. Because of its worldwide importance the introduction of standard methods by atomic absorption started first with iron ores [67] and refractories. The ISO/TC 102 is presently occupied with draft proposals for the atomic absorption determinations of Na/K Ca/Mg, Zn, V, Pb and Al.

References pp. 246—249

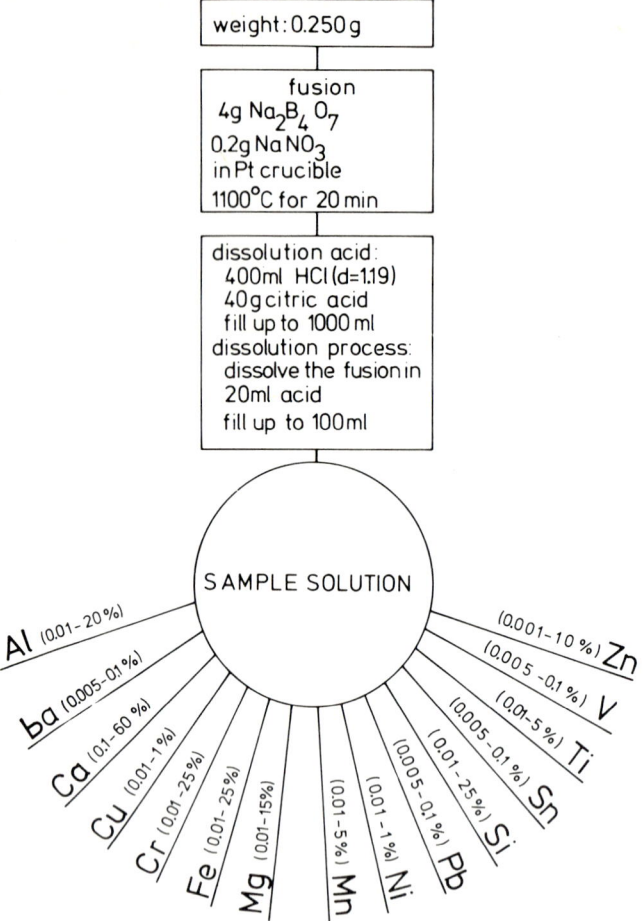

Fig. 12. General fusion technique.

B. Trace-element analysis

In general, for trace analysis, very small dilutions are preferred [54], in which case only small solution quantities are available for atomic absorption. These can be brought into the flame either with the Hoesch injection technique [3, 13, 14, 31, 99, 130, 142] or with the boat technique [76] or they can be determined in the furnace atomiser [75, 90, 91, 120] so that even such elements as La in ores [109] or the rare earths generally [36, 37] as well as Pb [48, 50], Bi [49], As, Sb, and Sn [116] can be determined.

Because of the variety of oxide materials there are a number of special methods, for example for the determination of Cr [26, 44] and Cu, Mn, V and Ni [28]. Also, sampling of the solid material, as in the determination of Al and Sn [6]. An attempt has been made to determine fluorine with the aid of the AlF molecule [145]. A universally applicable method is elusive.

1. *Graphite tube methods (low salt content)*

The furnace atomiser is readily used for trace determinations and also in the case of oxide materials it provides extremely low detection limits. Because of the inhomogeneity of the sample materials, the sample may frequently be no smaller than 1 g (or with extremely small concentrations up to 5 g), so the solution volume is limited as far as possible.

Furthermore, it is useful to limit the salt content of the solution. This generally means that the furnace technique should only be used to determine trace elements soluble in acids — possible after pressure bomb decomposition. In this case the time of analysis becomes irrelevant. An example is the determination of Cd-traces in iron ores and related oxides.

(i) *Determination of cadmium in iron ores and related oxides*

Basis. Samples are dissolved in the acid mixture by a long reaction. The solution evaporates in the furnace atomiser where it is transformed into atomic vapour and the absorption of light intensity is measured. (Addition method.)

Concentration range. 0.01—1 mg Cd kg^{-1}.

Reagents. HCl Suprapur ($\delta = 1.15$ g/ml). HNO$_3$ Suprapur ($\delta = 1.4$ g/ml). H$_2$SO$_4$ Suprapur (1 + 1). Cd stock solution: 1.000 g Cd dissolved in 5 ml HNO$_3$ and 25 ml H$_2$O and topped up to 1000 ml with H$_2$O (1 ml = 1 mg Cd). Cd standard solution (I): 50 ml of the stock solution diluted with H$_2$O to 500 ml (1 ml = 100 µg Cd). Cd standard solution (II): 10 ml of standard solution (I) diluted with H$_2$O to 1000 ml freshly every day (1 ml = 1 µg Cd). Cd solutions for the addition method: 10.0, 25.0 and 50 ml of standard solution (II) (10, 25 and 50 µg Cd accordingly) are topped up with H$_2$O to 1000 ml.

Operating conditions. AAS-device with graphite tube furnace and deuterium background corrector. Wave-length, 228.8 nm; EDL, Cd (6 W).

Procedure. Approximately 5 g of the sample are weighed exactly and placed in a 250 ml beaker. A second, similar beaker is used for the dummy solution. To both are added, consecutively, 40 ml H$_2$O, 20 ml HCl and 20 ml HNO$_3$. To dissolve the sample it is boiled for 1—2 h. After it has cooled 50 ml of H$_2$SO$_4$ are added and evaporation encouraged until H$_2$SO$_4$ vapour escapes. After cooling, 100 ml H$_2$O are added. The solution should be left overnight on a warm hotplate. The solution in the beaker without the sample is treated in the same way. Afterwards each is filtered through an appropriate filter, washed out with diluted H$_2$SO$_4$ and hot H$_2$O, cooled, and filtered into a 250 ml flask, which is then topped up. Diluted sample solution, 40 ml, is now pipetted into each of four 50 ml graduated flasks (A, B, C, D). Then, 5 ml of the Cd solutions with 10 ng ml^{-1}, 25 ng ml^{-1} and

References pp. 246—249

50 ng ml^{-1} are pipetted into the graduated flasks B, C, D, in this sequence. All graduated flasks are topped up to 50 ml and stored in polythene bottles. The dummy solutions are produced by corresponding additions to 40 ml of the initial solution. At 3 minute intervals 5 µl of the test solutions at a time are injected into the furnace by a flask ejector pipette.

The sequence of measurements is important. It has proved useful to measure C, B, A, and D again in the same way after solution D has been measured 2—4 times, to recognise possible equipment changes during the series.

In addition, the 4 dummy solutions are measured in the same sequence.

Evaluation. The results of the measurements are plotted on a graph against the concentration of Cd in 50 ml (addition solution) and are amended with the corresponding blank values.

The Cd-concentration of the unknown sample can be read off the result of solution A after it has been amended by the blank value A. This method is also very exact in terms of the error margin (RSD = 10% at 0.1 mg Cd kg^{-1}) and is therefore a typical example for this procedure.

Analysis by AAS of (most) trace elements in bulk materials from the field of ferrous metallurgy is still in its early stages.

2. *Injection technique (high salt content)*

Clearly, trouble-free universal methods are the ideal for trace analysis of oxide materials. Numerous materials can only be dissolved via melting fusion; they then cannot be diluted overmuch and therefore contain a relatively high salt content. However, many laboratories are not yet equipped with furnace atomisers so the flame method must be used.

During trace analytical analyses with the flame technique a separation and therefore concentration, e.g. via ion exchange, can be introduced if the concentrations during normal sample preparation are likely to be below the detection limit of the instrument [92].

The direct use of the Hoesch injection technique is useful for trace analysis of oxide substances after fusion [31, 99, 142]. It is possible to take up the fused substance in small quantities (max. 10 ml, usually 1 ml). It is best to use an acid mixture which contains 400 ml HCl ($\delta = 1.15$ g ml^{-1}) and 40 g citric acid in 1000 ml. Generally, 10 µl, and in special cases 50 µl, of the analytical solution are injected. It is thus possible to determine all important ascertainable elements in 1 ml of analytical solution several times over. Salt concentrations of 100 mg ml^{-1} are no problem for the injection technique.

The following general methods can also be recommended for the analysis of almost all oxide products.

(1) Fusion of an analytical sample. The sample is ground until it passes through a 100 µm sieve. A 0.250 g quantity is weighed and mixed with 4.0 g of sodium tetraborate and 0.2 g of sodium nitrate. The fusion occurs in a platinum-crucible either at 1350°C with a coil which is heated induc-

tively or in a muffle furnace at 1100°C. The period of fusion is at the most 5—20 min.

(2) X-ray fluoresence spectroscopy. The melt is poured into a pre-heated platinum-mould, producing a tablet of approximately 30 mm diameter. Cracking of the tablet is prevented by slow cooling. The level surface is measured directly.

(3) AAS-analysis. The same tablet or an aliquot part of it is dissolved directly in acid (400 ml conc. HCl + 40 g citric acid in 1000 ml) and topped up to 100 ml with H_2O. All elements which can be ascertained by AAS can be determined directly from this solution with the injection technique, if corresponding standard solutions have been used for calibration each time.

A combination of the rapid procedures of spectroscopy and corresponding control methods is presently the only analytical procedure with the necessary precision and accuracy and at a reasonable cost. In the field discussed AAS is an essential technique for routine work.

C. Interference of sample matrix

The chemical make-up of oxide substances is often very complex and the matrix frequently contains two or three compounds of almost equal concentration. The determination of individual interference effects is very complicated and time consuming, because the composition of these substances can be subject to wide variations. As in the case of steel, pig iron and cast iron the interference effects are generally not determined.

By using suitable calibration systems the interferences can be compensated for without being determined in detail. For this purpose calibration using model solutions or the use of the addition method have proved useful.

In the case of the addition method an analytical-grade solution of the element to be determined is added to the sample solution in graded concentration steps. The concentrations of the additional solutions are measured so that only a few millilitres have to be added. Thus the change in the sample matrix is so insignificant that normally neither physical nor chemical interference effects are altered. The prepared solutions are measured and the determined absorption values are plotted against the concentration of the added solution. The values are always made up of two components, one from the sample and the other from the added analytical-grade element solution. If the straight line obtained is extended beyond the ordinate it will intersect the abscissa. This value always lies to the left of the origin and is thus a negative value. The absolute value at this point corresponds with the concentration of element in the sample. The addition method can also be applied if an atomic absorption spectrometer plus microprocessor is being used. In this case the sample solution is aspirated and the concentration adjusted to zero. Then the sample with the highest addition is aspirated and the microprocessor is fed the concentration of the element solution. If a dummy solution containing all acids and solvents is then used, the micro-

References pp. 246—249

processor will print out the required concentration of the sample with a 'minus' sign.

This method of internal calibration may only be used for linear calibration curves. For a curved calibration plot a linear extrapolation against zero would be very inaccurate.

D. Precision and accuracy

Statistical data on the analysis of oxide products should be based on the same criteria as for iron and steel analysis. Reproducibility limits of about 1—2 rel.% are valid in the case of direct analysis with AAS. The use of the injection technique leads to reproducibilities of 2—5 rel.% for high salt content and 10—20 rel.% for the graphite tube technique (low salt content) used as a trace method.

While analysis in an iron matrix ranges from the trace range to 20%, element concentrations of up to 60% are determined in oxide substances.

By using an electrically controlled 'diluter' reproducibility of 1—3 rel.% can be achieved even for concentrations not normally suitable for determination by AAS.

Since classical techniques such as gravimetry and photometry produce similar variations of reproducibility in these ranges of concentration, the use of AAS for these determinations is justified. The gain in time compared with traditional methods is considerable. However, the accuracy of the analysis of oxide products has to be checked by the reference method (XRF, classical chemistry).

IV. SPECIAL AAS METHODS

There have been attempts to develop special methods for some elements which cause problems with solvent analysis when using the flame or furnace atomiser techniques.

Two basic trends can again be recognised, i.e. variations in the equipment or in the associated chemical reactions.

The direct introduction of atomic vapour into the AAS flame can also be seen as an equipment variation [148, 152]. Here, the evaporation of the steel samples can be carried out with the help of the glow-discharge lamp [148] or an aerosol generator with a low current d.c.-arc discharge [152]. Another example is the combination of gas chromatography and AAS [92], where AAS is used as an element detector.

A. Development of methods for gases

Chemical reactions whose product is gaseous and contains the element to be determined have proved useful. The determination of Hg as metal vapour

[79, 150] and the determination of elements which form volatile hydrides, for example As [30, 43] and Sb, Ge [19, 132], Se [23, 98] and Te, Bi, Pb and Sn [42], have been described. The Hg determination and those of As and Sb relevant to ferrous metallurgy are now part of the routine analysis of materials. The combination of EDL and the evolution of gas method have proved especially useful for this. For reasons of environmental control the determination of Hg in bulk materials has become very important, for example in iron ores, the procedure for which is now given as an example.

1. *Determination of mercury in iron ores, dust and slurry*

Basis. Samples are dissolved by pressure fusion, are then reduced in a wash bottle with $SnCl_2$ and analysed flameless as cold vapour.

Concentration range. 0.005—1 mg Hg kg^{-1}.

Reagents. HCl, Suprapur ($\delta = 1.15$ g ml^{-1}); H_2SO_4, Suprapur $(1 + 1)$; HNO_3, Suprapur ($\delta = 1.4$ g ml^{-1}); HF, 40 mass % p.a.

$K_2Cr_2O_7$ solution, 10 mass % p.a. NH_2OH solution, 5 mass % p.a. $SnCl_2$ solution, 100 g $SnCl_2 \cdot 2H_2O$ in 50 ml HCl to be topped up to 1000 ml with H_2O. N_2 to be passed through for 30 min.

Hg stock solution: 108.0 mg HgO are dissolved in 5 ml HNO_3 and topped up to 1000 ml with H_2O: 1 ml = 0.1 mg Hg. Hg standard solution I: 5 ml of $K_2Cr_2O_7$ solution and 10 ml HNO_3 are added to 10 ml of stock solution and topped up to 1000 ml with H_2O: 1 ml = 1 μg Hg. Hg standard solution II: 5 ml of $K_2Cr_2O_7$ solution and 10 ml HNO_3 are added daily to 10 ml of standard solution I and topped up to 1000 ml with H_2O: 1 ml = 10 ng Hg.

Operating conditions. AAS device with background corrector, absorption cuvette open (15 cm length, 15 mm diameter) and closed system with crystal windows circulating pump, 2.5 l min^{-1} pressure fusion vessel. Wavelength, 253.7 nm; EDL, Hg (5 W); deuterium compensator.

Procedure. According to the Hg concentration, 0.2 to 1.0 g of sample are weighed into the pressure fusion vessel, mixed with 4—10 ml H_2SO_4, 1—5 ml HNO_3 and 0.2 ml HF, close and heat for 45 min at 140°C stirring constantly with a magnetic mixer. After cooling transfer to a gas wash bottle (250 ml), add 5 ml of $K_2Cr_2O_7$ solution and top up to 100 ml with H_2O. The excess oxidiser is reduced with NH_2OH solution (decolourising). Then 5 ml of $SnCl_2$ solution is added, the wash bottle is closed and Hg-free air is passed through the reagent vessel with the help of the circulation pump and through the absorption cuvette, until the maximum value appears (closed system). In the case of the open system wait for 15 s after the reduction with $SnCl_2$ solution before air is passed through with the pump.

Calibration. With the standard solutions a calibration curve is produced by following the same sequence; this curve should be linear for the closed system for the concentration range 10—250 ng Hg. A linear calibration curve is produced for the concentration range 200—1000 ng for the open system. In the case of samples of unknown composition the addition method is

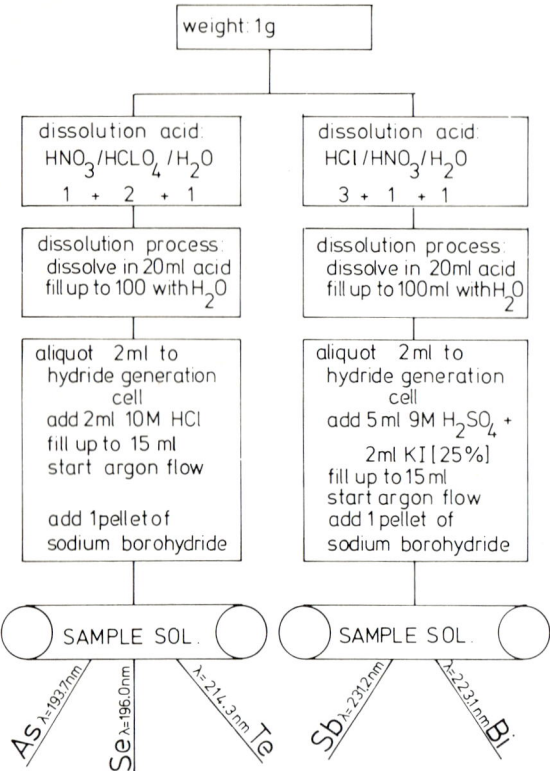

Fig. 13. General gas generation method.

recommended for best results. In this case certain quantities of the standard solutions can be added after pressure fusion.

Today, metal hydrides can be synthesised in purchasable equipment. Principally a solution which contains $AsCl_3$ or $SbCl_3$ is pre-reduced, with for example KI and ascorbic acid, and then reduced with sodium borohydride to arsine or stibine. This is done in a closed system and the method is simple (Fig. 13). The determination of As via arsine in iron ores is now available as draft proposal ISO/TC102. The method described for the concentration range of 0.0001—0.05 mass % is as follows.

Decompose the test portion by sintering with sodium peroxide and leaching with water and HCl. Transfer the solution to a distillation flask and evaporate a portion of the solution. Reduce arsenic to arsenic(III) by treatment with potassium bromide and hydrazine sulphate. Adjust the acidity and distill $AsCl_3$, collecting the distillate in water. Reduce an aliquot portion of the solution with potassium iodide and ascorbic acid, and treat with sodium borohydride to reduce arsenic ions to the volatile hydride, arsine. Rapidly sweep the arsine into a hydrogen—nitrogen (or argon)-entrained air flame and measure the transient peak absorbance at 197.2 nm in an atomic absorption spectrometer.

The determination of As in iron and steel for the concentration range 0.002—0.01 mass % is possible with the hydride method with an RSD of approximately 2—5%.

B. Metal determination in organic solvents

A concentration procedure with the help of extraction or chromatographic separation of the elements to be determined from the matrix is often available using the organic phase.

Disregarding possible losses during extraction or elution it is a common procedure to aspirate organic solutions with metal ions into the AAS flame since this generally improves the measuring sensitivity [78, 92, 151].

In industry analysis for metals in lubricating oils, greases or fuels is important for checking characteristics on delivery or for testing the quality of used lubricants for the purpose of preventing maintenance. In the first case one talks of the determination of metal ions in oils among others, in the second of wear metals, such as Cd, Cr, Cu, Fe, Mn, Ni, Sb, Sn and Ti. Metals which may be present in lubricants or fuels, are Al, Ba, Ca, K, Li, Mg, Mo, Na, P, Pb, S, Si, V and Zn [74]. This analytical area is very important because considerable financial considerations are related to the use of the correct lubricants, to the possibility of harmful substances developing in fuel oils (S or V) and also to maintenance.

This is also shown by the existence of a large and increasing number (Table 5) of standardised AAS-methods [7, 35, 66] relating to metal ions in lubricating oils or greases (Al, Ba, Ca, and Zn). In the literature, methods can be found which check additive content (Ba, Ca, Zn) with the help of AAS [16, 20, 61, 63, 64, 112, 133].

There are also methods for V in fuel oil [88], as well as Hg [80, 144] and Cd [118] and the AAS determinations of Ti [121], Pb [61, 95] and Mn [96] in petrol. The use of the graphite tube furnace in the determination of metal ions in oils and greases, including wear metals, has been described [9, 22, 69, 86].

Other reports deal with individual elements, such as Ni [1, 86, 87] or Fe [11, 84]. The efficiency [71—73] of flame methods (AAS) has been compared with flameless techniques (NFAAS) (Table 6). Because of their significance there have been attempts to determine the elements P [38] and S [78] directly with AAS. This, however, requires a device which can measure ultraviolet lines (ca. 180 nm) with sufficient sensitivity. Good results can also be achieved by gas chromatographic separation and successive AAS determination [92] and simultaneous multielement analysis with a Vidicon-detector has been tried [68] because the speed with which the information is gained can be very important in practice. Some work [39, 53] reports on the problem of molecular bands which can appear when working with

TABLE 5
STANDARDISATION ACTIVITIES

Standard	Lubricants, fuels	Elements	AAS method Fuel/oxidant	
ASTM D 2788-69 T [7]	Gas turbine fuels	Ca, P, Mg, Na, K V	C_2H_2/air N_2O/C_2H_2	Dilution with organic solvent Direct nebulisation
IP 308/74 [66]	Lubricating Oils	Ba, Ca, Mg, Zn	N_2O/C_2H_2	Dilution with organic solvent Direct nebulisation
DIN 51 391, Teil 1	Oils	Ba, Ca, Zn	C_2H_2/air	Dilution with petrol Direct nebulisation
DIN 51 393, Teil 1	Lubricants	Pb	C_2H_2/air	Dilution with petrol Direct nebulisation
DIN 51 395, Teil 1	Oils	Mo	C_2H_2/air	After incineration Direct nebulisation
DIN 51 395, Teil 2				
DIN 51 397, Teil 1	Oils	Fe	Non-flame	After incineration dissolution in HCl
DIN 51 401 [35]	—	—	General working principles	
DIN 51 431	Oils	Mg	C_2H_2/air	Dilution with petrol Direct nebulisation
DIN 51 769, Teil 7	Petrol	Pb	C_2H_2/air	Dilution with isooctane Direct nebulisation
DIN 51 769, Teil 8	Petrol	Pb (5—25 mg l^{-1})	C_2H_2/air	Direct nebulisation
DIN 51 790, Teil 3	Fuels, oils	Na	C_2H_2/air	After incineration dissolution in HCl
DIN 51 793, Teil 3	Oils, fuels Fuels, oils	V Na	Non-flame C_2H_2/air	After incineration dissolution in H_2O
DIN 51 797, Teil 3	Oils	Na	C_2H_2/air	After incineration
DIN 51 797, Teil 4			C_2H_2/air	Dilution with ethanol/xylene direct nebulisation
DIN 51 815	Greases	Li, Na, Ca	C_2H_2/air	Dilution with cyclohexene, HCl, H_2O-extraction

TABLE 6

DETECTION LIMITS AND LIMITS OF DETERMINATION OF DIRECT AAS ANALYSIS OF WEAR METALS FROM ORGANIC SOLUTION [74]

Element	AAS		NFAAS	
	Detection limit (mg kg^{-1})	Limit of determination (mg kg^{-1})	Detection limit (mg kg^{-1})	Limit of determination (mg kg^{-1})
Ag	0.1	0.4	0.01	0.15
Al	3	10	0.2	0.6
Co	0.2	0.6	0.02	0.06
Cr	0.2	0.6	0.02	0.06
Cu	0.2	0.6	0.01	0.03
Fe	0.2	0.6	0.01	0.03
Mg	0.008	0.025	0.0002	0.0006
Mn	0.1	0.3	0.002	0.01
Ni	0.2	0.6	0.02	0.06
Pb	0.8	2.5	0.005	0.02
Si	3	10	0.05	0.2
Sn	5	15	0.01	0.03
Ti	0.4[a]	2[a]	—	—
W	1[a]	20[a]	—	—
Zn	0.01	0.05	0.001	0.005

[a] Only from water solution after incineration.

organic solutions. This demands specially selected lines or background correction.

Owing to the difficulty in extracting a homogeneous sample from used oils, or to the uncertain distribution, types and sizes of particles, an early incineration is generally undertaken, which should ideally be without waste. The treatment of greases is similar. The analysis of the ash is basically similar to the analysis of oxide substances in an aqueous medium.

Fresh oil and petrol are normally analysed directly by AAS; oils of differing viscosity should be diluted with organic solvents such as MIBK, xylene, etc. until the rate of atomisation is no longer influenced by the viscosity.

Two basic methods of sample preparation can be distinguished. The direct method includes all bound or completely dissolved ions. No agreement has been reached concerning particles or their diameters determined in the process. The indirect method includes all metal or non-metal ions which are not volatilised during incineration.

To measure the increase or the decrease of a certain type of ion, which corresponds to contamination, wear or thinning (consumption of additives), it is necessary to determine even slight variations of concentration with high precision [125]. The efficiency and precision of the AAS method is

References pp. 246—249

such that slight variations can be detected even for concentrations around 1 mg kg^{-1}. In practice, far reaching decisions often depend on this. For example, the checking of accuracy and construction of calibration curves is now done world-wide, with oil-soluble standards available for all possible types of ions (e.g. CONOSTAN). Due to the increasing importance of preventive maintenance this analytical field will expand considerably.

V. CALIBRATION TECHNIQUES

There are various calibration systems for AAS which differ mainly in the time needed.

For standardised methods the calibration curve is always produced with a pure-element solution. Hence, interferences have to be considered and eliminated, leading to time consuming checks of variable parameters such as matrix effects, anion or cation influences or interferences caused by acids.

In the analytical procedures of ferrous metallurgy calibration methods have quickly proved useful, interferences being compensated for from the outset although they are not known individually.

A. Calibration using standard solutions

Atomic absorption spectrometers fitted with microprocessors allow calibration points to be stored. The necessary data can be made available to the microprocessor by aspirating suitable calibration solutions whose concentrations should be such that they cover the entire concentration range to be analysed, at appropriate intervals. After storing the calibration points the microprocessor constructs the calibration curve, inspection of which shows whether the plot is linear or curved.

With the help of microprocessors, recalibrations have become so easy and so quick, that a new calibration curve can be plotted using different parameters (a different analytical line, slit width, position of burner or dilution of sample solution).

When using a dual-channel spectrometer the calibration is easier since one concentration range can be covered by channel A and another by channel B. Standards for all categories of substances which are being used routinely in the laboratories of steel works can be purchased.

For calibration, a series of standard solutions is prepared which differ in their concentration of the element to be determined but which are equal in their concentration of all other elements. When calibrating with a pure-element solution a stock solution which contains a high concentration of the element to be determined and which only has to be diluted in certain ratios is used. However, in the case of an analytical sample interferences have to be expected. It is recommended that a model solution is prepared and adjusted to the necessary concentration by diluting with a separately prepared solution which contains all matrix elements and accompanying elements but not the element to be analysed. Thus, standard solutions with different concentrations of the element to be analysed but with the same matrix are produced. For steel, pig iron and cast iron it is mostly sufficient to prepare a stock solution. Dilution of the stock solution for calibration to various concentrations is done with an analytical-grade iron solution.

The concentration of iron in the standard has to correspond with that of the sample solution. In the case of alloy steel of varying iron concentration, pure-iron solution is added to the stock solution and to the sample solution in excess. Thus, the difference in the concentration of iron in the sample compared with that in the standard is negligible. Interferences due to the matrix then remain constant during calibration and analysis. Calibration with model solutions is carried out by following the same pattern as with standard solutions.

B. Recalibration procedures

The standardised sequence for measuring calibration solutions and solutions to be analysed is: (1) Measuring of calibration solutions in order of increasing concentration. (2) Measuring of sample solution, twice. (3) Measuring of calibration solutions in sequence of decreasing concentration.

Measuring the calibration solution in order of decreasing concentration, after the solution to be analysed has been measured, has no influence on the results for the sample solutions. It is better to calibrate in order of increasing concentration before measuring the sample solutions and to check the calibration values by measuring in order of decreasing concentration; memory-effects can then be recognised when calibration points deviate.

Once the data for the calibration solution have been stored it is recommended to measure a recalibration solution after every fifth sample solution. This solution must be prepared as for a calibration solution but may not be used for calibration. The concentration of the element in the sample solution to be determined should lie in about the middle of the concentration range of the calibration solutions and must be known. If the calibration curve does not change while measuring the sample solution, then the recalibration value must correspond with the known nominal value. Should deviations occur the calibration curve has to be checked with the calibration solutions and, if necessary, the calibration data fed in again. This should not, however, be done without having checked for causes such as blocking of the burner slit, twisting of the burner port or blocking of the atomiser.

If a measurement has to be interrupted because of time, or if the same problem occurs again after a few hours or the next day (same element in the same matrix, e.g., Al in steel), then the calibration curve may be stored. With the spectrometer switched to stand-by, data are preserved in the micro-

References pp. 246—249

processor while other functions, such as the fuel gas supply or the chopper motor are switched off. The lamp current is adjusted to a low setting so that the hollow-cathode lamps are kept pre-warmed and ready to operate. To balance slight alterations in equipment parameters the lowest concentration calibration sample is aspirated and the new value fed into the microprocessor by passing on the nominal concentration. The remaining calibration points of the curve are then related to the new recalibration value. Thus the complete curve is recalibrated using one calibration sample. After approximately two minutes, the spectrometer will be operational again.

When analysing low concentrations, in the range of a few $\mu g\,ml^{-1}$, or even less, it has to be remembered that solutions of very low concentrations are not very stable. It is essential that fresh solutions are prepared for calibration as well as for recalibration. Thus, it is recommended to prepare a solution of a higher concentration, such as a stock solution, since it will remain stable longer.

Just before measuring the calibration solutions are diluted to their nominal concentrations with the appropriate solutions.

C. Background correction

Background correction is carried out with a continuum source, e.g. a hydrogen hollow-cathode lamp or a deuterium-arc lamp.

If the background absorption is wide band as in the case of dissociation continua of alkali halides, non-specific loss of light can be compensated for using the background compensator. However, in the case of finely structured molecular bands, e.g. electron stimulation spectra, use of background correction leads to overcompensation and consequent inaccurate measurements. Therefore, the background correctors, now part of nearly all modern atomic absorption spectrometers, should only be used if a thorough knowledge of the spectral background has been gained. The operational range of the background corrector lies between 190 and 350 nm.

In ferrous metallurgy background correction is mainly used after fusion, when light absorption occurs on salt particles because of high total salt concentrations. By dissolving the fusion substance in inorganic acids, such as HCl, alkali halides develop in high concentration (NaCl, KCl with soda—potassium fusions; LiCl with lithium tetraborate fusions etc). These result in the molecular absorption already described.

Even if the non-specific light losses do not lead to analytical mistakes when balancing the calibration solution and the sample solution, the signal specific to the element can be increased by using the background absorption adjustment and thus the determination limit is lowered.

When analysing solid substances using graphite-tube absorption, signals arising from formation of smoke from the matrix have often been observed. The solid sample method, with steel chips as described in section II.B.1, shows neither smoke or fog phenomena nor other unspecified absorptions.

The direct analysis of oxide products such as iron-sinter, slag or refractories, leads to considerably non-specific loss of light due to the large silicate matrix. It has yet to be investigated whether the use of background correction will make direct oxide analysis possible.

First experiments indicate that a direct determination of trace elements in oxide substances is possible but the determination of matrix elements, because of their high concentrations, should however be left to other methods of analysis, such as solution analysis with AAS or direct X-ray fluorescence spectrometry.

D. Calculations and data handling

The increasing use of advanced computing machinery can be applied to the processing of AAS-data. Some modern atomic absorption spectrometers allow calibration curves to be stored. The adaptation of the equipment parameters to a base calibration curve can then be carried out with just one recalibration solution. This, however, is only valid for the calibration curve of one element in a specific matrix.

The microprocessor can store calibration curve data, compute statistical data such as the mean value, standard deviation and relative standard deviation and compensate for equipment parameters. In addition it is easily possible to couple it to a large computer via an interface. Thus, the functions of the equipment as well as calibration values and analytical values can be controlled on-line and evaluated. Direct coupling has been in use for years for simultaneously registering emission and X-ray fluorescence spectrometers. On-line coupling has proved useful when similar analytical problems with a fixed number of elements to be analysed occur. It leads to fewer errors in operation and to quicker analyses. However, the value of on-line coupling for sequential analytical methods such as polarography or AAS is questionable. Personnel have to be present continuously to operate the equipment (changing lamps, feed samples), so that no economy of effect can be achieved. When using automatic sample feeders, direct coupling to a computer is only feasible if the same elements in the same matrix are always determined.

In ferrous metallurgy analysis, where AAS is used as a special procedure for special analyses, direct coupling is not profitable [58]. To install a spectrometer controlled by a computer, analysis of a far larger number of samples or possible problems would be necessary. The ratio of computer location to frequency of use would be very unfavourable and would increase the cost of the individual analysis. It is more useful to feed the analytical data, issued by the microprocessor and printed by an automatic printer, to the laboratory computer via an off-line system which would then include the data among that from on-line analytical systems.

If the data are stored on floppy-discs, they are always available for further statistical investigations concerned with reproducibility, ring experiments or survey statistics, without taking up storage space.

References pp. 246—249

REFERENCES

1 J. F. Alder and T. S. West, Anal. Chim. Acta, 61 (1972) 132—135.
2 E. Althaus, Perkin-Elmer Analytical Report, 1964.
3 A. D. Ambrose, Proc. Chem. Conf., 27 (1974) 59.
4 E. Ametrano and B. Grassi, Scan., 5 (1974) 35.
5 Analytical Methods Committee, Analyst, 103 (1978) 643—647.
6 M. A. Ashy, J. B. Headridge and A. Sowerbutts, Talanta, 21 (1974) 649—652.
7 American Society for Testing and Materials, 1916 Race St., Philadelphia, PA 19103.
8 American Society for Testing and Materials, Part 32, 1973, pp. 1147—1159. (Proposed Recommended Practices for AAS.)
9 W. B. Barnett, H. L. Kahn and G. E. Peterson, At. Absorpt. Newsl., 10 (1971) 106—110.
10 W. B. Barnett and E. A. McLaughlin, Jr., Anal. Chim. Acta, 80 (1975) 285—296.
11 T. T. Bartels and M. P. Slater, At. Absorpt. Newsl., 9 (1970) 75—77.
12 B. Bernas, Anal. Chem., 40 (1968) 1682—1685.
13 H. Berndt and E. Jackwerth, Spectrochim. Acta, Part B, 30 (1975) 169—177.
14 H. Berndt and E. Jackwerth, At. Absorpt. Newsl., 15 (1976) 109—113.
15 P. L. Bertolaccini, Metallurgica Ital., 61 (1969) 396—403.
16 H. Binding and H. Gawlick, Deutsche Forschungsberichte, 1969 (TIB Hannover).
17 H. Boesch, Mikrochim. Acta, (1976) 49.
18 H. Bosch, E. Büchel, H. Grygiel and K. Lohau, Arch. Eisenhüttenwes., 45 (1974) 699—704.
19 R. S. Braman and M. A. Tompkins, Anal. Chem., 50 (1978) 1088—1093.
20 M. P. Bratzel, Jr. and C. L. Chakrabarti, Anal. Chim. Acta, 61 (1972) 25—32.
21 F. Brivot, I. Cohort, G. Legrand, J. Louvrier and I. A. Voinovitch, Analusis, 2 (1973) 570—576.
22 K. G. Brodie and J. P. Matoušek, Anal. Chem., 43 (1971) 1557—1560.
23 K. G. Brodie, Int. Lab., (1977) 65—74.
24 W. M. G. T. van den Broek and L. de Galan, Anal. Chem., 49 (1977) 2176—2186.
25 W. D. Cobb, W. W. Foster and T. S. Harrison, Lab. Pract., 24 (1975) 143.
26 W. D. Cobb, W. W. Foster and T. S. Harrison, Analyst, 101 (1976) 39—43, 255—259.
27 W. D. Cobb, W. W. Foster and T. S. Harrison, Anal. Chim. Acta, 78 (1975) 293—298.
28 J. Collin, J. Sire and I. A. Voinovitch, Bull. Liaison Lab. Ponts Chaussées, 79 (1975) 78.
29 A. Condylis and B. Mejean, Analusis, 3 (1975) 94—109.
30 E. A. Crecelius, Anal. Chem., 50 (1978) 826—827.
31 M. S. Cresser, Anal. Chim. Acta, 80 (1975) 170—175.
32 M. Damiani, M. G. Del Monte Tamba and F. Bianchi, Analyst, 100 (1975) 643—647.
33 M. B. Denton and H. U. Malmstadt, Anal. Chem., 44 (1972) 241—246.
34 H. A. von Derschau and H. Prugger, Fresenius Z. Anal. Chem., 247 (1969) 8—12.
35 Deutsches Institut für Normung (DIN), P.O. Box 1107, D-1000 Berlin 30.
36 K. Dittrich, E. John and I. Rohde, Anal. Chim. Acta, 94 (1977) 75—81.
37 K. Dittrich and K. Borzym, Anal. Chim. Acta, 94 (1977) 83—90.
38 D. J. Driscoll, D. A. Clay, C. H. Rogers, R. H. Jungers and F. E. Butler, Anal. Chem., 50 (1978) 767—769.
39 Y. Endo and Y. Nakahara, Tetsu To Hagané, 59 (1973) 800—807.
40 Y. Endo and Y. Nakahara, Trans. Iron Steel Inst. Jpn., 16 (1976) 396—403.
41 T. A. Eroshevich and D. F. Makarov, Zavod. Lab., 41 (1975) 186.
42 H. D. Fleming and R. G. Ide, Anal. Chim. Acta, 83 (1976) 67—82.
43 D. E. Fleming and G. A. Taylor, Analyst, 103 (1978) 101—105.
44 A. G. Fogg, S. Soleÿmanloo and D. T. Burns, Talanta, 22 (1975) 541—543.
45 P. Foster, R. Molins and H. Bozon, Analusis, 1 (1972) 434—438.

46 P. Foster and J. Garden, Analusis, 2 (1974) 675.
47 W. Frech, Talanta, 21 (1974) 565—571.
48 W. Frech, Anal. Chim. Acta, 77 (1975) 43—52.
49 W. Frech, Fresenius Z. Anal. Chem., 275 (1975) 353—357.
50 W. Frech and A. Cedergren, Anal. Chim. Acta, 82 (1976) 83—102.
51 T. W. Freudiger and C. T. Kenner, Appl. Spectrosc., 26 (1972) 302—305.
52 L. de Galan and G. F. Samaey, Anal. Chim. Acta, 50 (1970) 39—50.
53 L. de Galan and E. J. Benes, Met. Ital., 64 (1972) 343—348.
54 C. A. Gomez and L. M. T. Dorado, Rev. Metal. (Madrid), 10 (1974) 355.
55 R. J. Guest and D. R. McPherson, Anal. Chim. Acta, 71 (1974) 233—253.
56 R. K. Hansen and R. H. Hall, Anal. Chim. Acta, 92 (1977) 307—320.
57 J. B. Headridge and A. Sowerbutts, Analyst, 98 (1973) 57—64.
58 W. Heinemann and W. Prinz, Fresenius Z. Anal. Chem., 289 (1978) 17—23.
59 R. Herrmann, Z. Instrum., 75 (1967) 101—111.
60 R. Höhn and F. Umland, Fresenius Z. Anal. Chem., 258 (1972) 100—106.
61 H. -E. Hoffmann, H. Nathansen and F. Altmann, Jenaer Rundsch., 6 (1976) 302—304.
62 M. E. Hofton, British Steel Corp., Report 1976 (GS/TECH/558/2/76/C).
63 S. T. Holding and P. H. D. Matthews, Analyst, 97 (1972) 189—194.
64 S. T. Holding and J. J. Rowson, Analyst, 100 (1975) 465—470.
65 I. D. Ingle, Jr., Anal. Chem., 46 (1974) 2161—2171.
66 IP Standards for Petroleum and its Products, Part I: Methods for Analysis and Testing, Sect. 2, IP Method 308 (1975), Applied Science Publishers, Barking, Gt. Britain.
67 International Organization for Standardisation (ISO), TC 102/SC 2 (Iron Ore/Chemical Analysis).
68 K. W. Jackson, K. M. Aldous and D. G. Mitchell, Appl. Spectrosc., 28 (1974) 569—573.
69 E. Jantzen, Schmiertech. Tribol., 22 (1975) 31—37.
70 A. H. Jones and W. D. France, Jr., Anal. Chem., 44 (1972) 1884—1886.
71 S. H. Kägler, Erdoel/Kohle, 27 (1974) 514—517.
72 S. H. Kägler, Erdoel/Kohle, 28 (1975) 232—237.
73 S. H. Kägler, Erdoel/Kohle, 29 (1976) 362.
74 S. H. Kägler, Schmiertech. Tribol., 25 (1978) 84—88.
75 H. L. Kahn and S. Slavin, At. Absorpt. Newsl., 10 (1971) 125.
76 H. L. Kahn, G. E. Peterson and J. E. Schallis, Perkin-Elmer Analytical Report No. 17 (1969).
77 I. P. Kharlamov and G. V. Eremina, Zavod. Lab., 40 (1974) 385—391.
78 G. F. Kirkbright, M. Marshall and T. S. West, Anal. Chem., 44 (1972) 2379—2382.
79 G. F. Kirkbright and P. J. Wilson, Anal. Chem., 46 (1974) 1414—1418.
80 H. E. Knauer and G. E. Millman, Anal. Chem., 47 (1975) 1263—1268.
81 K. H. Koch and K. Ohls, Arch. Eisenhüttenwes., 39 (1968) 925—928.
82 P. König, K. H. Schmitz and E. Thiemann, Arch. Eisenhüttenwes., 40 (1969) 553—556.
83 J. Kožušnikova and A. Kolářová, Hutn. Listy, 32 (1977) 810—812.
84 J. Komárek, J. Jambor and L. Sommer, Fresenius Z. Anal. Chem., 262 (1972) 91—94.
85 G. R. Kornblum and L. de Galan, Spectrochim. Acta, Part B, 28 (1973) 139—147.
86 J. J. Labrecque, J. Galobardes and M. E. Cohen, Appl. Spectrosc., 31 (1977) 207—210.
87 I. Lang, G. Šebor, V. Sychra, D. Kolihová and O. Weisser, Anal. Chim. Acta, 84 (1976) 299—305.
88 I. Lang, G. Šebor, O. Weisser and V. Sychra, Anal. Chim. Acta, 88 (1977) 313—318.
89 L. L. Lewis, Anal. Chem., 40 (1968) 28A—47A.
90 D. Littlejohn and J. M. Ottaway, Analyst, 103 (1978) 595—606.

91 D. Littlejohn and J. M. Ottaway, Anal. Chim. Acta, 98 (1978) 279—290.
92 J. C. van Loon, B. Radziuk, N. Kahn, J. Lichwa, F. J. Fernandez and J. D. Kerber, At. Absorpt. Newsl., 16 (1977) 79—83.
93 R. P. Lucas and B. C. Ruprecht, Anal. Chem., 43 (1971) 1013—1016.
94 H. -M. Lüschow, D. Lindenberger and G. Kraft, Fresenius Z. Anal. Chem., 279 (1976) 347—349.
95 R. J. Lukasiewicz, P. H. Berens and B. E. Buell, Anal. Chem., 47 (1975) 1045—1049.
96 R. J. Lukasiewicz and B. E. Buell, Appl. Spectrosc., 31 (1977) 541—547.
97 O. Luzar and V. Sliva, Hutn. Listy, 30 (1975) 55.
98 D. C. Manning, At. Absorpt. Newsl., 10 (1971) 123—124.
99 D. C. Manning, At. Absorpt. Newsl., 14 (1975) 99—102.
100 T. Maruta, M. Suzuki and T. Takeuchi, Anal. Chim. Acta, 51 (1970) 381—385.
101 H. Massmann, Fresenius Z. Anal. Chem., 225 (1967) 203—213.
102 H. Massmann, Spectrochim. Acta, Part B, 23 (1968) 215—226.
103 H. Massmann, Angew. Chem., 86 (1974) 542—552.
104 G. L. McPherson, At. Absorpt. Newsl., 4 (1965) 180—191.
105 J. Musil, Hutn. Listy, 30 (1975) 292.
106 J. Musil and J. Doležal, Anal. Chim. Acta, 92 (1977) 301—305.
107 G. Nonnenmacher and Fr. -H. Schleser, Fresenius Z. Anal. Chem., 209 (1965) 284—293.
108 N. Oddo, Met. Ital., 64 (1972) 359—362.
109 W. Ooghe and F. Verbeek, Anal. Chim. Acta, 73 (1974) 87—95.
110 W. Ooghe and F. Verbeek, Anal. Chim. Acta, 79 (1975) 285—291.
111 B. D. Pederson and R. W. Taylor, Electr. Furnace Conf. Proc. 1974, 32 (1975) 168.
112 G. E. Peterson and H. L. Kahn, At. Absorpt. Newsl., 9 (1970) 71—74.
113 G. E. Peterson and J. D. Kerber, At. Absorpt. Newsl., 15 (1976) 134—143.
114 A. A. Petrov and G. V. Skvortsova, Zh. Prikl. Spektrosk., 22 (1975) 991.
115 W. J. Price and P. J. Whiteside, Analyst, 103 (1978) 643—647.
116 D. B. Ratcliffe, C. S. Byford and P. B. Osman, Anal. Chim. Acta, 75 (1975) 457—459.
117 R. D. Reeves, B. M. Patel, C. J. Molnar and J. D. Winefordner, Anal. Chem., 45 (1973) 246—249.
118 W. K. Robins and H. H. Walker, Anal. Chem., 47 (1975) 1269—1275.
119 R. C. Rooney and C. G. Pratt, Analyst, 97 (1972) 400—404.
120 C. J. Rowe and M. W. Routh, Res. Develop., (Nov. 1977) 24—30.
121 C. S. Saba and K. J. Eisentrant, Anal. Chem., 49 (1977) 454—457.
122 Z. Sámsoni, Mikrochim. Acta, Suppl. II, (1978) 177—190.
123 A. Sato and T. Ito, Tetsu To Hagané, 56 (1970) 144.
124 K. -H. Sauer and M. Nitsche, Arch. Eisenhüttenwes., 40 (1969) 891—893.
125 W. Schmidt, F. Dietl, G. Schadow and D. Ade, Motortechn. Z., 37 (1976) 307—310.
126 P. H. Scholes, Proc. Soc. Anal. Chem., 5 (1968) 114—115.
127 P. H. Scholes, Proc. BISRA Conf. 1968 (MG/D/Conf. Proc./688/68).
128 P. H. Scholes, British Steel Corp., Report 1972 (MG/CC/588/72).
129 P. H. Scholes, British Steel Corp., Report 1973 (CAC/Conf. Proc./115/73).
130 E. Sebastiani, K. Ohls and G. Riemer, Fresenius Z. Anal. Chem., 264 (1973) 105—109.
131 I. L. Shresta and T. S. West, Bull. Soc. Chim. Belg., 84 (1975) 549.
132 R. K. Skogerboe and A. P. Bejmuk, Anal. Chim. Acta, 94 (1977) 297—305.
133 S. Skujins, Techtron Appl. Notes, No. 4, 1970.
134 W. Slavin, Appl. Spectrosc., 20 (1966) 281—288.
135 W. Slavin and S. Slavin, Appl. Spectrosc., 23 (1969) 421—433.
136 B. G. Stephens and H. L. Felkel, Jr., Anal. Chem., 47 (1975) 1676.
137 K. A. Stewart, Proc. Soc. Anal. Chem., 5 (1968) 116—117.
138 R. E. Sturgeon, Anal. Chem., 49 (1977) 1255A—1267A.
139 D. H. Svedung, Jernkontorets. Ann., 155 (1971) 295—297.

140 D. R. Thomerson and W. J. Price, Analyst, 96 (1971) 825—834.
141 W. Thomich, Arch. Eisenhüttenwes., 42 (1971) 779—781.
142 K. C. Thompson and R. G. Godden, Analyst, 101 (1976) 96—102.
143 M. Tomljanovic and Z. Grobenski, At. Absorpt. Newsl., 14 (1975) 52.
144 W. -Ch. Tsai and L. -J. Shiau, Anal. Chem., 49 (1977) 1641—1644.
145 K. -J. Tsunoda, K. Fujiwara and K. Fuwa, Anal. Chem., 49 (1977) 2035—2039.
146 G. L. Vassilaros, Talanta, 21 (1974) 803—808.
147 Verein Deutscher Eisenhüttenleute, Handbuch für das Eisenhüttenlaboratorium, Vol. 5, Verlag Stahleisen, Düsseldorf, 1977.
148 A. Walsh, Appl. Spectrosc., 27 (1973) 335—341.
149 H. C. Wagenaar and L. de Galan, Spectrochim. Acta, Part B, 28 (1973) 175—177.
150 R. J. Watling, Anal. Chim. Acta, 94 (1977) 181—186.
151 B. Welz, CZ Chem. Tech., 1 (1972) 373—380 and 455—460.
152 R. K. Winge, V. A. Fassel and R. N. Kniseley, Appl. Spectrosc., 25 (1971) 636—641.
153 R. Whitman, Analyst, 100 (1975) 555—562.
154 J. D. Winefordner, J. J. Fitzgerald and N. Omenetto, Appl. Spectrosc., 29 (1975) 369—383.
155 R. Woodriff, M. Marinkovic, R. A. Howald and I. Eliezer, Anal. Chem., 49 (1977) 2008—2012.

Chapter 4f

The analysis of non-ferrous metals by atomic absorption spectrometry

F. J. BANO
*Linksfield Laboratories Ltd., Chalon Way, St. Helens, Merseyside
(Gt. Britain)*

I. INTRODUCTION

Atomic absorption spectrometry has been applied to the analysis of over sixty elements. The technique combines speed, simplicity and versatility and has been applied to a very wide range of non-ferrous metal analyses. This review presents a cross section of applications. For the majority of applications flame atomisation is employed but where sensitivity is inadequate using direct aspiration of the sample solution a number of methods using a preconcentration stage have been described. Non-flame atomisation methods have been extensively applied to the analysis of ultra-trace levels of impurities in non-ferrous metals. The application of electrothermal atomisation, particularly to nickel-based alloys has enabled the determination of sub-part per million levels of impurities to be carried out in a fraction of the time required for the chemical separation and flame atomisation techniques.

II. DETERMINATION OF IMPURITIES IN ALLOYS BASED ON VARIOUS NON-FERROUS METALS

A. Aluminium

The analysis of aluminium alloys by atomic absorption was one of the earliest applications of the technique to non-ferrous metal analysis. Wallace [1] has described the determination of magnesium in aluminium alloys using an air—propane flame. Because of the strong interference of aluminium on magnesium in low temperature flames, 8-hydroxyquinoline was used as a releasing agent. The use of the nitrous oxide—acetylene flame by Wilson [2] eliminated the interference of aluminium on magnesium.

Aluminium alloys may be dissolved using the following procedure. Weigh 1.000 g of sample into a PTFE beaker in a cooling bath. Add 20 ml of concentrated hydrochloric acid dropwise, allowing the reaction to subside between additions. (Note that the reaction is vigorous and the beaker should be cooled during the addition of acid.) Add 4 ml hydrofluoric acid and allow to stand for ten minutes, taking care that the temperature does not exceed 50—60°C. Add 4 ml of nitric acid dropwise, again taking care to cool the

References p. 259

TABLE 1

STANDARD CONDITIONS FOR THE ANALYSIS OF ALUMINIUM ALLOYS

Analyte	Wavelength (nm)	Flame	Notes
Aluminium	237.3	nitrous oxide/acetylene	Aluminium is ionised in the nitrous oxide/acetylene flame; add 2000 $mg\, l^{-1}$ potassium to the sample solution
Magnesium	285.2	nitrous oxide/acetylene	
Copper	324.7	air/acetylene	
Iron	372.0	nitrous oxide/acetylene	
Tin	224.6	nitrous oxide/acetylene	
Zinc	213.9	air/acetylene	At the 213.9 nm wavelength non-atomic species absorb strongly; background correction is therefore necessary
Silicon	251.6	nitrous oxide/acetylene	At the 217.0 nm wavelength non-atomic species absorb strongly; background correction is therefore necessary

mixture during the reaction. Cool and add 0.5 g A.R. boric acid to complex the excess fluorides. Transfer to a 100 ml volumetric flask and make up to volume with distilled water. Standard conditions for the analysis of aluminium alloys are shown in Table 1.

Table 2 shows results obtained by Campbell [3] for the determination of silicon in aluminium alloys. In this procedure the sample was dissolved in 20% sodium hydroxide/hydrogen peroxide. After dissolution the solution was treated with a mixture of nitric and hydrochloric acids and diluted to volume.

B. Cobalt

Welcher et al. [4] have described the analysis of cobalt-based alloys. Cobalt alloys may be dissolved according to the following procedure. Weigh 0.5000 g of sample into a PTFE beaker. Add 2 ml of water, 10 ml of concentrated hydrochloric acid and 5 ml of concentrated nitric acid. Warm the mixture for approximately thirty minutes. Add 2 ml of concentrated nitric acid and 5 ml of concentrated hydrochloric acid and reheat for a further thirty minutes. Cool, add 5 ml of hydrofluoric acid and fume to dryness. Add 5 ml of concentrated nitric acid and heat for ten minutes. Repeat with a further 5 ml of nitric acid. Add 10 ml of hydrochloric acid and reheat to boiling, cool and dilute to 100 ml. Standard conditions for the analysis of cobalt alloys are shown in Table 3.

TABLE 2

DETERMINATION OF SILICON IN ALUMINIUM ALLOYS

Sample	A.A.	Chemical	% Relative error
1	12.93	13.10	1.3
2	13.86	13.79	0.2
3	13.20	13.40	1.5
4	18.35	18.28	0.4
5	14.12	14.05	0.5
6	11.15	11.07	0.7
7	11.55	11.64	0.8
8	13.23	13.41	1.3
9	10.99	11.07	0.7
10	11.66	11.64	0.2
11	12.30	12.18	1.0
12	17.59	17.50	0.5
13	17.47	17.55	0.2
14	17.67	17.80	0.7

TABLE 3

INSTRUMENTAL PARAMETERS FOR THE ANALYSIS OF COBALT-BASED ALLOYS

Analyte	Wavelength (nm)	Flame	Notes
Cobalt	304.4	nitrous oxide/acetylene	High levels of chromium can lead to interference in the presence of hydrochloric acid; the use of the nitrous oxide/acetylene flame avoids this interference.
Chromium	520.8	nitrous oxide/acetylene	
Iron	386.0	nitrous oxide/acetylene	The use of the nitrous oxide/acetylene flame is recommended in order to eliminate interference

C. Copper

The analysis of copper-based alloys by atomic absorption has been described by Johns and Price [5].

For the sample preparation, weigh 0.5000 g of sample into a 250 ml beaker and add 20 ml of a mixed acid solution containing 250 ml of concentrated hydrochloric acid, 250 ml of concentrated nitric acid and 500 ml of water per litre of solution. Warm gently until dissolution is complete, cool and transfer the solution to a 100 ml volumetric flask. Make up to the mark.

Instrumental conditions for the analysis of copper alloys are shown in Table 4.

TABLE 4

INSTRUMENTAL PARAMETERS FOR THE ANALYSIS OF COPPER-BASED ALLOYS

Analyte	Wavelength (nm)	Flame	Notes
Aluminium	209.3	nitrous oxide/acetylene	
Antimony	217.6	air/acetylene	
Arsenic	193.7	air/acetylene	
Copper	327.4	air/acetylene	For the determination of major alloying elements the use of Class A volumetric glassware is recommended
Iron	372.0	air/acetylene	
Lead	283.3	air/acetylene	
Manganese	403.1	air/acetylene	
Nickel	352.5	air/acetylene	
Silicon	251.6	nitrous oxide/acetylene	
Tin	286.3	nitrous oxide/acetylene	
Zinc	213.9	air/acetylene	

Results obtained for four samples of BCS copper alloys are shown in Table 5. All the results indicate good agreement with certified values. The results obtained for copper indicate that in favourable cases it is possible to obtain satisfactory results for major alloying components. The excellent results which can be obtained for the analysis of large percentage amounts of copper is further confirmed by the work of Sattur [6], whose results for five copper-based alloys are shown in Table 6.

D. Lead

A major problem in the analysis of lead-based alloys is the selection of a suitable solvent mixture. Price [7] has described a method based on hydrobromic acid/bromine. The disadvantage of this method is that the solvent mixture is unpleasant to handle and precipitation often occurs on dilution. Hwang [8] has described a method for the analysis of trace and minor elements in lead/tin solders using a solvent mixture of fluoboric acid and nitric acid.

For fluoboric acid preparation, to 200 ml 40% w/v hydrofluoric acid at 10°C, add, in small quantities, 75 g boric acid. Allow to dissolve and store in a polythene bottle. This reagent should be prepared freshly each day.

Weigh 1.000 g of finely divided sample into a 250 ml beaker and cover with a watch glass. Add 40 ml of water, 6 ml of fluoboric acid and 12 ml of concentrated nitric acid. Warm the solution gently. Agitating the solution using a magnetic stirrer will aid dissolution. Cool to room temperature. Add 5 ml of 1% tartaric acid solution. Dilute to 100 ml with water.

TABLE 5
RESULTS FOR FOUR SAMPLES OF BCS COPPER ALLOYS

BCS No.	Aluminium		Antimony		Arsenic		Copper		Iron		Lead		Manganese		Nickel		Silicon		Tin		Zinc	
	Cert.	Obt.	Cert.	Obt.	Cert.	Obt.	Cert.	Obt.	Cert.	Obt.	Cert.	Obt.	Cert.	Obt.	Cert.	Obt.	Cert.	Obt.	Cert.	Obt.	Cert.	Obt.
179/2	2.22	2.20					58.5	58.7	1.02	1.02	0.35	0.35	0.86	0.87	0.56	0.56	0.044	0.04	0.70	0.71	35.8	35.3
		2.15						59.0		1.01		0.36		0.87		0.56		0.04		0.72		35.7
183/2			0.24	0.24	0.14	0.16	85.0	85.2			3.35	3.40			0.51	0.51			5.03	5.00	5.16	5.14
				0.25		0.15		34.7				3.40				0.52				5.02		5.12
304	8.92	8.78					80.4	80.2	4.71	4.65			0.51	0.495	4.75	4.75	0.19	0.185			0.60	0.607
		8.82						80.1		4.68				0.500		4.70		0.19				0.604
364			0.18	0.18			80.6	81.0			9.25	9.30			0.28	0.28			9.35	9.31	0.13	0.128
				0.18				80.5				9.35				0.28				9.22		0.126

References p. 259

TABLE 6

RESULTS OBTAINED [6] FOR THE ANALYSIS OF FIVE COPPER-BASED ALLOYS

Sample	Type	A.A.	% Cu Cert. Value
NBS37d	Brass	70.6	70.78
63c	Bronze	80.7	80.48
124d	Bronze	83.6	83.6
157	Cu–Ni–Zn(nickel silver)	71.3	72.14
164a	Aluminium bronze	82.3	82.25

TABLE 7

INSTRUMENTAL PARAMETERS FOR THE ANALYSIS OF LEAD-BASED ALLOYS

Analyte	Wavelength (nm)	Flame
Aluminium	309.3	nitrous oxide/acetylene
Antimony	217.6	air/acetylene
Arsenic	193.7	air/acetylene
Bismuth	223.1	air/acetylene
Cadmium	228.8	air/acetylene
Copper	324.8	air/acetylene
Gold	242.8	air/acetylene
Iron	248.3	air/acetylene
Lead	283.3	air/acetylene
Nickel	232.0	air/acetylene
Silver	328.1	air/acetylene
Tin	224.6	nitrous oxide/acetylene
Zinc	213.9	air/acetylene

Instrumental parameters for the analysis of lead-based alloys are shown in Table 7 and results obtained by Hwang [8] for the analysis of impurities in lead solder are shown in Table 8.

E. Nickel

Welcher and Kriege [9] have described a procedure for the analysis of the major alloying elements in nickel alloys.

Prepare the sample by dissolving 1.000 g of turnings in a mixture of 10 ml of hydrochloric acid and 10 ml of nitric acid. Heat gently until the cessation of nitrogen dioxide fumes. Transfer to a PTFE beaker and add 5 ml of hydrofluoric acid dropwise. Ensure that the solution temperature does not exceed 30°C. Transfer the solution to a plastic 100 ml volumetric flask and make up to volume with water.

TABLE 8

RESULTS OBTAINED [8] FOR THE ANALYSIS OF IMPURITIES IN LEAD SOLDER

Element	Reported value (%)	Found (%)	Element	Reported value (%)	Found (%)
Ag	0.019	0.019	Cu	0.05	0.05
	0.036	0.035		0.10	0.11
Al	0.05	0.05	Fe	0.006	0.004
	0.07	0.03		0.016	0.014
As	0.019	0.018	Ni	0.0025	0.0025
	0.031	0.032		0.033	0.032
Au	0.010	0.010	Sb	0.12	0.11
	0.040	0.038		0.30	0.31
Bi	0.038	0.039	Zn	0.005	0.0008
	0.089	0.094		0.0013	0.0020
Cd	0.0057	0.0059			
	0.010	0.010			

Instrumental parameters for the analysis of nickel alloys are shown in Table 9.

F. Zinc

To prepare the sample, dissolve 1.000 g of the sample in 10 ml of hydrochloric acid. Oxidise with 2 ml of 100 vol. (30% w/v) hydrogen peroxide. Boil to destroy excess peroxide. Cool, transfer to a 100 ml volumetric flask and make up to volume with distilled water.

Instrumental parameters for the analysis of zinc alloys as shown in Table 10.

III. ELECTROTHERMAL TECHNIQUES

The methods already described have illustrated the wide applicability of flame atomisation techniques to the analysis of non-ferrous alloys. The introduction of electrothermal atomisation has enabled the direct determination of sub-part per million levels of impurities. The presence of very low levels of lead, bismuth and other low melting point metals is known to have a deleterious effect on the metallurgical properties of nickel alloys.

Welcher et al. [10] have described the direct determination of trace quantities of lead, bismuth, selenium, tellurium and thallium in high-temperature nickel alloys using electrothermal atomisation.

References p. 259

TABLE 9

INSTRUMENTAL PARAMETERS FOR THE ANALYSIS OF NICKEL-BASED ALLOYS

Analyte	Wavelength (nm)	Flame	Notes
Chromium	520.8	nitrous oxide/acetylene	Chromium is subject to interference from nickel. It is advisable to adjust the fuel flow until the absorbance obtained from a 1500 μg ml^{-1} chromium solution and a 1500 μg ml^{-1} chromium plus nickel at the sample level produce the same absorbance
Bismuth	223.1	air/acetylene	Sample preparation. Dissolve 2.000 g of sample in 12 ml of hydrochloric acid, 6 ml of nitric acid and 20 ml of water in a PTFE beaker. Evaporate the solution to 10—12 ml, cool, add 1 ml of hydrofluoric acid dropwise and boil for five minutes. Cool, add 5 ml of 1% boric acid solution and dilute to 50 ml. Background correction using a UV continuum source is advisable for bismuth
Iron	386.0	nitrous oxide/acetylene	Nitrous oxide/acetylene flame is recommended to avoid interference
Cobalt	304.4	nitrous oxide/acetylene	
Lead	217.0	air/acetylene	Sample preparation as for bismuth. Background correction recommended
Silicon	251.6	nitrous oxide/acetylene	

The samples were dissolved by the following procedure. Dissolve 1.000 g of metal chips in 30 ml of a 1:1:1 mixture of nitric acid, hydrofluoric acid and water in a PTFE beaker. When dissolution is complete, reduce the volume to approximately 5 ml by evaporation. Cool, add about 20 ml of water and heat to dissolve all salts. Cool, transfer to a 50 ml plastic volumetric flask and dilute to the mark.

Prepare standards by making standard additions to previously characterised samples. Operating parameters and detection limits are shown in Table 11. The results obtained by Welcher indicate that electrothermal atomisation is capable of yielding a very rapid means of analysing trace impurities in non-ferrous alloys. In many cases lengthy separations are not necessary and the technique is clearly applicable to a wide range of alloys.

TABLE 10

INSTRUMENTAL PARAMETERS FOR THE ANALYSIS OF ZINC-BASED ALLOYS

Analyte	Wavelength (nm)	Flame	Notes
Aluminium	309.3	nitrous oxide/ acetylene	Add 2000 μg ml^{-1} potassium to sample solutions to suppress ionisation. Use a standard additions procedure to avoid interference from the zinc matrix
Magnesium	285.2	nitrous oxide/ acetylene	
Iron	248.3	air/acetylene	
Copper	327.4	nitrous oxide/ acetylene	The nitrous oxide/acetylene flame is used to minimise the effect of the high zinc/copper ratio
Cadmium	228.8	air/acetylene	
Lead	217.0	air/acetylene	Background correction should be used

TABLE 11

OPERATING PARAMETERS AND DETECTION LIMITS FOR THE ELECTROTHERMAL ATOMISATION TECHNIQUE OF WELCHER ET AL. [10]

Element	Lead	Bismuth	Tellurium	Selenium	Thallium
Wavelength (nm)	283.3	223.1	214.3	196.0	276.9
Spectral band width (nm)	0.7	0.2	0.7	0.7	0.7
Solution volume (μl)	50	20	50	50	50
Sample concentration (mg ml^{-1})	20	20	20	20	4
Drying temperature (°C)	150	150	150	600	150
Drying time (s)	20	20	20	20	20
Char temperature (°C)	400	800	600	1000	500
Char time (s)	60	45	60	45	60
Atomisation temperature (°C)	2000	2200	2200	2400	2000
Atomisation time (s)	5	5	5	5	5
Detection limit in sample (ppm)	0.1	0.1	0.2	0.1	0.1

REFERENCES

1 F. J. Wallace, Analyst, 88 (1963) 259.
2 L. Wilson, Anal. Chim. Acta, 40 (1969) 503.
3 D. E. Campbell, Anal. Chim. Acta, 46 (1969) 31.
4 G. G. Welcher and O. H. Kriege, At. Absorpt. Newsl., 9 (1970) 61.
5 P. Johns and W. J. Price, Metallurgia, 81 (1970) 75.
6 T. W. Sattur, At. Absorpt. Newsl., 5 (1966) 37.
7 W. J. Price, Analytical Atomic Absorption Spectrometry, Heyden, London, 1972.
8 J. Y. Hwang and L. M. Sandonato, Anal. Chem., 42 (1970) 744.
9 G. G. Welcher and O. H. Kriege, At. Absorpt. Newsl., 8 (1969) 97.
10 G. G. Welcher, O. H. Kriege and J. Y. Marks, Anal. Chem., 46 (1974) 1227.

Chapter 4g

Atomic absorption methods in applied geochemistry

MICHAEL THOMPSON and SHIRLEY J. WOOD
Applied Geochemistry Research Group, Department of Geology, Imperial College, London SW7 (Gt. Britain)

I. INTRODUCTION

A. Analytical requirements in applied geochemistry

Applied geochemistry consists essentially of two aspects, namely exploration for mineral reserves and environmental studies. Both aspects employ the same basic technique, which is the execution and interpretation of surveys of the concentration of various elements in samples taken at or near the surface of the earth. The primary purpose is to identify geochemical 'anomalies', that is, areas which have unusually high concentrations of one or more elements. Such surveys may cover many thousands of square kilometres and produce comparably large numbers of samples. As a second stage the anomalies found are investigated in detail, with much greater sampling density. The purpose of this is to determine whether the anomaly is due to mineralisation, to naturally-occurring concentration processes, metalliferous but unmineralised rock strata, or to pollution. Sampling media most frequently used are (in decreasing order) soil, stream and lake sediment, rock, water and herbage.

Geochemical surveys impose upon the analytical chemist certain special constraints [1]. The over-riding factor is cost-effectiveness, as can be gauged from the fact that for about 95% of all analyses (in this field) the sole outcome is to eliminate localities from further consideration. Large numbers of samples have to be analysed with an optimally low cost and a fast turn-around time, sometimes at a temporary camp remote from normal services. A reasonable standard of accuracy and precision must be consistently obtained. The precision required depends on the contrast, which is the ratio of the anomalous concentration to the background level, and also on the sampling precision which is often the limiting factor. As a general rule a coefficient of variation of 10% is completely acceptable. Even higher relative variations are often usable, and have sometimes to be tolerated, as when the only available analytical method is being used near the detection limit.

B. Atomic absorption spectrometry in applied geochemistry

Most of this analytical requirement is met by atomic absorption spectrometry (AAS) methods: about 70 percent of all geochemical samples are

References p. 284

analysed by conventional flame techniques, usually for several elements. Metals most frequently determined by AAS in the exploration industry are copper, nickel, lead and zinc in soils and sediments. Iron and manganese are also frequently determined to assist interpretation, and in all about ten elements are regularly required. Several important elements are determined by methods other than AAS, notably tin, molybdenum, tungsten and arsenic. This is because flame AAS cannot provide the low detection limits required without a costly pre-concentration stage. In water and herbage samples, the concentration of the analytes may be lower by between 1 and 3 orders of magnitude and concentration procedures may have to be employed.

In a typical exploration laboratory the average output from an AAS instrument is 800—1000 determinations in an eight hour period. To meet this capability and minimise cost special sample preparation methods are adopted. The main emphasis is on simplifying the analytical procedure so that large batches of samples can be handled by a single person, typically with 50—300 samples per batch. As an example of the efficiency obtainable, in the AGRG a two-person team analysed stream sediments for zinc and cadmium at an average rate of 300 samples per working day over a ten month period. They carried out every aspect of the task themselves including weighing, sample digestion, dilution, instrumental analysis, washing up, and elementary data quality control.

This rate of working is achieved by using, as far as possible, a number of special techniques, including: (i) carrying out the whole procedure in a single vessel, which avoids time consuming processes such as quantitative liquid transfer and filtration; (ii) the use of accurate liquid dispensers to provide a known final volume, rather than volumetric flasks; (iii) use of a fixed sample weight so that a single dilution factor applies to a whole batch; (iv) use of partial sample digestion procedures which may extract only about 90% of the analyte from the sample; (v) use of optimised glassware cleaning procedures; and (vi) use of a rapid automatic diluter for out-of-range samples. Space is conserved by using as far as possible test tubes as digestion vessels rather than the wider beakers or flasks. Test tubes, in addition, provide a maximum depth of liquid, which facilitates nebulisation for several elements from a small volume, and they can be conveniently heated in large numbers in aluminium blocks or air baths.

Flameless AAS methods have so far made little impact in applied geochemistry, except for methods for the element mercury, for which cold-vapour AAS is uniquely effective [2]. A plethora of methods are now in use ranging from simple attachments to standard instruments, to fully portable specific mercury meters of high sensitivity incorporating ingenious methods of background correction [3].

Graphite furnace atomisers are only used in special cases, i.e. when the analyte concentration is very low, (a) because the sampling rate is typically about 5—10 times slower than flame methods, (b) because interference effects tend to be severe and (c) because more skill is required for manual operation. Interference effects are especially important in applied geochemistry because chemical separation methods are usually too time consuming to be cost-effective. A graphite furnace method for gold is used in some laboratories, but generally flame methods requiring large samples with solvent extraction/concentration are preferred. Perhaps the most obvious application of the graphite furnace is in the analysis of fresh water samples, where heavy-metal concentrations are typically too low for direct determination by flame methods (e.g. about $10\,\mu g\,l^{-1}$ for Cu and Pb). Even in this field, however, concentration procedures by ion-exchange resins or solvent extraction are preferred, especially if a number of elements are to be determined. In such cases the extra time of the solvent extraction step is offset by the quicker throughput and greater reliability of the flame methods.

At the present time the technique of forming the volatile hydrides of certain elements (Ge, Sn, As, Sb, Bi, Se and Te), as a method of separation and rapid introduction of these elements into an atomiser (flame or hot tube), has had little impact in applied geochemistry. A few applications have been reported but are not yet widely used despite the very low detection limits which are obtainable. The main problems with the method are an abundance of interference effects, mainly from transition elements, and short linear calibration ranges. However Bedard and Kerbyson [4, 5] have shown that it is possible to separate in advance traces of As, Sb, Bi, Se and Te from pure copper, (the most serious interferer) by co-precipitating the elements on lanthanum hydroxide. It has further been shown that this precipitation method is applicable to the majority of interfering elements, and can be adapted to provide a rapid large batch method suitable for geochemical analysis of soil and sediment [6].

There is great interest in the volatile hydride forming elements as mineral 'pathfinders', as pollutants and, for selenium at least, as an essential dietary element. There are no existing alternative methods for the elements which are both accurate and cost-effective and there is little doubt that hydride methods will be used extensively in the near future.

C. Instrumental requirements

Almost any standard AAS instrument can be used for basic work in applied geochemistry, but there are a number of features which can improve the scope and speed of analysis.

Semi-automatic features. Facilities such as autozero, automatic linearization of the calibration curve, digital readout in concentration units, and a printer are all highly desirable, and jointly they greatly improve the cost-effectiveness of analysis, because the cost of the operator and the instrument can be distributed among a disproportionately larger number of samples. However, it has yet to be demonstrated that, over a long period, an automatic sample introduction system is as quick or reliable as a motivated human operator in flame AAS. Fully automatic systems are rarely

References p. 284

encountered in applied geochemical laboratories at the time of writing.

Stability. Base line stability is an important feature in AAS as in conjunction with sensitivity, it determines the detection limits which can be obtained. This is important where elements are present at concentrations close to the normal practical detection limits, e.g. Pb and Cd in soils and sediments, and where a small improvement would widen the applicability. For instance a 5-fold decrease in noise (or increase in sensitivity) would enable Mo to be determined in normal soils and sediments by a flame method. Long term stability is also important as it cuts down the frequency with which the instrument needs to be recalibrated.

Background interference. Correction for non-specific absorption is sometimes essential in geochemical work (section I.D.) and is most conveniently carried out by means of a continuum source. However, not all systems available are completely effective, and this feature should be checked before purchase of an instrument. It is essential that the beams from the hollow-cathode lamp and the continuum lamp traverse exactly equivalent paths through the flame, and can be accurately balanced for energy.

D. Interference effects

In the context of geochemical analysis, i.e. the determination of traces of heavy metals in the presence of large concentrations of major constituents (Na, K, Ca, Mg, Fe, and Al), numerous instances of interference can be found. These interferences have either to be rendered insignificant directly (e.g. by the addition of a further reagent to the sample solution, or instrumentally), or ignored. If they are too large to be ignored alternative analytical methods are used. Chemical separation techniques are generally inadmissible on cost grounds.

Two distinct types of interference can be recognised, namely background (translational) effects and enhancement/suppression (rotational) effects, sometimes occurring together in the same interference/analyte combination. Background effects can be overcome (with varying degrees of success depending on the instrument and the elements) by the use of continuum lamp background correction, and this method is widely used in geochemical analysis especially in the determination of Cd, Co, Ni and Pb in the presence of calcium [7]. If a continuum background corrector is not available, adequate compensation can be made by subtraction of a correction based on the concentration of calcium in the sample solution, which is determined separately. Rotational effects can, in principle, be overcome by the method of standard additions, but this would not be practicable in applied geochemistry because of the additional cost of this labour-intensive technique. Consequently rotational effects are largely ignored.

In a recent comprehensive survey [8] most of the important interference effects relevant to the analysis of stream sediments by flame AAS have been identified. A large number of significant effects were detected, but only those effects judged likely to adversely affect interpretation of the data were distinguished as important. Important translational (background) effects were those of calcium on Cd, Co, Ni and Pb, and aluminium on Pb.

Rotational effects were found to be important only in the instances of calcium on Li and aluminium on Co, Li and Ni. However, it was found that all of these effects could be adequately corrected by a linear function of the interferent concentration.

E. Sample attack methods

Applied geochemistry seldom requires the complete chemical attack of the sample. In soil and sediment, for example, the trace elements incorporated in the crystal lattice of quartz particles are usually of little interest. Consequently sample attacks which liberate trace elements only from less resistant minerals such as clays or hydrolysates (Fe and Mn oxides) are completely satisfactory. Sometimes methods designed to attack a specific constituent are used: examples are cold reducing buffers for extracting trace metals from precipitated manganese(IV) oxide, or cold ammonium acetate to extract metal ions bound to ion-exchange sites on clay minerals. Results of such methods need to be interpreted with caution. At the other extreme are methods designed to bring into solution metals which are present in discrete resistant minerals, such as cassiterite, which are not amenable to attack with mineral acids.

Fusion methods are avoided if at all possible in applied geochemistry, especially in atomic absorption work. Fusions are not convenient for rapid large batch methods because of the manipulative difficulties. The alkali metal salt introduced into the sample solution is invariably a drawback in atomic absorption methods because the high solid content of the solution can cause blockages of the nebuliser and burner slot, as well as giving poorer detection limits caused by background absorption and noisy signals. Fluxes tend to be very expensive because of the high degree of purity required, as they need to be used in at least a five-fold excess over the sample. Mineral acids, by contrast, are easy to purify, quick to dispense, and, apart from sulphuric acid, do not produce large matrix effects in AAS.

II. GENERAL ASPECTS OF SAMPLE PREPARATION METHODS

A. Digestion vessels

Most chemical attack on samples is undertaken in glassware. Except where special equipment is required or constant shaking is necessary, all such attacks on rock, soil or sediment samples can be carried out in test tubes, the most convenient type being medium-wall rimless borosilicate tubes, size

References p. 284

19 × 150 mm. These can be handled in batches of 50 in nylon-coated wire racks. Cold extractions are best carried out in small polypropylene bottles (50 ml or 100 ml). Hot attacks involving hydrofluoric acid can be carried out in large batches in open 50 ml PTFE beakers on PTFE-coated or "Sindanyo" topped hot plates (from S.J. Juniper Ltd., Harlow, Essex, U.K.). Cleaning of test-tubes in regular use is straightforward. After any unused liquid has been poured away, the solid residue is washed out with a jet of tap water. The tube is then rinsed once with 1% nitric acid in demineralized water and dried in an inverted position in a wire rack in an oven at about 50°C. This procedure is effective, and no measurable contamination of a sample occurs even when the sample previously attacked in the tube was 1000 times more concentrated in trace elements. This is true of both new and of heavily-used tubes. Wire stemmed test tube brushes must not be used, because of the danger of contamination from zinc and other metals. PTFE beakers can be rendered free from trace elements derived from previous samples by thorough rinsing in dilute nitric acid or by boiling in a dilute solution of a laboratory glassware detergent such as Decon.

B. Heating equipment

The most convenient and effective method of heating large batches of test tubes is undoubtedly by means of heating blocks. A simple heating block (see Fig. 1) can be cheaply made by obtaining a 2 in. thick block of aluminium or Dural, of the same size as a standard hot plate. This is drilled at about 25 mm intervals to a depth of about 30 mm with holes about 1 mm larger than the test tubes to be accommodated. By use of the normal hot-plate controls, a temperature steady to within ± 5°C can be obtained at any point on the block, over the temperature range 50—200°C. These shallow blocks can be used for a variety of purposes where heating below the boiling point or gentle refluxing is required.

Large commercially-made blocks can be obtained from Scienco Western Ltd. (Toft, Cambridge, CB3 7RL, U.K.). These are much deeper blocks (see Fig. 2) and are designed to accommodate over two hundred 150 mm test tubes to a depth of 130 mm. The temperature control is very fine, and temperature programmers are also available if required. Deep blocks are ideal for critical work and suitable for the gentle evaporation of acids such as perchloric acid without excessively heating the residue. Tops of aluminium blocks can be made resistant to corrosion by acid fumes if they are given a heavy coating of PTFE from an aerosol spray.

A cheaper but less controllable device for steady evaporation of acids to dryness is the air bath shown diagramatically in Fig. 3. It consists of an aluminium sheet box with the vertical sides insulated with Sindanyo sheet. The bottom is open and rests on an ordinary hot plate. The top is drilled to hold test-tubes suspended by the rims. The temperature can be controlled by judicious use of the hot plate control for the gentle evaporation of

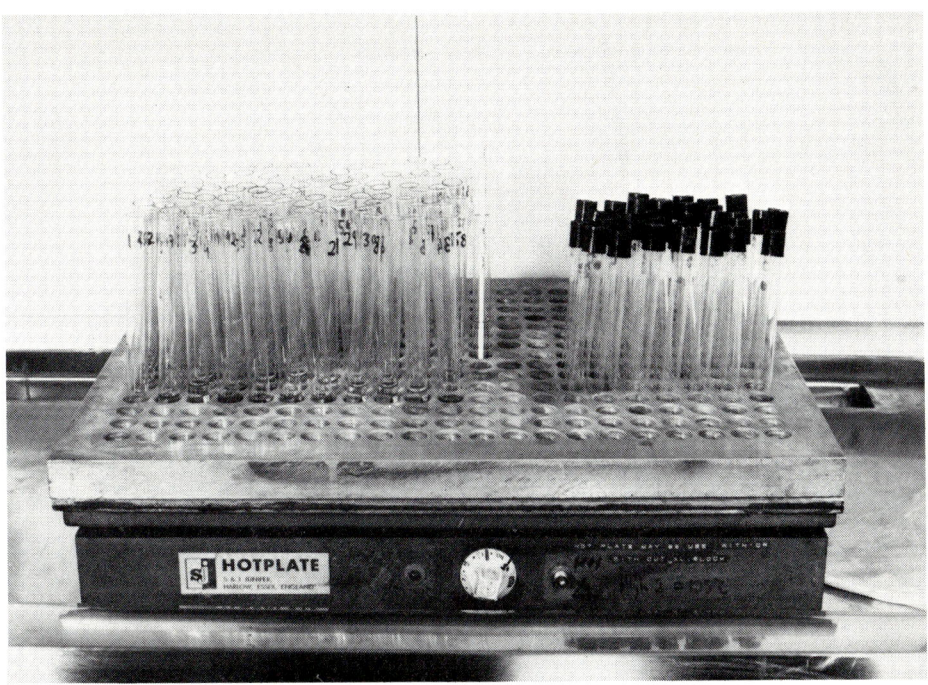

Fig. 1. Shallow aluminium block-bath for heating many test-tubes.

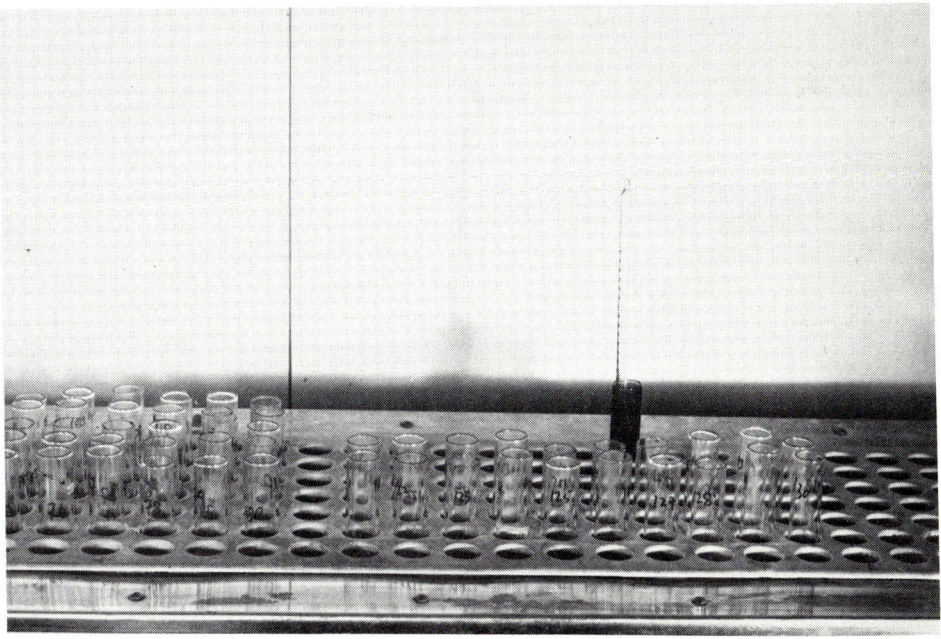

Fig. 2. Top surface of deep aluminium block-bath for evaporation to dryness in test-tubes.

References p. 284

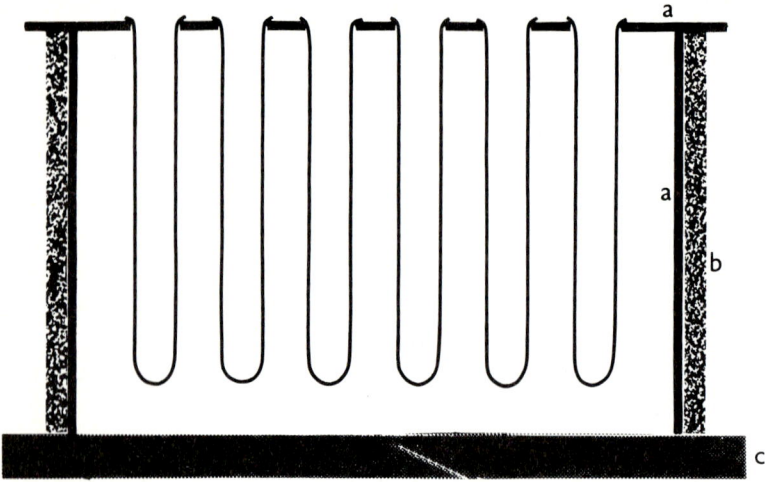

Fig. 3. Cross section showing construction of an air bath for evaporation to dryness in test-tubes: (a) sheet aluminium; (b) sheet Sindanyo; and (c) steel top surface of a hot plate.

perchloric and other acids over extended periods (such as overnight), but there is little control over the final temperature to which the residue is subjected.

C. Dispensers and diluters

The repetitive measuring of small ($< 20 \text{ cm}^3$) volumes of liquid is best undertaken by means of equipment such as the Oxford all-glass dispenser (The Bohringer Corporation Ltd., Bell Lane, Lewes, Sussex BN7 1LG, U.K.). Even concentrated nitric acid or perchloric acid can be permanently kept in such a device, the volume dispensed is remarkably precise, and the action very rapid. Other manufacturers equipment may be suitable but those with metal parts (such as springs) must be avoided because of the inevitable corrosion caused by acid fumes. No dispenser suitable for hydrofluoric acid has yet been found.

In geochemical analysis the concentration range of an analyte often spans several orders of magnitude, and dilutions are frequently required to bring the sample solution within the calibration range. An automatic variable-ratio diluter is essential for this purpose. Many manufacturers produce suitable models, but corrosion problems are almost inevitable with the acidic solutions which are regularly used. Small amounts of acid can escape between the piston and the barrel, and cause corrosion, especially where metal parts are enclosed.

D. Reagents and calibrators

Analytical grade reagents (i.e. 'Analar' or equivalent) are suitable and should be used for most geochemical analysis, but for very low concentrations of analytes, as may be encountered in water analysis, high purity reagents (e.g. 'Aristar') are preferable. Demineralised water is sufficiently pure for most purposes.

Calibration solutions (calibrators) should always be made up in the same medium as the final sample solution, and can be conveniently prepared in one dilution from stock solutions of $1000\,mg\,l^{-1}$ by using pipettes of the 'Eppendorf' type. These pipettes provide a relative error of less than 1%. The stock solutions of standards supplied by chemical houses are cost-effective in commercial analysis.

E. Safety

Most of the procedures involving attacks with strong acids must be carried out in adequately constructed fume cupboards. Great care in the design and choice of constructional materials is required for cupboards used for hydrofluoric acid or perchloric acid. In particular, perchloric acid vapours or condensate must not be allowed to come into contact with wood, which can form a self-igniting or explosive mixture. In handling hydrofluoric acid, a face mask, rubber gloves and a plastic apron are *essential*, because irreversible tissue damage can be caused very quickly by this acid.

Perchloric acid mixtures can be used with complete safety for attacks on virtually all rocks, soils and sediments, as in methods III.B, III.C and III.H. Even peaty soils can be safely treated by these methods. The only explosive hazard which we have encountered is with rocks containing appreciable quantities of oil or bitumen. These types, however, are easily recognised during sample preparation and weighing. The acid can also be used with complete confidence for herbage analysis so long as the procedure described is followed closely, and not attempted on fatty or oily samples. The Society for Analytical Chemistry's guide to the use of perchloric acid should be studied before attempting *any* attacks with perchloric acid [9]. See also Chapter 3 of this book.

F. Mechanical sample preparation

Rocks. Rock samples are reduced to 1—2 mm fragments by a jaw crusher or percussion mortar. They are then reduced to a fine powder to pass a 200 mesh (63 μm aperture) sieve by means of a swing-mill such as the one produced by Tema Machinery Ltd., Banbury, Oxon, U.K. Agate mortars for this device are expensive but do not contaminate the sample with trace metals. A hard steel mortar is suitable if contamination with traces of chromium can be tolerated. Small hammer mills are suitable for the fine grinding, but as

References p. 284

the hammers are commonly stainless steel, they contaminate the sample with nickel and chromium. Non-metallic sieves can be cheaply constructed from perspex tubing (10—15 cm diameter) and monofilament polyester bolting cloth, which is supplied in a wide range of mesh sizes by Henry Simon Ltd., P.O. Box 31, Stockport, Cheshire, SK3 0RT, U.K.

Soils. The dried soil is gently disaggregated (not crushed) and the particles greater than 2 mm sieved out. The <2 mm fraction is then crushed in a swing mill or hammer mill as described for rocks.

Sediments. Sediments are usually prepared by simple sieving of the natural material after drying and disaggregation. A number of size fractions are used for special purposes, but the most widely used is the —80 mesh (ca. 200 μm aperture) fraction, which is analysed without further preparation. Coarser fractions may need to be crushed before analysis.

Herbage. Herbage samples must be milled to pass a 1 mm aperture before analysis. This can be achieved by means of a beater mill with a carbon steel construction (as produced by Christie and Norris Ltd., Broomfield Road, Chelmsford, Essex, U.K.). This device introduces no detectable concentration of foreign metals into the sample. A small hammer mill is also satisfactory, but may contaminate the sample slightly with traces of nickel and chromium.

G. Analytical quality control

A prominent feature of applied geochemical analysis is the regular use of control methods to ensure that a satisfactory level of accuracy and precision is maintained from batch to batch of samples. The methods are based on the planned use of standard samples and/or duplication, as much as 5—10% of the analyses being devoted to the control system. For example, in a batch of 300 samples, it would be usual to insert 5 each of 2 standard samples (representing low and medium levels of the analyte), 10 samples (selected at random) in duplicate, and 10 reagent blanks. The control samples should be inserted at random positions in the analytical sequence. While the statistical principles behind the control methods are elementary, their application needs a considerable amount of close attention to ensure that realistic (rather than optimistically biassed) results are obtained.

III. SAMPLE ATTACK METHODS

A. Nitric acid attack for soil or stream sediment

Reagent
(a) Nitric acid (70%).

Equipment
(a) Test-tubes and racks.
(b) Liquid dispensers (2).
(c) Shallow heating block sited in a suitable fume cupboard.

Procedure
(a) Weigh each sample (0.250 g, −80 mesh) into a clean, dry, numbered test-tube.
(b) Weigh standard and duplicate samples, and leave empty test-tubes for blank determinations at random intervals.
(c) Add nitric acid (1.0 ml) to each test-tube.
(d) Place the tubes in the heating block at $105 \pm 5°C$ and leave for one hour.
(e) Transfer the test-tubes to a wire rack and allow to cool.
(f) Add water (9.0 ml) to each test tube and mix the contents thoroughly.
(g) Allow the solid residue to settle to the bottom of the tube (at least 4 h), and aspirate the solution directly from the test-tube.

Remarks
(a) This attack is designed for the digestion of clay minerals, usually the major constituent of the −80 mesh fraction of soils and sediments, and is completely effective (i.e., > 90%) in solubilising trace elements such as Cu, Pb, Zn, Cd, Mn, Fe, Co, Ni, Hg and Ag. For other rock forming minerals the attack may be virtually complete (feldspar, olivine), partial (e.g. about 50% for pyroxine biotite, amphibole), or negligible (quartz). Lateritic soils are only partially attacked.
(b) Samples with a high organic matter content (e.g. peaty soils) may react vigorously and char. Such samples should be left standing overnight in nitric acid (procedure step c) prior to hot digestion.
(c) The dilution factor for this attack is 40.

B. Nitric acid—perchloric acid attack for rock, soil, or stream sediment

Reagents
(a) Perchloric acid (60%).
(b) Nitric acid (70%).
(c) Hydrochloric acid (6 M). Dilute hydrochloric acid (516 ml, 36% acid) to 1 l with water.

Equipment
(a) Test-tubes and racks.
(b) Liquid dispensers (4).
(c) Deep heating block sited in a suitable fume cupboard.

Procedure
(a) Weigh each sample (0.250 g, −80 mesh) into a clean, dry, numbered test-tube.

References p. 284

(b) Weigh standard and duplicate samples, and leave empty test-tubes at random intervals for blank determinations.
(c) Add nitric acid (4.0 ml) to each test-tube.
(d) Add perchloric acid (1.0 ml) to each test-tube.
(e) Place the test-tubes in the cold block and raise the temperature to 150°C ± 5°C over 2–3 h. Leave at this temperature until copious evolution of fumes ceases.
(f) Increase the temperature of the heating block to 185°C ± 5°C and transfer the test-tubes to wire racks when the residue is dry.
(g) Allow the tubes to cool. Add hydrochloric acid (2.0 ml) to each test-tube.
(h) Place the tubes in a shallow heating block (60°C) and leave for one hour. Transfer the tubes to wire racks and allow to cool.
(i) Add water (8.0 ml) to each tube and mix thoroughly.
(j) Allow the residue to settle (at least 4 h) and aspirate the supernatant liquid directly from the test-tube.

Remarks
(a) This is a more powerful attack than nitric acid alone. In addition to the minerals completely attacked by the method 3.1, pyroxenes, biotite, limonite and some amphiboles are almost completely attacked. Trace element extraction from lateritic soil is almost complete. However, some common minerals containing important metals are attacked to a negligible extent (e.g. rutile, chromite, cassiterite, zircon, beryl), or to a minor degree (barite).
(b) Samples with a high organic-matter content may react vigorously with nitric and perchloric acids. Such samples should be kept overnight in the heating block at 50°C after the addition of the acids at stage (d). The normal procedure can be resumed at stage (e).
(c) This method must not be attempted on samples containing oil or bitumen.
(d) The dilution factor for this method is 40.

C. Hydrofluoric, nitric and perchloric acid attack for rock, soil, or sediment

Reagents
(a) Hydrofluoric acid (40%).
(b) Perchloric acid (60%).
(c) Nitric acid (70%).
(d) Hydrochloric acid (6 M). Dilute hydrochloric acid (516 ml, 36% acid) to 1 l with water.

Equipment
(a) PTFE beakers (50 ml).
(b) Graduated flasks (25 ml) or graduated test tubes (10 ml).
(c) Hotplate sited in a suitable fume cupboard.

(d) Polythene measuring cylinder and plastic tray for dispensing hydrofluoric acid.
(e) Liquid dispensers (4).

Procedure
(a) Weigh each sample (0.250 g, −200 mesh) into a clean, dry, numbered PTFE beaker.
(b) Weigh standard and duplicate samples, and leave empty beakers at random intervals for blank determinations.
(c) Add nitric acid (3.0 ml) followed by perchloric acid (3.0 ml) to each beaker.
(d) Then add hydrofluoric acid (10 ml) to each beaker.
(e) Heat the beakers on a hotplate until dense white fumes are seen (1—1.5 h).
(f) Heat for a further 20 min and then allow the beakers to cool.
(g) Add further hydrofluoric acid (2 ml) to each beaker.
(h) Heat the beakers on the hotplate until the solution is gently evaporated to dryness (about 4 h) and allow the beakers to cool.
(i) Add further perchloric acid to each beaker (2.0 ml).
(j) Heat gently, evaporate to dryness and allow the beakers to cool.
(k) Add hydrochloric acid (2.0 ml if the final volume is 10 ml, 5.0 ml if the final volume is 25 ml) to each beaker and warm gently.
(l) Transfer the solutions from the beakers to either graduated flasks (25 ml for a dilution factor of 100) or to graduated test tubes (10 ml for a dilution factor of 40) and dilute to volume with water.

Remarks
(a) This method will completely digest most constituents of rocks, soils and sediments. A few minerals will partly or completely resist attack, e.g., barite, chromite, cassiterite, tourmaline, kyanite, some spinels and magnetites, rutile, zircon and wolframite.
(b) When using PTFE beakers on a hotplate, care should be taken not to exceed the temperature at which PTFE becomes plastic (240°C).
(c) A shortened form of this method can be used for less resistant samples by omitting steps f, g, i and j from the above method. The double fuming with perchloric acid is necessary for calcareous samples (i.e., > 10% Ca) to destroy the insoluble calcium fluoride residue.
(d) This method must not be attempted on samples containing oil or bitumen.

D. Hydrofluoric acid—boric acid attack for silicon determination in rock, soil, or sediment

This method is useful for the determination of silicon, but the solution can also be used for other major constituents. It is based on decomposition of the sample with hydrofluoric and hydrochloric acids in a polypropylene bottle. Boric acid solution is added to dissolve precipitated fluorides [10, 11].

Reagents
(a) Hydrofluoric acid (40%).
(b) Hydrochloric acid (36%).
(c) Saturated boric acid solution. Weigh out 200 ± 5 g boric acid into a beaker, add 1000 ± 50 ml water, cover the beaker and heat until the acid has dissolved. Cool to $40 \pm 10°C$ and decant into a bottle.

Equipment
(a) Polypropylene bottles with screw caps (125 ml).
(b) Water bath or an oven.
(c) Plastic tray and plastic dispenser for hydrofluoric acid.
(d) Liquid dispenser.
(e) Measuring cylinder, 50 ml.

Procedure
(a) Weigh each sample (0.100 g, —80 mesh) into a dry, numbered polypropylene bottle.
(b) Weigh standard and duplicate samples, and leave empty bottles at random intervals for blank determinations.
(c) Add hydrochloric acid (1.0 ml) to each bottle, wetting the sample thoroughly.
(d) Add hydrofluoric acid (5.0 ml) to each bottle and close firmly.
(e) Place the bottles in either an air oven or a water bath at $95 \pm 5°C$ and leave for one hour. Allow the bottles to cool.
(f) Add boric acid solution (50 ml) to each bottle, close firmly and replace it in the air oven for a further hour. Allow the bottles to cool.
(g) Add water (44.0 ml) to each bottle and mix thoroughly.
(h) Use this solution for the determination of silicon.

Remarks
(a) Ensure that the bottles used are of polypropylene or other plastic material that will withstand temperatures up to about $130°C$. If the screw caps do not give a tight seal, this can be improved by using 'washers' cut from thin plastic film.
(b) This method is suitable for the same range of minerals as method III.C.
(c) The hydrofluoric—boric acid solutions should not be left in contact with glass apparatus for more than two hours to evoid etching the glassware and contaminating the sample solutions with silicon.
(d) Other elements can be determined on the same solution. Make the calibrators for aluminium with the same concentration of hydrofluoric—boric acid as the sample solution. Determine magnesium and calcium by using the nitrous oxide—acetylene flame, making a dilution of the sample solution to contain $1000\ \mu g\ ml^{-1}$ of potassium as an ionisation suppressant. Determine sodium and potassium by using the air—acetylene flame making an appropriate dilution of the sample solution to contain $1000\ \mu g\ ml^{-1}$ of caesium as an ionisation suppressant.

(e) For samples with low (< 5%) silicon content, use higher sample weights (up to 0.5 g).
(f) The dilution factor for this method is 1000.

E. Lithium metaborate fusion attack for silicon determination in rock, soil or sediment

This method is useful when silicon determinations are required on samples which are not decomposed by hydrofluoric—boric acids at 95°C. The finely ground sample is fused with lithium metaborate in a graphite crucible and the melt is dissolved in dilute nitric acid [12].

Reagents
(a) Lithium metaborate, high purity flux.
(b) Nitric acid (4% v/v).

Equipment
(a) Graphite crucibles supplied by Heydon & Son Ltd., 24 Ninion Avenue, Hendon, London NW4 3XP, U.K.
(b) Magnetic stirrer with plastic-coated stirrer bar.
(c) Plastic bottles (125 ml wide mouth) with screw caps.
(d) Graduated flasks (100 ml).
(e) Beakers (100 ml, polypropylene).
(f) Muffle furnace.

Procedure
(a) Pre-ignite a graphite crucible for 30 min at 950°C and then cool taking care not to disturb the powdery inside surface.
(b) Mix the sample (0.200 g, −200 mesh) with lithium metaborate (1.0 g) in a porcelain crucible, transfer to the graphite crucible, and heat in a muffle furnace at 900°C for 15 min.
(c) Add nitric acid (50 ml) to a polypropylene beaker.
(d) Remove the crucible from the furnace and immediately pour the melt into the nitric acid. Introduce a plastic coated stirrer bar into the solution and stir to dissolve the melt (about 10 to 15 min).
(e) Transfer the solution into a graduated flask (100 ml) and dilute to volume with nitric acid (4% v/v).
(f) Immediately, transfer the solution to a clean, dry numbered plastic bottle, and use to determine silicon.

Remarks
(a) Samples with an appreciable organic matter content should be ignited prior to the fusion.
(b) Aluminium can also be determined on the solution directly, by using the nitrous oxide—acetylene flame.
(c) Calcium and magnesium are usually determined on dilutions of the

References p. 284

original sample solution. The diluted solutions and calibrators are prepared to contain potassium (1000 μg ml^{-1}) as ionisation suppressant and the determinations carried out in the nitrous oxide—acetylene flame.
(d) Sodium and potassium are determined in an air—acetylene flame.
(e) All calibration solutions must contain an appropriate concentration of lithium metaborate and nitric acid.
(f) The dilution factor for this method is 500.

F. Chelation—solvent extraction method for determining trace metals in water samples

This method is for the determination of cadmium, cobalt, copper, iron, manganese, nickel, lead and zinc, which are solvent extracted and concentrated as their diethyldithiocarbamate chelates. After destruction of the organic complexes dissolution of the residue in dilute acid gives a solution suitable for atomic absorption analysis [13].

Reagents
(a) Acetic acid (100%).
(b) Ammonia solution, isothermally distilled from the low-lead reagent S.G. = 0.880. Allow the reagent to equilibrate in a desiccator at room temperature, with an equal volume of water.
(c) Chloroform ('Aristar' grade).
(d) Deionised water.
(e) Hydrochloric acid (1 M). Dilute 89 ml of hydrochloric acid (36%, 'Aristar') to 1 l.
(f) Nitric acid (70% 'Aristar').
(g) Nitric acid (2 M). Dilute 125 ml of nitric acid to 1 l.
(h) Sodium acetate trihydrate.
(i) Sodium diethyldithiocarbamate (SDDC).
(j) SDDC/buffer solution. Dissolve sodium acetate trihydrate (250 g) in water (500 ml), add acetic acid (6 ml) to the solution and mix well. Add SDDC (50 g) to this solution and mix. Dilute the solution to 1 l with water. If necessary adjust the pH of the solution to between 8.0 and 9.0. Extract any SDDC—metal complexes by successive treatments in a separating funnel with 30 ml aliquots of chloroform, until the extract is colourless.

Equipment
(a) pH narrow range test paper.
(b) Separating funnels (1 l).
(c) 'Quickfit' conical flasks (100 ml).
(d) Hotplate sited in a suitable fume cupboard.
(e) Membrane filters (0.45 μm) and a suitable filtering assembly.
(f) Sample bottles, high density polyethylene, 1 l capacity.
(g) Glass rods.

(h) Liquid dispenser.
(i) Measuring cylinder (1 l).

Procedure

A. *Sampling procedure*
(a) Fill sufficient sampling bottles with nitric acid (2 M) and leave for 24 h.
(b) Rinse each bottle out three times with water.
(c) Rinse the bottle out with the water being sampled, and take a 1-l sample.

B. *Preliminary treatment — 'soluble metal fraction'*
(a) Wash a membrane filter with hydrochloric acid and rinse with water.
(b) Filter the sample through the membrane as soon as possible after collection.
(c) Add nitric acid (2 ml) to the filtrate, mix well, and store in a polyethylene bottle.

C. *Alternative preliminary treatment — 'total metal fraction'*
(a) Add nitric acid (2 ml) to the sample in the sample bottle and mix well.
(b) Set the sample aside for at least four days, shaking the bottle each day.
(c) Wash a membrane filter with hydrochloric acid and rinse with water. Filter the sample through this washed membrane, and store in a sample bottle.

D. *Extraction procedure*
(a) Place the pretreated sample (500 ml) in a separating funnel (1 l) and adjust the pH to approximately 7 with ammonia solution.
(b) Add SDDC/buffer solution (20 ml) and if necessary, adjust the pH to between 5.8 to 6.1 by dropwise addition of either ammonia solution or nitric acid.
(c) Shake the funnel for 5 min and then add chloroform (30 ml).
(d) Shake the funnel for 5 min and then allow the phases to separate. Run the chloroform layer into a conical flask.
(e) Add chloroform (20 ml) to the funnel and repeat the extraction. Add the separated chloroform phase to the conical flask.
(f) Add nitric acid (1 ml) to the combined extracts and place the flask on a hotplate.
(g) Gently evaporate the chloroform extracts to dryness. Repeat the addition of nitric acid until a white or pale-yellow residue remains.
(h) Add hydrochloric acid (5.00 ml of 1 M) to the residue, stopper the flask and leave to dissolve.
(i) After mixing determine the metal concentrations by atomic absorption analysis.
(j) Carry out a blank determination by repeating steps (a) to (i) on deionised water (500 ml) which has already been stripped of any trace metals by the extraction procedure.

Remarks
(a) Cleanliness of apparatus and precautions against contamination at all stages are essential. Soak glassware in nitric acid (2 M) followed by a solution

of 'Decon' (5% v/v) solution in deionised water. Wash out five times in deionised water before use. Equipment required immediately can be cleaned by shaking with the SDDC/buffer solution and rinsing five times with deionised water. Separating funnels should be cleaned in this way before beginning a series of extractions.

(b) Where high trace-metal concentrations are indicated by the formation of an immediate deep colour or a precipitate on the addition of the SDDC/buffer solution, further extractions with chloroform will be necessary.

(c) The SDDC solution is unstable under acid conditions and decomposes rapidly. At a pH of 9, the SDDC solution is stable for at least one month.

(d) Acidification of the sample as described will stabilise the concentration of extractable trace elements for at least 35 days.

(e) A duplicate and a blank determination should be carried out after every tenth sample. Spiked samples should be analysed periodically, particularly at the beginning and end of each batch of SDDC/buffer solution.

(f) In this method 'soluble' refers to those metal species capable of passing through a $0.45\,\mu$m membrane whilst 'total' refers to that metal fraction of the particulate solubilised by dilute nitric acid and capable of passing through a $0.45\,\mu$m membrane plus the soluble fraction. A genuine 'total' fraction would include metals still not leached from the particulate matter.

(g) The concentration factor for this method is 100.

G. Nitric acid — perchloric acid attack for herbage samples

CAUTION. Fatty or oily samples must not be digested by this procedure.

Reagents
(a) Nitric acid (70%).
(b) Perchloric acid (60%).
(c) Hydrochloric acid (6 M). Dilute hydrochloric acid (36%, 516 ml) to 1 l.

Equipment
(a) Conical flasks (250 ml) or Phillips beakers each with a watch glass cover.
(b) Hotplate sited in a suitable fume cupboard.
(c) Either graduated flasks (25 ml) or graduated test-tubes (10 ml), depending on the dilution factor required.
(d) Liquid dispensers.

Procedure
(a) Weigh each sample (2.0 g dried at $105°C$ and ground) into a clean, dry, numbered conical flask.
(b) Weigh standard and duplicate samples and leave blank flasks at random intervals.
(c) Add nitric acid (40 ml) to each flask, cover with a watch glass and set aside in a fume cupboard overnight.
(d) Place the covered flasks on a hotplate and warm gently until frothing ceases.

(e) Allow the flasks to cool and add perchloric acid (3 ml).
(f) Replace the flasks on the hotplate, remove the covers, and cautiously heat just to dryness.
(g) Allow the flasks to cool, add hydrochloric acid (2.0 ml) and deionised water (2 to 3 ml) to each flask. Warm gently to dissolve the residue.
(h) Transfer the cooled solution to either a graduated flask or to a graduated test tube.
(i) Dilute to volume with water and use the solution for atomic absorption analysis.

Remarks
(a) On completion of the attack, most samples exhibit a residue of insoluble silica.
(b) The method is suitable for a large number of metals.
(c) The dilution factor for this attack is 12.5 if 25 ml flasks are used, or 5.0 for 10 ml test tubes.

H. Mercury in rocks, soils and sediments

Reagents
(a) Nitric acid (70% m/v).
(b) Tin(II) chloride solution; dissolve 25 g of $SnCl_2 \cdot 2H_2O$ in 100 ml diluted (1 + 1) hydrochloric acid. Pass oxygen-free nitrogen through the solution for 3 h or until the mercury blank is negligible.
(c) Magnesium perchlorate (anhydrous).
(d) Soda asbestos.

Equipment
(a) Shallow heating block.
(b) Eppendorf pipettes.
(c) Atomic absorption cell.   (The equipment is set up as shown in Fig. 4.)
(d) Bubbler.
(e) Dispensers, 2.
(f) Test tubes (100 × 12 mm), two for each sample.
(g) Nitrogen cylinder.

Procedure
(a) Weigh each sample (0.300 g) into a test tube.
(b) Add nitric acid (1.00 ml) and heat at 80°C for 1 h in the heating block.
(c) Cool, add water (2.00 ml), mix, and allow the solid residue to settle.
(d) Add water (2.00 ml) and the sample solution (1.00 ml) to a second tube, and then tin chloride solution (200 µl). (Use Eppendorf pipettes for the sample and the tin chloride solutions).
(e) Pre-set the nitrogen carrier gas flow rate at 0.5 l min^{-1}.
(f) Attach the second tube to the carrier gas line and record the peak absorption on a chart recorder.

References p. 284

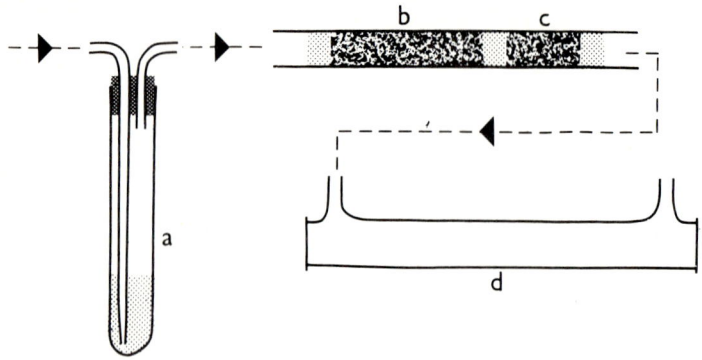

Fig. 4. Schematic diagram of equipment for the determination of mercury showing: (a) tube for reduction with $SnCl_2$ and outgassing of mercury vapour; (b) magnesium perchlorate desiccant; (c) soda-asbestos; and (d) silica-windowed cell to fit in light path of instrument.

(g) If the absorption is greater than the maximum of the calibration range, immediately repeat the determination with a smaller aliquot of sample solution at stage (d).

(h) Calibrate the system, starting at stage (d), by adding 1.00 ml of a solution of mercury(II) chloride in diluted (1 + 3) nitric acid. Suitable calibration solutions are made at concentrations of 0, 10, 20, 40, 60, 80 and 100 ng ml^{-1}, and should be prepared fresh daily. Check the calibration and blank after every 10 samples.

Remarks

(a) The detection limit is about 0.5 ng Hg, depending on the internal volume of the apparatus and the stability of the spectrometer. The upper limit of the useful calibration range is about 100 ng Hg. For the weights and volumes given the working range is 5—1000 ppb. A lower range can be obtained by using 0.600 g sample and by using the whole of the solution at stage (c).

(b) Special care is required in pre-handling of samples to avoid loss of mercury. Rocks should be subjected to a minimum of grinding: a particle size of —200 μm is suitable. Soil samples should be air-dried only, the water content being determined on a separate sub-sample. Sediments should be deep-frozen until just before analysis and weighed wet, to avoid loss of the volatile organo-mercury compounds which may account for a substantial proportion of the mercury present in sediments.

(c) The internal dimensions of the bubbler, cell and tubing are kept as small as possible for good sensitivity.

(d) The absorption cell should be as narrow as possible without occluding the light beam, and as long as can be accommodated in the instrument.

(e) Samples can be analysed at the rate of about one per minute.

(f) Water used in the sample solutions should be demineralised, and outgassed to remove traces of mercury metal.
(g) Background correction should not be necessary in this method.

I. Tin in rocks, soils and sediments

Cassiterite-bearing samples are attacked by volatilisation with ammonium iodide, and the sublimate containing the tin is dissolved in dilute tartaric acid. The solution can be used for normal nebulisation into an air—acetylene flame for high levels, or for hydride formation for low levels of tin. Interfering elements are not volatilised [14].

Reagents
(a) Ammonium iodide, ground to $-200\,\mu m$.
(b) Tartaric acid solution (1% m/v). Dissolve 10 g of tartaric acid in 1 l of water.
(c) Sodium tetrahydroborate solution. Dissolve 10 g of reagent in 1 l of 0.1 M sodium hydroxide solution.

Equipment
(a) Modified 3-slot burner for argon—hydrogen diffusion flame as shown in Fig. 5.
(b) Volatilisation tubes as shown in Fig. 6.
(c) Heating block (600°C).
(d) Hydride generator as shown in Fig. 7.
(e) Peristaltic pump, Watson Marlow MHRE 200 or equivalent with silicone rubber tubing (0.8 mm i.d. for reagent, 0.5 mm i.d. for sample solution).

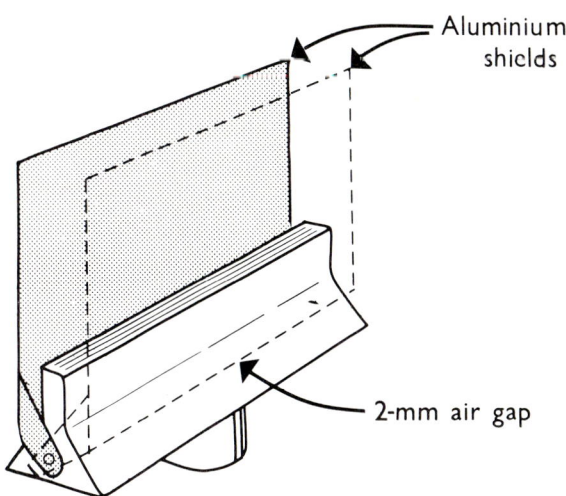

Fig. 5. A three-slot burner modified with aluminium shields for tin determination in an entrained air—hydrogen flame.

References p. 284

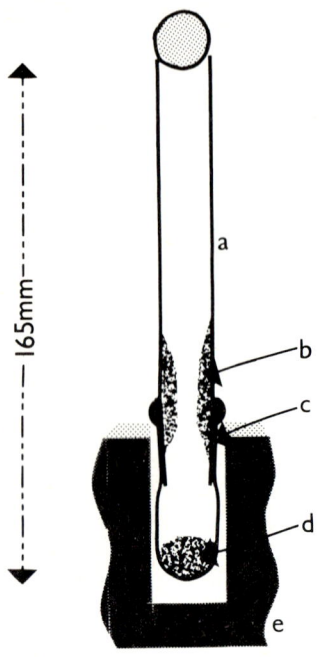

Fig. 6. Apparatus for the volatilisation of tin with ammonium iodide showing; (a) 14/23 cone used as an air condenser; (b) sublimate above hot zone; (c) 14/23 socket adapted to form a short tube; (d) sample residue; and (e) part of a multi-hole hot-block.

Procedure A (nebulisation)
(a) Grind the sample to pass a 200 mesh (75 µm) sieve.
(b) Weigh the sample (0.200 g) and the ammonium iodide (0.30 g) into a heating tube and mix.
(c) Heat the tube at 600°C for 10 min in the aluminium block, with the condenser and glass sphere in place, the condenser being cooled by a forced draught from, e.g., a vacuum cleaner.
(d) When cool detach the condenser and place it in a test tube. Add 20.0 ml of tartaric acid solution and heat at 50°C for 20 min.
(e) When cool, nebulise the solution into the air—acetylene flame, and record the absorption.
(f) Prepare calibration solutions in the range 2 to 250 µg ml^{-1} of tin(II) chloride in a solution containing tartaric acid (1% m/v) and ammonium iodide 1.5% m/v. The calibration is linear up to 230 µg ml^{-1} and serviceable up to at least 250 µg ml^{-1}.

Procedure B (hydride generation)
(a) Follow procedure A up to stage (d).
(b) Block the nebuliser capillary and attach the hydride generator to the auxiliary oxidant inlet. Light the argon—hydrogen diffusion flame supported on the shielded 3-slot burner (argon flow 13 l min^{-1}, hydrogen flow 8 l min^{-1}).

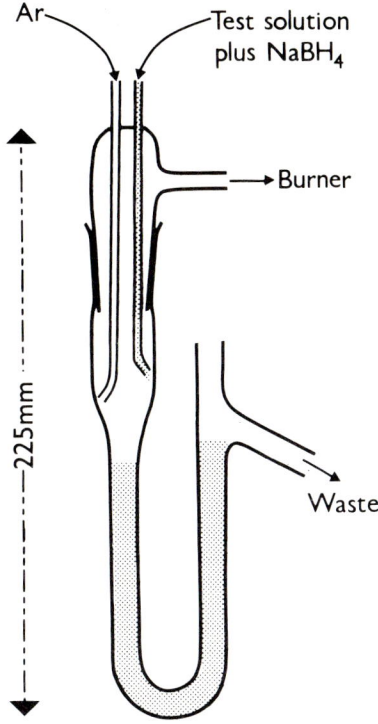

Fig. 7. Hydride generator for the determination of tin.

(c) Operate the hydride generator with the following flow rates: carrier gas (argon) $1 l min^{-1}$; reagent flow rate $4.5 ml min^{-1}$; sample solution flow rate $9.2 ml min^{-1}$.
(d) After about 5 s for signal stabilisation, integrate for 10 s (using "100 average" on the PE403) and record the absorption.
(e) Prepare calibration solutions in the range 0.02 to $2.0 \mu g ml^{-1}$, as before containing tartaric acid and ammonium iodide.

Remarks
(a) Fine grinding of the sample is necessary to ensure complete attack and homogeneous distribution of the cassiterite throughout the sample.
(b) The heating can be carried out with a bunsen burner if a 600°C heating block is not available.
(c) No interfering elements are volatilised with the tin.
(d) The equipment can be set up so that nebulisation or hydride generation can be employed in quick succession depending on the concentrations of tin encountered in the samples. The appropriate ranges are: nebulisation: from about $200 \mu g g^{-1}$ (detection limit) to $25000 \mu g g^{-1}$; hydride generation: from about $0.80 \mu g g^{-1}$ (detection limit) to $200 \mu g g^{-1}$. Higher ranges could be

References p. 284

accommodated by dilution of the solution obtained at stage (d) in procedure A.

(e) Dilute tin standards in tartaric acid solution are stable for at least one week.

References

1. J. S. Webb and M. Thompson, Pure Appl. Chem., 49 (1977) 1507.
2. W. R. Hatch and W. L. Ott, Anal. Chem., 40 (1968) 2085.
3. J. C. Robbins in M. J. Jones (Ed.), Proc. 4th Int. Geochemical Exploration Symp., London, April 17—20, 1972, IMM 1973, pp. 315—323.
4. M. Bedard and J. D. Kerbyson, Anal. Chem., 47 (1975) 1441.
5. M. Bedard and J. D. Kerbyson, Can. J. Spectrosc., 21 (1976) 64.
6. M. Thompson, B. Pahlavanpour, S. J. Walton and G. F. Kirkbright, Analyst, 103 (1978) 705.
7. J. R. Foster, Can. Min. Met. Bull., 60 (1973) 85.
8. M. Thompson, S. J. Walton and S. J. Wood, Analyst, 104 (1979) 229.
9. Analytical Methods Committee Report, Analyst, 84 (1959) 214.
10. F. J. Langmyhr and P. E. Paus, Anal. Chim. Acta, 43 (1968) 397.
11. B. Bernas, Anal. Chem., 40 (1968) 1682.
12. J. C. Van Loon and C. M. Parisis, Analyst, 94 (1969) 1057.
13. H. Watling, D. I. C. Thesis, Imperial College, London, 1974.
14. D. Gladwell, M. Thompson and S. J. Wood, J. Geochem. Explor., 16 (1981) 41.

Chapter 4h

Applications of atomic absorption spectrometry in the petroleum industry

W. C. CAMPBELL
I.C.I. Petrochemicals Division, R & D Department, New H.Q., P.O Box 90, Wilton, Middlesbrough (Gt. Britain)

I. INTRODUCTION

Atomic absorption spectrometry (AAS) is a technique of particular utility in the determination of trace elements in petroleum feedstuffs and products [1, 2]. It combines the virtues of simplicity, sensitivity, wide elemental coverage and relatively low cost.

Problems of trace element analysis in the petroleum industry are many and varied in their nature. For this reason it is not possible to present here detailed methods for every possible combination of element of interest and sample type. Methods are presented for the most common analytical problems but many other determinations may be made with only minor changes to the methods given.

A. Flame atomisation

The flame is still by far the most popular and convenient atomisation source employed in AAS. It provides sufficient sensitivity for most trace metal analysis requirements met in the petroleum industry. Methods are described for use with both air—acetylene and nitrous oxide—acetylene flames. The properties of these flames are described in Chapter 2.

B. Electrothermal atomisation

Electrothermal atomisers come in many shapes and sizes, ranging from tube furnaces to metal ribbons. At present, the most popular form is the graphite tube furnace of which a number of designs are commercially available. The electrothermal devices vary markedly in their atomisation characteristics and a method suitable for one design will not necessarily work on another without some modification.

Methods given here are written in as general a format as possible to allow, it is hoped, application to a range of atomisers. The analyst may resort to electrothermal atomisation as a result of one of two factors: insufficient sample size to allow nebulisation into a flame or insufficient sensitivity when using flame atomisation. Where neither of these factors apply then the flame should be used, since it is more convenient. An exception to this rule may be where it is desired to atomise solid samples directly.

It is usual with electrothermal atomisers to pipette between 5 and 100 μl samples into the device using a micropipette. With petroleum samples dissolved in organic solvents this may be a problem. Due to the low surface tension of many of these solvents they do not pipette easily and often dry irregularly in the atomiser, both factors giving rise to poor reproducibility. The problem of poor drying characteristics may be overcome with many solvents by pipetting into a pre-heated atomiser at approximately 80°C. The solvent is removed immediately, leaving the analyte on a reproducible spot each time. This technique, however, requires some care so as not to melt or contaminate the pipette tip.

Automatic injection and sampling systems are now available from a number of manufacturers and these undoubtedly help to relieve the tedium of the technique and will in general lead to an increase in precision through more consistent sample injection.

II. SAMPLING

The reader is referred to ASTM Method D270-65 [3] for details of proper procedures in the sampling of petroleum products. It is obviously vital that the sample being subjected to analysis by AAS is representative of the material of interest.

The practice in many industrial analytical service laboratories is to present the analytical chemist with samples and some form of written request for analysis. Often discrepancies occur between expected values and those obtained by analysis and the blame for this is often placed on the analytical chemist. In many cases little or no thought is given to the sampling or handling of the material prior to submission for analysis. The analytical chemist is well advised to enquire into the history of samples and where possible to maintain some form of control over sampling, sample handling and storage.

A. Sample contamination

Sample contamination from sampling apparatus, storage media or by introduction of foreign elements during analytical procedures must be given consideration. This is a particular problem in those determinations carried out using electrothermal atomisers, due to the high sensitivity of these devices.

It is not possible to precisely define sampling conditions and storage media for all situations. Certain general recommendations can, however, be made. Solder seams in metal containers often used for sampling are to be avoided. Exposure of samples to metal surfaces may be reduced by coating these with inert polymeric material. A particularly troublesome source of contamination may be the atmosphere in the analytical laboratory itself. In many industrial environments the presence of dust and fumes in the atmosphere can cause problems normally manifesting themselves in high and erratic blank values. For these and other reasons a blank value should always be obtained using a process identical to that used for the sample itself. Where this value is high the source of contamination must be identified and removed.

B. Reagent impurities

In the analysis of metals in petroleum and petroleum products one of the most common sample preparation procedures is the dilution of the sample with an organic solvent such as xylene, methyl isobutyl ketone (MIBK) or white spirit. It is of great importance that the solvent system chosen is as free as possible from metallic contamination. Elements such as sodium and zinc are commonly found in many organic solvents. Similarly, other reagents such as mineral acids must be investigated for metal content before use. Where ultra-trace level determinations are to be attempted the reagents used may need to be purified. For solvents, the use of redistillation or extraction with mineral acid may improve the blank levels.

C. Sample storage

Although it is well known that solutions of metals at trace levels may deteriorate with time it is not possible to specify general storage conditions for all combinations of sample and element of interest. Metals in storage may undergo precipitation, volatilisation, adsorption on container walls or diffusion through container walls. The stability of the sample will depend on the solvent or sample matrix, the element of interest and its form of association, the nature of the sample container and the conditions under which it is kept. As a general rule glass bottles and polyethylene made by the high-pressure free-radical process, may be used for storage. The use of rubber stoppers should be avoided as these are attacked by hydrocarbons.

The storage container must be free from contamination before use. A solution of strong non-metallic detergent may be used for most applications. After soaking in detergent for a time, rinse the container with tap water, then distilled or deionised water. Dry in an oven at approximately 50°C before use. The use of air-jet drying after application of a volatile solvent, such as acetone, often gives rise to contamination.

D. Sample preparation

The petroleum sample must be prepared for analysis prior to its presentation to the AAS instrument. The choice of sample preparation method is

References p. 306

varied, each method having its own particular advantages and limitations. The main methods of interest to the petroleum analyst are dry ashing, reagent aided dry ashing, wet digestion and dilution with solvent.

Where the element of interest is present in the sample as a soluble complex then it is often sufficient to dilute the sample with a suitable organic solvent. The sample may be diluted to the desired concentration level and any viscosity effects reduced. Not all solvents are suitable for flame AAS. Some, such as n-hexane and benzene, give smokey, unstable flames and should be avoided. Solvents such as white spirit, MIBK, n-heptane and xylene are acceptable. A dilution ratio should be chosen to bring the metal concentration into the region 0.2—0.8 absorbance units (or approximately fifty times the manufacturers quoted sensitivity).

Petroleum oils often contain suspended or colloidal inorganic materials. In the case of used lubricating oils small particles of metals are present. In many cases these suspended solids are of sufficiently small particle size that efficient breakdown occurs in the flame and a simple dilution procedure may be used. However, where it is suspected that this is not the case then it is recommended that an ashing technique is used to prepare the sample.

A general dry ashing procedure is to weigh 5—50 g of sample into a platinum or silica crucible. The sample is subjected to temperatures of between 150—200°C (usually an IR heater), to remove low boiling components, and then ignited in a muffle furnace at 500—550°C. The ash is normally taken up in hydrochloric or nitric acid. A major disadvantage of dry ashing is the possible loss of volatile elements or compounds. Elements such as Zn, Cd and Pb are subject to volatilisation at low temperatures and Ni and V, when present as volatile porphyrins in oil, are easily lost during ashing. The problem may be reduced by the use of ashing aids such as benzene sulphonic acid or *p*-toluene sulphonic acid. The acid used to dissolve the ash should also be present in the standards at the same concentration level. In practice it is found that hydrochloric acid is best for flame analysis and nitric acid for electrothermal analysis.

Wet digestion techniques have been advocated, similar to those used for biological and food analysis. However the technique is slow and only a limited number of samples may be handled.

III. STANDARDS FOR PETROLEUM ANALYSIS

It is vital that the standards used are as well characterised as possible, since the accuracy of any measurement is a function of the comparison made. Standard materials may be obtained in two forms, as solid organometallic compounds of known metal content or as ready made metal in oil-based standards of known concentration.

Table 1 illustrates the organometallic compounds suitable for use as standards in AAS. These or similar materials may be obtained along with

TABLE 1

ORGANOMETALLIC STANDARDS FOR AAS

Element	Compound
Al	Aluminium 4-cyclohexanebutyrate
Ag	Silver 2-ethylhexanoate
Ba	Barium 4-cyclohexanebutyrate
Ca	Calcium 2-ethylhexanoate
Cd	Cadmium 4-cyclohexanebutyrate
Co	Cobalt 4-cyclohexanebutyrate
Cr	Tris (1-phenyl-1,3-butanedione)chromium(III)
Cu	Copper 4-cyclohexanebutyrate
Fe	Iron 4-cyclohexanebutyrate
Hg	Mercuric 4-cyclohexanebutyrate
K	Potassium 4-cyclohexanebutyrate
Li	Lithium 4-cyclohexanebutyrate
Mg	Magnesium 4-cyclohexanebutyrate
Mn	Manganese 4-cyclohexanebutyrate
Na	Sodium 4-cyclohexanebutyrate
Ni	Nickel 4-cyclohexanebutyrate
Pb	Lead 4-cyclohexanebutyrate
Si	Octaphenylcyclotetrasiloxane
Sn	Dibutyltin bis(2-ethylhexanoate)
V	Bis (1-phenyl-1,3-butanedione)-oxovanadium(IV)
Zn	Zinc 4-cyclohexanebutyrate

specification analysis and recommended methods of standard preparation from the following sources: The Office of Standard Reference Materials, Room B314, Chemistry Building, National Bureau of Standards, Washington, D.C. 20234, U.S.A. and Eastman Kodak Company, Eastman Organic Chemicals, 343 State Street, Rochester, New York 14650, U.S.A.

To prepare a $500 \mu g\,g^{-1}$ standard from solid organometallic compounds proceed as follows: calculate the weight of material containing 50 mg of the element of interest. Dry the material in a low temperature oven and then weigh the necessary amount. Transfer the weighed material to a weighed 200 or 250 ml volumetric flask. Add 5 ml of 2-ethyl hexanoic acid, 4 ml of 6-methyl heptane-2,4-dione and 2 ml of xylene. Gently heat on a hot plate and swirl until a clear gel forms, then add 2 ml of 2-ethyl hexylamine. Heat and swirl until a clear solution is obtained. Immediately add approximately 80 ml of metal-free lubricating oil, mix and allow to cool to room temperature. Make the contents of the flask up to 100 g with lubricating oil, stopper and shake.

Calibration standards in the chosen solvent may be prepared by subsequent dilution of the above standard. As an alternative to lubricating oil, solvents such as MIBK, xylene or white spirit may be used directly to prepare the master standard. It is not always necessary to follow rigidly the above procedure as many of the organometallics dissolve easily in a range of

References p. 306

solvents. Dissolution of the salt in 2-ethyl hexylamine in xylene followed by dilution with the desired solvent is often sufficient. When in doubt consult the manufacturer of the salt being used.

A range of single and blended organic sulphonate standards in oil bases is available. These appear to offer excellent stability and may be diluted with paraffinic and aromatic hydrocarbons as well as ketones. These may be obtained from: Conostan Division, Continental Oil Company, P.O. Box 1267, Ponca City, Oklahoma 74601, U.S.A.

IV. APPLICATIONS

A. Crude and residual fuel oils

Crude oil contains a number of metallic elements which are of interest either due to the undesirable effects they cause in the refining process or as an indication of the origin of the oil. The concentration levels encountered will vary with the type of crude oil. Crudes originating from different oil fields may vary markedly in trace-metal content. Also some crude oils will become contaminated in transport from the oil well to the refinery by, for example, pipeline material or seawater. The levels of such metals as Ni, V and Na must be carefully controlled in order to reduce production problems such as plant corrosion or catalyst poisoning.

Residual fuel oils, consisting mainly of the residue from the distillation of the more asphaltic crude oils, are used largely as fuels for ships, locomotives and various heating purposes. It thus contains much of the inorganic constituents originally present in the crude oil. Deposits from the fuel oil may reduce boiler efficiency. Vanadium in the deposits may catalyse the conversion of SO_2 to SO_3 resulting in sulphuric acid production in cooler regions, giving rise to corrosion.

Two methods are given here for the determination of metals in crude and residual fuel oils; the dilution—flame analysis method and an ashing procedure. Because of the nature of these oils simple dilution with solvent may leave undissolved solids in the solvent to be presented to the spectrometer. This could give rise to a decrease in the precision of the method. The ashing procedure overcomes this difficulty but is more time consuming. The method of choice will depend on whether the analytical emphasis is placed on speed of analysis or precision. Where very low levels of Ni or V are important it may be possible to modify the method given under fuel oils (Section IV.B.2.).

1. The determination of metals in crude and residual fuel oil by dilution and flame analysis

(i) *Summary*

The method may be used for the determination of Ca, Fe, V, Ba, Ni, Na and Mg in crude and residual fuel oils. After dilution with the chosen solvent the solution is aspirated into either the air—acetylene or nitrous oxide—acetylene flame. Organometallic standards are prepared in the same solvent. Limits of detection are generally in the range $1-5\ \mu g\ g^{-1}$.

(ii) *Sample preparation*

Should the oil be very viscous it is recommended that it is placed in an oven at 60°C for at least one hour prior to sampling. All samples must be stirred or vigorously agitated in order to homogenise them prior to sampling.

The exact ratio of sample to solvent depends on the concentration of the element to be determined in the oil. However a dilution of 1 in 10 (1 + 9) in the chosen solvent is a good general rule and further dilution may be made if required. The solvent used is to some extent a matter of individual choice. Solvents found to be useful in this application are MIBK, xylene and a mixture of 10% isopropanol and 90% white spirits. Weigh accurately approximately 10 g of the oil sample and dissolve with shaking in the chosen solvent. Make up to 100 ml and stopper. Dilute further if necessary to match calibration standards.

(iii) *Standard preparation*

Using either commercially available metal in oil standards or a stock solution, prepared as described in Section III, prepare calibration standards for the elements of interest by dilution with the same solvent as used for the oil samples. The concentration range spanned by the standards should as far as possible give a linear response and will be a function of the particular AAS instrument used and the conditions chosen. The following may be used as a rough guide to the concentration range best used and will be approximately correct for most modern instruments. At least four standards, including a blank, should be used to cover the suggested range.

Barium	$0-10\ \mu g\ ml^{-1}$
Calcium	$0-2\ \mu g\ ml^{-1}$
Iron	$0-5\ \mu g\ ml^{-1}$
Magnesium	$0-1\ \mu g\ ml^{-1}$
Nickel	$0-5\ \mu g\ ml^{-1}$
Sodium	$0-2\ \mu g\ ml^{-1}$
Vanadium	$0-25\ \mu g\ ml^{-1}$

Standards at these concentration levels have doubtful long-term stability and they should therefore be prepared as and when necessary and not stored for more than a few days.

(iv) *Instrumental*

Adjust the instrumental parameters of wavelength, spectral band pass and lamp current in accordance with the manufacturer's recommended conditions. The flame conditions should be established while aspirating a solvent blank. It is recommended that a fuel lean nitrous oxide—acetylene flame be

References p. 306

used for Mg, Ca and Ba, a fuel rich nitrous oxide—acetylene flame for V and a fuel lean air—acetylene flame for Fe, Ni and Na. For other elements of interest consult the manufacturers handbook. It will normally be necessary to decrease the acetylene flow rate to compensate for the fuel contribution of the solvent. It may in some cases be helpful to increase oxidant flow and/or adjust nebuliser uptake rate.

Check that there are no background (scatter or non-specific absorption) effects by using a hydrogen or deuterium lamp or a nearby non-absorbing line.

(v) *Procedure*

Prepare sample and standard solutions as described. Optimise the instrumental conditions for the element of interest and aspirate first standards and then samples obtaining the relevant absorbance values. Aspirate the standards again and check that the absorbance values are unchanged. Calculate the concentration in the samples by preparing a calibration graph and comparing the sample absorbances with this.

$$\text{Concentration in oil} = (x \times 100/y)\,\mu g\,g^{-1}$$

where $x = \mu g\,ml^{-1}$ in final solution and y = weight of oil sample (g).

(vi) *Notes*

(1) One source of poor precision in this method is the differing viscosities found after dilution of the oil samples with solvent. These differences give small changes in sample uptake rate with respect to both other samples and to standards. The method of standard additions may be used to overcome this problem but with a considerable increase in analysis time.

(2) The alkali and alkaline earth elements are easily ionised in the flame. It is therefore necessary to add an ionisation suppressant when analysing for these elements. The potassium salt of naphthasulphonic acid or a commercially available potassium standard solution must be added to give a final potassium concentration of approximately $1000\,\mu g\,ml^{-1}$ in both samples and standards.

2. *The determination of metals in crude and residual fuel oils by ashing and flame analysis*

(i) *Summary*

The method is suitable for the determination of Ca, Fe, V, Ba, Ni, Na and Mg in crude and residual fuel oils. It may also be applicable to other elements. The oil is ashed to obtain the inorganic residue and the ash taken up in hydrochloric acid. Standards are prepared in hydrochloric acid media. Limits of detection will lie between 0.1 and $1\,\mu g\,g^{-1}$ for most elements.

(ii) *Sample preparation*

Very viscous oil samples should be placed in an oven at 60°C before sampling is attempted. All oils must be vigorously agitated before sampling. Accurately weigh approximately 20 g of the oil sample into a silica crucible and add 4 ml of a 20% solution of benzenesulphonic acid in butanol. Place the crucible under an IR heater or similar apparatus at approximately 150—200°C and allow slow charring to occur. Slowly introduce the crucible in stages into a muffle furnace at approximately 550°C until all carbon has been removed. Allow the crucible to cool and add 2 ml of 50% reagent grade hydrochloric acid to dissolve the ash. Quantitatively transfer to a 25 ml volumetric flask and make up to the mark with distilled or deionised water. Prepare a blank by taking an empty crucible through the ashing procedure.

(iii) *Standard preparation*

Commercially available aqueous standard stock solutions or solutions prepared from analytical reagent grade chemicals may be used. Dilute these to give calibration standards approximately in the range given in Section IV.A.1.

At least four standards including a blank should be used to cover the range. Each calibration standard must contain 4% v/v hydrochloric acid in order to match the samples. Prepare the calibration standards as required; do not store.

(iv) *Instrumental*

Adjust the instrumental parameters of wavelength, spectral band pass and lamp current in accordance with the manufacturers recommended conditions. It is recommended that a fuel lean nitrous oxide—acetylene flame be used for Mg, Ca and Ba, a fuel rich nitrous oxide—acetylene flame be used for V and a fuel lean air—acetylene flame be used for Fe, Ni and Na. For other elements of interest consult manufacturer's handbook.

Check that there are no background (scatter or non-specific absorption) effects by using a hydrogen or deuterium lamp or a nearby non-absorbing line.

(v) *Procedure*

Prepare the sample and standard solutions as described. Optimise the instrumental parameters for the elements of interest and aspirate standards, then samples, obtaining the relevant absorbance values. Aspirate the standards again and check that the absorbance values are unchanged. Calculate the concentration in the samples by preparing a calibration graph and comparing the sample absorbances with this.

Concentration in oil $= (x \times 25/y) \mu g \, g^{-1}$

where $x = \mu g \, ml^{-1}$ in aqueous solution, $y =$ weight of oil sample (g).

References p. 306

(vi) *Notes*

(1) When determining the alkali and alkaline earth elements it is necessary to add an ionisation suppressant, usually potassium, at approximately 1000 μg ml^{-1} to both standards and samples.

(2) It has been reported that the addition of approximately 10 000 μg ml^{-1} lanthanum to vanadium solutions increases the vanadium sensitivity. This can be useful where it is necessary to determine vanadium at very low levels.

(3) When the more volatile elements are of interest care must be taken not to lose these by, for example, inadvertently subjecting the sample to a higher temperature than is necessary.

B. Fuel and gas oils

The presence of certain elements in fuel and gas oils may accelerate the oxidative deterioration of refined products or otherwise reduce their stability during storage. Vanadium, for example, in boiler firing oils may produce corrosion and toxic emissions.

1. *The determination of metals in fuel and gas oils by flame analysis*

(i) *Summary*

The method may be used to determine Pb, Na, V, Ni and Zn in fuel and gas oils. It may also be applicable to other elements. The sample is diluted with MIBK or xylene and comparison made to organometallic standards in the same solvent. For most elements the limits of detection will be in the range 1—5 μg g^{-1}.

(ii) *Sample preparation*

Accurately weigh approximately 5 g of the oil into a 50 ml volumetric flask and dilute with MIBK or xylene. Shake to dissolve and mix.

(iii) *Standard preparation*

Use either commercially available oil base standards or organometallic standards prepared as described in Section III. Dilute these with the same solvent as used for the samples to prepare calibration standards. At least four standards including a blank should be prepared to cover an approximately linear range. This will vary from instrument to instrument but the following is a useful guide.

Lead	0—10 μg ml^{-1}
Sodium	0—2 μg ml^{-1}
Vanadium	0—25 μg ml^{-1}
Nickel	0—5 μg ml^{-1}
Zinc	0—2 μg ml^{-1}

Prepare the calibration standards as required; do not store.

(iv) *Instrumental*

Adjust the instrumental parameters of wavelength, spectral band pass and lamp current in accordance with manufacturer's recommended conditions. A fuel rich nitrous oxide—acetylene flame should be used for the determination of vanadium and may be useful for other elements if interference effects are suspected. The air—acetylene flame may be used for the determination of Pb, Na, Ni, and Zn. For other elements consult the manufacturer's handbook. Establish the flame conditions while aspirating the blank solvent. It will normally be necessary to reduce the acetylene flow to compensate for the solvent contribution. In some cases the use of auxiliary air in the air—acetylene flame may prove advantageous.

Check that there are no background (scatter or non-specific absorption) effects by using a hydrogen or deuterium lamp or a nearby non-absorbing line.

(v) *Procedure*

Prepare the samples and standards as described and optimise the instrumental parameters. Aspirate the standards then samples and record the absorbance readings. Aspirate the standards again and check that the absorbance values are unchanged. Prepare a calibration graph and compare sample absorbance values with this.

Concentration in oil $= (x \times 50/y)\,\mu g\,g^{-1}$

where $x = \mu g\,ml^{-1}$ in final solution, and $y =$ weight of oil sample (g).

(vi) *Notes*

(1) Where the sample concentration is above the calibration range used then a further dilution with solvent is necessary. This further dilution must be accounted for in the calculation.

(2) In the determination of sodium it is necessary to add an ionisation suppressant such as potassium at approximately $1000\,\mu g\,ml^{-1}$ to both samples and standards. The potassium salt of naphthasulphonic acid or a commercially available standard may be used for this purpose.

2. *The determination of low levels of nickel and vanadium in fuel oils using electrothermal atomisation*

(i) *Summary*

The method is suitable for the determination of low levels of Ni and V (down to approximately $0.01\,\mu g\,g^{-1}$) in fuel oils. Samples are diluted with xylene and comparison made with organometallic standards prepared in the same solvent. Electrothermal atomisation is used to achieve increased sensitivity.

(ii) Sample preparation

Accurately weigh the fuel-oil sample in a volumetric flask and dilute to the mark with xylene. The weight taken will depend on the concentrations of Ni and V in the fuel oil but in any case the dilution should never be less than 1 : 2, sample : solvent. Shake well to homogenise.

(iii) Standard preparation

Standards for the analysis may be prepared from organometallic standards, analysed samples or the NBS (GM-5) Heavy Oil Standard. The most satisfactory results are likely to be obtained using the second or third options. The sensitivity available is critically dependent on the electrothermal device to be used. This and the size of aliquot chosen for injection into the atomiser (normally 5—100 μl) will determine the selection of the concentration ranges chosen for the standards. Refer to the manufacturers information on Ni and V sensitivity and linear range and prepare calibration standards accordingly. Always prepare a blank solution and at least three standards to cover the chosen range.

Prepare the standards as required; do not store.

(iv) Instrumental

Adjust the spectrometer variables of spectral band pass, wavelength and lamp current according to the manufacturer's recommended conditions for Ni and V. It will be necessary to employ background correction for the determination with most electrothermal atomisers, especially in the case of Ni. Where available on the spectrometer, set up the deuterium or hydrogen lamp background correction system as recommended by the manufacturer. Otherwise use a nearby non-absorbing line to estimate the background intensity.

The following conditions, for use with electrothermal atomisation, are generally applicable.

Purge gas	argon or nitrogen
Dry time	10—50 s (volume dependent)
Dry temperature	100—200°C (ramped if possible)
Pre-atomisation heating time	10—50 s
Pre-atomisation heating temperature	700—1000°C (ramped if possible)
Atomisation time	5—15 s
Atomisation temperature	maximum available

(v) Procedure

The choice of measurement mode, peak height or peak area, will depend on the instrumental facilities. All injections should be made in duplicate and an average value taken. Where repeat injections give large variations in signal a third injection should be made. Obtain peak height or peak area values for standards and then samples. Recheck standard values periodically to ensure that the sensitivity is unchanged. Prepare a calibration graph for the standards and read the sample concentrations from this.

Concentration of V or Ni in fuel oil $= (x \times z/y) \mu g\, g^{-1}$

where $x = \mu g\, ml^{-1}$ in diluted sample, z = volume of sample solution, and y = weight of fuel oil sample (g).

(vi) *Note*

This method may be applicable to other elements in fuel oil, for example Pb.

C. Unused lubricating oils

The most commonly encountered additive elements in lubricating oils are Ca, Ba, Mg and Zn. They are normally present at relatively high concentrations such that a simple dilution—flame analysis procedure may be used to determine their concentrations. The control of the additive concentrations is important in the control of the physical and chemical properties of the lubricant. Problems associated with metallic particulate matter are not generally encountered with unused lubricating oils.

1. *The determination of calcium, barium, magnesium and zinc in unused lubricating oil by dilution and flame analysis*

(i) *Summary*

The method is applicable to the determination of Ca, Ba, Mg and Zn in unused lubricating oils and may also be of use for the determination of other elements of interest. Limits of detection will normally be in the region of 0.005% w/w. The lubricating oil is diluted with white spirit and aspirated into the nitrous oxide—acetylene flame. Organometallic standards in white spirit are used for calibration.

(ii) *Sample preparation*

Accurately weigh the lubricating oil in a volumetric flask and dilute to the mark using high quality white spirit. The exact dilution required will depend on the sample concentration and the calibration range chosen. Where Ca, Mg or Ba are to be determined K must be present in the final solution at $1000\, \mu g\, ml^{-1}$ (see note 3).

(iii) *Standard preparation*

Using either commercially available metal in oil standards or organometallic standards, prepared as described in Section III, prepare calibration standards by dilution with white spirit. Standards for Ca, Ba and Mg must contain $1000\, \mu g\, K\, ml^{-1}$ as ionisation suppressant. Choose a concentration range for each element which exhibits an approximately linear response.

References p. 306

This will vary from instrument to instrument but the following guide will hold for most modern spectrometers.

Calcium	0–2 $\mu g\,ml^{-1}$
Barium	0–10 $\mu g\,ml^{-1}$
Magnesium	0–2 $\mu g\,ml^{-1}$
Zinc	0–2 $\mu g\,ml^{-1}$

Use at least four calibration standards, including the blank, to span the concentration range. Prepare these standards as required; do not store.

The preparation of combined calibration standards containing all four elements is acceptable. Again a potassium concentration of 1000 $\mu g\,ml^{-1}$ must be maintained.

(iv) *Instrumental*

Adjust the instrumental parameters of wavelength, spectral band pass and lamp current in accordance with the manufacturer's recommended conditions. On certain instruments it may be found that a problem occurs when attempting to determine Ba in the presence of high Ca concentrations. In the nitrous oxide—acetylene flame high concentrations of Ca give rise to an increase in spectral emission from the flame at and around the Ba wavelength of 553.6 nm. It is thought that this gives rise to a high D.C. noise level in the detection electronics, which in turn results in high and erratic values for Ba absorbance. The problem may be minimised by using the smallest possible high voltage setting on the photomultiplier. This is best achieved by increasing the Ba lamp current to the highest value allowed by the manufacturer, so increasing the light throughput to the photomultiplier. Do not increase the spectral band pass.

It is recommended that the nitrous oxide—acetylene flame be used for all four elements (see note 1). Establish the flame conditions while aspirating a solvent blank.

Check whether background correction is necessary, especially when determining Mg and Zn, by using a hydrogen or deuterium lamp or a nearby non-absorbing line.

(v) *Procedure*

Aspirate standards, then samples and note absorbance readings. Recheck standards to ensure that calibration drift has not occurred. Prepare a calibration graph and read the sample concentration from this.

Concentration in lubricating oil = $[(x \times z \times D)/(y \times 10^4)]$ %w/w

where x = $\mu g\,ml^{-1}$ in final dilution of sample, z = volume (ml) of final solution, D = dilution factor (if any) prior to final solution, and y = weight of lubricating oil sample (g).

(vi) *Notes*

(1) All four elements are best determined using the nitrous oxide—acetylene flame although Ca, Mg and Zn may be determined with adequate sensitivity in the air—acetylene flame. The nitrous oxide—acetylene flame gives greater freedom from interference effects and also gives rise to improved calibration linearity.

(2) The general purpose solvent, 10% isopropanol—90% white spirit, may also be used for this analysis. Solvents such as toluene and MIBK have been found to give poor solubility for Ba additives.

(3) The alkali and alkaline earth elements are easily ionised in the nitrous oxide—acetylene flame. Potassium is therefore added to both samples and standards to give a final concentration of $1000\,\mu g\,ml^{-1}$. The potassium salt of naphthasulphonic acid or a commercially available potassium standard may be used.

D. Used lubricating oils

In all engines there is frictional contact between moving and stationary parts. This inevitably leads to small particles of metal or alloy becoming suspended in the lubricating oil. Examination of the concentration of various metals in used lubricating oils often gives an early indication of component failure. Thus, piston-ring wear is indicated by Cr, bearing wear by Sn, Ag and Cu, blowby by Pb and inefficient air filtration by Si.

The analysis of used lubricating oils is accepted as an effective and practical means of monitoring engine wear. It allows preventative maintenance to be carried out at convenient times when needed and avoids the expense of major breakdowns and excessive lay-offs. Navies, air-forces, air-lines, rail companies and many companies with large truck fleets make use of this diagnostic aid.

Because of the nature of the metallic constituents in the oil, suspended particles of metal or alloy, agreement between laboratories and individual instrumentation is often poor. However it is not the absolute concentration level of an element which is important but the change in concentration with time. A history of the particular engine being monitored is required, and a sudden change in wear metal content is indicative of excessive wear and indicates that maintenance is required. It is therefore useful if the same instrumentation is used for a particular analytical procedure such that values obtained may be compared over a period of time.

In used lubricating oils the wear metals are normally present in an extremely finely divided form and thus flame chemistry, similar to that for metals in true solution, is normally found. However, in some oils, such as those from heavy duty equipment or where excessive dust has been accumulated, larger particles may be present. In these cases more accurate results may be obtained by employing an ashing procedure to ensure proper dissolution of the metals (see Section II.D).

Finally, in order to obtain representative samples of used lubricating oil for analysis it is necessary to run the engine for some time prior to sampling. This is of prime importance since all subsequent operations will be invalidated by poor sampling techniques.

1. *The determination of wear metals in used lubricating oils*

(i) *Summary*

The method is applicable to the determination of Al, Ag, Cr, Cu, Fe, Ni, Si, Sn, Mg, Pb and Zn in used lubricating oils and may also be applicable to other elements of interest. Samples are diluted with MIBK and compared with organometallic standards in MIBK. Limits of detection will range from approximately $0.1\ \mu g\ ml^{-1}$ for elements such as Zn and Cu, approximately $1\ \mu g\ ml^{-1}$ for elements such as Sn and Al and greater for Si.

(ii) *Sample preparation*

Pipette 5 ml of the used lubricating oil, which has been previously agitated, into a 25 ml volumetric flask and make up to the mark with MIBK. Shake well to dissolve and homogenise.

(iii) *Standard preparation*

Using either commercially available standards or those prepared as described in Section III, prepare calibration standards by dilution with MIBK. Choose a concentration range for each element which exhibits an approximately linear response. This will vary from instrument to instrument but as a rough guide the following is useful.

Aluminium	$0-10\ \mu g\ ml^{-1}$
Silver	$0-5\ \mu g\ ml^{-1}$
Chromium	$0-5\ \mu g\ ml^{-1}$
Copper	$0-5\ \mu g\ ml^{-1}$
Iron	$0-5\ \mu g\ ml^{-1}$
Nickel	$0-5\ \mu g\ ml^{-1}$
Silicon	$0-40\ \mu g\ ml^{-1}$
Tin	$0-20\ \mu g\ ml^{-1}$
Magnesium	$0-1\ \mu g\ ml^{-1}$
Lead	$0-10\ \mu g\ ml^{-1}$
Zinc	$0-2\ \mu g\ ml^{-1}$

Prepare at least four calibration standards, including the blank, to cover the chosen concentration range. Prepare the calibration standards as required for use; do not store.

(iv) *Instrumental*

Adjust the instrumental parameters of spectral band pass, wavelength and lamp current in accordance with the manufacturer's recommended conditions. The nitrous oxide—acetylene flame should be used when determining

Al, Si and Sn and may also prove advantageous for the determination of Cr, Fe and Ni. The air—acetylene flame should be used when determining the other elements mentioned. For elements not covered here refer to the manufacturer's handbook.

Establish the flame conditions while aspirating a blank MIBK solution. As is normal when aspirating organic solvents a reduction in the acetylene flow rate will be required to obtain an acceptable flame. When using the air—acetylene flame the use of auxiliary air or an increase in air flow rate may help to produce an improved flame.

Check to see if background correction is necessary, especially for elements with resonance wavelengths in the 200—300 nm range, by using a deuterium or hydrogen lamp or a nearby non-absorbing line.

(v) *Procedure*

Aspirate standards, then samples, noting the absorbance readings. Aspirate the standards again to check that no drift has occurred. Prepare a calibration graph for the standards and read the sample concentrations from this.

Concentration in lubricating oil $= x \times 5 \,\mu g \, ml^{-1}$

where $x = \mu g \, ml^{-1}$ in the diluted sample.

(vi) *Note*

Where the concentration of the element of interest in the diluted sample exceeds the range of the calibration, a further dilution should be performed and the appropriate factor included in the above formula.

E. Gasoline

Alkyl Pb compounds, mainly tetramethyl and tetraethyl Pb, are added to gasoline as anti-knock agents. The amount added varies from country to country but is generally in the range 200—1500 $\mu g \, ml^{-1}$ Pb and this may be present as one or both of the Pb alkyl species mentioned. Unfortunately, it has been found that the Pb response in atomic absorption spectroscopy is dependent on the particular alkyl with which it is associated. Thus, unless it is known that only one particular Pb alkyl is present in a gasoline sample, it is not possible to employ a simple dilution procedure. Since the exact nature of the Pb species is seldom known for sure, then a general analytical method must ensure that the response from the various Pb alkyls is equalised in some manner. This is achieved by stabilisation with iodine and a quaternary ammonium salt.

In unleaded gasoline the Pb, present naturally, may exist in a variety of organometallic complexes. These may exhibit differing pre-atomisation and atomisation characteristics during the determination of Pb using electrothermal atomisation. The addition of iodine is again used to overcome this problem.

References p. 306

The use of cyclopentadienyl-carbon monoxide compounds of Mn, as anti-knock agents in fuels, has been known for some time. In recent years the use of methylcyclopentadienyl manganese tricarbonyl (MMT) as an anti-knock agent for gasoline has become more widespread, due to the decline in use of Pb alkyl compounds.

1. *The determination of lead in leaded gasoline*

(i) *Summary*

The method is suitable for the determination of Pb in gasoline in the range $1-1000\,\mu g\,ml^{-1}$ ($0.001-1\,g\,l^{-1}$). The gasoline sample is diluted with MIBK and shaken with a solution of iodine in toluene. Iodo lead alkyl species are formed and these are stabilised with a liquid anion exchanger. Thus, the problem of the variable Pb response is overcome. This solution is compared with standards prepared in a similar fashion from Pb chloride in MIBK.

(ii) *Sample preparation*

The sample dilution required will depend on the concentration of Pb present in the gasoline. This should be chosen to fall within the range defined by the calibration standards. Pipette the chosen volume of gasoline into 100 ml volumetric flasks in duplicate. To this add approximately 50 ml of MIBK and 0.2 ml (200 µl) of a solution of iodine in toluene (see note 2). Swirl the mixture and allow to stand for two minutes. Add 5 ml of a 1% v/v solution of Aliquat 336 (see note 1), swirl to react and dilute to the mark with MIBK.

(iii) *Standard preparation*

A stock Pb standard must first be prepared. Place reagent-grade Pb chloride in an oven at approximately 100°C for a few hours. Accurately weigh 0.3355 g of this salt and dissolve in 200 ml of a 10% v/v solution of Aliquat 336 in a 250 ml volumetric flask. Dilute to the mark with 10% v/v Aliquat 336 solution. This standard contains $1000\,\mu g\,ml^{-1}$ Pb. Store in a darkened bottle, tightly stoppered.

Calibration standards are prepared by dilution of the stock standard. Choose an approximately linear range for Pb on the instrument to be used (normally $0-10\,\mu g\,ml^{-1}$). Into four 100 ml volumetric flasks place the following: a volume of Pb free iso-octane (see note 3) equal to the volume of gasoline in the diluted sample; 0.2 ml of the iodine in toluene solution; 5 ml of the 1% v/v solution of Aliquat 336; and such volumes of the stock solution (after dilution, with MIBK, to $100\,\mu g\,ml^{-1}$ Pb if necessary) as are necessary to give the desired calibration concentrations. One standard must contain no Pb. Dilute to the mark with MIBK. Prepare these calibration standards freshly as required; do not store.

(iv) *Instrumental*

Adjust the instrumental parameters of spectral band pass, wavelength and lamp current in accordance with the manufacturer's recommended conditions. Use an air—acetylene flame and, aspirating the blank solution, adjust the acetylene flow rate to obtain a fuel-lean flame.

(v) *Procedure*

Aspirate the standards, then samples, and record the absorbance values. Aspirate the standards again to check for instrumental drift. Plot a calibration graph for the standards and from this read the concentration of Pb in the samples.

Concentration of Pb in gasoline $= (x \times D/1000)\, \text{g l}^{-1}$

where x = averaged Pb concentration in diluted sample ($\mu\text{g ml}^{-1}$) and D = dilution factor (including further dilution of sample).

(vi) *Notes*

(1) Aliquat 336 is tricapryl methyl ammonium chloride. Two diluted solutions of this are required, a 1% v/v and a 10% v/v solution in MIBK. Aliquat 336 is available from a number of sources, including: Phasesep R & D Chemical Company, Deeside Industrial Estate, Queensferry, Flintshire, Gt. Britain, and General Mills Corporation, Minneapolis, MN 55415, U.S.A.

(2) Iodine solution (3% w/v): dissolve 3 g of iodine in toluene and dilute to 100 ml.

(3) Pb free gasoline: use 2,2,4-trimethyl pentane (iso-octane).

2. *The determination of lead in lead free gasoline using electrothermal atomisation*

(i) *Summary*

The method is applicable to the determination of Pb in unleaded gasoline, in the range 0.001—1 $\mu\text{g ml}^{-1}$. The gasoline is diluted with MIBK and iodine added to overcome the problems of varying response from different Pb species. Furnace, rod or cup type atomisers may be used for the determination. Only general conditions for the determination are given as these will vary markedly from one instrument to another. Samples are compared with standards prepared by the dilution of oil-based standard material with MIBK.

(ii) *Sample preparation*

Dilute the Pb free gasoline with MIBK in order to bring the concentration within the chosen working calibration on the instrument being used. Add iodine from the stock iodine solution (see note 1) to make the final solution 1000 $\mu\text{g ml}^{-1}$ in iodine.

(iii) *Standard preparation*

Use Pb in oil standards (see note 2) and dilute to the established linear range (generally 0—0.1 µg ml^{-1}) with MIBK. Do not dilute by greater than a factor of twenty in any one dilution stage. Add iodine from the stock iodine solution to make the final solutions 1000 µg ml^{-1} in iodine. Prepare at least four calibration standards to cover the range, including a blank. Prepare these standards freshly as required; do not store.

(iv) *Instrumental*

Adjust the spectrometer parameters of spectral band pass, wavelength and lamp current in accordance with the manufacturer's recommended conditions for Pb. It will be necessary to correct for background effects with most instruments. Use automatic background correction facilities where available, otherwise make use of a nearby non-absorbing line. The use of the 283.3 nm Pb wavelength will reduce background effects compared with the 217.0 nm wavelength.

The following are general conditions for the electrothermal atomisation device.

Purge gas	argon or nitrogen
Dry time	10—50 s (volume dependent)
Dry temperature	100°C
Pre-atomisation heating time	20—50 s
Pre-atomisation heating temperature	up to 550°C
Atomisation time	5—10 s
Atomisation temperature	1800—2200°C

Where the gas flow may be interrupted during the atomisation period (gas stop facility) then this may be used to advantage in increasing the sensitivity of the determination.

(v) *Procedure*

The sample volume for injection into the electrothermal atomiser will vary between approximately 5 and 100 µl depending upon the particular instrument being used. The choice of measurement mode, peak height or peak area will depend on the facilities available, both being applicable. Inject standards, then samples in duplicate and obtain peak height or peak area measurements for each. Where repeat injections give large variations in signal, a further injection should be made. An average peak height or peak area value may then be obtained for each standard and sample. While obtaining sample data it is advisable to periodically inject a standard to check for instrument drift. Prepare a calibration graph for the standards and from this read the concentration of Pb in the sample.

Concentration of Pb in gasoline = $x \times D$ µg ml^{-1}

where x = µg ml^{-1} Pb in dilute sample, and D = dilution factor.

(vi) *Notes*
(1) Stock iodine solution (5% w/v): dissolve 5 g of iodine in MIBK and make up to 100 ml.
(2) Pb in oil standard: use the commercial standard described in Section III, or similar.

3. The determination of manganese in gasoline

(i) *Summary*
The method is suitable for the determination of Mn, present as methylcyclopentadienyl manganese tricarbonyl, in gasoline in the range 0.5–40 μg ml^{-1}. The sample is reacted with bromine then diluted with MIBK and the standards prepared in a similar manner. This ensures an equal response from the Mn species in both sample and standard solutions.

(ii) *Sample preparation*
Dilute the sample with MIBK so as to fall within the chosen calibration range. Add 1 ml of the bromine solution (see note 1) prior to making up to the mark in a 100 ml volumetric flask. Shake well to mix.

(iii) *Standard preparation*
Use a commercially available Mn in oil standard or an organometallic Mn standard prepared as described in Section III (see note 2). Dilute the stock standard with MIBK and add 1 ml of the bromine solution (see note 1) to give four calibration standards, including the blank, to cover the range approximately 0–4 μg ml^{-1} Mn in 100 ml volumetric flasks.

(iv) *Instrumental*
Adjust the instrumental parameters of spectral band pass, wavelength and lamp current in accordance with the manufacturer's recommended conditions. Use an air–acetylene flame and, aspirating an MIBK blank solution, adjust the acetylene flow rate in order to obtain a lean, blue flame.

(v) *Procedure*
Aspirate standards, then samples, and record the absorbance values. Aspirate the standards again to check that no drift has occurred. Prepare a calibration graph for the standards and from this read the concentration of Mn in the samples.

Concentration of Mn in gasoline = $(x \times D/1000)$ g l^{-1}

where x = concentration of Mn in final dilution (μg ml^{-1}) and D = dilution factor.

References p. 306

(vi) *Notes*

(1) Bromine solution: add reagent grade bromine to an equal volume of carbon tetrachloride. Shake well to mix.

(2) Manganese standard solution: use manganese in oil standard as supplied by Conostan or manganese 4-cyclohexanebutyrate (see Section III).

REFERENCES

1 B. Welz, Atomic Absorption Spectroscopy, Verlag Chemie, Weinheim, New York, 1976, p. 215.
2 B. E. Buell, in E. L. Grove (Ed.), Applied Atomic Spectroscopy, Vol. 2, Plenum Press, New York, 1978, p. 53.
3 Sampling Petroleum and Petroleum Products, Annual Book of ASTM Standards, Part 18, Method D270-65.

Chapter 4i

Methods for the analysis of glasses and ceramics by atomic spectrometry

W. M. WISE, J. P. WILLIAMS and R. A. BURDO
Research and Development Laboratories, Corning Glass Works, Corning, New York 14830 (U.S.A.)

I. INTRODUCTION

The implementation of modern spectrometers with stable excitation sources and electronics has engendered atomic emission (AES) and atomic absorption spectrometry (AAS) methods for determining major ($>10\%$) and minor (0.1 to 10%) amounts of selected elements in glasses, ceramics and other similar materials. Since each measurement is done using a spectral line which is characteristic only for the analyte, separations are usually unnecessary, leading to quick results. In most cases AES and AAS signals are sufficiently stable, for major and minor components, to produce final results which are as accurate and precise as those acquired using the classical titrimetric and gravimetric procedures. It is fortunate that rather simple AES and AAS methods are available, because the classical procedures for some of the elements are so arduous that in the past results were often obtained by difference.

Recent reviews have appeared on the analysis of glasses and related materials using AES and AAS [1—3]. It is the purpose of this chapter to present to the laboratory technician with experience in AES and AAS some selected methods for determining the elements common to such substances. Since AES and AAS methods are based upon comparisons with standards, it is desirable to have reasonable estimates of the elemental concentrations in a sample before beginning an analysis. Sometimes this information is obtainable from batch compositions. Frequently, the composition of the material is unknown. A commonly used technique for obtaining semiquantitative estimates is optical emission spectroscopy, for which a procedure has been previously published [2].

II. APPARATUS

A. Flame emission and absorption

Spectrometers comparable to those available from Perkin-Elmer Corporation, Instrument Division, Main Avenue MS-12, Norwalk, Connecticut-06856,

Varian Associates, 611 Hansen Way, Palo Alto, California-94303, and Instrumentation Laboratories, Inc., Jonspin Road, Wilmington, Massachussets-01887, will provide adequate signals for those elements that can be determined by AES or AAS using air + C_2H_2 or $N_2O + C_2H_2$ flames.

B. Plasma emission

Since boron produces weak flame absorption signals, an echelle spectrometer equipped with a d-c plasma excitation source similar to that procurable from Spectrametrics, Inc., 204 Andover Street, Andover, Massachusetts-01810, can be used to determine this element by AES.

III. CHEMICALS

Only established analytical reagent grade chemicals should be employed for the preparation of standards and reagent solutions and the treatment of samples.

Anhydrous Li_2CO_3, $LiBO_2$ (−200 mesh), $BaCO_3$, $CaCO_3$, Na_2CO_3 (−200 mesh), K_2CO_3, $SrCO_3$, $Na_2B_4O_7$ (−200 mesh) and SiO_2 (99.9+%, −100 mesh). Fe, Zn, Cd, Al and Mg metals (all 99.99%). $Pb(NO_3)_2$ (99.5+%). H_3BO_3 (99.99%). $(NH_4)_6Mo_7O_{24} \cdot 4H_2O$. NaCl (< 0.001% K). KCl (< 0.002% Na). $LaCl_3 \cdot 6H_2O$ (< 0.001% Mg + Ca + Sr + Ba). EDTA (99.6% free acid). HF (29 M). $HClO_4$ (12 M). HCl (12 M). HNO_3 (16 M). H_3PO_4 (15 M). NH_4OH (15 M, filtered).

Reference to water implies the use of distilled or deionized water of equal quality. Unless otherwise specified, all dilutions to volume are done with water.

IV. REAGENT SOLUTIONS

(1) KCl flame buffer (2500 µg KCl ml^{-1}). Dissolve 2.5 g of KCl in water and dilute to one liter (1 l).

(2) NaCl flame buffer (25 000 µg NaCl ml^{-1}). Dissolve 25 g of NaCl in water and dilute to 1 l.

(3) LaCl$_3$ flame buffer (25 000 µg LaCl$_3$ ml^{-1} + 0.3 M HClO$_4$). Dissolve 36.0 g of $LaCl_3 \cdot 6H_2O$ in water, add 25 ml of 12 M $HClO_4$ and dilute to 1 l.

(4) Methyl orange indicator (0.1%). Dissolve 0.1 g of methyl orange in 100 ml of ethyl alcohol.

(5) LaCl$_3$ + EDTA flame buffer (25 000 µg LaCl$_3$ ml^{-1} + 0.25 M EDTA). Add 73.1 g of EDTA and 10 drops of methyl orange indicator to 800 ml of water containing 67 ml of 12 M $HClO_4$. Stir and add 15 M NH_4OH until all of the EDTA is dissolved. Introduce 36.0 g of $LaCl_3 \cdot 6H_2O$ and continue stirring and adding NH_4OH until the solution is clear and yellow. Cool and dilute to 1 l.

(6) $Ca(ClO_4)_2 + HClO_4$ flame buffer (25 000 μg Ca ml^{-1} + 1.8 M HClO$_4$). Introduce 31.25 g of calcium carbonate into a 500-ml volumetric flask which is half-filled with water. With slow stirring add 125 ml of 12 M HClO$_4$. After the effervescence ceases, heat the solution to expel excess CO$_2$, cool to room temperature and dilute to volume.

(7) $NaClO_4$ flame buffer (10 000 μg Na ml^{-1}). Dissolve 11.54 g of Na$_2$CO$_3$ with 100 ml of water and 20 ml of 12 M HClO$_4$. Heat to boiling to expel the CO$_2$, cool to room temperature and dilute to 500 ml.

(8) Molybdate complexing solution (39 000 μg (NH$_4$)$_6$Mo$_7$O$_{24}$ · 4H$_2$O ml^{-1}). Dissolve 78 g of (NH$_4$)$_6$Mo$_7$O$_{24}$ · 4H$_2$O in 700 ml of water. Dilute to 2 l after filtering through a Whatman No. 41 filter paper.

(9) Molybdate diluting solution. Dissolve 0.5 g of Na$_2$CO$_3$ and 0.5 g of Na$_2$B$_4$O$_7$ in 100 ml of 0.39 M nitric acid. Add 100 ml of molybdate complexing solution and dilute to 500 ml with 0.126 M HNO$_3$. This solution can be used until a scale of hydrated molybdenum oxide develops on the container walls (several days to a week). Then a fresh solution should be prepared. The scale can be removed with ammonium hydroxide solution.

(10) Silicon buffer solution (10 000 μg Si ml^{-1}). Place 21.4 g of SiO$_2$ into a 1-l plastic jar. Add 300 ml of cold water and then slowly add 250 ml of chilled 29 M HF. Allow the solution to approach room temperature slowly in order to dissipate the heat of reaction safely without the loss of silicon as the fluoride and subsequently stand at room temperature (jar loosely capped) until dissolution is complete. Dilute to 1 l in a plastic volumetric flask and store in plasticware.

(11) Silicon diluting solution. To a 1000-ml plastic volumetric flask, add 30 ml of silicon buffer solution and 42.5 ml of 29 M HF. Dilute to volume with water to yield a solution containing 300 μg Si ml^{-1} and 1.45 M HF. Store in plasticware.

V. STANDARDS

(i) Lithium, sodium, potassium, calcium, strontium and barium stock solutions

To obtain 1.000 g l^{-1} solutions, place the following quantities of each element carbonate into a 1-l volumetric flask, and mix with 100 ml of water (Li$_2$CO$_3$, 5.324 g; Na$_2$CO$_3$, 2.3042 g; K$_2$CO$_3$, 1.7674 g; CaCO$_3$, 2.4973 g; SrCO$_3$, 1.685 g; BaCO$_3$, 1.4371 g). Slowly add 10 ml of 12 M HClO$_4$. After the bubbling has ceased, heat to boiling to expel the excess CO$_2$, cool to room temperature and dilute to volume.

Prepare 50 μg ml^{-1} stock solutions by diluting 50 ml of each 1.000 g l^{-1} solution to 1 l.

Lithium, sodium and potassium AES calibration standards. Pipet 5, 10, 20, 30, 40, 60 and 80-ml portions of the 50 μg ml^{-1} stock solutions into 1-l volumetric flasks and add 2.5 ml of 12 M $HClO_4$. Dilute the lithium calibration standards to volume. Place 40 ml of the KCl flame buffer into the flasks containing the sodium calibration standards, and 40 ml of the NaCl flame buffer into the potassium calibration standards and then dilute each to volume. Lithium, sodium or potassium: 0.25, 0.50, 1.00, 1.50, 2.00, 3.00 and 4.00 $\mu g\,ml^{-1}$.

Calcium, AES and AAS, strontium AES and barium AES calibration standards. Pipet 5, 10, 15, 20, 25, 30, 40 and 60-ml aliquots of the 50 $\mu g\,ml^{-1}$ stock solutions into 1-l volumetric flasks. Add 500 ml of water, 10 drops of methyl orange solution, 40 ml of the $LaCl_3$ + EDTA flame buffer and sufficient 15 M NH_4OH (dropwise) to each flask to turn the color of the solution yellow and dilute to volume. Calcium, strontium or barium: 0.25, 0.50, 0.75, 1.00, 1.25, 1.50, 2.00 and 3.00 $\mu g\,ml^{-1}$.

Barium AAS calibration standards. Pipet 5, 10, 15, 20 and 25-ml portions of the 1.000 $g\,l^{-1}$ solution into 500-ml volumetric flasks. Add 250 ml of water, 5 drops of methyl orange solution, 20 ml of the $LaCl_3$ + EDTA flame buffer and sufficient 15 M NH_4OH (dropwise) to each flask to turn the color of the solution yellow and dilute to volume. Barium: 10, 20, 30, 40 and 50 $\mu g\,ml^{-1}$.

(ii) *Magnesium stock solution*
Dissolve 1.000 g of clean magnesium metal in a 1-l volumetric flask with 100 ml of water and 20 ml of 12 M $HClO_4$. After all of the metal has dissolved (sometimes heating is necessary) cool the solution to room temperature and dilute to volume. Prepare a stock solution containing 20 $\mu g\,Mg\,ml^{-1}$ by diluting 20 ml of the 1.000 $g\,l^{-1}$ solution to 1 l in a volumetric flask.

Magnesium calibration standards. Pipet 5, 10, 20, 30, 40, 50 and 60-ml aliquots of the stock solution into 1-l volumetric flasks, add to each 40 ml of the $LaCl_3$ flame buffer and dilute to volume. Magnesium: 0.10, 0.20, 0.40, 0.60, 0.80, 1.00 and 1.20 $\mu g\,ml^{-1}$.

(iii) *Aluminum stock solution*
Prepare a 5.000 $g\,Al\,l^{-1}$ solution by dissolving clean aluminum metal in 100 ml of water containing 50 ml of 12 M HCl and 10 ml of 16 M HNO_3. Cool the solution to room temperature and dilute to volume in a 1-l volumetric flask. Make a 500 $\mu g\,Al\,ml^{-1}$ stock solution by diluting a 100-ml aliquot to 1 l in a volumetric flask.

Aluminum calibration standards. Pipet 5, 10, 15, 20, 25, 30, 35 and 40-ml aliquots of the stock solution into 1-l volumetric flasks, add 40 ml of the $Ca(ClO_4)_2$ + $HClO_4$ flame buffer and dilute to volume. Aluminum: 2.5, 5.0, 7.5, 10.0, 12.5, 15.0, 17.5 and 20.0 $\mu g\,ml^{-1}$.

(iv) *Iron stock solution*

Dissolve 1.000 g of clean iron wire in 100 ml of 6 M HClO$_4$ and dilute to 1 l in a volumetric flask. Prepare a 50 μg Fe ml^{-1} stock solution by diluting a 25-ml aliquot to 500 ml.

Iron calibration standards. Pipet 5, 10, 20, 30, 40 and 50-ml aliquots of the stock solution into 250-ml volumetric flasks, add 10 ml of 1.25 M H$_3$PO$_4$ and dilute to volume. Iron: 1.0, 2.0, 4.0, 6.0, 8.0 and 10.0 μg ml^{-1}.

(v) *Zinc stock solution*

Prepare a 1.000 g Zn l^{-1} solution by dissolving the clean metal in 100 ml of 3 M HCl, and diluting to volume in a 1-l volumetric flask. Dilute a 20-ml aliquot to 1 l to obtain a 20 μg Zn ml^{-1} stock solution.

Zinc calibration standards. Pipet, 5, 10, 20, 30, 40 and 50-ml aliquots of the stock solution into 1-l volumetric flasks, add 2.5 ml of 12 M HClO$_4$ and dilute to volume. Zinc: 0.10, 0.20, 0.40, 0.60, 0.80 and 1.00 μg ml^{-1}.

(vi) *Cadmium stock solution*

Dissolve 1.000 g of mossy cadmium metal in 20 ml of 5 M HNO$_3$ and dilute to 1 l in a volumetric flask. Dilute a 20-ml aliquot to 500 ml to prepare a 40 μg Cd ml^{-1} stock solution.

Cadmium calibration standards. Pipet 5, 10, 20, 30, 40 and 50-ml aliquots of the stock solution into 1-l volumetric flasks, add 2.5 ml of 12 M HClO$_4$ and dilute to volume. Cadmium: 0.20, 0.40, 0.80, 1.20, 1.60 and 2.00 μg ml^{-1}.

(vii) *Lead stock solution*

Dissolve 1.5985 g of Pb(NO$_3$)$_2$ in 200 ml of water, add 0.5 ml of 16 M HNO$_3$ and dilute to 1 l in a volumetric flask. Prepare a 100 μg Pb ml^{-1} stock solution by diluting 50 ml to 500 ml in a volumetric flask.

Lead calibration standards. Pipet 5, 10, 20, 30, 40 and 50-ml aliquots of the stock solution into 500-ml volumetric flasks. Add 1.25 ml of 12 M HClO$_4$, 5 ml of the NaClO$_4$ flame buffer and dilute to volume; Lead: 1.0, 2.0, 4.0, 6.0, 8.0 and 10.0 μg ml^{-1}.

(viii) *Silicon stock solution*

Thoroughly mix 0.1070 g of SiO$_2$ with 0.5 g of Na$_2$CO$_3$ and 0.5 g of Na$_2$B$_4$O$_7$ in a 30-ml platinum crucible. Cover and place in a muffle furnace at 1000°C for 30 min. After cooling, insert a Teflon-coated magnetic stirring bar inside the crucible and immerse the crucible into a 250-ml Teflon beaker containing 100 ml of molybdate complexing solution and 100 ml of 0.39 M HNO$_3$. Stir magnetically until the melt is dissolved (10 to 30 min) and dilute to 500 ml with 0.126 M HNO$_3$ yielding a stock solution of 100 μg Si ml^{-1}.

References p. 319

Silicon calibration standards. Pipet 10, 20, 30, 50, 60 and 80-ml aliquots of the stock solution into 100-ml volumetric flasks and dilute to volume with the molybdate diluting solution. Silicon: 10, 20, 30, 50, 60 and 80 $\mu g\,ml^{-1}$. Standards containing less than 40 μg Si ml^{-1} are stable for about 10 days until a scale of hydrated molybdenum oxide forms on the container walls. Then fresh standards are prepared. Standards of higher concentrations are stable for months with no accumulation of scale. Note that all standards are stored in plastic containers.

(ix) *Boron stock solutions*

Dissolve 5.7194 g of H_3BO_3 in water and dilute to 1 l to yield a 1.000 g B l^{-1} stock solution. Prepare a 100 μg B ml^{-1} stock solution by diluting 100 ml of the 1.000 g B ml^{-1} stock solution to 1 l in a volumetric flask.

Boron calibration standards. Place 10, 20, 30, 40, 50, 60 and 80-ml aliquots of the 100 μg B ml^{-1} stock solution and 10, 20, 30, 40, 60 and 80-ml aliquots of the 1.000 g B l^{-1} stock solution into 1-l plastic volumetric flasks, add 30 ml of the silicon buffer solution, 42.5 ml of 29 M HF and dilute to volume. Boron: 1.0, 2.0, 3.0, 4.0, 5.0, 6.0, 8.0, 10.0, 20.0, 30.0, 40.0, 60.0 and 80.0 $\mu g\,ml^{-1}$. Store in plasticware.

VI. SAMPLE PREPARATION

In order to obtain products with prescribed properties, most commercial glasses and ceramics are prepared under conditions designed to produce homogeneity, i.e., pre-mixed batch materials and/or stirring during the melting process. Consequently, in the majority of cases it is not difficult to acquire a representative sample for analysis by crushing, then grinding to a suitable particle size with an agate or alundum mortar and pestle. It is important to first clean the sample, particularly if it was obtained as a result of some grinding or cutting operation. Also, crushing and grinding containers must be carefully selected to prevent the introduction of contaminants. Mortars and pestles can be precleaned by thoroughly grinding a portion of the sample which is then discarded. For most materials that are decomposed by acids, comminution to −100 mesh is satisfactory. Reducing the particle size to −200 mesh may be necessary for materials that require fusions. Samples should not be ground any finer than necessary for decomposition, to minimize contamination from the mortar and the pick-up of water and CO_2 from the atmosphere. When it is suspected that the sample has picked up water and/or CO_2 during comminution it may be necessary to dry it at 120°C or ignite at 1000°C for 1 h. Then, in order to prevent the sample from acquiring additional water and/or CO_2 prior to the analysis, it can be stored in a desiccator containing BaO.

Small crucible melts are sometimes inhomogeneous, and glass-ceramics and certain opal glasses normally consist of discrete phases with variable chemical compositions and different degrees of hardness. AES and AAS procedures for determining major constituents normally require only about 0.1 to 0.2 g of sample. Consequently, it is extremely important that if any question exists concerning the homogeneity of a material, the entire portion selected for grinding be comminuted fine enough to pass the sieve followed by thorough mixing of the sample.

VII. DETERMINATION OF Li, Na, K, Mg, Ca, Sr, Ba, Al, Fe, Zn, Cd AND Pb BY AES OR AAS

A. Procedure for acidic decomposition

This procedure is recommended for materials that are decomposed by attack with HF and $HClO_4$. (See also the section in Chapter 3 giving precautions to be taken when using perchloric acid.)

Weigh a 0.1 to 0.2-g portion of the ground sample (about −100 mesh) to the nearest 0.1 mg directly into a 50-ml platinum dish and moisten with several ml of water plus 4 ml of 6 M $HClO_4$. Gently swirl the mixture to prevent the formation of lumps and add 5 ml of 29 M HF. After all visible reaction has ceased, place the dish on a steam bath, and evaporate the solution until just $HClO_4$ remains. Place the dish on a hotplate and heat until fumes of $HClO_4$ are evolved. (At this point, insoluble fluorides may begin to decompose resulting in a loss of solution by effervescence. Boric acid (about 40 mg) can be added to effect non-violent decomposition of these fluorides, and gentle swirling of the dish and contents until all bubbling ceases helps prevent any loss.) Wash the sides of the dish down with a stream of water, and evaporate the solution again to $HClO_4$ fumes and fume hard to incipient dryness. Repeat the washing down of the sides and fuming two more times with the addition of more $HClO_4$ when required to prevent evaporation to dryness. Moisten the residue with 10 drops of 6 M $HClO_4$ and add 25 ml of water. If the solids do not dissolve with gentle heating on a hotplate, cover the dish with a platinum lid and heat at almost boiling for an hour to effect complete dissolution. It may be necessary to add water periodically to keep the volume at about 20 ml. Transfer the solution to a 100 or 200-ml volumetric flask. Add about 20 ml of water to the dish along with 0.5 ml of 6 M HCl. (Omit the HCl if lead is present.) Heat gently for several minutes and then quantitatively transfer to the flask. Cool to room temperature and dilute to volume. Pipet appropriate aliquots and dilute to volumes that will place the concentrations of the analytes in the optimum ranges of the calibration curves in the presence of the same amounts of flame buffers as are in the calibration standards.

References p. 319

B. Procedure for fusion

Fusions are used for materials that are resistant to attack by acids, i.e., refractories and abrasives. Platinum or platinum + rhodium-alloy crucibles and lids provide excellent containers for performing fusions. Since silicon interferes with the absorption and emission responses of some elements such as the alkaline earths and aluminum, it is common practice to remove major amounts of it by adding HF to the sample and evaporating to dryness before fusion. (Indeed, if silicon is absent, the preliminary treatment is omitted.) The most widely used fluxes are carbonates or carbonate + borate mixtures. The choice depends on which one will effectively decompose the sample at a suitable temperature within a reasonable period of time and what elements are being determined. Examples are: a 1:1 mixture of $Na_2CO_3 + Na_2B_4O_7$ or $LiBO_2$. Usually, a 5 to 10:1 ratio of flux to sample produces a clear melt on fusion making it easy to detect visually when decomposition of the sample has been completed. However, preliminary experiments should be performed to ascertain the minimum temperature and amount of flux required to facilitate rapid dissolution of the cooled melt, maintain low blanks and keep large amounts of extraneous salts from clogging burners and causing the scattering of radiant energy. Sometimes it is necessary to include in the standards the same quantity of salt introduced by the fusion into the analyte solution to compensate for any enhancement or depression of AES and AAS signals. Loss of sample components can occur during a fusion. Lead and cadmium can be volatilized from the system with prolonged heating. Iron can be lost in the following manner. At high temperatures carbonates used in the batch can decompose to the oxide and CO_2. The CO_2 decomposes to produce appreciable amounts of CO, and the CO reduces the iron to the metal which subsequently alloys with the platinum. About 25 mg of $NaNO_3$ added to 0.5 g of flux helps to prevent the reduction of iron and the loss of lead and cadmium by maintaining an oxidizing atmosphere.

Weigh a 0.1-g portion of the −200-mesh sample to the nearest 0.1 mg into a 30-ml platinum crucible, and mix with 5 ml of H_2O and 5 ml of 29 M HF. Place the crucible on a low-temperature hotplate and evaporate the solution to dryness. Heat the residue until HF fumes cease to evolve. Thoroughly mix the flux with the residue using a platinum stirring rod. Heat the mixture, slowly at first, until the flux is melted, then cover and fuse in a furnace or over a blast burner with an oxidizing flame until all of the sample is decomposed. Dissolve the cooled melt in 25 ml of hot 1.2 M $HClO_4$ or HNO_3 with stirring, taking precautions to prevent the loss of sample through effervescence. (Often it is advisable to begin the treatment of carbonate cakes with cold dilute acid. Sometimes it may be necessary to employ additional or more concentrated acid to effect dissolution.) Cool the solution to room temperature and dilute to a tractable volume that will keep all solids dissolved. Then perform the dilutions for determining the various analytes.

C. Measurement of analytes

Handbooks supplied by manufacturers contain valuable information for selecting instrument settings. However, in order to optimize the analyte's signal, burners should be clean; and flame stoichiometry and burner position should be adjusted before each group of measurements is taken on standards and samples.

The recommended conditions for determining the elements by AES and AAS spectrometry are given in Table 1. After a suitable burner warm-up time, one AES or AAS reading is obtained on each calibration standard and sample solution using an 8 to 10-second integration time. The process is repeated at least two or three more times to acquire an average of three or four readings for each solution.

Calibration curves are based on AES or AAS measurements on at least five (preferably six) standards. The calibration curve, relating the average AES or AAS reading to concentration, is graphed either manually or by a suitable computer program.

Below is shown a typical formula for calculating the percent oxide in a sample.

$$\frac{(\mu g\ ml^{-1}\ of\ element) \times (sample\ volume,\ ml) \times (dilution\ factor) \times (gravimetric\ factor)}{(sample\ weight,\ mg) \times (10)}$$

$$=\ \%\ oxide$$

D. Discussion

1. *Alkalies*

Alkali oxides are present in glasses to function as modifiers, lowering the liquidus temperature, and with the exception of silicon are probably the most frequently determined elements in glass analysis. With modern instrumentation the stabilities of the alkali resonance-line responses with the air + C_2H_2 flame are about equal using either the AES or the AAS mode. Consequently, to obviate an unnecessary potential variable, i.e. the hollow-cathode lamp, the alkalies are best measured by AES. The air + C_2H_2 flame, being hotter than air + propane, produces emission signals that are less sensitive but more stable and reproducible which are the necessary requirements for determining a major or minor constituent. With a stoichiometric air + C_2H_2 flame lithium does not need an ionization suppressor. However, the strontium hydroxide band at 670 nm may interfere with the lithium signal, requiring wavelength scanning. With a stoichiometric to oxidizing flame sodium and potassium require 100 μg KCl ml^{-1} and 1000 μg NaCl ml^{-1}, respectively, as ionization suppressors to produce signals that are unaffected by other elements. It has been found that the removal of radiant

TABLE 1

RECOMMENDED CONDITIONS FOR AES OR AAS MEASUREMENTS FOR Li, Na, K, Mg, Ca, Sr, Ba, Al, Fe, Zn, Cd AND Pb

Element	Mode	λ (nm)	Oxidant + fuel	Final buffer concentration
Li	AES	670.8	air + C_2H_2[a]	0.03 M $HClO_4$
Na	AES	589.0	air + C_2H_2[b]	100 μg KCl ml^{-1} + 0.03 M $HClO_4$
K	AES	766.5	air + C_2H_2[c]	1000 μg NaCl ml^{-1} + 0.03 M $HClO_4$
Mg	AAS	285.2	N_2O + C_2H_2[c]	1000 μg $LaCl_3$ ml^{-1} + 0.01 M $HClO_4$
Ca	AES, AAS	422.7	N_2O + C_2H_2[c]	1000 μg $LaCl_3$ ml^{-1} + 0.01 M EDTA, pH = 4.5
Sr	AES	460.7	N_2O + C_2H_2[c]	1000 μg $LaCl_3$ ml^{-1} + 0.01 M EDTA, pH = 4.5
Ba	AES, AAS	553.5	N_2O + C_2H_2[c]	1000 μg $LaCl_3$ ml^{-1} + 0.01 M EDTA, pH = 4.5
Al	AAS	309.3	N_2O + C_2H_2[c]	1000 μg Ca ml^{-1} + 0.07 M $HClO_4$
Fe	AAS	248.3	air + C_2H_2[b]	0.05 M H_3PO_4
Zn	AAS	213.8	air + C_2H_2[b]	0.03 M $HClO_4$
Cd	AAS	228.8	air + C_2H_2[b]	0.03 M $HClO_4$
Pb	AAS	216.9	air + C_2H_2[b]	100 μg Na ml^{-1} + 0.03 M $HClO_4$

[a] Stoichiometric flame (burner slot = 10 cm).
[b] Stoichiometric to oxidizing flame (burner slot = 10 cm).
[c] Oxidizing flame (burner slot = 6 cm).

energy below 625 nm with a Corning Code 2408 red filter results in more stable lithium and potassium signals.

2. Alkaline earths

Major and minor amounts of alkaline earth oxides can also function as glass modifiers. In addition, they act as stabilizers to improve chemical durability, mechanical strength and produce desired electrical properties. The alkaline earths are determined with an oxidizing N_2O + C_2H_2 flame. Using the AAS mode, 1000 μg $LaCl_3$ ml^{-1} is employed as an ionization suppressor for magnesium. Calcium can be determined either by AES or AAS, but in addition to the $LaCl_3$ needs 0.01 M EDTA at pH = 4.5 as a releasing agent. The $LaCl_3$ + EDTA flame buffer is also required to obtain satisfactory AES strontium and barium signals. When barium is present as a major constituent, the analyte solution is compared with the more concentrated standards using the AAS mode. However, if an appreciable amount of calcium is present in the analyte solution, the calcium hydroxide band at 554 nm interferes with the barium AAS signal; and the barium must then be done by AES and wavelength scanning.

3. Aluminum

Aluminum oxide is another stabilizer that is included in glass compositions to improve the resistance to breakage from thermal shock. Because of the refractory nature of the element, the oxidizing $N_2O + C_2H_2$ flame is used for excitation with the AAS mode. It has been found that a satisfactory ionization suppressor is the $1000\,\mu g\,Ca\,ml^{-1} + 0.07\,M\,HClO_4$ solution.

4. Iron

Usually, iron oxide is added to glasses as a coloring agent; or it can be present because of impure batch components. If the analyte solution is $0.05\,M$ in H_3PO_4, iron produces a stable and reproducible AAS signal with a stoichiometric to oxidizing air $+ C_2H_2$ flame.

5. Zinc and cadmium

Zinc and cadmium oxides are included in glass compositions as stabilizers and to improve chemical durabilities. With a stoichiometric to oxidizing air $+ C_2H_2$ flame the zinc and cadmium AAS signals are quite satisfactory if $0.03\,M\,HClO_4$ is present in the analyte solution. Background correction may be necessary at the wavelengths given in Table 1 if the analyte solution is high in salt concentration.

6. Lead

The oxide of lead acts as a glass stabilizer, lowers the melting temperature of the batch and produces glasses with high indexes of refraction necessary for some optical applications. The $100\,\mu g\,Na\,ml^{-1} + 0.03\,M\,HClO_4$ flame buffer acts as an ionization suppressor to enhance the AAS lead signal using the stoichiometric to oxidizing air $+ C_2H_2$ flame. If scattering becomes a problem, background correction may be required.

VIII. DETERMINATION OF SILICON BY AAS

A. Procedure for sample decomposition and AAS measurement

Mix the $0.1000\,g$ sample with $0.5\,g\,Na_2CO_3 + 0.5\,g\,Na_2B_4O_7$. Fuse, cool and dissolve the melt in the molybdate complexing solution $+ HNO_3$ as described in Section V.viii. Then dilute to $500\,ml$ with $0.126\,M\,HNO_3$ in a plastic container. Silicon signals are obtained using the conditions shown in Table 2(a).

References p. 319

TABLE 2

(a) CONDITIONS FOR SILICON AAS MEASUREMENT

Mode	λ (nm)	Oxidant + fuel	Flame buffer concentration
AAS	251.6	$N_2O + C_2H_2$[a]	7.8 g l^{-1} ammonium molybdate + 2 g l^{-1} flux + 0.16 M HNO$_3$

(b) CONDITIONS FOR BORON AES MEASUREMENT

Mode	λ (nm)	Excitation source	Final buffer concentration
AES	249.7	Ar plasma jet[b]	300 µg Si ml^{-1} + 1.45 M HF

[a] Stoichiometric flame (burner slot = 6 cm).
[b] An argon d-c plasma between two tungsten electrodes.

B. Discussion

The most commonly employed glass-former is SiO$_2$. Since silicon has poor flame emission sensitivity, normally it has not been determined by AES. The stability of the silicon AAS signal is inferior to that of other elements but its linearity is good in the range of interest. Therefore, it is possible to alternate measurements between the sample solution and a single standard solution which closely matches it in silicon concentration. The concentration of the sample solution is then calculated as $C_u = C_s(I_u/I_s)$ where (I) is signal intensity (absorbance) and the subscripts (u) and (s) refer to the sample and standard respectively. This "one sample — one standard" approach requires that the samples be pre-scanned to determine their approximate silicon concentrations in order that they may be compared against the proper silicon standards. At least six standards must be available in the range of 5—100 µg ml^{-1}. At least 11 alternate measurements between the sample and standard solutions should be obtained in order to compensate for imprecision and to bring the accuracy to within 1% relative. The integration time should be in the range of 10—20 s per measurement.

IX. DETERMINATION OF BORON BY AES

A. Procedure for sample decomposition and AES measurement

Place 0.1000 g of sample into a 30-ml plastic jar having a screw cap. Add 5 ml of cold water and then 5 ml of cold 29 M HF. Cap tightly and set aside at room temperature until dissolution is complete (1—24 h). Nearly all silicate glass samples do not require heat for dissolution, but if heat is applied then a viable seal must be assured in order to prevent the loss of boron as the fluoride. (Insoluble fluorides and unreactive oxides may remain undissolved but this does not affect the complete dissolution of boron. The formation of fluoroboric acid at room temperature prevents the loss of boron as the fluoride.) After complete boron solubilization, dilute to 100 ml in a plastic volumetric flask using water. Then, as a general procedure, dilute this solution 1 : 1 with the silicon diluting solution already described in Section IV.11. This latter dilution can be omitted for very low boron sample levels (0.3% B_2O_3). Boron signals are obtained using the conditions shown in Table 2(b).

B. Discussion

Boric oxide is also a glass-former, but the chief reason for including it in glass compositions is to lower the thermal expansion and make the final product resistant to thermal shock. The 249.7 nm boron line has sufficient sensitivity for most determinations and is also free from a possible iron spectral interference which can occur at 249.8 nm. However, this latter line has twice the sensitivity and is useful for the lower boron concentrations. The "one sample — one standard" method of measurement as described in Section VIII.B is also used for boron determinations. The alteration of sample and standard measurements compensates for the response drift of the d-c plasma jet (inverted V-type with two tungsten electrodes) and brings the precision of measurement into the 1% relative range. The signal integration time should be 20 s with at least 11 alternate measurements of the sample and standard solutions.

The authors express their appreciation to Mr. G. A. Machajewski, Mr. D. O. Robinson, Mr. S. D. Solsky, Ms. M. Snyder and Mr. W. C. Thresher who performed laboratory experiments that helped make this publication possible.

REFERENCES

1 W. M. Wise and J. P. Williams in I. L. Simmons and G. W. Ewing (Eds.), Progress in Analytical Chemistry, Vol. VIII, Plenum, New York, 1976, p. 29.
2 W. M. Wise, R. A. Burdo and J. S. Sterlace, Prog. Anal. Atom. Spectrosc., 1 (1978) 201.
3 C. B. Belcher, Prog. Anal. Atom. Spectrosc., 1 (1979) 299.

Chapter 4j

Clinical applications of flame techniques

B. E. WALKER
St. James's University Hospital, Leeds LS9 7TF (Gt. Britain)

I. INTRODUCTION

The early 1960's saw the introduction of flame atomic absorption spectrometry (AAS) to clinical laboratories and this provided a sensitive yet simple analytical technique for the estimation of some trace elements in biological fluids. "Trace" elements is now perhaps an inaccurate description for those metals previously detectable in only small amounts when using older and less sensitive analytical techniques. Many such elements can now be estimated with precision and this has proved a great stimulus to trace-element research in clinical medicine. The number of trace elements known to be essential to man has doubled in the last twenty years [1]. Specific and treatable diseases are now known to be associated with excess or deficiency of trace elements and even in advanced societies dietary intake of some elements may be less than ideal [2]. Thus, monitoring of biological fluids for trace element levels in both health and disease can contribute towards major advances in nutritional management.

The principal aim with regard to trace-element analysis in clinical medicine is the earliest detection of deficiency or toxic states. This necessitates access to biological specimens with a minimum of danger and inconvenience to the patient, but the analysis of which will furnish an accurate indication of body stores of the element. By far the most popular and easily accessible biological fluid used for this purpose is plasma (or serum), but this, as with many other biological specimens, presents special analytical problems when using flame AAS. Although there are numerous papers demonstrating abnormal levels of trace elements in a wide variety of diseased states, it is not always appreciated that methodological and physiological factors can profoundly affect trace element levels. As a result many publications have drawn false conclusions with regard to the possible causative role of abnormal trace-element levels in diseased states. Rightly, much has been written about the simplicity and sensitivity of flame AAS but insufficient consideration has been afforded to other factors likely to cause difficulty in the interpretation of results, particularly when using biological specimens. Representative sample collection and data interpretation can be as important as analytical technique in providing meaningful results for the clinician. Manufacturers recommended instrumental parameters when using aqueous samples do not necessarily represent optimal

References pp. 338—339

conditions for plasma and physiological variation may far exceed analytical variation for some elements [3]. The management of acutely ill patients often requires urgent investigation and methods used in a clinical laboratory must avoid lengthy and intricate procedures and be suitable for rapid analysis of large numbers of samples [4].

This chapter considers the optimisation of both physiological and analytical conditions for the estimation of those elements present in biological fluid best suited to analysis by flame AAS.

II. SAMPLE COLLECTION

Particular care must be taken in the collection of blood samples. In spite of the difficulty in obtaining representative samples, many biological studies concerning trace-element work depend upon the comparison of mean levels in "normals" with those in diseased states. The amount of specimen is often limited, particularly when investigating children, and every effort must be made to produce a reliable result without having to resort to further sampling. In some circumstances it may be necessary to use special techniques for obtaining blood specimens from very young subjects [5]. Of the elements present in plasma some show very little variation, being accurately maintained within narrow physiological limits. Others, however, show considerable variation between individuals when factors such as age, sex, pregnancy, time of day, and even the technique of blood collection [6] are taken into account. Such physiological factors are particularly important in clinical work and can cause large variation in plasma values, possibly masking or exaggerating changes due to disease states. In normal subjects geographical location may influence plasma values [7] and it is essential that a standardised collection system is employed and that each laboratory establish it own normal range with their particular analytical technique.

A. Contamination

Avoidance of contamination and stability of specimens is of the utmost importance in clinical work, since for some elements contamination can be expected in almost all steps of the analytical procedure [8]. Blood should be drawn using plastic disposable syringes and disposable needles. Although these are less liable to contamination than glass all makes should be carefully checked before they are accepted as suitable. Similarly, containers for collecting blood and subsequent storage of plasma or serum should be rigidly checked. When plasma is required the anticoagulant used should also be checked for metal content. For storage purposes disposable plastic containers are best though rubber stoppers are a potential cause of contamination. Polypropylene tubes and polythene stoppers are probably best [8]. Some containers allow trace metals to leech from the walls and increasing concentrations may be found following several days contact. Adsorption from the solution may also occur, particularly with metals in very low concentration [9]. These problems of contamination are potentially most serious for those metals present in plasma in very low concentrations, such as zinc and copper, and in these circumstances it is probably wise to collect the blood into two different containers and store each separately. However, even for metals present in relatively high concentrations, i.e. magnesium, reducing specimen handling to a minimum will increase precision [10]. Glass is a poor material for storage of high-purity solutions and plastics are better. Even with the most meticulous technique occasional contamination cannot always be avoided and every effort must be made to detect its occurrence in all stages of preparation and analysis [9].

B. Effect of occlusion

Venous blood samples are usually obtained after occlusion is applied to the arm. Such obstruction to venous return whilst making collection easier causes ultrafiltration and this can be a source of considerable error in trace-element work. Following venous occlusion fluid easily passes through the vein wall whilst large molecular weight substances such as proteins cannot. Most metals present in plasma are to some extent bound to protein and consequently their concentration also increases following venous occlusion. Figure 1 shows the mean results obtained in six normal individuals when blood samples were taken from both arms simultaneously (0 minutes) following which occlusion was applied to one arm. The results shown are expressed as the percentage increase in the plasma concentration of the occluded arm compared with the non-occluded arm.

About half the calcium and magnesium present in plasma is protein bound, the remainder being the ionised form. During occlusion the ionised

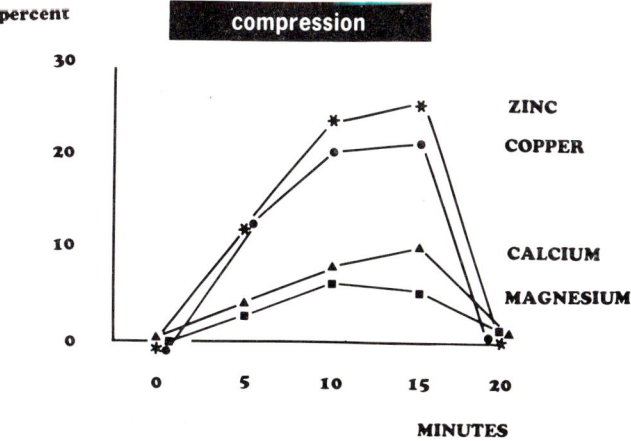

Fig. 1. Mean results for six normal individuals with one arm occluded (see text for details).

References pp. 338—339

part can pass out of the vein but the protein bound part cannot and so the total plasma concentration increases five to ten percent after fifteen minutes occlusion. Zinc and copper are almost totally protein-bound and so their concentration within plasma increases even more dramatically, amounting to 20—30% after fifteen minutes occlusion. Five minutes after the occlusion is released the concentration of all metals has returned to values similar to the non-occluded arm. Inappropriately low concentration of protein bound elements will occur when protein concentrations are low due to disease states. The difficulty of interpreting plasma calcium values in these circumstances is well documented, and adjustment of total plasma values for a lowered protein concentration improves the accuracy of diagnosis [11]. However, this is because only the ionised fraction of plasma calcium is of clinical importance and it is not at present justifiable to apply similar corrections for other elements. In particular it is inappropriate to make any adjustment for the increase in protein concentration following venous occlusion and to avoid the considerable errors likely to be introduced specimens have to be collected without occlusion [12].

C. Diurnal variation

The time of day the specimen is obtained may have considerable influence upon the accuracy of results for some elements. Many plasma constituents are known to vary in concentration throughout the day. This diurnal variation is illustrated in Fig. 2. In these studies blood was obtained from twelve healthy individuals at two hourly intervals throughout the day. The 0800 hours sample was obtained following an overnight fast, after which normal meals were taken. There is very little change throughout the day for magnesium, calcium and copper. Plasma iron however starts to fall about lunch-time and by 2200 hours is 8% lower than the early morning sample. The changes are even more dramatic for zinc which falls progressively from early morning until lunch time when it is almost 20% lower. The cause of this diurnal variation is not known but for elements such as zinc and iron normal ranges or apparent differences between groups of normals and patients will have little meaning unless collection times are standardised. Recent dietary intake may also cause alteration of plasma values and for the most precise results [13] it is recommended that specimens are obtained early in the morning following an overnight fast and before breakfast.

III. ANALYSIS

A. Effect of protein content

The protein content of biological fluids is not only a potential source of variation during the collection of samples, but also causes considerable

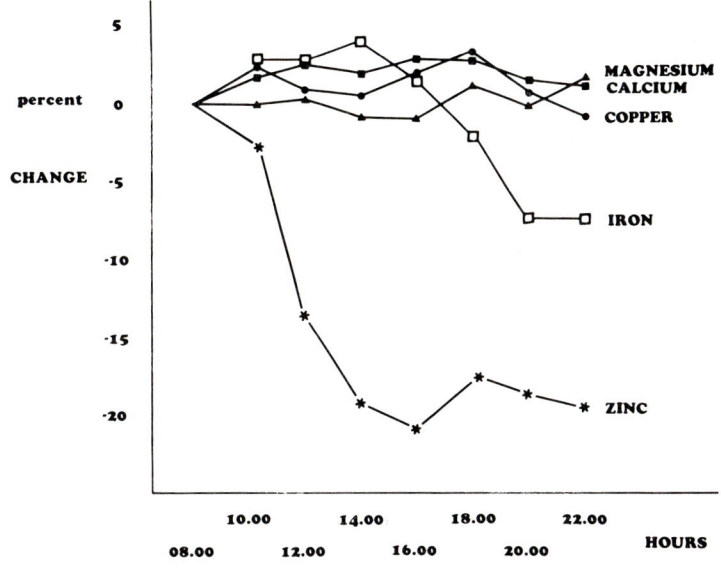

Fig. 2. Diurnal variation of some plasma constituents.

analytical difficulty when using flame AAS. The viscosity of plasma due to its protein content causes variation in aspiration rate to the atomiser and consequently inaccurate results. The effect of increasing plasma dilution on apparent zinc content is shown in Fig. 3. Due to the high protein content of undiluted plasma, aspiration rates are slow and the apparent zinc content depressed. This error would apply to all metals estimated and requires a dilution of at least 1 in 12 before an accurate result is obtained. For some elements this presents no problem since the concentration in plasma is sufficiently high to allow dilution factors considerably greater than 12.

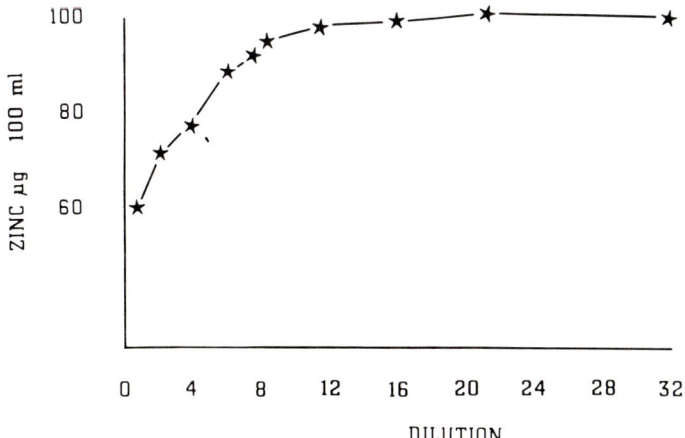

Fig. 3. Effect of plasma viscosity on zinc estimation.

References pp. 338—339

Zinc and copper, however, are present in low concentration and sufficient dilution to overcome viscosity and aspiration difficulties may so reduce the concentration that sensitivity is unsatisfactory. This is an important consideration and a number of techniques for resolving the problem have been suggested.

Perhaps the most popular method is to de-proteinise the sample and compare with an aqueous standard solution. This is usually a satisfactory procedure but it negates the prime advantage of AAS, that of simplicity and maintaining sample handling to a minimum to avoid contamination. Positive pressure sampling may be used [14] or standards may be "loaded" with a protein-like material to create a similar viscosity to that of plasma [15]. It might be anticipated that problems with aspiration rate would then apply equally to sample and standard. Unfortunately, this is not satisfactory since considerable variation in the protein concentration of plasma occurs even between normal subjects and very wide differences in concentration often occur between normal and pathological specimens. In most chronic disease the albumin content falls substantially and to obtain accurate results separate standards of appropriate concentration have to be prepared. Thus the technique of loading standards cannot be considered suitable for clinical work and simple sample dilution is recommended. Since significant alterations in aspiration rate can occur even when plasma is diluted ten-fold [16], a minimum dilution of twelve-fold is needed. Reduced signal strength may require scale expansion leading to an unstable baseline, but this can generally be overcome by employing a strictly timed sequence of measurements. Distilled water is aspirated for ten seconds followed by standard solution or unknown for a similar period, and subsequent analysis of the trace from a chart recorder [17].

B. Interference

Blood and plasma contain many organic ions in varying concentrations and these are a potential source of inaccuracy when using AAS, since the recovery of many elements is influenced by the medium in which the standards are dissolved [16]. Sodium is present in plasma in high concentration and although this undoubtedly causes interference problems with many elements the results of detailed studies are conflicting. For example, some workers have reported that sodium has a suppressive effect upon the determination of zinc [18] whilst others have found enhancement of absorption for zinc [19]. It is likely that the degree of inorganic interference is dependent upon many factors, such as the type of nebulizer used, composition of flame gases and the zone of flame used for measurements, and it is perhaps not surprising that conflicting reports appear. The complex nature of plasma makes it impractical to undertake a detailed study of all possible interference effects by the many inorganic ions present and the best approach is to prepare standards in a solution which contains the major ions and dilute in a similar way to the unknown.

IV. INDIVIDUAL ELEMENTS

A. Calcium

The average adult contains more that 1 kg of calcium, this being the most prevalent metallic element. The major proportion is contained within the bone (99%) the remainder being in teeth and soft tissues [20]. Plasma calcium represents only a minute fraction (0.03%) of total body calcium yet alteration in the concentration of this fraction can usually accurately reflect underlying disorders of calcium metabolism which are of great importance in clinical medicine.

Calcium is necessary for control of many biochemical processes at a cellular level and acts as a coupling agent between muscle excitation and contraction. Its presence is necessary for normal heart rhythm and contraction and effective blood clotting. Such functions require a mechanism for the accurate control of both intra- and extra-cellular concentrations. As for sodium the homeostatic control takes place not only at the cell membrane but also at a subcellular level. Daily intake for an adult is 250 mg this being derived largely from milk products. Proportional to body weight, the intake is greatest during the period of active growth and bone formation.

Not all calcium present in the diet is absorbed by the small intestine and mechanisms are present to ensure only amounts appropriate to body needs are absorbed. These processes are complex and involve the interaction of special transport protein, vitamin D and parathormone. Thus, abnormalities of calcium metabolism may result from many different disease processes. Diseases affecting the bowel may prevent normal absorption, diseases of the parathyroid gland may result in inappropriate levels of parathormone for calcium requirement and a nutritionally inadequate diet may cause vitamin D deficiency with consequent disordered calcium absorption.

Many pathological processes may be detected by assessing the calcium concentration of plasma. Calcium is present in plasma in two forms, one bound to albumin (which is less than half the total) and the other ionised. It is only alterations of the ionised fraction which provide an accurate indication of underlying pathology but, because of the difficulty in measuring this fraction, measurements of total calcium continue to be used. The bound part of plasma calcium is attached to albumin and many diseases cause a non-specific reduction in the plasma concentration of this protein. Consequently, a reduced total plasma calcium may be due to a reduction in albumin concentration whilst the more important ionised fraction remains normal. Measurement of total plasma calcium could therefore be misleading unless an adjustment is made for the reduced albumin levels. This has been the subject of much controversy but the technique does seem to be clinically useful in avoiding diagnostic errors when plasma albumin levels are low [11]. Very high levels of calcium represent a medical emergency demanding urgent investigation. There are many causes of increased plasma

References pp. 338—339

calcium the commonest being parathyroid tumour and malignancy [21]. Although several methods are available for the determination of plasma calcium atomic absorption spectrometry is to be preferred [22]. Errors associated with preparation and dilution of standard solutions and specimens are perhaps more important than instrumental performance [23].

1. *The determination of calcium in plasma and urine*

Calcium is determined in plasma following a fifty-fold dilution with lanthanum chloride.

Standards and reagents
1. Calcium: $1000\,\mu g$ Ca ml^{-1}.
2. Sodium/potassium: 140 meq Na, 5 meq K l^{-1}.
3. Lanthanum diluent: 0.1% La, 0.1 N HCl.

Working solutions
1. Intermediate standard: Ca 10 mg% — dilute 10 ml of stock standard to 100 ml with 140 meq Na, 5 meq K l^{-1} solution.
2. Working standard: dilute the calcium 10 mg% solution fifty-fold with lanthanum diluent.
3. Blank: dilute a convenient volume of 140 meq Na, 5 meq K l^{-1} solution fifty-fold with the lanthanum diluent.

Sample preparation
Take an appropriate volume of plasma and dilute fifty-fold with lanthanum diluent. For the determination of calcium in urine the sample should be diluted with lanthanum chloride to produce a concentration of about 100 mg/100 ml.

B. Copper

Copper has long been of interest in clinical medicine because, although an essential element, it is better known for its toxic effects which result in the syndrome called hepato-lenticular degeneration or Wilson's disease. This disease is rare but of considerable clinical importance because its early recognition and treatment can prevent the development of serious and progressive brain and liver damage. The average adult contains 100 to 150 mg of copper, the highest concentrations being found in the hair, liver, muscle and lung [24]. Absorbed copper is first attached to plasma albumin and then taken up by the liver. Subsequently it is released, bound to the specific copper-carrying protein ceruloplasmin. Like many other trace elements copper is an essential component of several metallo-enzymes, including cytochrome C oxidase.

Hepato-lenticular degeneration occurs predominantly in young people and was first described in 1912 by Kinnier Wilson. It is inherited as an autosomal recessive trait and the basic underlying abnormality is a deficiency of ceruloplasmin, a protein which acts as a carrier for copper in the plasma. In its absence copper is carried loosely bound to albumin which allows excessive copper to be deposited in tissues and large amounts to be excreted in the urine. The serum copper is often low because of this rapid transfer from the blood. The high tissue levels predominantly affect the liver resulting in cirrhosis and the brain causing progressive involuntary movement and disability. The diagnostic biochemical abnormalities of this disease are low serum ceruloplasmin, normal or low serum copper and high urinary copper. Provided the condition can be recognised before irreversible damage occurs the prognosis is good since it is possible to remove large amounts of copper from the body via the urine using chelating agents such as penicillamine. Treatment may have to continue for some years and urine copper levels monitored to demonstrate effectiveness of the treatment. Recognition of cases of Wilson's disease should demand an assessment of other members of the family to detect possible asymptomatic cases.

Whilst from the clinical point of view the chief interest in copper is in relationship to Wilson's disease, other toxicity and deficiency states have also been recognised. Most cases of copper deficiency have occurred following prolonged gastro-intestinal loss or malabsorption [25] but premature infants may also be at risk due to the low body stores. In these circumstances hypochromic microcytic anaemia occasionally develops, probably due to an effect of copper on iron metabolism [26]. Toxicity due to copper has been reported in patients with renal failure maintained on artificial dialysis. In these circumstances the use of water with a high concentration of copper can result in copper being dialysed into the patient and toxicity symptoms [27]. Plasma copper is not influenced by meals and diurnal variation is slight [28]. Red blood cell levels are slightly less than plasma levels. Although plasma copper may be a better index of body copper than whole blood levels [29] plasma levels are raised in many apparently unrelated disease states [24]. The wide range of these diseases suggests that the raised copper levels are a non-specific effect, probably secondary to a raised ceruloplasmin concentration. In many conditions the increase in plasma copper is accompanied by a reduced plasma zinc and the increase in the copper/zinc ratio may be valuable in estimating the extent of some disease processes [30].

Copper in plasma can be estimated following twenty-fold dilution with 0.1 normal HCl. Although few metals cause interference when using flame AAS [31] standards containing 6 μg Cu per 100 ml should be made up in the plasma or urine equivalent [32]. Only very small amounts of copper are excreted in the urine of normal subjects and specimens are best aspirated undiluted. However, in Wilson's disease and especially during treatment urine copper may be high and appropriate dilution will be needed.

References pp. 338—339

1. The determination of copper in plasma and urine

Copper is determined in plasma following a twenty-fold dilution with 0.1 N HCl. The use of butanol (6% v/v butanol) as a diluent may improve precision.

Standards and reagents
1. Copper: 1000 μg ml^{-1}.
2. Sodium/potassium: 140 meq Na, 5 meq K l^{-1}.

Working solutions
1. Diluent 0.1 N HCl or butanol.
2. Intermediate standard; 100 μg Cu ml^{-1}. Take 10 ml of stock and dilute to 100 ml with distilled-deionised water.
3. Intermediate standard; 1 μg Cu ml^{-1}. Take 1 ml of the 100 μg Cu ml^{-1} solution and dilute to 100 ml with 140 meq Na, 5 meq K l^{-1}.
4. Working standard. Take a convenient volume of the 1 μg Cu ml^{-1} solution and dilute twenty-fold with 0.1 N HCl or butanol.
5. Blank. Take a convenient volume of the 140 meq Na, 5 meq K l^{-1} solution and dilute twenty-fold.

Sample preparation
Dilute plasma twenty-fold with 0.1 N HCl or butanol. Urine dilution depends on copper concentration. The same dilution equipment should be used for preparation of standards and samples.

C. Iron

Iron was one of the first of the minor elements to be recognised as being necessary for both plants and animals. Man contains a total of four to five grams of iron most of which is complexed with protein. A major proportion of body iron is present as haemoglobin [33]. The dietary intake of 15 to 20 mg per day is obtained mostly from meat, eggs and some vegetables. Only a small proportion of the dietary intake is absorbed, the uptake being regulated according to body needs. The only way significant amounts of iron are lost from the body is by blood loss; in menstruating females the loss may exceed dietary intake and cause iron deficiency. Dietary requirement during pregnancy is large and supplementation is usually recommended. Iron absorbed from the gut is transported in the plasma attached to the specific carrier protein transferrin and has a rapid turnover with a half life of only 100 min. Normally, only about a third of transferrin is bound with iron, representing 30% saturation of total iron binding capacity. Iron is stored mainly in the liver. Deficiency of iron may result from inadequate intake, poor absorption or excessive loss through bleeding. Poor dietary intake is probably more common than usually thought, particularly in the elderly and following gastric operations for peptic ulcer. Iron deficiency is manifest as a progressive hypochromic microcytic anaemia.

Although under normal circumstances iron absorption is accurately regulated according to bodily needs, excessive iron absorption occurs in alcoholics and patients with defective pancreatic function. Haemochromotosis (iron storage disease) is an inborn error of metabolism in which excessive amounts of iron are absorbed from the gut. Total body iron may increase ten or twenty times normal and the iron deposited in the tissues causes progressive damage. The organs principally involved are the liver (cirrhosis), the heart (heart failure) and the pancreas (diabetes). In the investigation of this disease diagnosis may be helped by finding a high iron binding capacity though it is also necessary to demonstrate excessive amounts of iron deposited in the liver.

Serum iron can easily be determined by simple dilution of plasma. However, the possibility of iron contamination due to the presence of non-visible haemolysis makes this procedure highly unreliable. Data indicate that a positive enhancement of 50% can occur from non-visible haemoglobin and greater than 100% from serum showing slight haemolysis. Since it is almost impossible to ensure that the sample is not haemolysed during collection a procedure must be employed which can guarantee that any haemoglobin present is removed. The recommended procedure is an adaptation of the photometric method suggested by the International Committee for Standardisation in Haematology as a reference method [34]. This highly selective atomic absorption technique eliminates most of the interference present in colorimetric procedure.

1. *The determination of plasma iron and iron binding capacity*

Standards and reagents
1. Iron: $1000 \, \mu g \, Fe \, ml^{-1}$.
2. Sodium/potassium: 140 meq Na, 5 meq K l^{-1}.
3. Magnesium carbonate: reagent grade powder.
4. Trichloroacetic acid.
5. Thioglycolic acid (mercaptoacetic acid).

Working solutions
1. Protein precipitant: 100 g trichloroacetic acid (TCA), 30 ml of thioglycolic acid and 166 ml of concentrated hydrochloric acid per litre. This should be stored in a dark-brown bottle to ensure stability for at least two months.
2. Intermediate standard: $100 \, \mu g \, Fe \, ml^{-1}$. Take 10 ml of the stock solution and dilute to 100 ml with distilled-deionised water.
3. Intermediate standard: $1 \, \mu g \, Fe \, ml^{-1}$. Take 1 ml of the $100 \, \mu g \, Fe \, ml^{-1}$ and dilute to 100 ml with 140 meq Na, 5 meq K l^{-1}. Prepare this solution daily just before use.
4. Working standard: Take 25 ml of the $1 \, \mu g \, Fe \, ml^{-1}$ and add 25 ml of the protein precipitant solution.
5. Blank. To 25 ml of 140 meq Na, 5 meq K l^{-1} add 25 ml of the protein precipitant solutions.

References pp. 338—339

Sample preparation (plasma iron)

Add 1.0 ml of the protein precipitant solution to 1.0 ml of plasma. Mix thoroughly and let stand for 5 min. Centrifuge until clear.

Total iron binding capacity

1. Pipette 1 ml of plasma into a small test tube and add 1 ml of 500 µg% Fe (0.5 ml of 100 µg Fe ml^{-1} diluted to 100 ml). Mix and let stand for 5 min.
2. Add 0.25 g of magnesium carbonate powder, stopper and shake vigorously for 10—15 s. Let stand for 30 min but remix thoroughly 4—5 times during this interval.
3. Centrifuge until clear.
4. Pipette 1 ml of clear supernatant to another tube and proceed exactly as with serum iron procedures.

D. Lithium

Lithium is not considered an important serum cation but its measurement is necessary to monitor plasma levels in patients given lithium carbonate for the treatment of depression. Although lithium has no known vital function in living organisms it is found in most tissues and fluids in very low concentrations [33]. Large doses of lithium carbonate are used in the treatment of depression and the therapeutic plasma concentration of lithium is only slightly less than that causing toxicity so that plasma levels must be maintained within narrow limits if maximum benefit is to be obtained whilst avoiding toxicity. Side effects are related to dosage and usually develop slowly over a period of days. Initially symptoms are a mild gastro-intestinal upset with nausea and vomiting but if excessive dosage continues tremor, confusion and fits may occur [35]. Cardiac arrhythmias, abnormal thyroid function and diabetes insipidus have also been reported.

Whilst plasma levels provide good evidence of impending toxicity (levels should be kept below 1.5 meq l^{-1}) it has been suggested that measuring erythrocyte levels is more reliable [36]. This is most easily accomplished by estimating whole-blood lithium and calculating erythrocyte levels using blood haematocrit. Lithium in plasma and whole-blood can be estimated following twenty-fold dilution using distilled-deionised water.

1. *The determination of lithium in plasma and whole-blood*

Lithium is determined in plasma or whole-blood following a twenty-fold dilution with 0.1 N HCl. This would be applicable to monitoring plasma levels in patients receiving lithium carbonate for the treatment of depression. Plasma lithium levels in normal subjects are very low and best determined by electrothermal atomisation.

Standards and reagents
1. Lithium: 1 meq ml^{-1}.
2. Sodium/potassium: 140 meq Na, 5 meq K l^{-1}.

Working solutions
1. Working standard. Dilute stock standard twenty-fold with 0.1 N HCl.
2. Blank. Dilute the 140 meq Na, 5 meq K l^{-1} solution twenty-fold with 0.1 N HCl.

Sample preparation
Dilute plasma or whole-blood twenty-fold with 0.1 N HCl.

E. Magnesium

The average adult contains 24 g of magnesium 99% of which is in bone and muscle, and in health plasma levels are kept remarkably constant. About one third is bound to protein and like calcium the remainder is ionised and represents the biologically active component. The concentration of magnesium in red blood cells is about four times that of plasma. Magnesium is a co-factor for all enzymes involved in phosphate transfer reactions that utilize adenosine triphosphate (ATP). Binding of messenger RNA to ribosomes is magnesium dependent and many enzymes are activated by magnesium [37]. In humans magnesium deficiency is characterised by abnormal muscular transmission and abnormal cardiac rhythms occur with both abnormally high and low plasma levels.

The kidneys, by controlling renal excretion of magnesium, play an important role in maintaining plasma levels within narrow limits. High plasma levels are associated with diminished kidney function and when excessive magnesium sulphate is used as a laxative. Magnesium deficiency is uncommon and usually a consequence of excessive loss from the gastrointestinal tract in the absence of adequate replacement [38]. Depletion can occur following diarrhoea, severe vomiting and when large amounts of bile, pancreatic juice and other body fluids are lost. In some circumstances hypocalcaemia may result from magnesium deficiency and magnesium plays a role in normal parathyroid gland function and in maintaining end organ response to parathormone.

Magnesium, as for calcium, is determined following simple dilution (1 in 50) with lanthanum diluent. Because red blood cells contain high levels of magnesium great care has to be taken to avoid haemolysis of red blood cells. It is convenient to measure both calcium and magnesium on the same sample.

1. *The determination of magnesium in plasma and urine*

Magnesium can be determined on the same sample as for calcium, following a fifty-fold dilution with lanthanum chloride.

References pp. 338—339

Standard and reagents
1. Magnesium: $1000\,\mu\text{g ml}^{-1}$.
2. Sodium/potassium: 140 meq Na, 5 meq K l^{-1}.
3. Lanthanum diluent: 0.1% La, 0.1 N HCl.

Working solutions
1. Intermediate standard: Mg 2 mg%. Dilute 2 ml of the stock standard to 100 ml with 140 meq Na, 5 meq K l^{-1} solution.
2. Working standard. Dilute the Mg 2 mg% solution fifty-fold with lanthanum diluent.
3. Blank. Dilute a convenient volume of the 140 meq Na, 5 meq K l^{-1} solution fifty-fold with lanthanum diluent.

Sample preparation
Take an appropriate volume of serum and dilute fifty-fold with lanthanum diluent. For magnesium in urine dilute with lanthanum diluent so that the concentration is about 2 mg/100 ml.

Haemolysis of the sample will cause a significant increase in plasma magnesium level. Lanthanum is added to samples and standard for convenience so that calcium and magnesium determinations may be made on each sample.

F. Potassium

Whereas sodium is predominantly extracellular, potassium is a major intracellular cation. An active energy requiring transport mechanism maintains this distribution and many diseases causing alteration of sodium levels also result in disturbances of potassium metabolism.

An adult contains about 3 g of potassium. The plasma level is maintained within narrow limits by the kidney which can efficiently conserve or excrete potassium according to need. Abnormally low plasma levels cause muscle irritability and weakness together with potentially dangerous cardiac arrhythmias. Similar changes can also occur with abnormally high plasma potassium levels, thus making potassium determination one of the more important investigations in clinical medicine.

High plasma levels occur when excretion is impaired and is most commonly seen as a consequence of kidney disease. Low levels are a consequence of excessive loss, usually through the gastro-intestinal tract following vomiting or diarrhoea.

1. *The determination of potassium in plasma*

Potassium is determined in plasma by flame emission using a fifty-fold dilution with distilled water.

Standards and reagents
1. Sodium/potassium: 140 meq Na, 5 meq K l^{-1}.

Working solutions
1. Working standard. Dilute the sodium/potassium solution fifty-fold with distilled water.
2. Blank. Distilled water.

Sample preparation
Dilute an appropriate volume of serum fifty-fold with distilled water.

G. Sodium

Almost all metabolic processes in the body are dependent in some way upon the presence of sodium. Sodium maintains osmotic pressure and hydration of all body tissues and helps to regulate pH. Changes in plasma levels of sodium occur in a wide variety of disease states and are often accompanied by changes in the fluid content of the various body compartments. The maintenance of normal blood volume is linked to that of sodium and is of crucial importance since the blood volume is a major factor in determining normal cardiac output [39]. The average adult contains about 70 g of sodium and is the major cation of extracellular fluids. About 97% of body sodium is extracellular, either in plasma, interstitial fluid or bone. Because sodium is so important in maintenance of body fluids there are sensitive homeostatic mechanisms directed towards maintaining sodium concentrations constant. Most dietary sodium is absorbed by the gastro-intestinal tract, the excess being excreted in the urine. Under hormonal control, the kidneys are able to regulate accurately the amount of sodium appearing in the urine, thus maintaining sodium balance. Excess sodium may be lost from the gastro-intestinal tract following severe vomiting or diarrhoea, and when this occurs urine sodium is reduced to a minimum.

There are many diseases which cause alteration in plasma sodium level. Low levels accompany hypoadrenalinism, prolonged diarrhoea or vomiting, diabetic acidosis and in some types of renal disease. High plasma sodium levels accompany severe dehydration and hyperadrenalinism. The management of patients with disturbances of sodium metabolism require accurate and frequent determination of the plasma sodium. Thus the measurement of plasma sodium is a common and important clinical investigation which allows the correct management of patients with disturbances of electrolyte metabolism. Although atomic absorption spectrometry is the method of choice for many metals, it is not usually employed for sodium since less expensive flame emission methods are adequate. Nevertheless those instruments providing emission facilities are ideally suited for the determination of sodium in plasma.

1. *The determination of sodium in plasma*

Sodium is determined in plasma following a two hundred and fifty-fold dilution with distilled water.

Standards and reagents
Sodium and potassium: 140 meq Na, 5 meq K l^{-1}.

Working solutions
1. Working standard. Dilute sodium/potassium solution two hundred and fifty-fold with distilled water.
2. Blank. Distilled water.

Sample preparation
Dilute a convenient volume of sample two hundred and fifty-fold with distilled water.

H. Zinc

The importance of zinc in animal nutrition has been well known for many years yet it is only in the last few years that the essential nature of zinc in human nutrition has been established. The introduction of AAS to clinical laboratories provided the stimulus to many studies and zinc is now established as an essential trace element of considerable importance in clinical medicine. The average adult contains 2—3 g of zinc, which is about half that of body iron but ten to fifteen times greater than for copper. Although zinc is present in all cells, particularly high concentrations are found in the eye, testes, prostate and liver. The metallo-enzyme carbonic-anhydrase was the first zinc containing enzyme to be recognised and now more than eighty other enzymes are known to contain zinc, including alkaline phosphatase and alcohol dehydrogenase. The metallo-enzyme retinene reductase accounts for the high zinc concentration of the retina and provides a link between the metabolism of vitamin A and zinc. Zinc is necessary for DNA synthesis and also plays a role in carbohydrate metabolism.

The concentration of zinc in red blood cells is 10 times that of plasma which implies scrupulous avoidance of haemolysis during specimen collection. Most red-cell zinc is tightly bound to the enzyme carbonic-anhydrase and not easily exchangeable with plasma zinc. Virtually the whole of plasma zinc is protein bound, mostly to albumin (60%), alpha 2 macroglobulin (30%) and the remainder to low molecular weight proteins including amino acids [40]. A syndrome of iron deficiency anaemia, hepatosplenomegaly and dwarfism has been described in association with abnormalities of zinc metabolism [41]. Although it has been suggested that this represents a human zinc deficiency syndrome definite conclusions are complicated because other nutritional deficiencies were also present in these patients. More recent evidence however suggests that human zinc deficiency may also be associated with symptoms and signs similar to those now established to accompany zinc deficiency in animals. Prolonged intravenous nutrition with fluids containing minimal quantities of zinc is complicated by diarrhoea and parakeratosis and shows a striking response to zinc replacement [42].

These features are very similar to those of acrodermatitis enteropathica a syndrome occurring in infants and associated with high mortality. This is known to represent a specific human zinc deficiency syndrome [43]. The recognition of clinical zinc deficiency emphasises the importance of establishing accurate biochemical criteria for early detection of such states. The measurement of plasma zinc might provide the simplest means for assessing zinc deficiency but special precautions are necessary for this to provide meaningful information. Specimens should be collected without venous occlusion and samples for analysis must be diluted at least twelve times to avoid errors due to viscosity effects. Plasma zinc undergoes a substantial diurnal variation necessitating accurately timed collections.

Care should be taken in automatically attributing low plasma-zinc values to zinc deficiency. As with calcium a major proportion of plasma zinc is attached to albumin and a reduction in plasma albumin will cause a reduction in total zinc values. In chronic diseases total zinc and albumin show a close relationship [44] and it is perhaps for these reasons that so many diseases are found to be associated with a reduced total plasma zinc. At present it is not clear which is the most important fraction of plasma zinc and consequently it is not justifiable to make any correction for protein alteration as has been suggested for calcium.

1. *The determination of zinc in plasma, whole-blood and urine*

Zinc is determined in plasma following a twenty-fold dilution with 0.1 N HCl. The use of butanol (6% v/v butanol) as a diluent may improve precision.

Standards and reagents
1. Zinc: $1000\,\mu g\,ml^{-1}$.
2. Sodium/potassium: 140 meq Na, 5 meq K l^{-1}.

Working solutions
1. Diluent: 0.1 N HCl.
2. Intermediate standard: $100\,\mu g$ Zn ml^{-1}. Take 10 ml of stock standard and dilute to 100 ml with distilled-deionised water.
3. Intermediate standard: $1\,\mu g$ Zn ml^{-1}. Take 1 ml of the $100\,\mu g$ Zn ml^{-1} and dilute to 100 ml with the 140 meq Na, 5 meq K l^{-1} solution.
4. Working standard. Take a convenient volume of the $1\,\mu g$ Zn ml^{-1} solution and dilute twenty-fold with 0.1 N HCl.
5. Blank. Dilute a convenient volume of 140 meq Na, 5 meq K l^{-1} twenty-fold with 0.1 N HCl.

Sample preparation
Dilute plasma twenty-fold with 0.1 N HCl. For whole-blood zinc dilute one hundred-fold and for urine ten-fold with 0.1 N HCl.

Sodium and potassium enhance zinc absorption and must be present in the standards and blank solutions.

REFERENCES

1. I. Lombeck and H. J. Bremer, Nutr. Metab., 21 (1977) 49—64.
2. H. H. Sandstead, Am. J. Clin. Nutr., 26 (1973) 1251—1260.
3. K. O. Pederson, Scand. J. Clin. Lab. Invest., 30 (1972) 191—199.
4. W. J. Price, Spectrochemical Analysis by Atomic Absorption, Heyden, London, 1979 p. 259.
5. D. D. Michie, N. H. Bell and F. H. Wirth, Am. J. Med. Technol., 42 (1976) 424—427.
6. F. Bjorksten, A. Aromaa, P. Knekt and L. Malinen, Acta Med. Scand., 204 (1978) 67—74.
7. J. Kubota, V. A. Lazar, B. S. Ithaca and F. Losee, Arch. Environ. Health, 16 (1968) 788—793.
8. R. W. Reimold and D. J. Besch, Clin. Chem., 24 (1978) 675—680.
9. V. D. Anand, J. M. White and H. V. Nino, Clin. Chem., 21 (1975) 595—602.
10. E. L. Pruden, R. Meier and D. Plaut, Clin. Chem., 12 (1966) 613—619.
11. R. B. Payne, A. J. Little, R. B. Williams and J. R. Milner, Br. Med. J., 4 (1973) 643—646.
12. B. E. Walker and R. B. Payne, J. Clin. Pathol., 32 (1979) 488—491.
13. M. R. Wills, J. Clin. Pathol., 23 (1970) 772—777.
14. R. Hicks and M. Haven, Am. J. Clin. Pathol., 54 (1970) 235—238.
15. B. M. Hackley, J. C. Smith and J. A. Halstead, Clin. Chem., 14 (1968) 1—5.
16. B. Momcilovic, B. Belonje and B. G. Shah, Clin. Chem., 21 (1975) 588—590.
17. J. B. Dawson and B. E. Walker, Clin. Chim. Acta, 26 (1969) 465—475.
18. A. S. Prasad, D. Oberleas and J. A. Halstead, J. Lab. Clin. Med., 66 (1965) 508—516.
19. R. I. Henkin, in W. Hertz and W. E. Cornatzer (Eds.), Newer Aspects of Copper and Zinc Metabolism in Newer Trace Elements in Nutrition, Marcel Dekker, New York, 1971 p. 255.
20. B. E. C. Nordin (Ed.), Calcium, Phosphate and Magnesium Metabolism, Churchill Livingstone, Edinburgh, 1976 p. 3.
21. L. Watson, Br. Med. J., 2 (1972) 150—152.
22. N. Weissman and V. J. Pileggi, in R. J. Henry, D. C. Cannon and J. W. Winkelman (Eds.), Clinical Chemistry, Harper and Row, New York, 1974 p. 651.
23. Association of Clinical Biochemists News Sheet No. 193, May 1979 p. 9.
24. I. J. T. Davies, Clinical Significance of the Essential Biological Metals, William Heineman, London, 1972 p. 48.
25. I. Sternleib and H. D. Janowitz, J. Clin. Invest., 43 (1964) 1049—1055.
26. C. J. Gubler, J. Am. Med. Assoc., 161 (1956) 530—535.
27. W. H. Lyle, J. E. Payton and M. Hui, Lancet, 1 (1976) 1324—1325.
28. G. E. Cartwright and M. M. Wintrobe, Am. J. Clin. Nutr., 14 (1964) 224—232.
29. I. E. Dreosti and G. V. Quike, Br. J. Nutr., 22 (1968) 1—7.
30. S. Inutsuka and S. Araki, Cancer, 42 (1978) 626—631.
31. F. W. Sunderman and N. O. Roszel, Am. J. Clin. Pathol., 48 (1967) 286—293.
32. J. B. Dawson, D. J. Ellis and H. Newton-John, Clin. Chim. Acta, 21 (1968) 359—365.
33. E. J. Underwood, Trace Elements in Human and Animal Nutrition, Academic Press, New York, 1977.
34. Proposed Recommendations for the Measurement of Serum Iron in Human Blood. The International Committee for Standardisation in Haematology. Blood, 37 (1971) 598.
35. Martindale, The Extra Pharmacopoeia, The Pharmaceutical Press, London, 1977, p. 1548.
36. G. H. Hisayasu, J. L. Cohen and R. W. Nelson, Clin. Chem., 23 (1977) 41—45.
37. J. K. Aikawa, in A. S. Prasad (Ed.), Trace Elements in Health and Disease, Vol. II, Academic Press, New York, 1976 p. 47.
38. The Lancet, 1 (1976) 523—524.

39 R. D. Cohen, in E. J. Campbell, C. J. Dickinson and J. D. H. Slater (Eds.), Clinical Physiology, Blackwell, London, 1974 p. 14.
40 R. R. Burns and G. S. Fell, Scott. Med. J., 21 (1976) 153—154.
41 A. S. Prasad, A. R. Schulert, H. H. Sandstead, A. Miale Jr. and Z. Farid, J. Lab. Clin. Med., 62 (1963) 84—89.
42 K. Weismann, N. Hjorth and A. Fischer, Clin. Exper. Dermatol., 1 (1976) 237—242.
43 B. Portnoy and M. Molokhia, Br. J. Dermatol., 91 (1974) 701—703.
44 B. E. Walker, J. B. Dawson, J. Kelleher and M. S. Losowsky, Gut, 14 (1973) 943—948.

Chapter 4k

Elemental analysis of body fluids and tissues by electrothermal atomisation and atomic absorption spectrometry

H. T. DELVES

Chemical Pathology and Human Metabolism, The University of Southampton, Southampton General Hospital, Tremona Rd., Southampton SO9 4XY (Gt. Britain)

I. INTRODUCTION

Elemental analysis of body tissues and fluids by atomic absorption spectrometry with electrothermal atomisation has advanced significantly the understanding of the role of trace elements in clinical biochemistry. All of those aspects of metabolic processes that are affected by changes in the concentrations of accessible trace elements have been studied. These include: deficiencies of essential trace elements as a result of inherited or acquired metabolic disorders, or from nutritional inadequacy; and excesses of trace elements producing toxicity states as a result of inherited metabolic disorders involving essential trace elements or from the inappropriate exposure to, or ingestion of, non-essential trace elements.

Three general headings are used in the discussion of the analytical measurements: essential trace elements; non-essential trace elements used therapeutically; and non-essential, toxic trace elements. The possibility that all trace elements might be both essential and toxic, depending upon their available concentrations has been discussed by Schwarz [1]. Although I would not argue against the toxicity of any element at excessively high concentrations it is difficult to accept that elements with no known functions, such as lead and cadmium, might be essential to man; or that because nickel is undoubtedly essential to the Japanese quail it could therefore be vital to non-avian species. It is not possible to discuss the analysis of every element that could be classified under the three general headings mentioned above. Those elemental analyses discussed here are intended to reflect recent, current and probable future interest in the application of atomic absorption spectrometry and electrothermal atomisation methods.

II. ELECTROTHERMAL ATOMISATION FOR ATOMIC ABSORPTION

Electrothermal atomization for atomic absorption is the most useful of all of the currently available analytical techniques for measuring trace elements in biological samples. The high sensitivity and element specificity that allow most measurements to be made using microsamples are matched only by techniques such as Isotope Dilution Mass Spectrometry (IDMS) and Proton Induced X-Ray Emission Spectroscopy (PIXIE) which are significantly more expensive and less amenable to routine analysis, or by Anodic Stripping Voltammetry (ASV) and related electrochemical techniques that have a limited range of application.

The disadvantages of electrothermal atomisation (ETA) — atomic absorption spectrometry (AAS) are the physical, chemical and spectral interferences, these being more severe than with flame atomic absorption spectrometry (FAAS), and which depend critically upon the experimental and operational conditions within the atomiser and the nature of the chemical pretreatment used. It is not intended to discuss here the theoretical aspects of these interferences which have been reviewed excellently elsewhere [2], but it is pertinent to consider briefly how these interferences affect the various stages of the analysis and how they may be minimised.

A. Physical interferences

Physical interferences occur as a result of the differences in surface tension and viscosity between body fluids or dissolved/digested tissues and aqueous standard solutions. They affect the accuracy and precision of injecting microlitre volumes of sample solutions into the atomiser, and the degree to which these solutions spread along and diffuse into the graphite atomiser. The rates of volatilisation and atom formation will vary with the position of the deposited sample within a tube atomiser because of the high temperature gradients during atomisation. Approximately 20% of the atomic vapour formed within a tube atomiser is lost by diffusion through the wall, a process which is also dominant with graphite-cup atomisers [3]. It is necessary therefore to control the position of sample injection within the tube and to minimise the diffusion of the deposited analyte solution, in order to obtain reproducible results. Xylene ($2 \mu l$) has been added to a carbon-rod atomiser prior to sample injection as a means of limiting sample diffusion into the rod [4]. With graphite-tube atomisers this effect may be made reproducible by injecting relatively large volumes ($10-20 \mu l$) of a sufficiently diluted sample solution, usually greater than 1 + 9 using as diluents either water or a surfactant solution, e.g. 0.1% v/v Triton X 100. A precision (RSD) of 0.03—0.05 might be expected using this procedure and manual injection, with a three to five-fold improvement using automatic sample injection (e.g., Fernandez [5] obtained an RSD of 0.01 for blood Pb analysis at $2.5 \mu mol\, l^{-1}$). It is probable that this improvement is not simply the result of a more precise measurement of microlitre volumes but a combination of this with a precisely controlled position and rate of sample injection.

Analyte solutions that are deposited onto solid support materials placed into the atomiser are not subject to positional and diffusional errors.

Originally, this method of sample introduction was devised for specific applications (e.g., Cernik's punched-disc of filter paper for blood lead analysis [6], and the analysis of metalloproteins separated on cellulose acetate membranes [7]), but there is now a commercially available system that uses carbon "boats" which has been recently applied to the direct analysis of cadmium and lead in blood and urine [8].

B. Chemical interferences

The rates and mechanisms of free analyte-atom production within electrothermal atomisers depend upon many factors, one of which is the chemical nature of the compounds present [2]. In the analysis of body tissues and fluids, the net effect of the matrix upon the analyte sensitivity is the resultant combination of an enhancement by the organic constituents and a suppression by the inorganic constituents present during atomisation. The suppressive interference on lead absorption signals from the inorganic constituents of biological matrices is shown in Fig. 1. The organic constituents had been destroyed prior to analysis by oxidation with $HNO_3/HClO_4/H_2SO_4$, and the residues dissolved in HCl. The resultant inorganic matrices from all five of the tissues and fluids studied produced lower sensitivities for Pb than that obtained with aqueous solutions containing the same concentration (0.05 M) of HCl. Evenson and Prendergast [9] have even noted different suppressive effects in blood lead analysis with samples preserved using heparin or oxalate as anti-coagulants. The nature and magnitude of the matrix effects upon analyte sensitivity will depend upon the thermal treatment of the sample in the atomiser during all stages prior to atomisation, and upon any chemical pretreatment that is used. (See Section III.) It is essential therefore to ensure complete reproducibility of the operating conditions during the drying and ashing stages to obtain a constant matrix composition for atomisation. Long, ramp-drying and ashing stages are preferable to abrupt temperature changes. Matrix matching and matrix modification can often lead to improved analyses and these are discussed in some detail in Section III.A.

C. Spectral interferences

Spectral interferences by light scattering or molecular absorption during atomisation are severe in most ETA—AAS determinations and must be compensated for by using background correction. Completely ashing biological samples prior to analysis is not sufficient to eliminate these interferences because the inorganic constituents of the matrix absorb radiation over a wide spectral range. This is illustrated in Fig. 2, the data for which was obtained using 0.05 M HCl solutions of the residues of the biological samples that remained after oxidation with $HNO_3/HClO_4/H_2SO_4$ and evaporation to dryness. A D_2 (deuterium) hollow-cathode lamp was used

References pp. 377—380

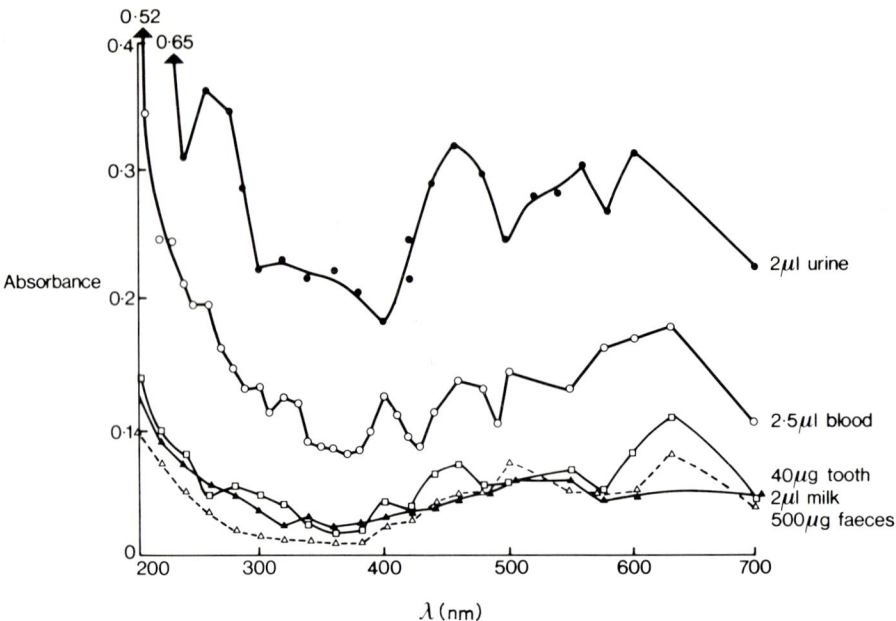

Fig. 1. Molecular absorption from the inorganic constituents of ashed biological samples.

as radiation source with a band-pass of 1 nm. The ETA conditions were: dry at 120°C for 20 s, ash at 500°C for 30 s, atomise at 2200°C for 5 s, clean at 2500°C for 5 s, followed by a "blank" atomisation at 2200°C for 5 s. The data points were obtained by subtracting the blank signal from that obtained during the initial atomisation. The curves for all five matrices (Fig. 1) show similar patterns, with peaks near 250—280 nm, slightly higher wavelengths than the 240 nm NaCl peak, and near 420—450 nm and 600—650 nm. Although there was very little radiation output from the D_2 lamp at wavelengths >400 nm, measurable signals were obtained. The very high absorption signals at <220 nm have been observed by others using simple halides and other inorganic salts [10], and is probably due to light scattering. The data presented in Fig. 2 suggest that molecular absorption/radiation scatter interferences exist over the whole of the analytically useful spectral region and that background correction is essential for the analysis of body tissues and fluids even after complete oxidation of the organic material.

It is possible to reduce these interferences by using higher temperatures during the ashing stage but this may lead to losses of some of the more volatile elements. The optimum ashing temperature for a given element will vary with the type of sample to be analysed and must be established or confirmed prior to analysis. For example, volatilisation of cadmium and lead from aqueous solution begins at 400°C and 500°C respectively, but these elements will volatilise at temperatures 50 to 100°C lower in the presence of a diluted blood or urine matrix. The use of matrix modification

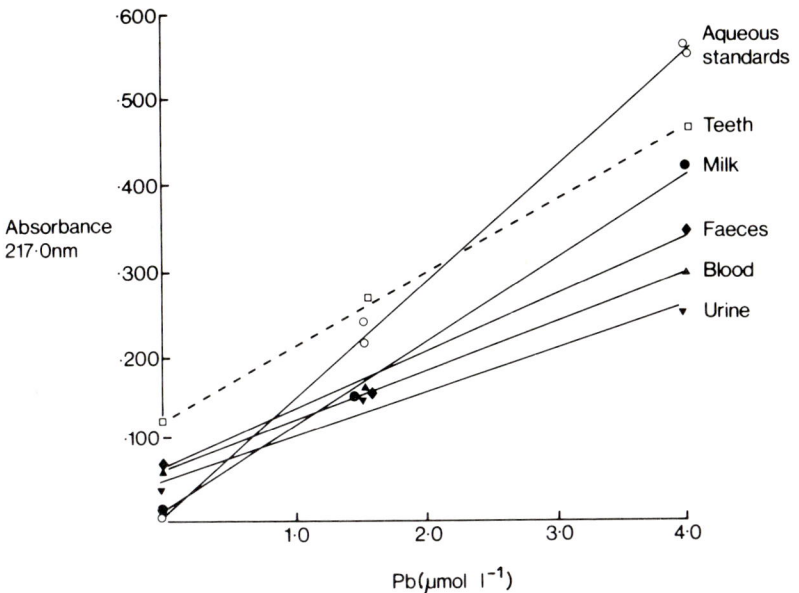

Fig. 2. Effect of matrix from ashed samples on sensitivity.

to allow higher ashing temperatures in the atomiser to minimise molecular absorption interferences is discussed later (Section III.A). Issaq and Zielinski [11] modified the design of a graphite tube atomiser to allow easier removal of the matrix during the ashing stage and thus reduce molecular absorption interferences in the direct analysis of whole-blood and serum. Although all of these various procedures can be used to minimise molecular absorption/ light scatter it is unlikely that they can be eliminated completely, and background correction is almost invariably necessary for elemental analysis of biological matrices.

The most commonly used methods of background correction measure the molecular absorption signal using either continuum sources (D_2/H_2), or line sources that provide a non-resonance line adjacent to the analytical wavelength. These signals are automatically subtracted from those obtained using the resonance line of the element to be determined. The disadvantage of the latter lies in the assumption that there is no change in molecular signal over the wavelength interval between the two lines. Continuum sources emit radiation poorly at wavelengths >300 nm and it is often difficult to balance the radiation output of the line and continuum sources. The two line method therefore becomes attractive at higher wavelengths e.g., 357.9 nm Cr line and 356.7 nm Lu line.

Recently, background correction systems employing the Zeeman effect have been described [12, 13]. This method of correction is more accurate

References pp. 377—380

than the others because it does not require any radiation source other than the analyte hollow-cathode lamp, and correction is made at the analytical line and not over the band-pass of the instrument. However, the extra instrumentation required is relatively complex and expensive. Koizumi and Yasuda [12] corrected for molecular absorbances of up to 1.5 A (A = absorbance unit), in the direct analysis of lead in blood without using any drying or ashing stages in the atomiser. Since the atomic signals would have been only approximately 0.05 A, an error of ± 5% at this level would require a measurement of difference between the molecular signal plus atomic signal (1.5 A) and molecular signal only (1.45 A) of better than 0.3%. Whether this could be maintained on a routine basis remains to be seen. An excellent discussion of the theoretical and practical aspects of background correction in AAS has recently been published by Newstead et al. [14].

III. SAMPLE PREPARATION FOR ELECTROTHERMAL ATOMISATION—AAS

Many of the earlier publications on the use of ETA—AAS to measure metals in bodyfluids advocated little or no sample pretreatment. The injection into the furnace of microlitre volumes of whole-blood or serum, either directly or after a simple aqueous dilution, can provide adequate clinical information and produce results of acceptable precision. However, accurate analyses will only be obtained under optimal conditions, and it is difficult to maintain these on a routine basis without some form of chemical pretreatment. For example, there are two time-dependent interferences in the analysis of lead in blood, when using a simple 1 + 19 dilution with water and taking $20\,\mu l$ volumes for analysis. Firstly, within 1—2 min. of dilution, a precipitate of the red-cell membranes appears and reaches a maximum after about 15 min. It is necessary therefore to ensure adequate mixing immediately before sampling for analysis. Secondly, the repeated injection of diluted whole-blood into the furnace results in a gradual deposition of a carbonaceous residue within the furnace that will affect analyte sensitivity if it is not removed at regular intervals. Both of these interferences are easily removed by simple pre-treatment of the samples with HNO_3 (see Section III.A).

According to Stoeppler et al. [15], severe errors up to a factor of two may result from ETA—AAS analysis of biological materials without some form of sample pretreatment. The approaches that will be discussed here are: (a) the use of diluent solutions to minimise matrix and molecular absorption interferences; (b) partial decomposition techniques in which metals are extracted from proteins with acids; (c) dissolution of tissue samples without complete oxidation; (d) complete oxidation procedures such as dry ashing, wet digestion at ambient and elevated pressures, and low temperature ashing with reactive gases at low pressures.

A. Diluent solutions for matrix modification

A few simple chemical reagents may be added to biological samples in order to minimise molecular absorption interferences by providing a more complete oxidation within the atomisation cell, or to minimise the effect of the matrix upon analyte sensitivity either by modifying the volatility of the analyte, or the matrix, or both.

Complete elimination of molecular absorption interferences in the analysis of serum for Cu was obtained by diluting the samples (1 + 9) with 0.1 M HNO_3 [16], but the greater protein content of whole blood and the greater molecular interference at lower wavelengths restricted the effect of conc. HNO_3 to merely reducing these interferences in blood Pb analysis [17]. The dilution of whole-blood (1 + 9) in a solution that finally contained 10% v/v $HClO_4$ in acetone was claimed to eliminate completely the molecular absorption interferences for blood Cd analysis [18]. The use of $HClO_4$ in organic solvents is potentially hazardous and is not recommended.

Ediger [19] discussed the use of chemical modification of either matrix or analyte volatility for ETA—AAS. The addition of NH_4NO_3 was recommended for the removal of NaCl by conversion to NH_4Cl and $NaNO_3$, both of which decompose below 400°C. A mixture of H_3PO_4 and NH_4NO_3 has proved successful for the elimination of molecular absorption and matrix interferences in the determination of Cu in urine [20]. Volatile elements such as Pb and Cd may be converted to the less volatile fluoride, sulphate or phosphate salts by addition of the appropriate ammonium salts all of which are easily decomposed. Lagesson and Andrasko [8] used NH_4NO_3 to remove NaCl for urine Pb analysis, $(NH_4)_2HPO_4$ to prevent the volatilisation of Cd during ashing blood samples at 525°C, and NH_4F to resolve the molecular and atomic signals obtained during urine Cd analysis.

Other volatile elements, e.g. As, Se, may be converted to the less volatile arsenides and selenides by the addition of Ag or Ni salts. Ashing temperatures of up to 1200°C for Se and 1400°C for As may then be used without volatilisation losses [19, 21, 22].

B. Acid extraction of metals from proteins

Trichloracetic acid (TCA) has been used to release Fe from serum proteins for many years, yet it is only recently that this kind of preparation has been used for the analysis of other metals. A comparison of TCA precipitation with simple dilution as preparation for serum Zn analysis was reported by Boyde and Wu [23] and by Kelson and Shamberger [24]. Both groups found that Zn was quantitatively recovered from serum using TCA and that the concentrations found were consistently higher than those obtained by direct dilution. The former group found a mean positive bias of +20% for the TCA preparation, but the latter found more consistent recoveries of 99.9% for TCA treatment compared with 99.3% for the dilution method, with RSD's of 0.014 and 0.022, respectively.

Zinc was quantitatively extracted from 0.5 g samples of homogenised diets and faeces by mixing with 6 ml of 1 M HCl and leaving to stand for 24 h [25]. The recoveries of added zinc were: $99.7 \pm 3.1\%$ ($n = 47$) for faeces and $98.9 \pm 2.9\%$ ($n = 50$) for diets. A comparison of the results obtained with this extractive preparation, with those obtained after complete digestion with $HNO_3/H_2SO_4/H_2O_2$ was excellent ($r = 0.998$). This most efficient yet simple procedure will no doubt be applied to the measurement of other metals in diets and faeces and perhaps moderately elevated temperatures of 60—80°C may reduce the analysis time. The quantitative release of Mn from plasma and from liver homogenates in 1% Triton X 100, was accomplished by heating in 2 M HCl solutions for 60 min at 60°C [26]. The analytical results for samples treated in this way agreed very well with those obtained after $HNO_3/HClO_4$ ashing, $r = 0.96$. Hinners [27] found that extraction of bovine liver with 1% v/v HNO_3 gave complete recovery of Cd, Cu, Mn and Zn.

De-proteinisation of diluted whole blood, and acid extraction of lead from proteins with 2 M HNO_3 is the preparation chosen for the National Bureau of Standards reference method for blood-lead analysis [28]. Centrifugation of blood samples treated in this way yields clear supernatant fractions that contain all of the lead present, and which are easily dispensed into electrothermal atomisers using auto-sampling techniques. This method gave results that compared very well with those obtained using anodic stripping voltammetry, $r = 0.975$, for concentrations of 10—900 $\mu g\,l^{-1}$ of lead in blood [15].

C. Dissolution of tissues

Jackson et al. [29] reported a simple method for dissolving tissue samples using a 25% alcoholic solution of tetramethylammonium hydroxide which contained 2% m/v APDC. Clear homogeneous solutions were obtained when 100—300 mg weights of liver, brain, or muscle tissues were heated at 60°C with 1—3 ml of the base/APDC solution, and the cooled solutions were diluted with toluene. The solutions were then analysed for Zn, Cu, Fe and Mn. This procedure has also been applied successfully to the analysis of Cu in brain [30] and Pb in human placental tissue and animal liver [31]. The addition of APDC ensured the stability of solutions of exogenous metal ions added for calibration as well as those of endogenous metals from the tissues for at least 24 h after dilution with toluene. Gross and Parkinson [32] ommitted APDC from their solution of the quaternary ammonium hydroxide and used water rather than toluene for the final dilution. They were able to dissolve successfully 28 different types of human tissue, and measured Cu, Pb, Zn, Cd and Mn in liver by ETA—AAS with RSD's of 0.04—0.06. Quantitative recovery, 94—104%, was observed for all metals except copper which was only 85%. It was also noted that standard solutions of metals formed precipitates with the base solution in the absence of tissue samples. These observations suggest that the ommission of APDC was a retrograde step and that procedures similar to that originally described by Jackson et al. [29] should be used.

D. Complete oxidation of tissues and fluids

1. *Wet and dry ashing*

The relative merits of completely oxidising biological samples using either hot mineral acids or by prolonged heating at elevated temperatures have been well documented since 1959 [33, 34]. The main disadvantages of wet ashing are the "blank" values from the acids used, even though these may be of the highest purity available; whereas dry ashing has the disadvantages of potential losses of metals either by volatilisation or by absorption onto the container. With care, both techniques can produce complete oxidation of biological samples with quantitative recoveries of the metals sought. In my own laboratory, wet ashing with a mixture of $HNO_3/HClO_4/H_2SO_4$ (10 + 6 + 1), or dry ashing at 460°C in transparent quartz beakers, have both proved successful for measuring 10—15 elements in blood, urine, diets and faeces [35, 36]. The wet-ashing method is no longer used because of the incompatibility of the fume-cupboard systems with the potentially hazardous $HClO_4$. Those who use or intend to use $HClO_4$ for sample oxidation on a routine basis should first establish that their fume cupboards and ducting are suitable for use with this reagent. An alternative mixture which gives complete oxidation is $HNO_3/H_2SO_4/H_2O_2$, but HNO_3/H_2O_2 produces only incomplete oxidation of biological samples.

The volatilisation losses of metals from tissues and fluids during dry ashing is widely known, but little is known of the losses during the preliminary drying stages. Iyengar et al. [37] investigated the losses of isotopically labelled endogenous metals from rat tissues and fluids with different drying temperatures, and showed that oven drying at 120°C resulted in losses of Sn (10—15%), Mn, Sc and Ce (up to 5%) from liver, muscle, heart and brain. Significant losses of Sn from muscle were observed as low as 80°C. These observations indicate that some form of extractive pretreatment of tissue homogenates, e.g. evaporation with HCl, is necessary before dry ashing.

The disadvantages of both wet and dry ashing which have only been outlined here, have led to the development of alternative methods for sample oxidation based on these two techniques. These are wet ashing using vapour phase attack or at elevated pressures, and dry ashing at low temperatures and pressures with reactive O_2.

References pp. 377—380

2. Vapour phase and pressure digestion

The exploration of gas-phase reactions to dissolve biological matrices has rarely been considered in spite of its well established success for silicates and other refractory inorganic materials. A simple apparatus for the vapour-phase oxidation of blood and other tissues with HNO_3, has been described by Roos [38]. The sample containers are placed inside a reaction vessel containing conc. HNO_3 and fitted with a reflux condenser. The sample attack, which takes 3—4 h, is from the vapour phase only, ensuring that no acid accumulates in the sample and that contamination is minimised.

Rapid digestion of inorganic matrices may be accomplished by heating to 110—170°C with oxidising acids in sealed pressure vessels. Most vessels are based on the original design by Bernas [39] of a crucible-shaped stainless-steel container fitted with a removable polytetrafluorethylene (PTFE) crucible and sealed with a stainless steel screw-cap with a PTFE disc inserted in the metal body. The gases produced during heating and decomposition raise the internal pressures to 30—50 bar, and the vessels must be cooled to room temperature before opening. Holak et al. [40] digested 1 g samples of fish for Hg analysis in only 30 min using HNO_3 and Paus [41] determined Hg, Cd, Zn, Cu, Fe and Pb in biological samples after pressure digestion with HNO_3/H_2SO_4. The main advantage over conventional wet ashing was the complete dissolution of fat. Stoeppler et al. [42] have given an excellent evaluation of pressurised digestion of biological matrices including some very important safety considerations. Pressure peaks of 92 bar, which could be hazardous in routine applications, were observed during the oxidation of animal feeds with conc. HNO_3. Procedures for minimising such pressure peaks are discussed.

A simplified pressurised digestion technique was used by Lutyen et al. [30]. Tissue samples (250 mg) were placed in Nalgene bottles, 4 ml of (1 + 1) $HNO_3/HClO_4$ was added and the bottles were sealed and kept at 100°C for 2 h. After cooling, the bottles were opened and the solutions diluted with water. This procedure did not work well with brain tissue because of the high lipid content.

3. Low-temperature ashing

Decomposition of biological tissues in a low pressure (<1 torr) stream of radio-frequency excited O_2 gas takes place at relatively low temperatures (70°C) with no volatilisation losses and 99—102% recovery of Sb, As, Cs, Co, Cr, Fe, Pb, Mn, Mo, Se, Na and Zn [43]. Appreciable losses of Hg, I, Ag, Au and Pt do occur, the latter three probably as a result of catalytic reaction with the excited O_2 [44]. The main disadvantage of this technique is the very long ashing time, which can be up to 32 h for 1 g samples. The use of more reactive gas mixtures can reduce the ashing time. Lopez-Escobar and Hume [45] described a mixed-gas technique in which 1.4 nmol min^{-1} of O_3 in O_2 effected the release of 98.5% of Hg from organic matrices in only

15 min. Carter and Yeoman [46] used an equimolar mixture of CF_4 and O_2 to oxidise 10 mg weights of whole-blood for Cd analysis in 12.5 min compared with 60 min required with O_2 alone. Pretreatment of blood samples with H_2O_2, can further reduce the sample time for complete oxidation with CF_4/O_2 to 6 min for 20 mg (20 µl) samples [47]. The application of mixed reactive gases for the rapid oxidation of biological matrices merits further study.

All of the sample preparation techniques that have been discussed may be used with solvent extraction and other separation procedures to concentrate the analyte and separate it from the matrix. Where such procedures are applicable they will be discussed together with the methods for determining the individual elements.

IV. ESSENTIAL TRACE ELEMENTS

Measurements of the concentrations of essential trace elements in body tissues and fluids can provide valuable information for the diagnosis and treatment of a wide range of human disorders. The physiological concentrations in body fluids of those elements generally considered to be essential to man are listed in Table 1, together with data on their analytical sensitivity by ETA—AAS. The elements fluorine and iodine are not readily amenable to measurements by ETA—AAS and are not included. Although the analytical sensitivities will vary with different types of atomiser the data given allow useful conclusions to be reached about the sample volumes that are required for accurate measurement of the elements. These data are important because the volumes of blood that may be taken from patients are limited, particularly where children are concerned. It may be seen from these data that the volumes of whole blood or serum that are required increase in the order $Zn \ll Fe = Cu < Mn \approx Se < Cr < Mo < V$. The dilutions of serum/blood that would be required to produce a moderate absorbance signal of 0.10 A with a typical injection volume of 20 µl, calculated from the data in Table 1 are: Zn (1 + 221), Fe and Cu (1 + 30), and Mn and Se (1 + 1). All other elemental analyses must be carried out either without dilution or at lower analyte signal strength. It is possible to measure Mn or Cr in serum with high scale expansion and low absorbance signals, but the analysis of Mo, Co and V requires some degree of preconcentration to produce absorbance signals of sufficient magnitude to allow precise measurements of the lower concentrations for reference populations.

The order of increase in volume requirements also reflects to a certain degree the difficulties of analysis, but this will be modified by matrix interferences and volatilisation problems. The calculations do indicate however that ETA—AAS should never be the method of choice for serum Zn analysis. The problems of contamination associated with such large dilutions can only lead to poor precision.

References pp. 377—380

TABLE 1

PHYSIOLOGICAL CONCENTRATIONS OF ESSENTIAL TRACE ELEMENTS IN BODY FLUIDS AND SAMPLE VOLUMES REQUIRED FOR ANALYSIS BY ETA—AAS

Element Concentration units		Reference concentrations for healthy controls		ETA—AAS sensitivity[a]	Sample volume required[b]
		μ/nmol l^{-1}	pg/μl	pg/0.0044 A	μl/0.0044 A
Zn (μmol l^{-1})	S	11—24	720—1570	3	0.004
	U	5—10	327—654		0.01
Fe[c] (μmol l^{-1})	S	11—36	614—2010	20	0.03
	U	0.2—1.0	11—56		1.8
Cu (μmol l^{-1})	S	12—26	762—1651	27	0.03
	U	0.2—0.8	13—51		2.1
Mn (nmol l^{-1})	B	18—180	1—10	4	0.4
	S	13—23	0.7—1.3		2.2
Se (μmol l^{-1})	S	1.2—1.4	92—108	44	0.44
Cr (nmol l^{-1})	S/B	3.8—135	0.2—7.0	13	65
	U	8—46	0.4—2.4		32
Mo (nmol l^{-1})	S	2—12	0.2—1.2	20	100
Co (nmol l^{-1})	B	1.7—17.	0.1—1.0	38	380
	U	1.7—34.	0.1—2.0		
V (nmol l^{-1})	B	9.2	0.47	200	426
	U	5.1	0.26		769

[a] Data from Price [2].

[b] Calculated for the minimum value of reference range where available, otherwise the mean value is used i.e. V.

[c] Serum Fe values vary widely with age and sex.

NOTE: The reference ranges for Mo, Co, and V and possibly Cr may prove to be lower than the quoted values (see Sections IV.F to IV.I).

(The data in Tables 1—3 were originally published in Prog. Anal. At. Spectrosc., 4 (1981) 1—48 and are reproduced here by permission of Pergamon Press.)

A. Zinc

The measurement of Zn in body tissues or fluids for clinical purposes does not require the sensitivity of ETA—AAS. The very early methods using carbon-rod atomisers required only 0.5 μl serum for direct analysis of Zn [48], and current techniques require <0.1 μl. The measurement of Zn in cerebrospinal fluid (CSF) by ETA—AAS [49] could have just as easily been determined using "pulsed"—nebulisation FAAS with <100 μl sample volumes.

The most useful role for ETA—AAS will be measuring the various low molecular weight and protein bound Zn species in body fluids and the cellular and sub-cellular distribution of Zn. The total Zn concentration and its distribution among the proteins of human parotid saliva has recently been measured using gel filtration and ETA—AAS [50]. Only 2 μl sample volumes were required for these studies and the precision at relatively low concentrations was good: RSD, 0.030 at 70 μg l^{-1} and 0.055 at 50 μg l^{-1}. The ETA conditions were: dry at 100°C for 20 s; ash at 600°C for 60 s; and atomise at 2100°C for 8 s. Only 87% of the Zn was recovered from the column, compared with 100 ± 5% for Cu and Cd, suggesting that the column had sequestered some low molecular weight Zn species.

B. Copper

The determination of Cu in serum/plasma is probably the easiest of all elemental analyses of biological fluids by ETA—AAS. Matousek and Stevens [48] analysed directly 0.5 μl volumes of serum and obtained an excellent correlation with FAAS. In my laboratory a simple 1 + 9 dilution of 100 μl volumes of serum with water has been found to be sufficient sample pretreatment. The diluted samples, 20 μl, are dried at 120°C for 20 s, ashed at 750°C for 30 s and atomised at 2500°C for 5 s. It is necessary to use background correction and to calibrate by matching the standard matrix with that of the samples to compensate for the suppressive interference from inorganic constituents. This is achieved easily by diluting the standards with a low Cu control serum and water (1 + 1 + 8). The best precision that can be achieved with this method using a manual sample injection is typically 0.05—0.06 (RSD) at 12 μmol l^{-1} (76 μg dl^{-1}) which is adequate for most clinical purposes. However the analytical performance may be improved considerably by using only a little sample pretreatment. Evenson and Warren [16] eliminated molecular absorption interferences during atomisation by diluting sera (1 + 9) with 0.01 M HNO$_3$, and obtained excellent RSD's of 0.021 at 12.3 μmol l^{-1} (78 μg dl^{-1}) and of 0.014 at 24.7 μmol l^{-1} (157 μg dl^{-1}). Wawschinek and Hofler [20] described a simple and attractive procedure for the routine analysis of serum and urine for copper, which they used as an aid to the diagnosis of Wilsons Disease. They formulated a diluent solution that helped to dissolve any urine precipitates, eliminated most but not all of the molecular absorption interferences, compensated for matrix effects on analyte sensitivity, and provided a uniform surface tension for diluted serum and urine samples. This solution contained 0.15 M H$_3$PO$_4$, 0.125 M NH$_4$NO$_3$ and 0.1% Triton X 100. Sera were diluted (1 + 9) and urines (1 + 1). A control serum was used for calibration for both matrices. The mean recovery of Cu added to serum at 10.4 μmol l^{-1} (660 μg l^{-1}) was 99%, range 95—104%, and that of Cu added to urine at 1.9 μmol l^{-1} (124 μg l^{-1}) was 100%, range 97—103%. The ETA conditions were dry at 100°C for 10 s, ash at 600°C for 30 s and atomise at 2700°C for 10 s. Background correction was necessary.

Few determinations of Cu in whole-blood or erythrocytes have been reported mainly because clinical interest has been concentrated on serum values and not because of any inherent difficulties in the analysis. Robbins et al. [51] analysed 0.5 ml volumes of erythrocytes, that had been washed free from plasma with isotonic saline, by ashing with conc. HNO_3 in a PTFE "bomb", diluting 50 times with water and injecting 5 μl volumes for ETA—AAS. Background correction was not needed but calibration required the method of standard additions. The RSD ranged from 0.02—0.08 and a mean value of 17 μmol l^{-1} (1100 μg l^{-1}) was observed for healthy subjects. The use of the Zeeman effect for background correction with minimal sample preparation has been reported by a number of workers [13, 52, 53].

The high sensitivity of ETA—AAS for Cu has stimulated the development of methods to measure concentrations of the Cu carrier species in biological fluids. Delves [7] analysed the Cu content of the protein fractions separated from 2 μl volumes of serum by cellulose acetate membrane (CAM) electrophoresis. The separated protein bands were cut from the CAM and placed directly into the ETA via a 6 mm × 1 mm hole cut in the wall of the graphite tube. Calibration was achieved by adding 2 μl volumes of aqueous standards to 8 mm × 6 mm strips of CAM. Background correction was essential. Approximately 94% of the Cu was located in the α_2 band, where caruloplasmin would run, whereas other fractions contained less than 5% of the total serum Cu. The recovery of Cu after electrophoresis was quantitative, 99%, and the RSD was 0.086 at 1.74 ng Cu. This method was applied to studies of Cu changes in patients with Menkes Syndrome receiving intramuscular injections of copper as the EDTA complex, and in children with acute lymphoblastic leukaemia.

Teape et al. [54] used a similar method to study the distribution of Cu among the serum proteins from patients with rheumatoid arthritis and found that 1—2% of their serum copper was bound to albumin. The same group of workers [55] measured Cu in plasma ultrafiltrate down to 63 nmol l^{-1} (4 μg l^{-1}) using 1000D MWT cut-off membrane to separate the high molecular weight species. They compensated for the 25% suppression from NaCl in the sample matrix by preparing standards in 0.9% NaCl solution. Ashing for 30 s at 800°C significantly reduced the molecular absorption interference, but background correction was still required. Although this temperature is approaching the maximum that can be used without volatilising Cu no losses of Cu were observed and the RSD of 0.04 at 0.8 μmol l^{-1} (50 μg l^{-1}) is excellent.

The total Cu content and the distribution of Cu among proteins of human parotid saliva was measured using gel-filtration and ETA—AAS [50]. The addition of 0.4 M HNO_3 or 0.1 M H_2SO_4 to portions of the Na_2HPO_4/NaH_2PO_4 eluate from the column helped remove NaCl during ashing for 20 s at 850°C. The presence of phosphate would allow the use of this temperature without volatilising Cu. The mean column recovery of Cu was 105% and the RSD was 0.055 at 0.8 μmol l^{-1} (50 μg l^{-1}).

Smeyers-Verbeke et al. [56, 57] analysed small sample weights of brain tissue, down to 1 mg, for Cu after ashing with H_2SO_4/H_2O_2. An RSD of 0.048 was obtained at a concentration of $20.9\,\mu g\,g^{-1}$ dry weight and agreement with the analysis by FAAS that required much larger sample weights was excellent. Small samples of duodenal mucosa which contained 7.7--$29\,\mu g\,g^{-1}$ dry weight were analysed directly using ETA—AAS without any chemical pretreatment [58], however it was necessary to use the method of standard additions for calibration.

C. Iron

The measurement of iron in serum using ETA—AAS is relatively simple because the two main sources of error (i.e., interference from haemoglobin and molecular absorption interference from NaCl) are easily overcome. The concentration of iron in erythrocytes is 1000 times greater than that in serum so that the slightest degree of haemolysis produces erroneously high results. Although the very earliest ETA—AAS methods involving direct analysis or the analysis of aqueous dilutions of serum were methodologically valid from an AAS viewpoint [48, 59], they did not allow for possible iron contamination from haemoglobin which is easily removed by precipitation with trichloracetic acid (TCA). Olsen et al. [60] precipitated proteins from $50\,\mu l$ volumes of serum by adding $50\,\mu l$ of 10% m/v TCA and heating at 90°C for 15 min, and measured iron in $10\,\mu l$ volumes of the supernate. The samples were ashed at 900°C for 25 s and atomised at 2400°C for 8 s. This method removed any haemoglobin by TCA precipitation and minimised the molecular absorption interferences by measuring at the 302.1 nm line where the NaCl absorption is negligible. Calibration was effected by treating the standards in a similar way to the samples, and a comparison with FAAS yielded an excellent correlation coefficient of 0.992 for the concentration range $20-250\,\mu g/100\,ml$. The within run RSD was 0.03 for physiological concentrations. Berman [61] has measured serum iron concentrations and total iron binding capacities routinely for many years using a similar procedure. She uses a larger dilution of serum with TCA (1 + 4) and measures Fe at the more sensitive 248.3 nm line. The iron binding capacity is measured by adding $1000\,\mu g$ Fe, as ferric ammonium citrate, to $200\,\mu l$ serum, equilibrating, removing the excess of Fe^{3+} by ion exchange and then measuring the increase in serum iron content.

Direct analysis of urinary iron concentrations resulted in poor sample injection precision and variable absorption signals during atomisation at 2700°C even after ashing for 30 s at 900°C. After low temperature ashing and dissolution of the residues in 1 M HCl the RSD improved to 0.038, with a mean recovery of 98.4% of Fe added at $0.36\,\mu mol\,l^{-1}$ ($20\,\mu g\,l^{-1}$) [62]. The analysis of $30-340\,\mu g\,g^{-1}$ Fe in rat aorta was accomplished by ashing with HNO_3 in a PTFE pressure vessel and ETA—AAS [63]. The detection limit using 10 mg samples was $4.2\,\mu g\,g^{-1}$.

References pp. 377—380

D. Manganese

Although it is possible to measure directly Mn in body fluids at normal concentrations of 18—180 nmol l^{-1} (1—10 µg l^{-1}) it is difficult to achieve a precision of better than 0.10 RSD. The elimination of the considerable molecular absorption interferences requires strict control of ETA ashing temperatures and a good background correction system, and the variable condensed phase matrix interferences from the inorganic constituents necessitates the use of standard additions for calibration [64]. Even this approach may not yield a viable method due to curvature of the calibration graph at very low absorbances, particularly with a diluted blood matrix [65].

D'Amico and Klawans [66] determined Mn directly with 50 µl volumes of serum or CSF. Careful control of the ETA ashing conditions was essential to avoid losses of Mn, which were significant at temperatures higher than 850°C. Ashing for 60 s at 800°C only reduced the molecular signal at atomisation to 0.5 A but this was within the instrument's correction capability. The detection limit of 1.3 nmol l^{-1} (0.07 µg l^{-1}) was more than adequate for the determination of the 'normal' concentrations which were found to be 13—23 nmol l^{-1} (0.74—1.25 µg l^{-1}) in serum and 10—16 nmol l^{-1} (0.55—0.89 µg l^{-1}) in CSF.

Buchet et al. [67] described excellent procedures for the determination of Mn in whole-blood and urine after extraction of the cupferron complex into MIBK. Urine samples required no treatment other than buffering at pH 7.1 with an imidazole—HCl buffer. Six different preparative treatments for whole blood samples were investigated, from simple (1 + 9) dilution with water or Triton X 100, TCA precipitation to acid oxidation and digestion. The best recovery of added Mn and most complete oxidation was obtained with a mixture of $HNO_3/HClO_4$. Analysis of the organic phase (20 µl ≡ 20 µl blood or 50 µl urine) was made without background correction using an ETA programme of drying for 5 s at 85°C, ashing up to 350°C and atomising at 2600°C. Calibration was by standard additions and linear from 18—455 nmol l^{-1} (1—25 µg l^{-1}). The Mn(II) cupferron complex was stable in MIBK for up to 24 h which is significantly better than the APDC chelate. The recovery of added Mn ranged from 104—108% and the RSD's were 0.033—0.053 for urine containing 73—255 nmol l^{-1} (4—14 µg l^{-1}) and 0.054—0.064 for whole-blood containing 164—236 nmol l^{-1} (9—13 µg l^{-1}).

The quantitative release of Mn from plasma and liver homogenates in 1% Triton X 100 was achieved by heating at 60°C in 2M HCl solutions for 60 min [68]. The results for samples treated this way agreed very well with those obtained using larger sample weights with $HNO_3/HClO_4$ oxidation and FAAS; correlation coefficient, $r = 0.96$. Calibration was by standard additions and the RSD for liver homogenates containing 474 nmol l^{-1} (26 µg l^{-1}) was 0.005 and that for plasma at 38 nmol l^{-1} (2.1 µg l^{-1}) was 0.035.

Dissolution of mg amounts of rat tissues in HNO_3 at 60°C was sufficient preparation to measure 18—109 nmol kg^{-1} (1—6 µg g^{-1}) levels of Mn [69]. The solutions were ashed at 800°C for 70 s prior to atomisation at 2400°C and background correction was essential. The sensitivity was 1.16×10^{-11} g for 1% absorption at 279.5 nm and the calibration, by standard additions, was linear over the range $1—8 \times 10^{-10}$ g. The precision at 5×10^{-10} g was 0.047 (RSD). Levels of Mn in brain 24—36 nmol kg^{-1} (1.3—2.0 µg g^{-1}) were two to three times lower than those in liver or kidney, but higher than in other tissues such as lung, spleen, heart and bone.

Pleban et al. [53] used a Zeeman background correction—ETA method to measure Mn in whole-blood and in serum samples, diluted 1 + 3, and 1 + 1 respectively with Triton X 100. Only 10—15 µl volumes of the diluted solutions were analysed and calibration was by standard additions over the range 9—180 nmol l^{-1} (0.5—10 µg l^{-1}). Within-run RDS's were 0.055—0.070 for whole-blood levels of 36—109 nmol l^{-1} (2—6 µg l^{-1}) and 0.050—0.10 for serum analyses at 18—55 nmol l^{-1} (1—3 µg l^{-1}). The values found for a control population expressed as mean ± SD were 33 ± 11 nmol l^{-1} for serum values, similar to the range reported by D'Amico and Klawans [66], and 164 ± 42 nmol l^{-1} (9 ± 2.3 µg l^{-1}) for whole-blood, which agrees with most published data. Without doubt this is an excellent demonstration of the background correction capability of the Zeeman system. The relatively low ashing temperatures of 350—450°C that were used resulted in molecular signals of 1.2 A during atomisation at 2400°C. The concurrent Mn atomic absorption signals were typically 0.005 A. The philosophy of using such low ashing temperatures and choosing to measure the difference between the molecular plus atomic signal (1.205 A) and the molecular signal (1.200 A) is questionable when with little extra effort the molecular signal could have been reduced to 0.2—0.3 A without losing Mn. Gross and Parkinson [32] obtained clear solutions after heating 1 g weights of a wide range of different human tissues with 2 ml of a 25% alcoholic solution of tetramethylammonium hydroxide at 70°C for 2 h. The solutions were diluted to 10 ml with water and were stable for three days. The analysis by standard additions gave a Mn content for NBS bovine liver of 9 µg g^{-1} which agreed with the reference value of 10 ± 1 µg g^{-1}. The RSD at this concentration was 0.041.

E. Selenium

Few simple direct procedures exist for the measurement of selenium in body tissues and fluids using ETA—AAS. Severe matrix and molecular absorption interferences usually require some form of separation procedure to be used. Saeed et al. [70] recently reported an excellent method for measuring Se in 20 µl volumes of serum. The samples were diluted with 20 µl volumes of 0.1% Ag or Ni nitrate solutions, which effected some degree of matrix modification prior to direct analysis by ETA—AAS. The Ag or Ni salts allowed atomisation temperatures of 1250°C and 1050°C respectively without losses of selenium. In the absence of these salts Se was lost at just over 400°C, and 6% of Se was lost during low-temperature ashing (LTA) at 150°C. An excellent agreement was obtained with results by neutron activation analysis, and the precision at normal Se concentrations in plasma of 1.2—1.4 μmol l^{-1} (92—108 μg l^{-1}) was 0.04 (RSD). Unfortunately this procedure is not directly applicable to whole-blood or other iron-rich tissues because of interference by Fe at the 196.0 nm line. The measurement of Se in whole-blood and tissues was accomplished by Ishizaki [71] who used oxygen-flask combustion, cation-exchange separation of interfering elements and then solvent extraction with dithizone/CCl$_4$ as sample pretreatment. Losses of Se during the ETA—ashing stage were eliminated by adding 50 μl of 0.1% Ni(NO$_3$)$_2$ to 20 μl of the organic extract in the tube. An RSD of 0.08 was obtained at a concentration of 0.13 nmol l^{-1} (0.01 μg g^{-1}).

F. Chromium

Volatilisation losses of organically bound Cr from body tissues and fluids make direct analysis difficult and most published methods use some form of oxidative pretreatment. It has even been suggested that such losses may occur with acid digestion at temperatures below 100°C [72]. Davidson and Sacrest [73] oxidised 200 μl volumes of blood/urine with HNO$_3$/HClO$_4$/H$_2$O$_2$ and dissolved the residues in 400 μl of 0.3 M HCl prior to analysis by ETA—AAS. The double ashing stage at 1100°C and then 1400°C gave no losses of Cr prior to atomisation at 2400°C. The method of standard additions was used for calibration and the overall procedure had an RSD of 0.04 at 38 nmol l^{-1} (2 μg l^{-1}) with a detection limit of 0.2 nmol l^{-1} (0.01 μg l^{-1}). A thorough investigation of the analytical parameters for ETA enabled Routh [74] to determine directly Cr in 20 μl volumes of urine. The samples were diluted 1 + 1 with either 0.1 M HNO$_3$ or 0.1 M HCl and aliquots were dried for 40 s at 100°C, then ashed by ramping the temperature at 600°C s^{-1} to 1000°C maintaining this temperature for 60 s, and finally atomised at 2300°C for 1.5 s. This procedure reduced the molecular absorption signals to approximately 0.2 A which were easily corrected. This interference could be reduced to less than 0.02 A by ashing at 1300°C but this resulted in significant losses of Cr from both nitrate and chloride matrixes. Calibration was by standard additions and the sensitivity of 6 × 10^{-12} g for 1% absorption at 357.9 nm was equivalent to 5.8 nmol l^{-1} (0.3 μg l^{-1}). Although the RSD was not better than 0.10 at 19 nmol l^{-1} (1.0 μg l^{-1}) the accuracy was excellent when the method was applied to NBS bovine liver reference material. The mean urinary Cr excretion for a normal population was 21 nmol (1.1 μg) per 24 h, range 8.6—46 nmol (0.45—2.4 μg) per 24 h. These values are similar to those reported by Vanderlinde et al. [75] of a modal urinary Cr level of 7.6 nmol l^{-1} (0.4 μg l^{-1}) with only a few levels in excess of 38 nmol l^{-1} (2 μg l^{-1}).

The analysis of Cr in hair following HNO_3 digestion has been reported by Rosson et al. [76] who obtained an RSD of 0.03 at concentrations of 4—5 nmol g^{-1}. Levels of Cr in pancreatic tissue were measured after low-temperature ashing of the samples with an RSD of 0.04—0.05 at the 200 pg level [77].

Solvent extraction of Cr from solutions of ashed tissues or body fluids is beset by the difficulties of maintaining the Cr(VI) oxidation state necessary for chelation. Recently Chao and Pickett [78], successfully extracted Cr(VI) at pH 2—3.5 with a 5% solution of methyltricaprylammonium chloride (MTCA) in HCl as the ion-association complex $[HCrO_4]^-[MTCA]^+$. A small excess of $KMnO_4$ prevented the reduction of Cr(VI) to Cr(III) by impurities in the solvent, MIBK. The extraction efficiency into MIBK of 97.1% was greater than that obtained with diphenylthiocarbazone (52%), HCl (75%), or APDC (87%). Bovine liver and other NBS reference materials were analysed for Cr down to 0.1 µg g^{-1} following $HNO_3/HClO_4/H_2SO_4$ oxidation, dissolution in HCl, and extraction with MTCA/MIBK. A relatively high ashing temperature of 1300°C was used prior to atomisation at 2700°C but without any losses of Cr. The RSD at 10 ng Cr was 0.06.

G. Molybdenum

The analysis of molybdenum in body tissues and fluids using ETA—AAS is extremely difficult because of the low physiological concentrations ($<$12 nmol l^{-1} ($<$1.2 µg l^{-1}) in blood/serum/urine), poor sensitivity, severe matrix interferences and carbide formation during atomisation. The role of the graphite tube in the latter interference can be minimised by lining the tube with Ta foil or by pre-forming a carbide layer on the inner wall of the tube by heating with added solutions of the salts of carbide forming elements. It is also necessary to oxidise the sample completely to avoid carbide formation from the organic matrix. Bentley et al. [79] used a low temperature ashing (LTA) procedure to oxidise samples of erythrocytes, serum and urine. The mean recovery of Mo added to serum at 10—40 µg l^{-1} was 101% (range 93—112%) whereas oxidation in a muffle furnace gave only 70% recovery. The residues after LTA oxidation were dissolved in (1 + 1) HNO_3 and analysed by ETA—AAS by drying for 50 s at 120°C, ashing for 40 s at 1800°C and atomising for 8 s at 2800°C. A prolonged high-temperature clean was carried out after each measurement to minimise memory effects. Calibration over the range 100—520 nmol l^{-1} (10—50 µg l^{-1}) required standard additions because of the considerable suppression of the Mo signal by solutions of the ashed matrix. The concentrations of Mo in sera from a control population ranged from 0—310 nmol l^{-1} (0—30 µg l^{-1}) whereas industrially exposed individuals had levels in excess of 52 nmol l^{-1} (50 µg l^{-1}). These levels are much higher than those obtained by neutron activation analysis [80] but are well within the range of the calibration graph of the ETA—AAS method. Further work is clearly needed to confirm these levels. (See also Co, Section IV.H.) The only reference materials available to the authors were NBS bovine liver and orchard leaves with mean concentrations of 3.2 and 0.3 µg g^{-1} respectively. The ETA—AAS values were in reasonably good agreement (3.1 and 0.2 µg g^{-1}) but these levels are considerably higher than the physiological concentrations of Mo in serum.

H. Cobalt

The measurement of Co in body fluids down to 1.7 nmol l^{-1} (0.1 µg l^{-1}) requires some form of sample oxidation, separation and preconcentration. Lidums [81] ashed whole-blood (2 ml) or urine (4 ml) with cHNO$_3$ followed by heating at 350°C. The residues were dissolved in HCl and separated on a cation-exchange column (Dowex 1XB). After elution with HCl and evaporation to dryness the Co containing residues were dissolved in 0.01 M HCl and analysed by ETA—AAS with ashing at 900°C and atomisation at 2600°C. This procedure could detect 1.7 nmol l^{-1} (0.1 µg l^{-1}) and had an excellent RSD of 0.032 at a concentration of 37 nmol l^{-1} (2.2 µg l^{-1}). The recovery of Co added to blood or urine at 32 nmol l^{-1} (1.9 µg l^{-1}) was 98%. The mean concentrations of Co in blood and urine from persons not occupationally exposed to this metal were 8.5 nmol l^{-1} (0.5 µg l^{-1}) and 6.8 nmol l^{-1} (0.4 µg l^{-1}) respectively, with an overall range of 1.7—37 nmol l^{-1} (0.1—2.2 µg l^{-1}).

Solvent extraction procedures have been used to measure comparably low concentrations of Co in body tissues and fluids. Ishizaki and co-workers [82] extracted Co from HCl solutions of ashed blood samples into tri-n-octylamine in CHCl$_3$ and then back-extracted into H$_2$O prior to analysis by ETA—AAS. The recovery of added Co was 97% and the method allowed determinations down to 1.7 nmol l^{-1} (0.1 µg l^{-1}). The mean "normal" concentration of Co in blood was found to be 19 nmol l^{-1} (1.1 µg l^{-1}). Barfoot and Pritchard [83] ashed 10-ml samples of blood or serum with HNO$_3$/HClO$_4$/H$_2$SO$_4$ prior to solvent extraction and separation of Co with α-nitroso-β-naphthol in CHCl$_3$. The overall recovery of Co added as ^{57}Co labelled vitamin B$_{12}$ was 95% and the within-run RSD was 0.037 for a concentration of 25 nmol l^{-1} (1.5 µg l^{-1}). The detection limit was approximately 1.7 nmol l^{-1} (0.1 µg l^{-1}). The "normal" ranges for Co in blood and serum were 34—47 nmol l^{-1} (2.0—2.8 µg l^{-1}) and 20—34 nmol l^{-1} (1.2—2.0 µg l^{-1}) respectively, which agree with the data presented above and some of the data reported using neutron activation analysis [83]. It is unlikely that this procedure would be used routinely because of the very large volumes of blood required, but it could be scaled down without too much difficulty. The direct analysis of Co in body fluids has been reported [84], but the concentration range studied was much higher than would be expected for non-occupationally exposed populations 85—170 nmol l^{-1} (5—10 µg l^{-1}).

The concentrations of Co in body fluids from populations not occupationally exposed to this metal reported using different ETA—AAS methods cover two orders of magnitude, from $1.7-170\,\text{nmol}\,l^{-1}$ ($0.1-10\,\mu g\,l^{-1}$). A similar but slightly lower range of disparate levels has also been reported by users of neutron activation analysis [83]. Many of the ETA—AAS procedures have detection limits that are less than one tenth of the concentrations found and have quantitative recoveries of added Co and good RSD's of 0.03—0.04 at the reported physiological levels. It is therefore difficult to criticise the validity of data obtained with such apparently analytically exact procedures. Sample collection and storage are obvious sources of contamination but most workers are aware of this. There is clearly an urgent need for reference samples of blood and serum with certified Co (and Mo see Section IV.G.) concentrations in the region of $3.4-17\,\text{nmol}\,l^{-1}$ ($0.2-1.0\,\mu g\,l^{-1}$).

I. Vanadium

Few atomic absorption methods can approach the sensitivity of neutron activation analysis (NAA) in measuring the low physiological concentrations of vanadium. The one notable exception is that reported by Ishizaki and Ueno [85], who used ETA—AAS to determine $<0.1\,\mu g\,l^{-1}$ in whole-blood, urine and tissues: large volumes of blood (10 ml) or urine (50 ml) were ashed with $HNO_3/HClO_4$ and the vanadium was preconcentrated 10—50 fold by solvent extraction with a 0.1% solution of N-cinnamoyl-N-2,3-xylylhydroxylamine (CXA) in CCl_4. This new reagent extracted quantitatively nanogram amounts of pentavalent vanadium from solutions of ashed tissues/fluids in 6 M HCl, and the extracted complex was stable for at least three days. The use of a pyrolytically coated graphite tube minimised the penetration of the organic extract (50 μl volumes) into the wall of the tube thus minimising carbide formation and giving a sensitivity of 400 pg for 0.0044 A at 318.4 nm which was 1.5 times greater than that obtained with conventional graphite tubes. The ETA—AAS parameters were: ramp dry over 25 s to 100°C, ash for 20 s at 1850°C and atomise for 10 s at 2800°C.

The mean recovery of $1\,\mu g\,l^{-1}$ vanadium added to blood was 107% and that of $0.2\,\mu g\,l^{-1}$ added to urine was 100%. The analytical precision at the mean concentrations found in blood and urine for healthy controls was excellent: 0.043 (RSD) for $0.47\,\mu g\,l^{-1}$ in blood and 0.117 (RSD) for $0.26\,\mu g\,l^{-1}$ in urine.

The increase in sensitivity from the pyrolytically coated tube has been noted by others [86] who cautioned against impregnation of the graphite tube with salts of elements which form stable carbides (Ta, Si, Nb, Zr, W, La). Decreased sensitivities for V have been obtained following such treatment which was attributed to the possible formation of ternary compounds between the impregnating element, vanadium and graphite, e.g. V—Ta—C.

References pp. 377—380

V. TRACE METALS USED THERAPEUTICALLY

The administration of compounds of certain metals can be effective in the treatment of a variety of disorders, for example; gastric ulcers (Al, Bi); chronic renal disorders (Al); rheumatoid arthritis (Au); and certain types of cancer (Ga, Pt). The effective therapeutic doses result in elevated concentrations of the metals in body tissues and fluids that can be two to three orders of magnitude higher than untreated levels (Table 2). Such high concentrations can produce toxic reactions so that their frequent monitoring is essential. Measurements of the total concentrations in plasma/whole-blood have been used for this purpose, but recent studies suggest that measuring the concentrations of the various metal-containing species can provide the information for better therapeutic management and will lead to a greater understanding of the role of these metals.

A. Aluminium

Aluminium hydroxide gels (Aludrox) have been widely used for the past fifty years to neutralise gastric acidity in patients with peptic and duodenal ulcers, and more recently to inhibit the gastrointestinal absorption of phosphate in patients with chronic renal failure. The apparent lack of toxicity, particularly for the former group of patients was attributed to the very low solubility of $Al(OH)_3$ and its supposedly poor intestinal absorption. However, a mean positive balance of up to 11 μmol (300 mg) Al per day by patients receiving Al-antacids has been reported recently [87] indicating the need to consider the accumulation of Al in tissues of patients receiving chronic administration of these antacids. Severe neurological disturbances — Dialysis Dementia — have been observed in some patients with chronic renal failure on haemodialysis. These have been associated with high Al concentrations in water used to prepare the dialysis fluid and with increased Al concentrations in post mortem samples of brains of affected patients [88]. It is worth noting that in 1973 Krishnan and Crapper [89] produced 'presenile dementia' in cats associated with increased concentrations of Al in brain tissue by the administration of aluminium salts.

Although the mechanism for the neurotoxicity of Al is not well understood, recent work by Gardiner [90] has shown that 10—30% of the plasma Al is ultrafiltrable. These low-molecular weight species could be more easily transported across membranes and thus reach the brain more easily than could protein bound Al. The higher levels of Al in the cerebrospinal fluid of patients on dialysis [90] would support this view.

The sensitivity of ETA—AAS for Al is approximately 2.5×10^{-11} g for 1% absorption at 309.3 nm. This amount of Al is equivalent to 2.5 μl serum or urine at 0.4 μmol l^{-1} (10 μg l^{-1}). Langmyhr and Tsalev [91] used only 2 μl of serum of haemolysed whole-blood for analysis and found that a direct analysis using a two-stage ashing procedure first at 500°C and then 1200°C

TABLE 2

REFERENCE AND THERAPEUTIC LEVELS OF TRACE METALS IN BODY FLUIDS AND TISSUES

Metal	Matrix	Concentrations in tissues/fluids μmol l^{-1}		ETA—AAS[a] sensitivity (pg/0.0044 A)	Volume required for analysis[b] (μl/0.0044 A)
		Reference levels	Therapeutic levels		
Al	Serum	<0.4	1.8—22	44	1.1
	Urine	<2.0	>15		0.1
	CSF	0.15—0.19	0.22—1.26		7.3
Au	Serum	<0.01	5—40	10	0.01
Bi	Serum Urine	<0.05	0.2—14	37	0.9
Ga	Serum Urine	<1	up to 14	400	5.7
Pt	Whole-blood Serum	<0.05	0.1—50	300	15

[a] Data from Price [2].
[b] At lower end of therapeutic range except for Ga which is calculated for 1 μmol l^{-1}.

was as effective in eliminating molecular absorption interferences during atomisation, as was a pretreatment stage of oxidation with hot conc. HNO_3. Gardiner [90] also found background correction to be unnecessary when he used a double ashing procedure; initially 700°C then a slow ramp to 1400°C. Losses of Al occurred at temperatures greater than 1400°C. The analyses of urine and bone were slightly less straightforward because of the variable Ca^{2+} and PO_4^{3-} concentrations which respectively suppressed and enhanced the Al atomic signals. These condensed phase interferences were compensated for by the addition of 4.7 mg l^{-1} Ca^{2+} as $CaCl_2$ and 4.0 mg l^{-1} PO_4^{3-} as KH_2PO_4. The higher salt matrix concentration necessitated the use of continuum source background correction even after ashing at 1600°C for 60 s [92].

All of these procedures used either standard additions or matrix matching of the standards for calibration, with linear working ranges up to 3.7 μmol l^{-1} (100 μg l^{-1}). The within batch precision of 0.053 (RSD), day-to-day precision of 0.059 (RSD) and recoveries of added Al of 98.5 ± 6.5% reported by Fuchs et al. [93] are representative of the quality of analysis achieved with the other procedures that have been discussed.

Alderman and Gitleman [94] overcame matrix interferences in the determination of Al in serum by diluting the samples, 1 + 1 to 1 + 5 depending upon Al concentration with a diluent containing 0.02 M Na_2H_2EDTA, 1% Triton X 100 and $NH_4OH/(NH_4)_2SO_4$. The recovery of Al added to serum ranged from 99—104% and the day-to-day precision was 0.034 at 8—9 μmol l^{-1} (220—250 μg l^{-1}). These levels are much higher than those observed for control subjects who did not receive Al therapy, 0—28 nmol l^{-1} (0—7.6 μg l^{-1}).

B. Gold

The complexes sodium aurothiomalate and sodium aurothioglucose provide effective therapy for rheumatoid arthritis but can produce unpleasant and toxic reactions because of the relatively large doses used (up to 250 μmol (50 mg) per week). The resultant high concentrations of Au in plasma of patients receiving chrysotherapy may be measured easily by solvent extraction and FAAS. The main advantages of ETA—AAS are the direct analysis and the sensitivity to measure the various Au containing species in body fluids.

Simple procedures that require only a dilution of serum or urine have been reported by Schattenkirchner and Grobenski [95], and by Ward et al. [96]. The former diluted samples of sera 1 + 4 with 0.1% Triton X 100 and urine samples 1 + 9 with 0.01 M HCl and the latter used a 1 + 9 dilution of serum in water. In both cases calibration was by standard additions to compensate for the considerable matrix interferences. Ward et al. [96] demonstrated an excellent correlation ($r = 0.98$) between neutron activation analysis (NAA) and ETA—AAS analysis of the total plasma Au and albumin bound Au, which contains up to 90% of the total. There was however, a bias towards higher values (10%) by NAA. This difference did not appear to be pre-atomisation losses during ETA—AAS, the recoveries of added Au ranged from 90—105%, nor over-correction by continuum source background corrector.

The separation into protein fractions was achieved by gel filtration using Sephadex G200 and it was interesting that the poor correlation between NAA and ETA—AAS for the globulin fraction, approximately 8% of the total Au, improved considerably when the NaCl was removed by dialysis against 0.02 M CH_3CO_2H. It is probable that the NaCl suppression of the Au atomic signal is much more pronounced at low total protein concentrations. Kamel et al. [97] have also measured Au in protein fractions using a similar separation procedure and using cellulose acetate membrane electrophoresis. In the latter case, portions of the cellulose acetate containing the various protein bands were inserted directly into the graphite tube furnace for analysis. Calibration was achieved by adding standards to "blank" portions of the membrane which were ashed at 780°C without losses of Au.

Samples of urine with variable Na content, and of synovial fluid of limited volume may be analysed for Au using solvent extraction ETA—AAS in order to avoid calibration and matrix problems. Wawschinek and Rainer [98] used dimorpholine thiuramdisulphide/MIBK extraction of Au from serum diluted with 7 M $HClO_4$ and 0.5% Triton X 100 (0.1 + 0.5 + 1.0 ml) and from 1.0 ml urine diluted only with 0.5 ml 7 M $HClO_4$. No other pretreatment was necessary. The recovery of Au added to serum at 50 μmol l^{-1} (10 mg l^{-1}) ranged from 97—98% and when added to urine at 0.5 μmol (100 μg) per 24 h the recovery was 98—99%. The excellent precision at 25 μmol l^{-1} (5 mg l^{-1}) 0.026 (RSD) was achieved using automatic injection of 20 μl portions of the organic phase. The sample was dried and ashed at 600°C and atomised at 2700°C. Background correction was essential and the less sensitive 267.6 nm line was used. This procedure appears to be much simpler than the more traditional extraction of Au chlor-ion association complexes from 6 M HCl following oxidation with $KMnO_4$ or $HNO_3/HClO_4$ [99].

C. Bismuth

Bismuth compounds have been used in the treatment of chronic constipation, peptic ulcers and as an aid to the control of odour and consistency of stomal discharge from colostomy and ileostomy patients. Bismuth compounds in current use are the subnitrate, subgallate and tripotassium-dicitrato-bismuthate. Severe neurological disturbances, and some fatalities have been observed in patients who have received prolonged oral administration of bismuth compounds. The severity of the encephalopathy did not correlate with the dose taken or duration of therapy, but there was a spontaneous and complete reversal of the condition on cessation of bismuth therapy: furthermore, symptomatic patients had higher blood Bi concentrations, median 0.7 μmol l^{-1} (146 μg l^{-1}), than did a group of asymptomatic patients with a median of 0.1 μmol l^{-1} (30 μg l^{-1}), which in turn was higher than the median value of untreated controls of 0.04 μmol l^{-1} (8 μg l^{-1}) [100]. Even higher blood Bi concentrations have been reported for other symptomatic patients, median 3.3 μmol l^{-1} (690 μg l^{-1}), range 0.24—13.6 μmol l^{-1} (50—2850 μg l^{-1}) [100]. These data indicate that with oral therapy there is a significant, albeit small, intestinal absorption of Bi that with prolonged therapy can lead to the accumulation of toxic levels within the central nervous system. Frequent monitoring of blood Bi concentrations would appear to be an essential part of effective therapy.

The measurement of Bi in body fluids and tissues may be achieved using either ETA—AAS or hydride generation techniques. Rooney [101] compared these procedures and reported that the latter was the method of choice. The severe molecular absorption interferences at 213.2 nm necessitate some form of chemical pretreatment for ETA—AAS. Thomas et al. [100] used a direct extraction with APDC/MIBK from blood samples (5 ml) diluted 1 + 1 with water and Triton X 100, to determine concentrations down to 0.005 μmol l^{-1} (10 μg l^{-1}). Allain [102] also used APDC to extract Bi from blood or cerebrospinal fluid (0.5 ml volumes diluted 1 + 5 with water) and from urine buffered at pH 6. The working concentrations

($0.05-9.6\,\mu\mathrm{mol\,l^{-1}}$) were much higher than those used by Thomas et al. [100] and were linear only for the initial part of the calibration. A precision of 0.05 (RSD) was obtained at $3.7\,\mu\mathrm{mol\,l^{-1}}$. The ashing temperature in the ETA was surprisingly low, at only 230°C. Djudzman et al. [103] used an ashing temperature of 470°C with atomisation at 2100°C for 2% H_2SO_4 solutions of tissues that had been previously oxidised in a low temperature asher. Calibration was achieved by standard additions.

D. Gallium

The potential use of Ga compounds as anti-cancer drugs has stimulated the development of methods suitable for measuring the pharmaco-kinetic parameters of their administration. Newman [104] analysed serum, urine and tissue samples from humans and animals receiving Ga compounds for the treatment of cancers. Oxidation with HNO_3 was required for the tissues, but a simple 1 + 9 dilution with EDTA was used for serum or urine. The latter reagent was necessary to overcome the severe suppression of the Ga atomic signals by Ca. A concentration of only 5×10^{-3} M Ca^{2+} caused a 50% suppression of the signal from 5×10^{-9} g Ga. The calibration graph was linear over the range $2-30 \times 10^{-9}$ g Ga using the 417.2 nm line. Although this wavelength is less sensitive than the 287.4 nm line, the molecular absorption interferences are considerably reduced. Ashing and atomisation temperatures were 900°C and 2500°C respectively, and the recovery of Ga added to tissues was 96—100%. Rannisteano-Bourdon et al. [105] also used ETA—AAS to determine Ga in serum and tissues and found serum levels of up to $14\,\mu\mathrm{mol\,l^{-1}}$ ($1000\,\mu\mathrm{g\,l^{-1}}$) following the intramuscular injection of $4.3\,\mu\mathrm{mol}$ (0.3 mg) Ga $\mathrm{kg^{-1}}$ body weight.

E. Platinum

Platinum coordination complexes have been investigated as potential anti-cancer agents since 1972. The most successful of these is the *cis* dichlordiammine platinum(II) complex which is particularly effective for the treatment of ovarian and testicular carcinomas. However, effective therapy requires high doses, typically $4.0\,\mu\mathrm{mol}$ ($825\,\mu\mathrm{g}$) $\mathrm{kg^{-1}}$ given intravenously, which produce unpleasant and toxic reactions. Pharmaco-kinetic studies of the various Pt species in plasma indicate that the active species is of low molecular weight and the protein bound Pt species is apparently inactive [106]. Further work is needed to identify the active species and to develop therapeutic procedures that produce maximum anti-tumour activity with minimum toxicity.

The total Pt concentrations of whole-blood, liver and kidney have been measured using ETA—AAS by Pera and Harder [107]. The considerable molecular absorption required the samples (0.2—0.5 ml blood, 1—2 g tissue) to be oxidised prior to analysis, which was accomplished by freeze-drying and digestion with conc. HNO_3. The sensitivity and detection limit, 1.8×10^{-9} g and 7.5×10^{-10} g respectively, were more than adequate for the observed concentrations of $1.5-50\,\mu mol\,kg^{-1}$ ($0.3-10\,\mu g\,g^{-1}$). The RSD obtained at the lower part of this concentration range was 0.05—0.10, and an excellent correlation ($r = 0.989$) was obtained between ETA—AAS and γ-scintillation counting of samples spiked with labelled Pt. Lower concentrations, down to $0.15\,\mu mol\,kg^{-1}$ ($0.03\,\mu g\,g^{-1}$) in blood and $0.015\,\mu mol\,kg^{-1}$ ($0.003\,\mu g\,g^{-1}$) in urine, have been measured following $HNO_3/HClO_4$ digestion and dissolution in HCl [108]. Calibration by standard additions or matrix matching was essential because of the suppression effect, which was greater with ashed urine than with ashed blood. The precision was good with RSD's of 0.03—0.05 at $0.5-4.0\,\mu mol\,kg^{-1}$ ($0.01-1.0\,\mu g\,g^{-1}$). Unfiltrable Pt species in serum have been measured down to $0.18\,\mu mol\,l^{-1}$ ($30\,\mu g\,l^{-1}$) by ETA—AAS after cation exchange collection [109].

VI. NON-ESSENTIAL, TOXIC TRACE ELEMENTS

The concentrations in body fluids of trace elements with no known biological functions for maintaining optimal human health, and which are toxic at relatively low concentrations are listed in Table 3. Measurements of these concentrations can be important in monitoring the health of populations occupationally exposed to these elements and in surveys to determine the extent of possible environmental contamination.

A. Lead

The measurement of lead in biological samples requires elimination, or more usually strict control, of the chemical effects of the matrix upon analyte sensitivity and of the considerable molecular absorption interferences during atomisation. Zeeman-effect background systems can compensate for very high absorbances, up to 1.5 A, and a direct analysis of Pb in blood was reported in which the pre-atomisation drying and ashing stages were omitted [12]. The accuracy of this method was demonstrated using NBS standard reference bovine liver, but it was necessary to modify the graphite tube to obtain a constant temperature zone at its centre in order to maintain analytical accuracy and precision. Others have used the Zeeman system with more traditional thermal pretreatments prior to atomisation [13, 53], but Hadeishi and McLaughlin [110] used a unique double-chamber furnace to improve the spatial resolution between the analyte and matrix vapours.

Molecular absorption interferences can be reduced considerably by ashing within the atomiser itself, although care must be taken to avoid pre-atomisation losses of lead which occur at 350—400°C with untreated blood samples. These losses may be avoided at higher temperatures by the addition of a solution of $(NH_4)_2HPO_4$ [111]. The net effect of the sample matrix on

TABLE 3

CONCENTRATIONS OF NON-ESSENTIAL, TOXIC TRACE ELEMENTS IN BODY TISSUES AND FLUIDS

Element (units)		Physiological concentrations		ETA—AAS sensitivity[a] (pg/0.0044 A)	Volume required for analysis[b] (μl/0.0044 A)
		μ/nmol l^{-1}	μg l^{-1}		
Pb (μmol l^{-1})	B	0.2—1.2	40—250	10	0.04
	U	0.05—0.4	10—80		0.13
Cd (nmol l^{-1})	B	0.9—27	0.1—3.0	1	0.33
	U				
Be (nmol l^{-1})	U	44—100	0.4—0.9	2	2.2
Ni (nmol l^{-1})	S	19±14	1.1±0.8	84	42
	U	7—68	0.4—4.0		21
As (nmol l^{-1})[c]	B	7—125	0.5—9.4	38	4.0
Hg (nmol l^{-1})[c]	B	5—50	1—10	200	20
	U				
Sb (nmol l^{-1})[c]	B	8	1.0	28	28
Te (nmol l^{-1})[c]	B	2	0.25	100	400

[a] Data from Price [2].

[b] Calculated for upper concentration limit where data available.

[c] Elements measured more easily by alternative techniques i.e. cold vapour (Hg) and hydride generation (As, Sb, Te).

analyte sensitivity will depend upon the chemical and thermal treatment of the sample prior to atomisation. The organic constituents enhance analyte sensitivity, whereas there is a suppression by the inorganic constituents, which results from interaction of Pb atoms with alkali, alkaline earth- and, for some samples, iron-chlorides [112]. It is therefore apparent that strict control of chemical and instrumental conditions is necessary for reproducible analysis, and it is not surprising that matrix interferences differ with different designs of atomisers and with only slight variations in chemical pretreatment of samples.

The most useful index of excessive exposure to lead is a blood-lead analysis, and the majority of methods discussed here are concerned with this measurement. Many of the earlier ETA—AAS methods used little or no sample pretreatment and in almost all cases the variety of matrix interferences encountered necessitated strict control of ETA conditions, and made calibration by standard additions mandatory. The problem of diffusion of liquid blood samples into the graphite atomisers was overcome as an additional advantage of the punched-disc method of Cernik [6]. Although automated injection of liquid samples has made this method somewhat less attractive, it has proved to be one of the most useful direct ETA—AAS methods of blood-lead analysis for monitoring industrially exposed personnel. Discs of 4 mm diameter are punched from the centre of a dried blood-spot on filter paper and placed directly into a carbon-cup atomiser. Calibration is achieved by preparing discs from blood samples to which known amounts of lead have been added. The results correlate well with other analytical methods, and the overall RSD of 0.056 at the concentrations important in industrial hygiene control ($2-5\,\mu\text{mol}\,l^{-1}$ ($400-1000\,\mu\text{g}\,l^{-1}$)) is sufficiently good for routine monitoring. Carter [113] has shown that there is an increased diffusion throughout the filter paper by blood samples with lowered haemoglobin values, leading to erroneously low blood-lead results. Therefore, if the method is to be applied to samples with abnormal haemoglobin values, a correction must be made for the variation in area of the dried blood spot.

Alternative procedures for minimising the errors from sample diffusion into graphite atomisers include using relatively large dilutions (1 + 9 to 1 + 19), or pretreatment of the graphite with organic solvents. Rosen and Trinidad [4] pretreated a carbon rod atomiser with xylene to minimise the diffusion of the sample into the rod. They obtained an excellent correlation of results using ETA—AAS (0.5 μl samples of heparinised blood) with results obtained by solvent extraction FAAS ($r = 0.990$), but found that samples preserved with EDTA gave low and variable recoveries of added lead unless they were diluted with Triton X 100. Evenson and Prendergast [9] also noted varying suppressive matrix effects from blood preserved with heparin or oxalate anticoagulants. Low and variable recoveries of added lead were found even when the samples were diluted 1 + 1 with 5% Triton X 100 or 1 + 9 with 0.01 M HNO_3. To overcome these effects they analysed diluted solutions of erythrocytes and obtained results that were in excellent agreement ($r = 0.998$) with those determined using Hessel's solvent extraction FAAS method [114]. The problems of the excessive accumulation of ashed residues of blood samples diluted 1 + 1 with Triton X 100 was reported by Kubasik and Volosin [115] who found that 0.5 μl was the maximum volume of blood that could be analysed without excessive accumulation of these residues. Del Castilho and Herber [116] reported that a cleaning atomisation was required after every fifth determination if 5 μl of a 1 + 5 dilution of blood in water was analysed. The accumulated residue would otherwise have caused an increasing suppression of the Pb atomic signal. Higher dilutions (1 + 19) eliminated the accumulation of ash residues. The accurate results that these and the previously mentioned workers obtained using direct ETA—AAS of untreated or simply diluted blood samples is evidence of the care and skill of the analysts rather than the exactitude of the methodology, since the problems discussed so far are additional to those of pre-atomisation losses and molecular absorption interferences.

References pp. 377—380

More recently Fernandez [5] has reported that aqueous lead standard solutions in 0.1% HNO_3 may be used for calibration in the direct analysis of whole-blood diluted 1 + 4 with 0.1% Triton X 100. The ETA conditions for 20 μl of the diluted samples were: ramp drying over 20 s to 100°C and maintain temperature for 15 s; ramp ash over 25 s to 525°C and maintain temperature for a further 25 s; atomise at 2300°C for 8 s. The calibration was linear to 5 μmol l^{-1} (1000 μg l^{-1}) and the RSD's were 0.039 at 1.5 μmol l^{-1} (300 μg l^{-1}) and 0.034 at 3.0 μmol l^{-1} (600 μg l^{-1}). The correlation of results with the group mean values in two external quality control schemes was excellent ($r = 0.968, 0.989$). In my opinion, it is unlikely that this direct method could be used for routine analyses with different types of ETA in different laboratories, since accurate analyses invariably involve some matching of the standard matrix with that of the blood sample.

One of the simplest and yet effective oxidative pretreatments of blood samples is the addition of HNO_3. This reduces molecular absorption interferences and affords some degree of matrix modification. Brodie and Stevens [117] heated equal volumes (usually 2 ml) of whole blood with conc. HNO_3 at 85—95°C in screw-capped glass tubes for about 30 min. Some lipid material remained undissolved so the solution was centrifuged before analysis of Pb (and Cd) using a carbon rod atomiser. This procedure reduced the molecular absorption signals with 2 μl of solution ($\equiv$ 1 μl blood) to < 0.10 A which was well within the capability of the background corrector and could even be resolved, in time, from the Pb atomic signal on a chart recording without background correction. The ETA conditions used were: dry at 110°C for 35 s; ash at 450°C for 20 s; and ramp atomise at 400°C s^{-1} to 2400°C. Using automated sample introduction the RSD was 0.032 at 1.0 μmol l^{-1} (200 μg l^{-1}) and ranged from 0.02—0.05 in routine analysis. Stoeppler et al. [15] deproteinised whole-blood by diluting 100 μl volumes with 150 μl conc. HNO_3 and 150 μl H_2O. After vigorous mixing the solutions were centrifuged to yield a supernate that was 0.75 M in nitrate and which contained all of the lead. Automated sample injection of 10 μl volumes ($\equiv$ 2.5 μl blood) into a graphite tube atomiser gave a day-to-day RSD of 0.035 at 2.3 μmol l^{-1} (480 μg l^{-1}) and of 0.084 at 0.5 μmol l^{-1} (100 μg l^{-1}) for routine analysis. Calibration was achieved either by means of standard addition or using matrix matched standards prepared from control-blood samples with low endogenous lead concentrations. These were linear up to 10 μmol l^{-1} (2000 μg l^{-1}). The ETA conditions were: dry at 100°C for 30 s; ash at 600°C for 30 s; atomise at 2200°C for 5 s; and clean at 2600°C for 2 s. This high ashing temperature did not result in any pre-atomisation losses of Pb — presumably because of the modification of the sample matrix. This procedure has been recommended as the basis of a reference procedure by the NBS, Washington.

Another attractive method using HNO_3 pretreatment is that of Nise and Vesterberg [118], who simply diluted blood 1 + 4 with 1% Triton X 100. After 30 min equilibration 10 μl of 0.01 M HNO_3 was added to 500 μl portions of the diluted blood solution, and $2\,\mu l$ ($\equiv 0.4\,\mu l$ blood) volumes of this solution were injected into a carbon rod atomiser. Calibration was by standard additions to the blood samples diluted in Triton X 100. The ETA conditions were: dry at 100°C for 30 s; ramp ash to 600°C over 35 s; and atomise at 1500°C for 3.5 s. The final ashing temperature is somewhat higher than would be expected for blood-lead analysis, but the authors stated that pre-atomisation losses of Pb were significant only at temperatures $> 700°C$ and that a minimum of 550°C was needed to reduce the background interferences during atomisation. The recovery of Pb added ranged from 95—102% over the concentration range 0.5—$2.9\,\mu mol\,l^{-1}$ (10—$600\,\mu g\,l^{-1}$) with most results with $100 \pm 2\%$. There was excellent agreement between this method and FAAS; $r = 0.998$ over the blood-lead concentration range 0.4—$4.0\,\mu mol\,l^{-1}$ (8—$800\,\mu g\,l^{-1}$). The within run precision ranged from 0.015—0.047 and on a day-to-day basis from 0.034—0.055 over a concentration range of 0.42—$2.99\,\mu mol\,l^{-1}$ (8—$600\,\mu g\,l^{-1}$). A maximum of 40 samples could be analysed in duplicate in an 8 h day.

Lagesson and Andrasko [8] oxidised $2\,\mu l$ volumes of blood at 500°C on graphite microboats which were later inserted into a rectangular cross-sectioned graphite tube for ETA—AAS. This technique does not have any advantages over the simple partial oxidation procedures discussed. For example, the RSD's ranged from 0.03—0.07, but it might prove useful for analysing solid tissue samples. A more precise method was described by Allain and Mauras [119] who extracted Pb and Cd directly from whole-blood (0.1 ml) diluted $1 + 3 + 2$ with H_2O and 2% APDC into 0.5 ml MIBK. Urine samples were adjusted to pH 5 prior to extraction. Calibration was achieved by standard additions to control samples prior to extraction. The ETA conditions for $40\,\mu l$ of extract were: dry by heating to 150°C over 10 s and maintaining this temperature for a further 10 s; ash by heating to 450°C over 10 s and holding at 450°C for a further 30 s; and atomise at 2400°C for 5 s. The RSD ranged from 0.015 at $2.3\,\mu mol\,l^{-1}$ ($478\,\mu g\,l^{-1}$) to 0.043 at $0.73\,\mu mol\,l^{-1}$ ($152\,\mu g\,l^{-1}$). There was no interference in the extraction efficiency from Na_2H_2EDTA at $2\,g\,l^{-1}$ so that the method may be used for patients on chelation therapy and for blood samples preserved with this anticoagulant.

Although there are procedures described for the analysis of Pb in tissues by solid-sampling techniques, such materials are best analysed following complete ashing or by dissolution and solvent extraction pretreatments. Legotte et al. [120] used wet or dry ashing of diets, faeces and urine samples, and dissolved the residues in 5% v/v HNO_3. There were considerable matrix interferences that could not be completely overcome. Even with the substantial dilutions, equivalent to $1 + 24$ of original sample, the method of additions was necessary to overcome the variable suppressions with different urine samples. It is possible that matrix modification with $(NH_4)_2HPO_4$ or the use of solvent extraction would have been advantageous.

Ward et al. [52] found a very good agreement between ETA—AAS and

References pp. 377—380

FAAS using a Zeeman correction system for the analysis of eight metals including Pb in whole-blood and tissue samples. Both procedures gave accurate analyses of NBS standard reference bovine liver samples. An interlaboratory comparison of blood-lead analysis by isotope dilution mass spectroscopy as a reference method and by 113 participating laboratories using six basic groups of methods showed that ETA—AAS procedures were low in an overall method ranking based on average accuracy and precision rankings for each method [121]. The ranking of methods was: anodic stripping voltammetry > Delves cup > extraction > tantalum strip > graphite furnace > carbon rod. In my experience of European inter-comparison programmes, the overall performance is independent of the analytical techniques that are used and that good analytical performances ensue from strict internal quality-control procedures regardless of the analytical method. The positive bias of AAS methods at low concentrations $<2\,\mu\text{mol}\,l^{-1}$ ($400\,\mu\text{g}\,l^{-1}$) and negative bias at higher concentrations observed by Boone et al. [121] is not an uncommon observation and is possibly the result of calibration graphs curving towards the concentration axis at regions erroneously assumed to be within the linear range. There is clearly a need for standard reference whole-blood samples for use in interlaboratory studies of blood-lead analysis.

B. Cadmium

The volatilisation of $1-2\,\mu l$ of untreated whole-blood or urine in a graphite tube atomiser produces molecular absorption signals in excess of 1.5 A. Direct analysis with minimal sample pretreatment is therefore only possible using Zeeman-effect background correction. Pleban et al. [53] diluted whole blood (1 + 9) with 1% Triton X 100 for direct ETA—AAS using this type of correction system and obtained a detection limit of $0.25\,\mu\text{g}\,l^{-1}$. Continuum source background correction, however, has a typical accuracy of 1% or 0.01 A at a level of 1.2 A [14] so that resultant atomic signals of 0.05 A would have errors of ±20%. It is necessary therefore to reduce the total molecular signal to 0.2—0.3 A in order to limit the correction errors to less than ±5% of the residual atomic signals. Wright and Riner [122] ashed $50\,\mu l$ volumes of blood or urine solutions diluted 1 + 49 with water at 400°C for 60 s in the ETA to reduce the background signals to levels within the capability of continuum-source correction. They did, however, encounter problems with the accumulation of carbonaceous residues even after atomisation at 2100°C. Cleaning the tube at 2600°C overcame these difficulties but considerably shortened the working-life of the tube.

An attractive instrumental procedure for reducing background absorption without losing Cd is to volatilise Cd selectively, from NaCl and the bulk matrix at relatively low temperatures (600—900°C). Cernik and Sayers [123] found pre-atomisation losses of Cd from blood spotted onto filter paper at temperatures >340°C, but were able to separate Cd signals from the resultant ashed matrix by atomisation at only 600°C. Higher atomisation temperatures of 850—900°C were used by others employing this principle [124, 125]. The latter workers conditioned the graphite tube by soaking in 8% m/v ammonium molybdate solution, heating it slowly to 1000°C and then rapidly to 2000°C. This procedure gave a reduced background absorption during subsequent analyses of urine samples which were ashed at 370°C and atomised at 900°C. The detection limit was 0.05 μg l^{-1} and the RSD's were 0.06 at 1 μg l^{-1}. Gardiner [90] analysed Cd directly in 5 μl volumes of urine after ashing at 400°C and selective volatilisation of Cd at 900°C. It was necessary to calibrate using standard additions because Cd was volatilised from aqueous solutions at lower temperatures than from diluted urine solutions. The mean recovery of Cd added to urine at only 0.5 μg l^{-1} was 107% and the comparison with results obtained using atomic fluorescence spectroscopy over the concentration range 0.2—11.0 μg l^{-1} was excellent (r = 0.994).

Langmyhr et al. [50] determined Cd in protein fractions separated from human parotid saliva by gel filtration. It was necessary to add either 0.1 M H_2SO_4 or 0.42 M HNO_3 to sample and remove excess NaCl by evaporation prior to Cd analysis. The RSD in the concentration range 10—20 μg l^{-1} was 0.034—0.041.

Prior ashing of the sample matrix to produce a substantial reduction in background interferences was accomplished by heating equal volumes (2 ml) of blood and conc. HNO_3 at 85—95°C [117]. Using automated sample injection an RSD of 0.027 at 13 μg l^{-1} was obtained. Legotte et al. [120] used wet or dry oxidation of diets, faeces and urine prior to Cd and Pb analyses. The sample matrices were dissolved in 5% v/v HNO_3 solution, and standard additions were used for calibration. An average RSD of 0.01 was obtained for Cd in urine over the range 0.16—1.65 μg l^{-1}.

Solvent extraction procedures to separate Cd from a whole-blood or urine matrix may be done directly, without wet or dry ashing treatments. Allain and Mauras [119] diluted 0.1 ml of whole blood with 0.3 ml H_2O and 0.2 ml 1% APDC solution and extracted Cd into 0.5 ml MIBK. Urine samples were buffered to pH 5 prior to extraction. The organic extract (40 μl) was dried by ramp-heating to 150°C over 10 s, ashed by heating to 350°C over a 10 s period and maintaining this temperature for a further 20 s, and atomised at 1800°C. The within batch RSD was 0.07 at 1.5 μg l^{-1} and 0.018 at 8 μg l^{-1}; with a day-to-day precision in the range 0.041—0.28. Direct extraction of Cd from blood with APDC has also been used by Ward et al. [126] and by many others. Ulluci and Hwang [127] used this reagent to extract Cd from ashed kidney samples. Alternative chelating agents such as NaDDC [115], dithizone [128] and dipivaloylmethane [129] have been used for biological samples after oxidative pretreatment.

C. Nickel

Direct measurements of the low concentrations of nickel found in serum or urine from non-occupationally exposed persons, 0.4—4.0 μg l^{-1} are difficult because of the high molecular absorption signals produced during atomisation. The inorganic constituents of the sample matrix do not begin to volatilise significantly at temperatures below 900°C and there is a danger of losing nickel if the temperature exceeds 1000°C. Gardiner [90] found that ashing for 60 s at 1000°C gave an efficient and reproducible oxidation of 10 μl volumes of plasma or urine with only a 4% loss of nickel. The samples were simply diluted 1 + 1 with water prior to injection into the furnace. The atomisation temperature was 2400°C and the sensitivity of 0.55 μg l^{-1} for 1% absorption at 232 nm obtained using a pyrolytically coated tube was 2—3 times better than that obtained with a conventional graphite tube. The detection limit of 0.3 μg l^{-1} was sufficient to allow the measurement of Ni in sera from healthy controls at 1.1 ± 0.8 μg l^{-1}. Patients on renal dialysis had higher levels of nickel in serum, 7.5 ± 3.1 μg l^{-1}, and it is interesting that at these levels some 20—80% of the nickel was shown to be bound to low molecular weight species.

Solvent extraction and separation procedures are frequently used to overcome the problems of molecular absorption interferences. Ader and Stoeppler [130] investigated the extraction efficiencies of APDC, dimethylglyoxime (DMG) and furildioxime (FD) using MIBK as the solvent in a comprehensive study of the recovery of ^{63}Ni added to urine. They also studied the relative merits of wet and dry ashing. Dry oxidation in quartz beakers resulted in 4—5% losses of Ni as insoluble silicon compounds whereas $HNO_3/HClO_4/H_2SO_4$ oxidation gave quantitative (> 98%) recovery. Extraction at pH 9 with either DMG or FD gave incomplete and variable recovery of Ni ranging from 80—90% but quantitative extraction (99.1—99.7%) was obtained using APDC at pH 2.5. The organic extracts (20 μl volumes) were dried at 120°C ashed at 1100°C and atomised at 2600°C. No pre-atomisation losses of Ni were observed. Mikac-Devic et al. [131] obtained quantitative recovery of Ni from ashed serum or urine samples using FD/MIBK extraction at pH 9 but with larger aqueous : organic phase volume ratios (5 : 1) to avoid the formation of precipitates which could occlude Ni. The recoveries ranged from 94—107% (mean 101%) at the 10—20 μg l^{-1} level and a good correlation ($r = 0.989$) was obtained with the mean values of seven other laboratories for the analysis of urine samples containing 4.5—81 μg l^{-1} Ni. The precision was reasonably good, with an RSD of 0.068 for 11 μg l^{-1} Ni in serum and 0.053 for 25 μg l^{-1} Ni in urine. These higher concentrations are representative of the levels which could be encountered with occupationally exposed persons. The main disadvantage of the FD/MIBK or APDC/MIBK extraction system is the interference by iron. This may be removed easily by a prior extraction with cupferron/MIBK from 2 M HCl [35]. An inter-laboratory comparison of the analysis of Ni in urine by 7—10 laboratories using ETA—AAS methods showed that procedures involving preliminary oxidation and extraction stages were generally superior to direct ETA methods in terms of inter-laboratory precision and recovery of added Ni [132].

D. Beryllium

In order to measure accurately the concentrations of Be in body tissues and fluids down to the low levels of $<0.5\,\mu g\,l^{-1}$ for unexposed persons, it is essential to overcome the severe matrix interferences that are the dominant features of this analysis. Cations in general (Ca^{2+} and Mg^{2+} in particular) enhance the Be absorption signals, whereas anions such as Cl^- and ClO_4^- cause severe suppression. The method of standard additions to diluted urine samples containing 0.5% HNO_3 and 4% H_2SO_4 has been used successfully to measure Be in the range 0.5—50 $\mu g\,l^{-1}$ [133]. However it is this author's experience that H_2SO_4 solutions severely degrade the graphite tube and limit its working-life.

An alternative and more attractive method of masking the matrix interferences is to add La^{3+} at 100 $\mu g\,ml^{-1}$ to sample solutions in 5% HNO_3 [134]. Urine samples were diluted directly for measurements down to 0.5 $\mu g\,l^{-1}$ but lower concentrations down to 0.05 $\mu g\,l^{-1}$ required preconcentration by coprecipitation with Ca using NH_4OH, dissolution in 5% HNO_3 before the addition of La^{3+}. Hair, faeces and fingernails were digested either with $HNO_3/HClO_4$ or by HNO_3/H_2O_2 and dry-ashing at 700°C and the residues dissolved in HNO_3. It was essential that all traces of Cl^- and ClO_4^- were removed from these samples before analysis. The ETA conditions were dry at 100°C for 50 s, ash at 1000°C for 60 s and atomise at 2600°C for 10 s. The calibration graph was linear over the range 1—25 $\mu g\,l^{-1}$ using 10 μl samples. The recoveries of Be added to all tissues were 90—110% with an average RSD of 0.15. The endogenous Be concentrations were 0.5 $\mu g\,kg^{-1}$ in all tissues except for faeces which contained 1—3 $\mu g\,kg^{-1}$. Similar concentrations, 0.8—4.6 $\mu g\,kg^{-1}$ were found in tissue samples analysed by ETA—AAS after ashing with HNO_3 in a PTFE pressure vessel and solvent extraction with acetylacetone in benzene [135]. Pretreatment of the graphite tube with $ZrOCl_2$ enhanced the Be signal by a factor of 3 for organic solvent phases and 9 for aqueous solutions. The detection limit for urine analysis of 0.6 $\mu g\,l^{-1}$ is near the upper concentration limits for persons not exposed to Be and will require improvement for routine screening.

E. Arsenic

Careful sample pretreatment is essential to prevent losses of volatile organo-As compounds from body tissues and fluids. Conventional wet-ashing techniques are satisfactory, but dry ashing requires the presence of an ashing aid i.e. $Mg(NO_3)_2$. Pre-atomisation losses of As from the atomiser may be eliminated by the addition of Ni salts that allow ashing at temperatures up to 630°C with atomisation at 2200°C [2]. Ishizaki [136] determined As down to 0.5 $\mu g\,l^{-1}$ in 2 ml samples of blood, after ashing at 550°C in the presence of $Mg(NO_3)_2$ and solvent extraction from 10 M HCl/KI into $CHCl_3$. The As was back-extracted into $Mg(NO_3)_2$ solution prior to

References pp. 377—380

ETA—AAS. A concentration range of 0.5—9.4 $\mu g\,l^{-1}$ (mean 3.8) was observed for an unexposed population ($N = 38$).

Lo and Coleman [137] used MgO/cellulose/Mg(NO$_3$)$_2$ as the ashing aid for the analysis of As in tissue samples. The ashing was continued at 550°C and the residues dissolved in 6 M HCl and diluted to 3 M. The solutions were further diluted 1 + 4 with H$_2$O and 10 μl volumes taken for analysis. The ETA conditions were; dry at 100°C for 20 s, ash at 250°C for 10 s and atomise at 2000°C for 10 s. Calibration was by standard additions.

F. Mercury, antimony, tellurium

At the present time there are no ETA—AAS methods that can compete with the cold vapour technique for Hg or with hydride generation methods for Sb and Te. Another attractive method for Sb and Te is low pressure microwave induced plasma (MIP) emission spectroscopy [138]. Using low-temperature ashing and solvent extraction as preparation, physiological concentrations of both elements ($<1\,\mu g\,l^{-1}$) were determined by this method.

VII. CONCLUSION

The development of ETA—AAS methods have enabled most clinically important elemental analyses to be done rapidly, with care and accurately, using sample volumes that cause the patient the minimum of discomfort. There are however, two important areas of work that require improvements in current techniques.

Firstly, there is an urgent need to develop methods for measuring abnormally low concentrations of certain essential trace metals, e.g. Co, Cr, Mn, Mo and V. Deficiencies of these elements would almost certainly accompany the deficiencies in Cu, Zn, and Se observed in patients receiving synthetic oral diets, or receiving intravenous nutrition with inadequate trace-element supplementation. It is also possible that inherited metabolic disorders involving deficiencies of these elements will be discovered, as has been the case for Cu (Menkes syndrome) and Zn (Acrodermatitis enteropathica). One of the major difficulties in measuring the "normal" concentrations of elements such as Co, Cr, Mn, Mo and V is the inadvertent contamination of the sample during collection. This problem has been discussed by Versieck and Cornelis [139] in their excellent appraisal of the variable literature values for 18 elements in human serum or plasma. Cornelis et al. [140] have developed a blood-collection technique that enables extremely low concentrations of V (17—66 ng l^{-1}) and of Mo (0.3—1.2 $\mu g\,l^{-1}$) to be measured. They collected whole-blood using a plastic cannula trocar and flushed this through with 40 ml of blood before the actual sampling. Clearly such enormous volumes are completely unsuitable for any clinical situation.

Alternative, contamination-"free" sampling procedures must be developed. It is possible that such procedures may be obtained using PTFE coated syringes and needles, or needles made from Ti.

Secondly there is a need to exploit fully the high sensitivity of ETA—AAS to measure the distribution of elements among proteins and other species separated from sera and from cellular material; and to measure the subcellular distribution of elements. These techniques will undoubtedly provide a most useful means of studying the role of trace elements in human disorders.

REFERENCES

1. K. Schwartz, in S. S. Brown (Ed.), Essentiality vs. Toxicity of Metals in Clinical Chemistry and Toxicology of Metals, Elsevier, North Holland, 1977, p. 3.
2. W. J. Price, in Spectrochemical Analysis by Atomic Absorption, Heyden, London, 1979.
3. F. J. Langmyhr, Analyst, 104 (1979) 993.
4. J. F. Rosen and E. E. Trinidad, J. Lab. Clin. Med., 80 (1972) 567.
5. F. J. Fernandez, At. Absorpt. Newsl., 17 (1978) 115.
6. A. A. Cernik, Br. J. Ind. Med., 31 (1974) 239.
7. H. T. Delves, Clin. Chim. Acta, 71 (1976) 495.
8. V. Lagesson and L. Andrasko, Clin. Chem. (NY), 25 (1979) 1948.
9. M. A. Evenson and D. D. Prendergast, Clin. Chem. (NY), 20 (1974) 163.
10. M. J. Adams, G. F. Kirkbright and P. Rienvatana, At. Absorpt. Newsl., 14 (1975) 105.
11. J. Issaq and L. Z. Zielinski, Anal. Chem., 47 (1975) 2281.
12. H. Koizumi and K. Yasuda, Anal. Chem., 48 (1976) 1178.
13. J. B. Dawson, E. Grassam, D. J. Ellis and M. J. Keir, Analyst, 101 (1976) 315.
14. R. A. Newstead, W. J. Price and D. J. Whiteside, in C. L. Chakrabarti (Ed.), Background Correction in Atomic Absorption Analysis in Prog. Analyt. Atom. Spectrosc., Pergamon Press, Oxford, Vol. I, 1979, pp. 267—298.
15. M. Stoeppler, K. Brandt and T. C. Rains, Analyst, 103 (1978) 714.
16. M. A. Evenson and B. L. Warren, Clin. Chem. (NY), 21 (1975) 619.
17. G. Hauck, Fresenius Z. Anal. Chem., 267 (1973) 337.
18. E. Schumacher and F. Umland, Fresenius Z. Anal. Chem., 270 (1974) 285.
19. R. D. Ediger, At. Absorpt. Newsl., 14 (1975) 127.
20. O. Wawschinek and H. Hofler, At. Absorpt. Newsl., 18 (1979) 97.
21. H. Freeman, J. F. Uthe and B. Flemming, At. Absorpt. Newsl., 15 (1976) 49.
22. B. Welz, Application of Graphite Furnace Atomic Absorption to the Analysis of Heavy Metals in Biological Materials, a Paper presented to CEC Course on Analytical Technology for Heavy Metals in Biological Fluids, Ispra, Italy, December 1978.
23. T. R. C. Boyde and S. W. N. Wu, Clin. Chem. (NY), 88 (1978) 49.
24. J. R. Kelson and R. J. Shamberger, Clin. Chem. (NY), 24 (1978) 240.
25. K. Ladefoged, Clin. Chim. Acta, 100 (1980) 149.
26. D. I. Paynter, Anal. Chem., 51 (1979) 2086.
27. T. A. Hinners, Fresenius Z. Anal. Chem., 277 (1975) 377.
28. T. C. Rains, personal communication.
29. A. J. Jackson, L. M. Michael and H. J. Schumacher, Anal. Chem., 44 (1972) 1064.
30. S. Lutyen, J. Smeyers-Verbeke and D. L. Massart, At. Absorpt. Newsl., 12 (1973) 131.

31 P. J. Barlow and A. K. Khera, At. Absorpt. Newsl., 14 (1975) 149.
32 S. B. Gross and E. S. Parkinson, At. Absorpt. Newsl., 13 (1974) 107.
33 T. T. Gorsuch, Analyst, 84 (1959) 135.
34 M. Zief and J. W. Mitchell, Contamination Control in Trace Element Analysis, ACS Monograph No. 47, John Wiley, New York, 1970, pp. 164—173.
35 H. T. Delves, G. Shepherd and P. Vinter, Analyst, 96 (1971) 260.
36 F. W. Alexander, B. E. Clayton and H. T. Delves, Q. J. Med., 43 (1974) 89.
37 G. V. Iyengar, K. Kasparek and L. E. Feinendegen, Analyst, 105 (1980) 794.
38 J. T. H. Roos, A Simple Vapour-Phase Dissolution Technique for Biological Materials, a paper presented at the Symposium on the Analysis of Biological Material, Spectroscopy Society of South Africa, October 1977, Pretoria, South Africa.
39 B. Bernas, Anal. Chem., 40 (1968) 1682.
40 W. Holak, B. Krimnitz and J. C. Williams, J. Assoc. Off. Anal. Chem., 55 (1972) 741.
41 P. E. Paus, At. Absorpt. Newsl., 11 (1972) 129.
42 M. Stoeppler, K. P. Muller and F. Backhaus, Fresenius Z. Anal. Chem., 297 (1979) 107.
43 C. E. Gleit and W. D. Holland, Anal. Chem., 34 (1962) 1455.
44 K. B. Cross, 1980, personal communication.
45 L. Lopez-Escobar and D. N. Hume, Anal. Lett., 6 (1973) 343.
46 G. F. Carter and W. B. Yeoman, Analyst, 105 (1980) 295.
47 H. T. Delves, (1980) unpublished.
48 J. P. Matousek and B. J. Stevens, Clin. Chem. (NY), 17 (1971) 363.
49 A. Mazzucotelli, M. Galli, E. Benassi, C. Loeb, G. A. Ottonello and P. Tangenelli, Analyst, 103 (1978) 863.
50 F. J. Langmyhr, B. Eyde and J. Jansen, Anal. Chim. Acta, 107 (1979) 211.
51 W. B. Robbins, B. M. DeKoven and J. A. Caruso, Biochem. Med., 14 (1975) 184.
52 N. I. Ward, R. Stephens and D. E. Ryan, Anal. Chim. Acta, 110 (1979) 9.
53 P. A. Pleban, K. H. Pearson and R. J. Shamberger, a paper presented at 17th ACS Meeting, Florida, 1978.
54 J. Teape, H. Kamel, D. H. Brown, J. M. Ottaway and W. E. Smith, Clin. Chim. Acta, 94 (1979) 1.
55 H. Kamel, J. Teape, D. H. Brown, J. M. Ottaway and W. E. Smith, Analyst, 103 (1978) 921.
56 J. Smeyers-Verbeke, P. Bell, A. Lowenthall and D. L. Massart, Clin. Chim. Acta, 68 (1976) 343.
57 J. Smeyers-Verbeke, G. Segebarth and D. L. Massart, At. Absorpt. Newsl., 14 (1975) 153.
58 D. M. Danks, B. J. Stevens and R. R. W. Townley, Science, 197 (1973) 1140.
59 M. T. Glenn, S. A. Savory, R. D. Fine, C. J. Reeves and J. D. Winefordner, Anal. Chem., 45 (1973) 203.
60 D. E. Olsen, L. L. Jarlow, F. J. Fernandez and H. L. Kahn, Clin. Chem. (NY), 19 (1973) 326.
61 E. Berman, Appl. Spectrosc., 29 (1975) 1.
62 M. E. Tatro, W. L. Raynolds and F. M. Costa, At. Absorpt. Newsl., 16 (1977) 143.
63 E. D. Seifert, Fresenius Z. Anal. Chem., 287 (1977) 317.
64 J. Smeyers-Verbeke, Y. Michotte, P. van den Winkel and D. L. Massart, Anal. Chem., 48 (1976) 125.
65 P. Del Castilho and R. F. M. Herber, Some Problems in the Determination of Mn in Blood by ETA—AAS, a paper presented at Euroanalysis III, Dublin, 1978.
66 D. J. D'Amico and H. L. Klawans, Anal. Chem., 48 (1976) 1469.
67 J. P. Buchet, S. R. Lauwerys and H. Roels, Clin. Chim. Acta, 73 (1976) 481.
68 D. I. Paynter, Anal. Chem., 51 (1979) 2086.
69 E. Bonilla, Clin. Chem. (NY), 24 (1978) 471.
70 K. Saeed, Y. Thomassen and F. J. Langmyhr, Anal. Chim. Acta, 110 (1979) 285.

71 M. Ishizaki, Talanta, 25 (1978) 167.
72 D. Behne, P. Bratter, H. Gessner, G. Hube, W. Mertz and U. Rosizk, Fresenius Z. Anal. Chem., 278 (1976) 269.
73 I. W. F. Davidson and W. L. Sacrest, Anal. Chem., 44 (1972) 1808.
74 M. W. Routh, Anal. Chem., 1 (1980) 182.
75 R. E. Vanderlinde, F. J. Kayne, M. J. Simmons, J. Y. Tson and R. L. Lavine, Clin. Chem. (NY), 24 (1978) 2151.
76 J. W. Rosson, K. J. Foster, R. J. Walton, P. P. Monro, T. G. Taylor and K. G. M. M. Alberti, Clin. Chim. Acta, 93 (1979) 299.
77 S. S. Chao, E. L. Kanabrocki, C. E. Moore, Y. T. Oester, J. Greco and A. Von Smolinski, Appl. Spectrosc., 30 (1976) 155.
78 S. S. Chao and E. Pickett, Anal. Chem., 52 (1980) 335.
79 G. E. Bentley, L. Markowitz and R. Meglen, in T. H. Risby (Ed.), Analysis of Molybdenum in Biological Materials, in Ultratrace Metal Analysis in Science and Environment, Am. Chem. Soc., Washington, DC, 1979, pp. 33—39.
80 J. Versieck, J. Hoste, F. Berbeir, L. van Ballenberghe, J. de Rudder and R. Cornelis, Clin. Chim. Acta, 87 (1978) 135.
81 V. V. Lidums, At. Absorpt. Newsl., 18 (1979) 71.
82 N. Oyamada, M. Ishizaki, S. Ueno, F. Kataoka, R. Murakami, K. Kubota and K. Katsamura, Ibaraki-Ken Eisei Kenkyusho Nempo, 15 (1977) 65.
83 R. A. Barfoot and J. G. Pritchard, Analyst, 105 (1980) 551.
84 R. A. A. Muzzarelli and R. Rochetti, Talanta, 22 (1975) 683.
85 M. Ishizaki and S. Ueno, Talanta, 26 (1979) 523.
86 A. Hulanicki, R. Karwowska and J. Stanczak, Talanta, 27 (1980) 214.
87 J. E. Gorsky, A. A. Dietz, H. Spencer and D. Osis, Clin. Chem. (NY), 24 (1978) 1485.
88 A. C. Alfrey, G. R. LeGendre and W. Kachny, N. Engl. J. Med., 294 (1976) 184.
89 S. S. Krishnan and D. R. Crapper, Al and Senility in Man, a paper presented to the 4th International Conference on Atomic Spectroscopy, Toronto, 1973.
90 P. H. E. Gardiner, Ph.D. Thesis, University of Strathclyde, 1979.
91 F. J. Langmyhr and D. L. Tsalev, Anal. Chim. Acta, 92 (1972) 79.
92 K. Garmestani, A. J. Blotcky and E. P. Rack, Anal. Chem., 50 (1978) 144.
93 C. Fuchs, M. Brashe, K. Paschen, A. Nordbeck, E. Quellnorst and U. Peek, Clin. Chim. Acta, 52 (1974) 71.
94 F. R. Alderman and H. J. Gitleman, Clin. Chem. (NY), 26 (1980) 258.
95 M. Schattenkirchner and Z. Grobenski, At. Absorpt. Newsl., 16 (1977) 84
96 R. W. Ward, C. J. Danpure and D. A. Fyfe, Clin. Chim. Acta, 81 (1977) 87.
97 H. Kamel, D. H. Brown, J. M. Ottaway and W. E. Smith, Analyst, 102 (1977) 645.
98 O. Wawschinek and F. Rainer, At. Absorpt. Newsl., 18 (1979) 50.
99 S. M. Pedersen and P. M. Graaback, Scand. J. Clin. Lab. Invest., 37 (1977) 91.
100 D. W. Thomas, T. F. Hartley, P. Coyle and S. Sobecki, in S. S. Brown (Ed.), Clinical Chemistry and Chemical Toxicology of Metals, Elsevier/North Holland, Amsterdam, 1977, p. 292.
101 R. C. Rooney, Analyst, 101 (1976) 749.
102 P. Allain, Clin. Chim. Acta, 64 (1975) 281.
103 F. Djudzman, E. van der Eeckhaut and P. de Moerloose, Analyst, 102 (1977) 688.
104 R. A. Newman, Clin. Chim. Acta, 86 (1978) 195.
105 S. Rannisteano-Bourdon, F. Prouillet and R. Bourdon, Ann. Biol. Clin. (Paris), 36 (1978) 39.
106 T. F. Patton, K. J. Himmelstein, R. Belt, S. J. Bannister, L. A. Sternson and A. J. Repta, Cancer Treat. Rep., 62 (1978) 1359
107 Pera and Harder, Clin. Chem. (NY), 23 (1977) 1245.
108 A. H. Jones, Anal. Chem., 48 (1976) 1427.
109 S. J. Bannister, Y. Chang, L. A. Sternson and A. J. Repta, Clin. Chem. (NY), 24 (1978) 877.

110 T. Hadeishi and R. D. McLaughlin, Anal. Chem., 48 (1976) 1009.
111 D. J. Hodges, Analyst, 102 (1977) 66.
112 H. Heinrichs, Fresenius Z. Anal. Chem., 295 (1979) 355.
113 G. C. Carter, Br. J. Ind. Med., 35 (1978) 235.
114 D. W. Hessel, At. Absorpt. Newsl., 7 (1968) 35.
115 N. P. Kubasik and M. T. Volosin, Clin. Chem. (NY), 19 (1973) 554.
116 P. Del Castilho and R. F. M. Herber, in S. S. Brown (Ed.), The Rapid Determination of Lead and Cadmium in Blood by Flameless Atomic Absorption, in Clinical Chemistry and Toxicology of Metals, Elsevier/North Holland, Amsterdam, 1977, p. 361.
117 K. G. Brodie and B. J. Stevens, J. Anal. Toxicol., 1 (1977) 282.
118 G. Nise and O. Vesterberg, Clin. Chim. Acta, 84 (1978) 129.
119 P. Allain and Y. Mauras, Clin. Chim. Acta, 91 (1979) 41.
120 P. A. Legotte, W. C. Rosa and D. C. Sutton, Talanta, 27 (1980) 39.
121 J. Boone, T. Hearn and S. Lewis, Clin. Chem. (NY), 25 (1979) 389.
122 F. C. Wright and J. C. Riner, At. Absorpt. Newsl., 14 (1975) 103.
123 A. A. Cernik and M. H. P. Sayers, Br. J. Ind. Med., 32 (1975) 155.
124 G. Lungren, Talanta, 23 (1976) 309.
125 G. D. Carmack and M. A. Evenson, Anal. Chem., 51 (1979) 907.
126 R. J. Ward, M. Fisher and M. Tellez-Yudilevich, Ann. Clin. Biochem., 15 (1978) 197.
127 P. A. Ullucci and J. Y. Hwang, a paper presented at the 17 th Colloquium Spectroscopium Internationale, Florence, Italy, 1973.
128 N. Lekehal, M. Hanocq and M. J. Helson-Cambier, Pharm. Tijdschr. Belg., 32 (1977) 76.
129 W. Oelschlager and W. Lautenschlager, Fresenius Z. Anal. Chem., 287 (1977) 28.
130 D. Ader and M. J. Stoeppler, J. Anal. Toxicol., 1 (1977) 252.
131 D. Mikac-Devic, F. W. Sunderman, Jr. and S. Nomoto, Clin. Chem. (NY), 23 (1977) 948.
132 D. B. Adams, S. S. Brown, F. W. Sunderman, Jr. and H. Zachariasen, Clin. Chem. (NY), 24 (1978) 862.
133 D. S. Grewal and F. Y. Kearns, At. Absorpt. Newsl., 16 (1976) 131.
134 J. A. Hurlburt, At. Absorpt. Newsl., 17 (1978) 121.
135 T. Stiefel, K. Schulze, G. Tolg and H. Zorn, Anal. Chim. Acta, 87 (1976) 67.
136 M. Ishizaki, Bunseki Kagaku, 26 (1977) 729.
137 Lo and Coleman, At. Absorpt. Newsl., 18 (1979) 10.
138 P. F. E. van Montfort, J. Agterdenbos and B. A. H. G. Jutte, Anal. Chem., 51 (1979) 1553.
139 J. Versieck and R. Cornelis, Anal. Chim. Acta, 116 (1980) 217.
140 R. Cornelis, L. Mees, J. Hoste, J. Ryckebusch, J. Versieck and F. Barbier, Neutron Activation Analysis of Vanadium in Human Liver and Serum, in Nuclear Activation Techniques in the Life Sciences 1979, International Atomic Energy Agency, Vienna, AEA-SM-227/25, p. 165.

Chapter 41

Forensic science

I. M. DALE

West of Scotland Health Boards, Department of Clinical Physics and Bio-engineering, Glasgow G4 9LF (Gt. Britain)

I. INTRODUCTION

In 1752 Mary Blandy was tried for the murder of her father by poisoning him with arsenic. During the course of the trial one of the chief medical witnesses for the prosecution, Dr. Anthony Addington, gave evidence regarding his examination of a white powder found in the possession of the accused. He stated that in his opinion the poison administered was white arsenic for the following reasons: (1) This powder has a milky whiteness, so has white arsenic. (2) This is gritty and almost insipid, so is white arsenic. (3) Part of it swims on the surface of cold water like a pale, sulphurous film, but the greatest part sinks to the bottom and remains there undissolved, the same is true of white arsenic, etc. The organs of the deceased were not analysed for the presence of arsenic and in fact appropriate chemical tests would not be available for some one hundred years when the Marsh (1836) and Reinsch (1841) tests were developed. In spite of the less than perfect scientific evidence, Mary Blandy was found guilty and executed [1].

It is fortunate, both for the accused and for the integrity of the legal system, that modern forensic science is more exact. However, this account does demonstrate one aspect of forensic science, that of comparing certain, hopefully characteristic, attributes of evidence found at or near the scene of the crime with similar items associated with the accused. The two most obvious examples of this type of investigation are in the use of fingerprints (physical characteristics of the fingers, first used as evidence in Britain in 1902) and blood grouping (biochemical characteristics). In recent years chemical techniques, particularly involving characteristic trace element analysis, have been successfully applied in the elucidation of forensic problems.

The forensic scientist employed in the analysis of specimens for metal concentrations is involved generally in two main areas of investigation. The first is in the determination of toxic metals in biological tissue in order to ascertain the cause of death or injury (homicidal or suicidal) in suspected poisoning cases. The second is to compare certain characteristic trace element concentrations in materials found at the scene of the crime with the same type of material found in the possession of the accused. A special case of this second approach is in the analysis of the elements barium, antimony and lead, deposited on people's hands after they have discharged a firearm. With the introduction in recent years of comprehensive legislation relating to health and safety in industry, the forensic scientist can be called upon to examine possibly hazardous concentrations of dust and fumes in the factory environment and to give evidence on the results.

II. BIOLOGICAL MATERIAL AND POISONING

A. Sampling

It is not normal for the analyst to be involved in the collection of specimens during the post mortem dissection of the victim. Accordingly it is imperative that the pathologist carrying out the examination is given clear instructions of the type of specimens required. A supply of clean plastic or glass jars is essential for the transport of portions of the organs to the laboratory. Most toxic heavy metals, such as mercury, arsenic, antimony and thallium, are known to accumulate in the kidney, liver and, to a lesser extent, the brain. If the poison was administered shortly before death substantial concentrations will also be found in the stomach. Specimens of kidney, liver, brain and stomach should certainly be obtained in any case of suspected exposure to toxic metals, together with samples of blood and urine if these can be collected.

If it is considered that the deceased had been poisoned over an extended period prior to death, it is important to collect hair (plucked from the roots with the ends tied with thread at the scalp end) and nail specimens. The growing hair and nail can absorb many trace elements from the blood stream and these elements are permanently bound to the keratin. Knowing that the average rate of growth of human hair is approximately 1 cm per month, sectional analysis of the hair from the root to the tip will produce a 'calendar' of exposure to the element.

B. Pretreatment

As in all fields of trace element analysis it is imperative that no contamination of the specimen is allowed to take place prior to examination. There is always the danger that during the dissection of the organs some contamination of the tissue could occur from the metal knives and scalpels; added to this are problems of dust and flaking paint and plaster from the mortuary walls and ceilings. Once the material has arrived in the laboratory it is advisable to remove the cut and exposed portions of the organs with a non-metallic knife. A suitable cutting edge can be prepared from broken pieces of clean laboratory glassware and this glass knife will cut most types of biological material with minimal risk of contamination.

It has to be admitted that decaying human organs have a rather unpleasant odour and for this reason, if for no other, some form of preservation is required. Placing the tissue in preserving liquids such as formalin is not recommended due to the contamination of the sample from trace elements found in commercial preserving fluids and to the possible leaching of elements from the sample into the liquid. An equivalent problem arises in the examination of bodies which have been embalmed with chemical solutions.

Dry ashing of the sample is not suitable because elements of interest such as mercury, arsenic, lead and antimony can be lost from the sample at the temperatures employed in the furnace, however the addition of magnesium and nickel nitrates has been shown to prevent the loss of arsenic and antimony by the formation of less volatile arsenides, arsenates and antimonides [2]. The two most suitable procedures for the pretreatment of biological material are freeze drying (lyophilisation) and low temperature ashing. The added advantage of such treatment is the concentration of the toxic metals in the final sample as shown in Table 1. A further advantage is in the ease of comparing results with previously reported data. Low-temperature ashing of biological material gives a white or grey powder soluble in hydrochloric acid, while the freeze-dried sample needs to be dissolved in concentrated nitric and sulphuric acids.

C. Freeze drying

De Goeij et al. [3] reported on an extensive investigation into the possible loss of trace elements from biological tissue during freeze drying at 10^{-5} atmospheres for 48 h. They found no significant loss of antimony, arsenic, bromine, cadmium, chromium, cobalt, copper, iron, mercury, molybdenum, selenium or zinc. Thus freeze drying of human tissue is a suitable treatment prior to analysis, together with the fact that most laboratories have access to appropriate facilities.

Thin slices of tissue are placed in clean plastic petri dishes, the loose fitting lids replaced and the dishes inserted into the chamber of the freeze drier. Blood specimens can be poured directly into the dish. It is important that the tissue be thoroughly frozen prior to evacuating the chamber. The time required for complete drying of the sample depends on the nature and weight of the material and the type of equipment employed. Drying is usually completed in 12—48 h. The dried tissue is not susceptible to decay and can be stored at room temperature in sealed plastic bags.

D. Low-temperature ashing

A radio-frequency coil is used to dissociate oxygen molecules in a suitable chamber for the ashing of the sample. A number of systems are commercially available. The sample is placed in a pure-quartz boat and introduced into the oxidation chamber. The chamber is kept under vacuum during the ashing, and oxidation products are vented through a suitable

References p. 394

TABLE 1

CONCENTRATION FACTORS FROM TRACE ELEMENTS IN DRIED AND ASHED ORGANS AND BODY FLUIDS (I.C.R.P.-23 REPORT, [32])

Sample	Concentration factor	
	Ashed	Dried
Blood, whole	100	5.0
Brain	67	4.7
Stomach	125	3.7
Hair	200	1.1
Heart	92	3.7
Kidney	91	4.4
Liver	78	3.6
Lung	91	4.5
Spleen	72	4.5
Urine	93	14.6

system to the outside atmosphere. The treatment of the sample takes 24—48 h, depending on the weight of material and the power and design of the low-temperature asher. A clean, dry powder easily dissolved in dilute nitric or hydrochloric acid is obtained and is entirely suitable for subsequent atomic absorption analysis by either flame or non-flame systems.

Locke et al. [4] have investigated the possible loss of trace elements during the low temperature ashing of human liver samples. No losses of magnesium, calcium, manganese, iron, copper, zinc, rubidium or cadmium were noted. Greater than 90% recovery of 50 $\mu g\,g^{-1}$ added elements were observed for antimony, barium, thallium, bismuth and lead, however, the recovery of arsenic was 64%, selenium 18% and tellurium 15%. Gleit and Holland [5] reported good recovery of selenium in alfalfa grass in low-temperature ashing.

E. Sample digestion

The acids and oxidising chemicals required for the dissolution of biological material will be discussed for each element separately.

III. TOXIC ELEMENTS

A. Arsenic

1. Introduction

This element can be considered to have the longest history of pharmaceutical and homicidal use of any substance; in fact, for many people the word poison is synonymous with arsenic. Its properties were described in detail by Greek physicians, the Roman physician Paulus Aegineta and in Chinese, Indian and Tibetan manuscripts. Arsenious oxide was employed on a grand scale during the middle ages as a murder weapon, and since the symptoms of acute poisoning (violent stomach pains and vomiting) are similar to the common disease of cholera, the homicide often went undetected [6]. A further advantage was that in amounts necessary for death, arsenic is practically tasteless, particularly when administered in food or drink. In the nineteenth and early twentieth century, arsenic was readily available to the poisoner either as a rat poison or as a constituent of fly paper, in which case the arsenic was extracted into hot water and then added to the victim's food.

Famous British trials of people accused of using arsenic include those of Mary Bateman (1807, home produced 'medicines'), Madeleine Smith (1857, rat poison), Elizabeth Frances Maybrick (1889, fly paper), Frederick Harry Seddon (1911, fly paper) and Herbert Rowse Armstrong (1921, weed killer). In recent years legislation restricting the use of arsenic in household products has reduced the incidence of arsenic poisoning. However the forensic scientist is often asked to examine specimens to eliminate the possibility of poisoning, and arsenic is one such element of interest to the pathologist. Various compounds of arsenic are used in industry, particularly electronics, wood processing and glass making. Arsenious oxide has been employed in taxidermy for the prevention of fungal attack of bird feathers and museum workers can be exposed to high concentrations of arsenic in dust [7]. Occasionally, chronic arsenic poisoning has been associated with the excessive ingestion of arsenic containing tonics [8]. Fatalities have occurred when workers have been exposed to arsine gas liberated during metal processing [9].

2. Symptoms

The symptoms of acute and chronic poisoning by arsenic have been described by Rentoul and Smith [10], Davidson and Henry [11] and Fowler [12]. Acute symptoms include gastrointestinal damage, convulsions and haemorrhage; at autopsy, fatty degeneration of the liver and kidneys is frequently noted. Acute inhalation of arsine is followed by extensive haemolysis, haemoglobinuria and death from renal failure.

In chronic arsenic poisoning anorexia, anaemia, disturbances in renal and hepatic function, dermatitis, skin hardening (hyperkeratosis) and pigmentation can occur. Arsine gas is a powerful haemolytic agent and symptoms of poisoning are due to its action on the blood.

3. Fatal dose

The smallest recorded fatal dose is 125 mg [10]. The action of vomiting has resulted in the recovery of a woman following the deliberate swallowing

References p. 394

of 14 g of arsenious acid. However in another case 13 g of arsenious acid caused death after seven days. Arsine gas is far more toxic and exposure to a concentration of over 30 ppm for an hour or more is dangerous.

4. Sampling

Arsenic is concentrated in the liver, spleen and kidney and is bound preferentially to sulphydryl groups in skin, hair and nails. Urine and gastric contents should also be obtained. The sequential analysis of hair sections has been shown to be of value in the examination of prolonged exposure to arsenic [8].

5. Analytical methods

Direct flame atomic absorption of the dissolved biological material has not been found suitable for the estimation of arsenic. Instead, arsenic has been determined following the generation of arsine gas by the action of sodium borohydride on the solution, and subsequent analysis of the arsenic either in an argon—hydrogen or nitrogen—hydrogen entrained air flame or in a heated quartz cell. Electrothermal atomic absorption determination of arsenic in biological samples requires the addition of nickel nitrate as a volatilisation suppressant.

6. Arsine-entrained air flame

Fiorino et al. [13] described the use of a shielded hydrogen (nitrogen diluted) entrained air flame for the analysis of food and animal tissue for arsenic.

Dry tissue (1—3 g) is digested with 30 ml of nitric acid, sulphuric and perchloric acids (4 : 4 : 1). After heating for approximately 1.5 h, the cooled solution is transferred to a 100 ml flask, 30 ml hydrochloric acid added and diluted to 100 ml with water. Suitable aliquots, up to 20 ml, of this solution are used for the subsequent arsenic analysis.

0.5 ml sodium iodide (10% w/v) is added as a pre-reductant, and the arsine is generated by the addition of 6—8 ml alkaline sodium borohydride solution (4 g in 100 ml sodium hydroxide, 10% w/v). The arsine gas is led to the burner and the transient absorption signal measured at 197.2 nm.

For a 5 g sample the detection limit was 7 ppb.

7. Arsine-heated quartz cell

Peats [14] described a method for the analysis of arsenic in dried algae and similar biological material. 100 mg of dried sample was digested with 1 ml of nitric, perchloric and sulphuric acids (23 : 23 : 1) by heating first to 150°C for one hour then to 500°C to near dryness and the material allowed to cool. A pre-reductant solution (0.5 ml) of 10% (w/v) potassium iodide in 5% hydrochloric acid, stabilised with 1% ascorbic acid was then added and the solution made up to 5 ml with 5% hydrochloric acid. The solution was left for at least four hours to assure quantitative reduction of As(V) to As(III). 200 µl of this solution was added to 10 ml of 5% hydrochloric acid in the reaction vessel and 3% sodium borohydride in 1% sodium hydroxide was used as the reducing solution. The generated arsine was carried to the quartz cell by high-purity nitrogen. The quartz cell was heated by the air/acetylene flame, an arsenic electrodeless discharge lamp used at 8 watts and the 193.7 nm line used for analysis (peak height). The absolute lower limit of detection was quoted as 0.45 ng arsenic, with the relative lower limit as 0.11 µg arsenic g^{-1}.

Peter et al. [15] used an electrically heated quartz cell for the determination of arsenic in urine. Urine, 2 ml, was digested with 2 ml of nitric and perchloric acids (1 : 1). Aliquots of this solution were used for the subsequent arsine generation by sodium borohydride. The normal level of arsenic in urine was found to be less than 10 ppb.

8. Arsenic-heated graphite atomiser

Lo and Coleman [16] analysed animal tissue for arsenic using a Perkin-Elmer HGA—2100 furnace. 5 g (wet weight) of tissue were mixed with 1.5 g magnesium oxide and 10 ml cellulose powder in a 100 ml beaker and charred carefully. The sample was cooled and 1.5 g magnesium nitrate added, heated to 550°C and ashed for 2 h. 5 ml water was added and the sample dissolved in 45 ml 6 N hydrochloric acid. This solution was diluted 1 : 5 for the atomic absorption analysis to eliminate the interference signal. 10 µl was used for analysis and the parameters were as follows:

Lamp:	electrodeless discharge arsenic lamp
Lamp power:	8 watts
Spectral slit-width:	0.7 nm
Wavelength:	193.7 nm
Background corrector:	on
Gas flow:	nitrogen
Drying:	20 s at 100°C
Ashing:	10 s at 250°C
Atomisation:	10 s at 2000°C

The sensitivity was reported to be 87 ng As g^{-1} of wet tissue.

Haynes [2] described a technique involving the dry ashing of 2 g sample with magnesium nitrate and nickel nitrate followed by heating with nitric (20—25 ml) and hydrofluoric acid (1—3 ml), taken to dryness and further heating with nitric acid (10 ml) and 30% hydrogen peroxide (10 ml). The resulting solution was filtered and made up to 100 ml with 0.5% (v/v) nitric acid. 20 µl was used for the arsenic analysis, employing the following instrument parameters:

References p. 394

TABLE 2

ARSENIC CONCENTRATIONS IN UNEXPOSED SUBJECTS PPM, DRY WEIGHT

Sample	n	Range	Arithmetic		Geometric	
			Mean	Standard deviation	Mean	Standard deviation
Brain	19	0.001–0.036	0.016	0.010	0.012	2.30
Blood, whole	12	0.001–0.920	0.147	0.270	0.036	7.04
Hair	52	0.01 –0.40	0.125	0.102	0.085	2.66
Heart	23	0.002–0.078	0.027	0.023	0.021	2.69
Kidney	25	0.002–0.363	0.050	0.075	0.026	3.45
Liver	27	0.005–0.246	0.057	0.059	0.034	2.84
Nail	124	0.02 –2.90	0.362	0.313	0.283	2.04
Spleen	23	0.001–0.132	0.032	0.035	0.017	3.62
Stomach	21	0.003–0.104	0.037	0.034	0.022	3.28
Urine ($\mu g\, ml^{-1}$)	25	0.006–0.77	0.114	0.164	0.053	3.53

Spectral slit-width:	0.2 nm
Wavelength:	193.7 nm
Background corrector:	on
Drying:	30 s at 110°C
Ashing:	30 s at 1200°C
Atomisation:	8 s at 2700°C

The sensitivity was given as 100 ng arsenic g^{-1} sample.

Modifications of the above techniques could be applied to the analysis of freeze dried and low temperature ashed tissue samples.

9. *Arsenic in human tissue — normal levels*

Cross et al. [7] reported on the arsenic concentrations in tissue samples from people who died as a result of violence and who had no known exposure to arsenic other than that in the general environment. Their results are given in Table 2 and are in general agreement with values reported by other workers (e.g., McKenzie and Neallie [17]; Larsen et al. [18]; and Gordus [19]).

10. *Arsenic in human tissue — exposed subjects*

The most popular era for homicidal arsenic poisoning occurred prior to the development of modern, accurate chemical analysis techniques. However, up-to-date information on the relative concentrations of arsenic in tissue samples is available in a number of cases of industrial, suicidal and homicidal exposure to arsenic.

Cross et al. [7] reported on cases of arsenic poisoning due to exposure to

TABLE 3

ARSENIC LEVELS IN EXPOSED SUBJECTS, PPM, DRY WEIGHT

Source	Subject	Hair		Nail		Urine (μg ml^{-1})
		Head	Pubic	Finger	Toe	
Steel bronze factory	A	71.5	31.2	15	51.9	—
	B	76.1	63.1	188	60.9	—
	C	10.8	8.6	18.6	10.8	—
	D	64.8	27.1	123	37.1	—
Zinc dross factory	E	2.1	6.3	2.2	—	—
	F	0.5	0.2	3.5	—	—
	G	14.0	4.4	3.2	—	—
Wood preservers	H	12.1	10.6	44.7	26.7	0.88
	I	37.4	16.7	—	32.2	0.39
	J	4.0	—	—	1.5	0.42
	K	25.2	17.9	22.5	0.8	—
	L	21.6	38.8	13.7	24.2	0.35

TABLE 4

ARSENIC LEVELS FOLLOWING INGESTION OF WOOD PRESERVATIVE SOLUTION PPM, DRY WEIGHT

Tissue	Concentration
Blood, whole	2.0
Brain	2.7
Heart	7.7
Kidney	17.5
Liver	13.9
Spleen	7.4
Stomach	21.1

wood preservative containing arsenic, to arsine gas liberated during metal working and to arsine gas leaking from gas cylinders.

Representative values are given in Table 3. Subject G died and the following arsenic concentrations found: brachial plexus 0.99 ppm (dry weight), heart 2.2 ppm, kidney 6.6 ppm and popliteal nerve 0.61 ppm.

A case of suicide due to the ingestion of a wood preservative solution was investigated by Cross et al. [20]. The subject swallowed an unknown quantity of a corrosive solution containing copper sulphate, sodium dichromate and an arsenic compound. Death occurred 36 h after ingestion and the cause of death was reported as poisoning by a corrosive substance. The results of the arsenic analysis are given in Table 4.

A case of murder by the administration of arsenic was reported by Barrowcliff [21]. On 8th September 1969, Mrs. Waite died of acute

References p. 394

gastro-enteritis with allergic polyneuropathy. Previous symptoms included diarrhoea, vomiting, peripheral neuritis, scaly skin, loss of hair and appetite, abdominal pain and shingles. Substantial concentrations of arsenic were found in the liver (40 ppm), duodenum (15 ppm) and small intestine (332 ppm). Sectional analysis of the hair showed variations in arsenic level corresponding with the history of her illness. Values ranged from 2 ppm (during stay in hospital) to 209 ppm (hair roots obtained at post mortem). Her husband was arrested, charged with the murder and found guilty.

IV. GUNSHOT RESIDUE ANALYSIS

A. Introduction

This subject is one of the more controversial topics in forensic science, with considerable debate as to the applicability and interpretation of the results. No matter how perfectly a firearm is constructed, there will always be some release of the products of the explosion onto the hands or the face of the person firing the weapon. The diphenylamine-sulphuric acid dermal nitrate test, first introduced in the 1930's, was supposed to detect the presence of nitrates and nitrites from gunpowder discharge residues [22]. In this test, successive layers of molten paraffin wax were applied to the hands of the suspect, the wax removed and the inner surface of the cast treated with an aqueous solution of diphenylamine and sulphuric acid. This reagent forms a blue colour with nitrites and nitrates, however, the test lacks specificity (it forms blue compounds with other nitrogen containing substances) and sensitivity. It was therefore found to give both false negative and false positive results in test firings [23]. Accordingly, in 1935 and 1940, the Federal Bureau of Investigation advised against the use of this test, although for want of any better technique it remained in use for a further twenty to thirty years. A possible alternative to the measurement of residues from the propellant charge of the cartridge was to determine residues from the primer of the ammunition. The primer charge is a shock-sensitive mixture which is detonated by the action of the hammer and subsequently fires the main propellant charge. Table 5 lists typical compositions of hand gun primer charges [24]. Harrison and Gilroy [25] developed colormetric spot tests for the detection of traces of barium, antimony and lead removed from the hand. However these tests also suffered from a lack of sensitivity and selectivity and were difficult to apply in the investigation of actual crimes as opposed to test firings.

Ruch et al. [26, 27] applied neutron activation analysis to the determination of barium and antimony deposited on the hands following the discharge of a gun. Table 6 lists the typical amounts of these elements removed from the back of the hand after a single firing of the weapon [23].

Electrothermal atomisation in atomic absorption spectroscopy has

TABLE 5

CHEMICAL COMPOSITIONS OF HANDGUN PRIMING MIXTURES

Component	Composition (%)				
	1	2	3	4	5
Lead styphnate	36	41	39	43	37
Barium nitrate	29	39	40	36	38
Antimony sulphide	9	9	11	—	11
Calcium silicide	—	8	—	12	—
Lead dioxide	9	—	—	—	—
Tetrazene	3	3	4	3	3
Zirconium	9	—	—	—	—
Pentaerythritol tetranitrate	5	—	—	—	5
Nitrocellulose	—	—	6	—	—
Lead peroxide	—	—	—	6	6

TABLE 6

TYPICAL AMOUNTS OF BARIUM AND ANTIMONY DEPOSITED ON THE HAND AFTER A SINGLE FIRING OF VARIOUS FIREARMS

Weapon	Barium (ng)	Antimony (ng)
0.22 calibre revolver	390	80
0.22 calibre automatic pistol	700	140
0.38 calibre revolver	1300	420
0.44 calibre revolver	1400	420
0.45 calibre automatic pistol	3600	600
0.25 calibre automatic pistol	4700	630
9 mm automatic pistol	7500	730

sufficient sensitivity for these elements, and techniques were developed for the determination of gunshot residues by Sherfinski [28], Goleb and Midkiff [29], Kinard and Lundy [22] and Instrumentation Laboratory [30], amongst others. For all of these methods, the gunshot residue is removed from the hand by rubbing the skin with cotton swabs moistened with 1 M nitric acid. These swabs are then leached with 2 ml of 1 M nitric acid and aliquots of this solution, with or without neutralisation by ammonium hydroxide, are used for the analysis.

The main problem associated with the application of gun-shot residue analysis is in the interpretation of the results. Lead is a ubiquitous contaminant of the environment, and little reliance can be placed on the presence of lead removed from the hand of a suspect. Barium and antimony are less commonly found in normal control samples; however these elements can be present in hand swabs due to metal working, painting and other normal occupations. Also, the suspect must be apprehended soon after the incident and the swabs taken before the hands are washed. Nesbitt et al. [24]

References p. 394

successfully applied scanning electron microscopy with an energy dispersive X-ray fluorescence detector to provide highly reliable identification of gunshot residue particles. This technique employed an examination of the characteristic morphology of the residue particles with direct simultaneous analysis of the distribution and quantity of lead, barium and antimony in these particles. This method would appear to be the most suitable for the unequivocal determination of the presence of gunshot residues, although the high cost of instrumentation has limited its routine application in forensic laboratories.

B. Cautionary note

On the 22nd November, 1963 the most infamous assassination in recent times took place when President J. F. Kennedy was shot in Dallas. One would expect that the most sophisticated techniques would have been employed in the examination of the evidence from the suspect, Lee Harvey Oswald. This was not the case. In spite of the F.B.I. reports on the unsatisfactory nature of the diphenylamine test, this procedure was used on a paraffin wax cast made from the hands and right cheek of Oswald. The test was positive for both hands and negative for the right cheek. The Warren Commission Report [31] stated "the test is completely unreliable in determining either whether a person has recently fired a weapon or whether he has not ... (the reagent) will react positively with nitrates from other sources and most oxidising agents, including dichromates, permanganates, hypochlorates, periodates and some oxides. Thus, contact with tobacco, Clorox, urine, cosmetics, kitchen matches, pharmaceuticals, fertilisers or soils may result in a positive reaction". However worse was to follow. The paraffin casts were examined by neutron activation analysis for the presence of barium and antimony. These elements were found on both surfaces of all the casts and in fact more barium was found on the outside surface of the cheek cast than on the surface next to the skin. It was clear that contamination of the paraffin wax had occurred at some stage prior to the analysis by neutron activation. That this could have happened during the most intensive scientific examination of evidence from a crime is a warning to all involved in forensic science.

C. Sampling

A representative sample of the gunshot residue is obtained by swabbing the thumb, web and forefinger with a cotton swab moistened with 1 M nitric acid. The barium and antimony is leached from the cotton tip by soaking it in 1 or 2 ml of 5% (v/v) nitric acid. Instrumentation Laboratory [30] recommend the neutralisation of the solution by the addition of 0.2 ml concentrated ammonium hydroxide solution.

TABLE 7

INSTRUMENT PARAMETERS FOR BARIUM ANALYSIS OF GUNSHOT RESIDUES (HEATED GRAPHITE FURNACE)

	Sherfinski [28]	Instrumentation Laboratory [30]
Wavelength	553.5 nm	553.5 nm
Band pass	—	1 nm
Sample size	50 μl	25 μl
Drying	25 s at 125°C	20 s at 60°C / 25 s at 120°C
Ashing	35 s at 600°C	5 s at 250°C / 15 s at 750°C
Atomisation	10 s at 2500°C	10 s at 2800°C

TABLE 8

INSTRUMENTAL PARAMETERS FOR BARIUM ANALYSIS OF GUNSHOT RESIDUES (TANTALUM STRIP FURNACE, GOLEB AND MIDKIFF [29])

Wavelength	553.6 nm
Spectrographic slit	50 μl
Sample size	10 μl
Drying	45 s at 100°C
Ashing	30 s at 450°C
Atomisation	2 s at 2500°C
Background correction	Use of neon line at 540.0 nm

TABLE 9

INSTRUMENTAL PARAMETERS FOR ANTIMONY ANALYSIS OF GUNSHOT RESIDUES

	Instrumentation Laboratory [30] Graphite furnace	Goleb and Midkiff [29] Tantalum strip
Wavelength	217.6 nm	217.6 nm
Bandpass	0.5 nm	—
Spectrographic slit	160 μm	50 μm
Sample size	25 μl	10 μl
Drying	20 s at 60°C / 25 s at 120°C	45 s at 100°C
Ashing	20 s at 300°C / 15 s at 400°C	30 s at 450°C
Atomisation	10 s at 3000°C	2 s at 2500°C
Background correction	On	Use of 217.9 nm antimony line

References p. 394

D. Analysis

The instrumental parameters for the analysis of barium and antimony in gunshot residues are given in Tables 7—9. Sensitivities for antimony are given as 0.09 ng and for barium 0.25 ng in the original gunshot sample [30].

REFERENCES

1. J. Glaister, The Power of Poison, C. Johnstone, London, 1954.
2. B. W. Haynes, At. Abs. Newsl., 17 (1978) 49.
3. J. J. M. de Goeij, K. J. Volkers and P. S. Tjiol, Anal. Chim. Acta, 109 (1979) 139.
4. J. Locke, D. R. Boase and K. W. Smalldon, Anal. Chim. Acta, 104 (1979) 233.
5. C. E. Gleit and W. D. Holland, Anal. Chem., 34 (1962) 1454.
6. S. Kind and M. Overman, Science Against Crime, Aldus Books, London, 1972.
7. J. D. Cross, I. M. Dale, A. C. D. Leslie and H. Smith, J. Radioanal. Chem., 48 (1979) 197.
8. A. C. D. Leslie and H. Smith, Med. Sci. Law, 18 (1978) 159.
9. J. E. Clay, I. M. Dale and J. D. Cross, J. Soc. Occup. Med., 27 (1977) 102.
10. E. Rentoul and H. Smith, Glaister's Medical Jurisprudence and Toxicology, 13th edn., Churchill Livingstone, Edinburgh, 1973.
11. I. Davidson and J. B. Henry, Clinical Diagnosis by Laboratory Methods, W. B. Saunders, Philadelphia, PA, 1974.
12. B. A. Fowler in R. A. Goyer and M. A. Mehlan (Ed.), Toxicology of Trace Elements, Wiley, New York, 1974.
13. J. A. Fiorino, J. W. Jones and S. G. Capar, Anal. Chem., 48 (1976) 120.
14. S. Peats, At. Abs. Newsl., 18 (1979) 118.
15. F. Peter, G. Growcock and G. Strunc, Anal. Chim. Acta, 104 (1979) 177.
16. D. B. Lo and R. L. Coleman, At. Abs. Newsl., 18 (1979) 10.
17. J. M. McKenzie and J. D. Neallie, Trace Substances in Environmental Health, Univ. Missouri, U.S.A., 8 (1974) 45.
18. N. A. Larsen, B. Neilson, H. Packenberg, P. Christoffersen, E. Damsgaard and K. Heydorn, Nuclear Activation Techniques in the Life Sciences, IAEA, Vienna, 1972.
19. A. A. Gordus, J. Radioanal. Chem., 15 (1973) 229.
20. J. D. Cross, I. M. Dale and H. Smith, Forensic Sci. Int., 13 (1979) 25.
21. D. Barrowcliff, Medico-Legal J., 39 (1971).
22. W. D. Kinard and D. R. Lundy, Am. Chem. Soc. Symp. Ser., 13 (1975) 97.
23. V. P. Guinn, Ann. Rev. Nucl. Sci., 24 (1974) 561.
24. R. S. Nesbitt, J. E. Wessel and P. F. Jones, Aerospace Report No. ATR-75 (7915)-2, El Segundo, California, U.S.A.., 1974.
25. H. C. Harrison and R. Gilroy, J. Forensic Sci., 4 (1959) 184.
26. R. R. Ruch, V. P. Guinn and R. H. Pinker, Trans. Am. Nucl. Soc., 5 (1962) 282.
27. R. R. Ruch, V. P. Guinn and R. H. Pinker, Nucl. Sci. Eng., 20 (1964) 381.
28. J. H. Sherfinski, At. Abs. Newsl., 14 (1975) 26.
29. J. A. Goleb and C. R. Midkiff, Appl. Spectrosc., 29 (1975) 44.
30. Instrumentation Laboratory Inc. Atomic Absorption Methods Manual, Vol. 2, Flameless Operations, Wilmington, MA, 1976.
31. Warren Commission Report, Report of the President's Commission on the Assassination of President John F. Kennedy, U.S. Government Printing Office, Washington, D.C., 1964.
32. I.C.R.P. -23 Report of the Task Group on Reference Man, Pergamon Press, London, 1975.

Chapter 4m

Fine, industrial and other chemicals

L. EBDON
*Department of Environmental Sciences, Plymouth Polytechnic,
Drake Circus, Plymouth PL4 8AA (Gt. Britain)*

I. INTRODUCTION

While many samples arrive in the analytical laboratory in a form ready for analysis by atomic absorption spectrometry, this is not usually so for the chemicals discussed in this chapter. Often the requirement is for quality control on trace metal impurities in a troublesome matrix (e.g. reagents, chemical starting materials, electronic components and medicines) to meet strict limits. The major component in the sample may cause chemical interference, non-specific absorption or salting-up of the burner, as well as potential dissolution problems. Thus, although some samples can be analysed by a straightforward dissolution and determination, a recurrent theme in this chapter will be methods of circumventing matrix problems. An attempt will be made to highlight particularly useful methods of matrix destruction and analytical procedures which avoid matrix problems. On occasion it is necessary to isolate the analyte from the matrix and a number of approaches to doing this will be described with an emphasis upon those which are simple and not time-consuming.

II. CHEMICALS

A. Inorganic

1. *Fine chemicals and analytical reagents*

In a number of cases, simple dissolution of a solid sample in an appropriate solvent is possible and some laboratory reagents may even be analysed without further treatment. Prior to flame analysis, the best solvent is dilute hydrochloric acid, provided of course that the major matrix elements are not silver, lead or another element which forms a sparingly soluble chloride. If additional oxidising ability is required, concentrated nitric acid may be added to the solvent. This acid is the preferred solvent when the analysis is to be completed by electrothermal atomisation. If the material contains large amounts of silica it may be necessary to add hydrofluoric acid after preliminary digestion with hydrochloric acid (see Chapter 4g). Care should of

References pp. 440—445

course be taken with this reagent to avoid all skin contact, to use plastic laboratory ware and to avoid losses of silicon in the form of volatile silicon tetrafluoride. It should be noted that such losses can be eliminated provided an excess of hydrofluoric acid is used and the solutions are not heated too strongly. In some cases the silicon may be deliberately removed by evaporation to dryness of the fluoride medium, although this will cause problems with other elements such as calcium and the lanthanoids which form insoluble fluorides. In the author's experience it is often preferable to use aqua regia as the solvent and to promote the precipitation of silica in those cases where the determination of silicon is not required. Provided the solution is kept highly acidic and the precipitate aged by prolonged boiling, it can be demonstrated that there is very little co-precipitation of analyte species onto the silica. The subsequent filtration of the sample removes the troublesome silica and a number of potential chemical interferences in the flame. Ways of digesting samples which require more oxidising treatments than outlined here can be treated as described in section II.B below, with the appropriate precautions.

Acid dissolution is a particularly favourable approach for carbonates and sulphides, where the matrix anion will be removed during the evolution of carbon dioxide or hydrogen sulphide, and for salts of organic acids, where the anion seldom causes interference problems. Conversely, sulphates can cause problems during flame atomisation and chlorides during furnace atomisation; ways of dealing with such problems are discussed below.

Inorganic chemicals not amenable to acid dissolution are unlikely to be successfully attacked by ashing procedures, but in these cases fusion procedures may be useful. Unfortunately, fusion increases still further the dissolved-solid content and may introduce contamination. Various fusion mixtures have been used by different workers. Silicates may be attacked using lithium metaborate (or a mixture of lithium carbonate and boric acid) [1]. Typically 0.2—0.5 g of finely ground sample is mixed well with 2 g of lithium metaborate in a platinum crucible and heated to 900°C in a muffle furnace for 15 min or until a clear melt is obtained. After cooling, the crucible is placed in a beaker containing concentrated nitric acid (8 ml) and water (150 ml). The dissolution is best completed at room temperature. Sodium carbonate fusions over standard Bunsen burners are favoured by many workers, but in my experience obtaining a sufficiently high temperature for a true fusion has proved problematical on air/natural gas flames. I have found fusions using sodium peroxide and sodium carbonate mixtures in a zirconium crucible much simpler and more reliable (zirconium crucibles have considerably longer practical lifetimes than nickel crucibles). The method, used successfully for many samples, is to mix 0.5 g of sample with 1.5 g of sodium carbonate (which acts to moderate the fusion) and 4 g of sodium peroxide. After fusion over a Bunsen burner, the zirconium crucible is cooled and immersed in a beaker containing concentrated hydrochloric acid (35 ml) and water (100 ml). Dissolution of the melt is completed by gentle warming if necessary and the final solution is made up to 500 ml.

The standards are best prepared with a matching sodium chloride content ($15\,\text{g}\,\text{l}^{-1}$) to minimise errors from differing viscosity and ionisation suppression.

Blank determinations following any of the above dissolution procedures are important and may prove significant at the trace level. While it is always advisable to prepare the standards in the same concentration of acid and major matrix elements as the sample, to overcome variations in uptake rate and atomisation, it is also preferable to take a blank through the whole dissolution procedure. Losses of trace metals will be minimised if dilute solutions are kept at a pH $<$ 2 and analysis is not delayed.

Little need be said about analytical procedures directly following dissolution. Using a flame as the atom cell care needs to be taken that large contents of dissolved solids do not cause the burner or nebuliser to partly block. A small Teflon cup, which accepts samples up to $100\,\mu\text{l}$, attached directly to the metal capillary of the nebuliser, the so-called injection-cup technique [2] can be of use here. Samples in the range $20-100\,\mu\text{l}$ are pipetted into the cup using a precision micropipette, and the resulting absorption peaks recorded using the peak height or area mode on the instrument or a chart recorder. In this way, total dissolved salt contents of 15% or so may be tolerated compared with the 5% normally recommended in the conventional flame mode. Using this technique, also known as discrete sample nebulisation, surprisingly little deterioration in detection limits is observed.

A warning note has already been sounded about possible interferences which may be encountered in the flame. Suppression of ionisation of easily ionised elements by large amounts of alkali metals (and alkaline earth metals in a nitrous oxide/acetylene flame) in the matrix may be overcome by adding excess potassium or cesium to all standards and samples ($1000\,\mu\text{g}\,\text{ml}^{-1}$ usually suffices). Physical interferences are commonly encountered in inorganic chemical analysis because of the high viscosity of the samples; this can be overcome by buffering the standards with equivalent amounts of high purity major matrix compounds and acids. Certain highly refractory matrices form stable compounds, usually oxides, in the air/acetylene flame, e.g. zirconium, trace metals otherwise fully atomised in this flame may be occluded in the microscopic particles and thus a negative error results. Therefore, the flame should be chosen with the matrix as well as the analyte in mind, and in such cases of occlusion the use of the hotter and reducing nitrous oxide/acetylene flame may be perferable. The addition of ammonium salts may, by rapid sublimation, also aid the break-up of clotlets in the flame. Chemical interferences, the formation of actual chemical compounds in the flame which atomise more slowly than other forms of the analyte, can be encountered. The classical example is the calcium/phosphate interference which may greatly reduce the population of calcium atoms in the flame compared with calcium in a chloride medium. This interference can be overcome by the use of releasing agents (such as lanthanum which

References pp. 440—445

preferentially complexes with the phosphate), sequestering agents (such as EDTA which preferentially complexes with calcium but does not retard atomisation) or a hotter flame (i.e. nitrous oxide/acetylene). Similar approaches can be used to overcome many other chemical interferences such as those of sulphate, phosphate, silicate and aluminate ions on alkaline earth metals. The method of standard additions will provide useful information on such interferences when results are compared with those obtained by direct calibration. Standard addition will, however, provide no information about the extent of non-specific or molecular absorption interferences. Fortunately, these can be minimised by the use of automatic background correction, via the use of a second beam of continuum radiation passing through the atom cell. In the analysis of solutions with high dissolved salt contents this is advisable whenever measurements are made at wavelengths below 300 nm.

The development of modern electrothermal atomisers has proved to be of immense value in the determination of low levels of trace elements in pure chemicals and reagents. Many analyses previously possible only after tedious, contamination-prone, pre-concentration techniques are now viable. This is especially so for involatile trace elements in volatile matrices, where the majority of the matrix can be removed during the ashing step without loss of the analyte element. Some elements which might be lost during a vigorous or prolonged ashing step can be retained by the addition of a stabilising agent. Examples of such stabilisations are: the addition of nickel to retain arsenic, selenium and tellurium; phosphate to retain cadmium; and sulphide for mercury. Others may be identified from the literature or from a study of the melting points of compounds of the analyte. While the use of coated furnace tubes, sophisticated temperature control and platform, or isothermal atomisation have reduced the extent of interferences observed in graphite furnace atomisation; these still exceed those observed in flames. Optimisation of the temperature programme, particularly the ash or char cycle, and the use of background correction are essential. The use of peak integration facilities and standard addition are advisable. Many interferences occur when the analyte atoms first leave the hot walls of the atomiser and condense with cooler vapour phase species, such problems can be ameliorated by iso-thermal atomisation. A practical approach to such conditions when using a modern small graphite furnace is to break up an old tube into about eight fragments. These fragments can be used as small boats and pushed into the centre of the furnace. The sample can then be pipetted onto the boat, from which atomisation will not take place until the tube is hot (i.e. the atoms enter a hot environment). A tube clean cycle is always advisable on the completion of atomisation to prevent the build up of matrix.

Chloride media present a particular problem in electrothermal atomisation. Manning and Slavin [3] have described the determination of lead in magnesium and sodium chlorides at the $0.1\,\mu g\,g^{-1}$ level, and their paper contains many points of relevance to other analyses. Pyrolytic carbon-coated tubes were further coated with molybdenum using ammonium molybdate dissolved in ammonia solution. Ammonium nitrate was added to the sample as matrix modifier. This is a particularly useful agent as it promotes the following reaction:

NaCl	+	NH_4NO_3	$\rightarrow$	$NaNO_3$	+	NH_4Cl
boiling point 1413°C		decomposes at 210°C		decomposes at 380°C		sublimes at 335°C

The temperature programme used was: dry for 20 s (110°C), ramp 15 s for a 15 s char (550°C) and ramp 9 s for a 9 s atomisation (2500°C). With determinations in chloride media, the ash or char stage is particularly critical in furnace atomisation, as hydrogen chloride may be formed which aids the removal of chloride. This might otherwise interfere by vaporising the analyte as the chloride before the atomisation temperature is reached. In more open rod-type systems this mechanism is not available, and chloride interferences may be more severe [4].

Despite these recent advances, for many determinations the level of interest is so low that the only feasible approach is to separate and perhaps preconcentrate the analyte prior to determination in a flame or furnace. Some of the possible methods of doing this for high purity mineral acids, such as volatilisation, precipitation and complexation, have been discussed by an IUPAC Commission [5]. An alternative way of categorising the possibilities and a discussion is presented below.

(i) *Evaporation*

An approach particularly suited to liquid samples such as mineral acids, ammonia solutions and volatiles such as silicon tetrachloride, is to evaporate the matrix carefully from a silica crucible. An inverted filter funnel can be usefully placed over the evaporating dish to reduce aerial contamination, for very stringent requirements a clean-air cupboard should be used. The residue should be taken up in 1—10 ml of hydrochloric acid for flame analysis or 0.5—1 ml of nitric acid for graphite furnace work. By evaporating large amounts of acid (e.g. several hundred ml) considerable concentration factors can be achieved. Kometani [6] determined chromium, cobalt, copper, iron, manganese, nickel and zinc at the ng g^{-1} level in silicon tetrachloride (used in optical glass fibre production) by evaporation at low temperature. In this case, the residue was taken up in high-purity hydrofluoric acid and the determination completed in a mini-furnace.

If suitable, samples can be evaporated directly in a graphite furnace. An elegant method has been described by Langmyhr and Hakedal [7] for the determination of cadmium, copper, iron, lead, manganese and zinc in reagent and technical grade acids and ammonia solution. Ammonia was evaporated directly in the furnace by adding 50—250 μl of the ammonia to a furnace preheated to 70—80°C. This procedure prevented the sample spreading. After evaporation to dryness the furnace was heated for 60 s at 200°C before

References pp. 440—445

TABLE 1

WAVELENGTHS AND ATOMISATION TEMPERATURES FOR DETERMINATION OF TRACE METALS IN ACIDS AND AMMONIA [7]

Element	Wavelength (nm)	Atomisation temperature (°C)
Cadmium	228.8	1800
Copper	324.7	2500
Iron	248.3	2540
Lead	283.3	1800
Manganese	279.5	1900
Zinc	213.9	1700

proceeding to the atomisation stage. The acids were analysed by injecting 20—150 μl onto a glassy carbon boat made to fit directly into the furnace. The boats were evaporated to dryness on a thermostatically controlled hotplate and were covered by an inverted filter funnel attached to a water pump. This latter arrangement removed the acid fumes as well as reduced the possibility of contamination. Sulphuric acid was evaporated initially at 180°C for 5 min with a finish at 220°C for 5 min, whereas hydrofluoric acid was evaporated at 115°C. The boats were inserted into the furnace using PTFE tipped forceps. The analytical conditions shown in Table 1 were used. A 60 s atomisation stage was used followed by a clean cycle of 30 s at 1950°C. Calibration was by the method of standard additions.

(ii) *Ashing and pyrolysis*

Although more widely applicable to organic chemicals, the matrix of some inorganic chemicals may be reduced or modified by heating to an elevated temperature. Care must be exercised not to lose volatile trace metals such as arsenic, cadmium and zinc; mercury will almost invariably be lost to some extent. Matrices which might be amenable to this treatment include ammonium salts, sulphates, nitrates and the salts of organic acids such as the oxalates.

(iii) *Electrodeposition*

Electrodeposition onto solid electrodes or mercury cathodes is a long established pre-treatment capable of large concentration factors, and provided the cathode potential is carefully controlled it is also of considerable selectivity. When atomic absorption is used as the finish, selective deposition is not usually required. There have been recent reports of electrodeposition of trace metals from water samples directly onto special graphite furnace tubes [8] and this technique should prove to be just as applicable to the analysis of reagents, where the chemical conditions can be more carefully controlled. The utility of electrodeposition for electrothermal atomisation has already been proven and, in particular, the determination of trace elements in commercial salts either using a hanging mercury drop cathode [9] or a solid graphite cathode [10]. The latter approach has a number of advantages. Analytical grade potassium and magnesium chloride were analysed for bismuth, cadmium, copper, indium, lead, mercury and thallium by acidifying the solution to pH 2 and carrying out electrolysis with a cathode of spectral grade graphite (previously cleaned by heating for 30 min at 2000°C) at a potential of -1.0 volts versus a silver/silver chloride reference electrode for 15 h. This has advantages over deposition onto a metal wire [11—13], as the active part of the cathode can be ground down, yielding about 250 mg for subsequent solid sample analysis in a graphite furnace. Sample (2—3 mg) was introduced into the furnace using a tantalum scoop, dried for 30 s at 300°C and then atomised at 1900°C using a $50°C\,s^{-1}$ ramp. The sample can usefully be moistened with 100 μl water to aid reproducibility. Graphite is inert, has a relatively high hydrogen overvoltage and offers an excellent solid sample matrix. The acidification with nitric acid prevents a rapid rise in pH during electrolysis and the use of -1.0 V enables indium, thallium, bismuth, antimony, manganese, silver, mercury, copper, lead and cadmium to be deposited. Manganese does not, however, give good deposits and silver cannot be determined using a silver/silver chloride reference electrode. Background correction was used to avoid problems from scattering. The graphite powder can also be safely stored before analysis.

(iv) *Adsorption*

Active carbon has been advocated by several groups of workers as a means of separating metal ions or their complexes. The latter approach being the more popular. Silver and thallium compounds have been dissolved in water, and xylenol orange solution added. The resulting chelates were adsorbed on activated charcoal and redissolved in nitric acid prior to a flame finish [14]. More recently Jackwerth and Berndt [15] have used diethyldithiocarbamate as the complexing agent when determining bismuth, cadmium, cobalt, copper, indium, iron, lead, nickel, silver, thallium and zinc during the purity control of several alkali and alkaline earth salts. A two percent solution of the ligand was used and the separation achieved by filtering the solution through a filter paper coated with 50 mg of activated carbon (sufficient for 10 g of sample). The carbon was treated with nitric acid to elute the metals for their determination in the air/acetylene flame. The injection-cup technique was used [2] and limits of detection in the range of 0.06—0.003 ppm were reported with a relative standard deviation of 4%. An interesting variation on these methods for non-volatile metals might be to ignite the carbon and dissolve the ash in hydrochloric acid for a flame, or nitric acid for a furnace. This would enable a very small amount of acid to be used and hence improve sensitivity.

References pp. 440—445

(v) Ion exchange

An obvious way to separate metal ions from acids, bases, digests and alkali metal salt solutions is to use an ion-exchange column or merely a batch of resin added to the analysis sample. The sample may be passed down a strong acid cation-exchange resin when divalent and trivalent metals will be strongly adsorbed or a chelating resin (such as Chelex—100) when transition metals will be adsorbed. Desorption can be by elution with acid or the resin may be directly added to a furnace. Unfortunately ion-exchange resins generally lack selectivity and this limits their application in the analysis of pure chemicals. A report of the determination of trace metals in high purity phosphorus has, however, been made [16]. The sample was dissolved in a mixture of hydrochloric and nitric acids and passed through a column of Dowex 50W—X8 resin to adsorb alkali, alkaline earth and heavy metal elements, which were later eluted with 50% hydrochloric acid.

(vi) Co-precipitation and precipitation

While trace levels of impurity elements may not be precipitated quantitatively from solutions of reagents, the addition of a suitable carrier usually ensures complete recovery. Such co-precipitation is a very simple and apt method for atomic absorption spectrochemical analysis. The author's laboratory has made extensive use of the precipitation of metal hydroxides by ammonia solution using lanthanum hydroxide carrier. A number of elements such as nickel and copper are not precipitated under such circumstances, as their ammine complexes are formed, thus some selectivity can be introduced. Valency control can further increase this selectivity; for example, vanadium and chromium form oxyanions in strongly oxidising media and are not precipitated. The alkali metals, which form soluble hydroxides, are similarly not precipitated. The precipitate is separated by centrifugation or filtration and redissolved in a preferred acid, the lanthanum additionally acting as a useful buffer in the subsequent analysis. Lanthanum (III) chloride solution (2 ml of a 5% w/v solution) is added in our method to an aliquot of solution (15—30 ml) in a centrifuge tube (50—100 ml capacity). Ammonia solution is then added to complete the precipitation, typically 10—20 ml of a concentrated solution suffices. The solution is mixed well, cooled in cold water and centrifuged at approximately 3000 r.p.m. for 3 min. A second aliquot of lanthanum chloride solution (2 ml) is added to the tube and the centrifugation repeated. After discarding the supernatant liquor, the precipitate is dissolved in hydrochloric or nitric acid, transferred to a volumetric flask and made up to volume (10 ml). Greater concentration factors can be introduced, if required, by repeating the process after discarding the supernatant liquor and precipitating the second aliquot upon the first. Standards matched for lanthanum content should be used. A similar method was reported by Young [17] for the determination of trace elements in alumina. Zirconium hydroxide was used as the carrier.

Twenty-one elements have been determined in high purity barium and strontium compounds after precipitation of the matrix as nitrates [18]. The compounds were dissolved in nitric acid and evaporated to a low bulk. The strontium and barium nitrates precipitated, as they were present in large amount and have a low solubility in the nitric acid. The injection-cup technique [2] was used to determine the metals in the range 1—0.01 ppm (aluminium, bismuth, cadmium, calcium, chromium, cobalt, copper, gallium, gold, indium, iron, magnesium, manganese, mercury, nickel, palladium, potassium, silver, sodium, thallium and zinc). The method is feasible as the solubility of lead, strontium and barium nitrates decreases rapidly with increasing nitric acid concentration (e.g. a hundred fold drop in solubility in 14 M acid compared with 1 M), yet there is no complex formation to increase solubility as for hydrochloric acid. There are problems with the co-precipitation of bismuth and lead, and losses of mercury (50% loss) and chromium (20%) upon evaporation. Other elements gave recoveries of about 95% and very low blanks were reported. Clearly the injection-cup technique is advantageous with such highly concentrated solutions and this was used with the air/acetylene flame, although aluminium and calcium were determined by continuous nebulisation into a nitrous oxide/acetylene flame. It would seem likely that many more problems would be encountered from co-precipitation using sulphate media. The method as described is obviously of value to the fireworks, explosives and electronics industries where high purity strontium and barium compounds are used.

(vii) *Solvent extraction*

Clearly the most popular separation and preconcentration technique for atomic absorption analysis is solvent extraction. In this case it is easy to identify extraction systems which will remove a broad range of impurities from matrices such as acids, bases and alkali metal salt solutions. In addition to the advantages to be gained from separation, especially valuable in furnace work, and the concentration factors available, solvent extraction confers an additional advantage (typically a factor of 3—5) in flame analysis arising from the favourable nebulisation characteristics of several organic solvents.

The most widely applied reagents have been chelating agents which will complex with many metals, e.g. dithizone and the various thiocarbamate derivatives such as diethyldithiocarbamate and pyrrolidine dithiocarbamate. The latter agent as the ammonium salt (APDC) has been shown to complex some thirty elements [19] most of which can be readily extracted into various solvents. 4-Methylpentane-2-one (methyl isobutyl ketone or MIBK) is usually the favoured solvent because of its excellent compatibility with flames. The solubility of MIBK in water is not negligible and this limits the available concentration factor to ten; higher molecular weight ketones (e.g. decan-2-one) offer better concentration factors and chloroform up to fifty times, but this latter solvent is only really suitable for electrothermal atomisation.

APDC complexes are formed over a wide pH range and generally extracted from solutions of pH 2—3. A wide range is however, permissible and a

TABLE 2

RECOMMENDED pH LIMITATIONS FOR APDC — METAL CHELATE EXTRACTIONS

Limit	Formation	Extraction
No limits pH range 1—14	Ag, Au, Bi, Cd, Co, Cu, Fe, Hg, Ir, Ni, Os, Pb, Pd, Pt, Re, Rh, Ru, Ti, Zn	Ag, Au, Cu, Ir
pH not less than		
2	Cr, Ga, In, Mn, Mo, Nb, Sb, Se, Te, U, V	In, Nb, Sb, Ti
3	Sn	Cr, Ga, Se, Te, U
4	Th	Mn, Sn, Th, V
pH not greater than		
3	W	Mo, W
4	Nb	Nb, U
5	U	As, Sb, Te
6	As, Mo, Te, V	Mn, Se, Sn, Th, V
7	Sn	Cr
8	Th	Ga, Pb
9	Cr, Sb	In
10	Ga, In, Se	Bi, Co, Fe, Hg, Ni, Os, Pd, Pt, Re, Ru, Zn
11		Cd
12	Mn	Rh, Ti

summary of typical limitations is given in Table 2. It should be noted that extractions should always be made at reproducible pH values. A suitable solution of APDC can be made by dissolving 1 g in 100 ml of water. This solution is not very stable and is best prepared freshly every few days and filtered before use. For a typical 50 ml of sample, 5 ml of this APDC solution is adequate. The pH can be adjusted as necessary with acetic acid or ammonia with measurement on a pH meter. Inspection of Table 2 shows that pH 5 is adequate for most metals and an indicator such as bromocresol green can be used to indicate this. In the case of manganese, a two minute excursion to pH 12 aids complexation and certain metals, notably chromium and molybdenum, need warming to 80°C for 5 min to overcome kinetic problems in complex formation. The solution is transferred to a 100 ml separating funnel and the complex (which may have precipitated) is extracted into 5 ml of MIBK. This is best achieved by vigorous inversions for 30 s. The extraction should be repeated with a second 5 ml aliquot to ensure quantitative recovery. The combined extracts can be aspirated into a fuel-lean air/acetylene flame, or if necessary a standard nitrous oxide/acetylene flame. For electrothermal atomisation chloroform may be used as the extractant and a larger volume of sample or smaller volume of solvent used. Standards must be taken through the same procedure.

Some extremely low detection limits in chemical samples have been reported by APDC/MIBK and other dithiocarbamate extractions. Further increased concentration factors can be obtained by back-extraction into a minimum of acid. Copper in ammonium fluoride solution used in semiconductor processing has been determined down to $4.5 \times 10^{-7}\%$ [20]. To a 40% ammonium fluoride solution (50 ml) was added ammonia solution (1 ml), 50–60 µg of a palladium(II) carrier solution (dissolved in 1 ml) and sodium diethyldithiocarbamate (5 ml). After extraction of the copper chelate into chloroform the solvent was evaporated, the residue dissolved in concentrated nitric acid (10 drops) and water (2 ml) and this was then evaporated. The final dissolution was in concentrated hydrochloric acid (2 drops) and concentrated nitric acid (1 drop). The solution was made up to 1 ml and the determination made by standard additions using an air/acetylene flame and the 324.7 nm copper resonance line.

An alternative complexing agent is 8-hydroxyquinoline (or oxine) which has the advantage of forming chelates with a number of elements, such as aluminium, calcium, magnesium and strontium, which do not react with APDC. Oxinates may be extracted into MIBK or chloroform, typically at pH 11 for calcium and magnesium. Traces of calcium have been determined in phosphoric acid and its salts by extraction as the oxine complex [21]. To a sample of phosphoric acid (9.8 g) was added 6 M hydrochloric acid (1 ml) and 10% sodium tartrate solution (20 ml). This was neutralised with 6 M sodium hydroxide and diluted to 100 ml. To an aliquot (10 ml), saturated sodium chloride solution (10 ml), butyl cellosolve (2 ml), oxine (3 ml of a solution of 5 g in 100 ml of ethanol) and potassium chloride/sodium hydroxide pH 13 buffer (5 ml) were added. The complex was extracted in 3-methyl-1-butanol which was sprayed into an air/acetylene flame. Measurement was at 422.7 nm. A similar method was used by the same authors to determine magnesium in brine, sodium carbonate and sodium bicarbonate [22]. The extraction in this case was at pH 11 and the method overcame otherwise pronounced salt effects. An interesting method for determining copper, iron, magnesium, manganese and zinc in high-purity silver chloride depended on an oxine/MIBK extraction followed by determination in the air/acetylene flame [23]. The sample (2 g) was first dissolved in concentrated ammonia solution (20 ml) with the aid of ultrasonic agitation for 15 min. Silver was not co-extracted.

Some elements can be extracted as simple salts into organic solvents, e.g. arsenic trichloride into benzene. For most metals alternative procedures are available. Cresser [24] has written a most useful guide to solvent extraction for flame spectrometry which can be consulted when developing new methods.

References pp. 440—445

(viii) *Vapour generation*

Mercury and those elements (antimony, arsenic, bismuth, germanium, lead, selenium, tellurium and tin) which form volatile covalent hydrides may be separated from the matrix by vapour generation. The use of tin(II) chloride to generate elemental mercury and its subsequent aeration into a long-path absorption cell with silica windows has been described elsewhere in this book, as has the use of sodium borohydride to produce hydrides which are swept to a flame or heated tube for atomisation. This approach is far more successful for mercury than for the other elements, as the hydride generation technique is subject to interference from a large number of transition metals and oxyanions.

Mercury has been determined in silver and silver nitrate by adding potassium bromide to the nitrate solution (for silver dissolve 0.1—1 g in 6 ml 8 M nitric acid) adjusted to pH 4—6 with ammonia (as indicated by the purple form of bromocresol purple) [25]. The silver bromide formed can be removed by filtration, the mercury remaining in solution as the $[HgBr_4]^{2-}$ complex. Tin(II) chloride is used to generate the mercury vapour. As the recovery obtained is only 70—85%, because of co-precipitation, a calibration curve was used. Arsenic has been determined in high-purity phosphoric acid by arsine evolution [26]. To phosphoric acid (1 ml), a 10% solution of potassium iodide (0.5 ml to reduce arsenic(V) to arsenic(III)) and 6 M hydrochloric acid (20 ml) were added. The arsine evolved upon injecting a 5% solution of sodium borohydride (5 ml) was led to a nitrogen/hydrogen/entrained air flame.

2. Industrial chemicals

Similar general remarks about inorganic industrial chemicals can be made as in section II.A.1. Additionally, such chemicals and technological chemical solutions may be more complex and their major constituents less well defined. Thus interferences may constitute a greater problem. Fortunately, however, the levels of trace metals present may be higher than those of interest in laboratory chemicals and flame analysis is often appropriate. Again there is a need to exercise care over blanks and to match standards for major constituents and acid content.

An on-stream atomic absorption method for the determination of zinc and manganese in flotation liquors containing calcium sulphate has been reported [27]. The nebuliser to an air/acetylene flame was continuously fed, by gravity, with a portion of the process stream buffered with EDTA to overcome calcium sulphate interference. The atomic absorption monitor operated continuously for 3 h before salting-up of the burner occurred.

Solid samples may frequently require fusions. Examples of such procedures in the literature are: (a) the determination of aluminium, silicon and sodium in molecular sieves using a nitrous oxide/acetylene flame following fusion (1 part to 10 of fusion mixture) with sodium carbonate, sodium tetraborate (2:1 mixture) and dissolution in nitric acid [28]; (b) the determination of chromium in electrode grade coke, using an air/acetylene flame, following ashing at 800°C, fusion with potassium hydrogen sulphate and dissolution in sulphuric acid [29].

The analysis of brines perhaps deserves special mention as the high sodium chloride concentrations are extremely unfavourable for electrothermal atomisation and most troublesome in flame analysis. The preferred approach is probably solvent extraction with either oxine or APDC to remove the trace metals into a small volume of MIBK for flame atomisation or chloroform for electrothermal cells. Care must be taken to avoid interference from chloro-complexes in the extraction, and if this is suspected an ion-association extraction of these complexes might be preferable.

B. Organic

1. *Organometallic*

The determination of the metallic components in organometallic compounds is an important adjunct in characterisation for the synthetic chemist and for quality control in industry. Atomic absorption offers an excellent means of doing this through a combination of good sensitivity with good precision. Often only small samples are available and electrothermal atomisers are preferred, but when precision is an important requirement the flame is likely to be optimal. Three approaches to organometallic analyses are frequently encountered: (a) dissolution in a suitable solvent, usually organic; (b) dissolution in a suitable solvent followed by ligand exchange; (c) digestion and dissolution. The first case is the simplest but since different organometallic compounds will nebulise and atomise in different ways, and probably very differently to inorganic standards, standards made from similar organic compounds in the same solvent must be used.

Duncan and Herridge [30] determined silicone fluid surfactants in polyurethane/polyether blends using the silicon 251.6 nm line and the nitrous oxide/acetylene flame. The concentration of the fluid is critical in foam manufacture and should be in the range 0.4—1.0 g in 100 g of blend. By diluting the sample (5 g) in ethanol/water (50 ml of 1:1) it could be sprayed into the flame. Standards were prepared by dissolving the pure silicone fluid in the same solvent. Carbonyl complexes of ruthenium, e.g. $[RuBr_2(CO)_3]_2$ and $RuBr_2(CO)_2 [P(C_6H_5)_3]_2$ have been analysed for ruthenium content by dissolving the organometallic compound (20—40 mg) in MIBK so as to give solutions in the range 1—10 μg Ru ml^{-1} [31]. Ruthenium tris-acetylacetonate in MIBK was used for standards in the subsequent assay with an air/acetylene flame.

After dissolution in a suitable solvent the analyte may be displaced from the organometallic compound by a stronger ligand. This has considerable

References pp. 440—445

advantages when either suitable standards are not available, or a wide range of organometallic compounds are being received for analysis. This ligand exchange method is not suitable for highly stable compounds. By dissolving nickel complexes in MIBK and then adding chloroform and diethyldithiocarbamate, Leonard and Swindall [32] developed a method independent of the bonding in organo-nickel complexes. This method also had the advantage of giving more stable solutions and allowing the use of the air/acetylene flame. Previously the nitrous oxide/acetylene flame had to be used to overcome the effect of the bonding on absorbance. The use of a buffer of lanthanum (2% in 0.5 M hydrochloric acid) and potassium phosphate (3% aqueous) has been found to overcome the different absorbances of cis- and trans-organo-platinum complexes in the air/acetylene flame [33, 34].

Clearly, if there is any uncertainty about the formulation, then destruction of the complex by ashing is preferable. Here, furnace atomisation offers a clear advantage as organometallic compounds may be destroyed in the charring stage, provided they are not volatile, and calibration performed by standard additions. For flame atomisers, ashing may be used again provided the compounds are not volatile as they often are. A number of metals may also be lost on ignition at temperatures as low as 500°C, therefore, wet digestion is frequently employed. A wide range of organo-tellurium compounds of the type $RTeX_3$ (e.g. $p\text{-MeC}_6H_4TeCl_3$) have been analysed by Thavornyutikarn [35] following digestion with nitric acid and perchloric acid. The digestate was evaporated to dryness and dissolved in hydrochloric acid prior to determination of tellurium at 214 nm in the air/acetylene flame. An approach to more volatile compounds is digestion in a PTFE lined pressure-vessel. Use of such digestion bombs is highly effective, if tedious, when several samples are to be analysed. This method has been used to digest organo-mercurial compounds of the type $(CH_3)_3SiCH_2HgX$, used in toxicity studies, in nitric and sulphuric acids [36]. The complex was dissolved in water (2—5 mg in 50 ml) and an aliquot of this solution (0.1 ml) sealed in the bomb with 16 M nitric acid (1.3 ml) and 18 M sulphuric acid (0.7 ml). The bomb was heated at 125°C for 4 h. After cooling, the contents were transferred to a flask and 4% potassium permanganate (7 ml) then 5% potassium persulphate (2 ml) were added. The solution was allowed to stand overnight before 6% hydroxylamine hydrochloride (10 ml) was added and mercury vapour generated with 20% tin(II) chloride solution (5 ml) for determination by the cold vapour technique. Lead salts such as resorcylate, salicylate, stearate and 2-ethylhexoate are used as modifiers of propellants and casting powders. After grinding and refluxing (2 g) with 10% acetic acid (10 ml) and acetone (50 ml) and dilution in acetic acid, the lead content of such powders can be determined in the air/acetylene flame [37]. A different approach to matrix destruction was used by Bigois and Levy [38]. They pyrolysed organo-silicon compounds in chlorine using a chlorine/nitrogen atmosphere at 1100°C. The vapour phase was transferred to a nitrous oxide/acetylene flame by a nitrogen stream for the determination of silicon at 251.6 nm.

Frequently coupled chromatography—atomic absorption systems can provide valuable information on organometallic speciation.

2. Impurities in organic compounds

At first sight the determination of trace metal impurities in organic compounds may appear to be refreshingly simple. The matrix can be more readily destroyed by wet digestion or ashing than some of the substances previously considered, and thus, several complications may be avoided. This is often the case, but words of warning must be given. Elements such as arsenic, cadmium, lead, selenium and zinc may be lost upon ashing at even moderate temperatures (e.g., 500°C is often recommended in the older literature) and there may be mechanical losses of analyte in any smoke evolved. If suitable precautions are taken, this can be a very useful approach, e.g. titanium has been determined in hydrazine by flash evaporation of the organic matrix and dissolution of the residue in acid for graphite furnace atomisation at 2800°C [39]. Vigler et al. [40] determined phosphorus in alkyl-phenyl phosphite, a polyester additive. The sample was diluted with N,N-dimethylformamide and ignited at 650°C with magnesium oxide in the presence of hydrogen peroxide. Similarly, tricresyl phosphate in petrol was ashed with magnesium sulphonate. The magnesium pyrophosphate formed in these ways was dissolved in nitric acid. A 25 μl aliquot plus 10 μl of 1% lanthanum nitrate solution were injected into a graphite furnace, measurement being at 213.6 nm using the programme: dry (30 s) at 120°C; char (40 s) at 1350°C; atomise (10 s argon interrupt) at 2700°C.

Wet ashing needs to be considered equally carefully. No one mixture will suffice for all compounds, sometimes organic sulphur may be precipitated as a troublesome suspension, above all, care must be taken with oxidising mixtures. The great value of peroxide and perchloric oxidants ensures their continued use, but when used to attack pure compounds additional care must be taken. As the temperature rises, the oxidation potential of different compounds is reached. This causes fewer problems with mixtures and more warning is given of the need to remove the source of heat temporarily, but with a pure, single compound this point may be reached rapidly and explosively. The reports of the Analytical Methods Committee of the Analytical Division of the Royal Society of Chemistry on the determination of small amounts in organic matter are authoritative guides to digestion procedures, these are published in the Analyst (e.g. for nickel [41] and selenium [42]).

The use of hydrogen peroxide (50% m/m) and concentrated sulphuric acid or nitric acid is often advocated and is successful, but the reader is referred to warnings on the use of peroxide published by the above committee [43, 44].

Ashing with mixtures of perchloric acid and nitric and/or sulphuric acid are again widely used, but extreme caution must be exercised. (See Chapter 3 of this book.) We have found few substances which withstand such an attack, but a large excess of acid to sample must be used, the sample should be well wetted before heating which should be applied gently and cautiously, e.g. with a micro-burner. Martinie and Schilt [45] have published the report of a very useful investigation of the effectiveness and hazards of two perchloric acid digestions of 87 organic substances. In one digestion, the sample (1 g) was attacked with a 2:1 perchloric acid/nitric acid mixture (15 ml). After observation for violent behaviour, the temperature was slowly raised to 120°C to distill off the nitric acid. If necessary, the heat was removed to moderate the reaction. Over a period of 15 min the temperature was raised to 140°C and then more rapidly to 203°C where it was boiled for 30 min, provided this was safe. A protective screen and fume hood were of course used. In the second digestion, the sample (1 g) was boiled with sulphuric acid (15 ml) for 15 min. After cooling, nitric acid (15 ml) was added and the temperature raised to 320°C slowly to allow for distillation of the nitric acid. After cooling, perchloric acid (15 ml) was added and the solution slowly heated so as to distill the perchloric acid from the sulphuric acid solution. Most of the 87 organic compounds investigated were completely oxidised by this treatment. Amongst compounds where more than 50% residual carbon were found in the nitric/perchloric digest were: alanine; proline; methionine; 2,2'-bipyridine; 8-hydroxyquinoline; 5-nitro-1,10-phenanthroline; 1,10-phenanthroline; pyridine; quinoline; and 2,4,6-trimethylpyridine. The following compounds gave violent reactions during the nitric/perchloric acid procedure: lanoline; cottonseed oil; lecithin; thiophene; furan; cholesterol; squalene; tygon; latex rubber; and Amberlite XAD—2. Other compounds giving vigorous reactions, and for which, therefore, more nitric acid should be used were: pyrrole; sodium tetraphenylborate; tetraphenylarsonium chloride; 1-nitroso-2-naphthol; 2-hydroxyquinoline coumarin; Amberlite CT—120; quinazoline; and anthranilic acid.

The determination of lead in organic colouring dyes by flame atomic absorption spectrometry [46] perhaps illustrates a typical approach to the problem of organic chemical samples. The dye (3 g) was digested in nitric acid and perchloric acid, unless water soluble. In order to meet the stringent regulations on the lead content of food colouring dyes and to overcome varying viscosity after digestion, solvent extraction using diethyldithiocarbamate from very acid solutions into xylene was employed. Ascorbic acid (5 ml of a 1% solution) was used to overcome interference of iodine in the extraction which was performed using phase separation paper.

C. Phosphors

Phosphors used in the production of fluorescent lamps are of the calcium halophosphate type, prepared by firing together calcium hydrogen phosphate, calcium carbonate and a lesser amount of calcium fluoride. The phosphor (0.25 g) can be dissolved in a small amount of concentrated hydrochloric acid, made up to 10 ml and impurities such as sodium determined using a flame [47].

Scott [48] has determined europium in yttrium phosphors. Yttrium vanadate (0.1 g) was fused with potassium carbonate (2 g) in a platinum crucible. Yttrium oxide and yttrium oxysulphide can be dissolved directly in the same solvent as the above melt, 50% hydrochloric acid. Standard flame conditions for yttrium were used.

III. PHOTOGRAPHIC FILM AND CHEMICALS

The measurement of various metals in photographic film is important in quality control, the presence of some metals at given concentration levels may be vital, especially in colour photography or deleterious. Often the levels in solutions of photographic chemicals can be determined by simple dilution and aspiration into a flame. Care must be taken to avoid the build up of silver in the cloud chamber and drain tube as explosive silver acetylide may be formed. Thorough washing out procedures are therefore advisable. Silver may be determined in photographic papers following wet ashing, with a detection limit of 600 ng cm^{-2} of paper [49], or after it has been stripped from the emulsion using potassium cyanide (0.2 M) [50]. The cyanide solution should, of course, not be acidified and the analysis can be completed at 328 nm using the air/acetylene flame. Potassium cyanide (0.04 M) can also be used to strip cadmium from photographic emulsion [51]. Dittrich and Mothes [52] used an 0.1% solution of the enzyme Mezymforte in 0.05 M ammonium chloride to digest photographic film in order to determine gold. The digestion was performed at 38°C. After evaporation to dryness and dissolution in nitric acid/hydrogen peroxide, the gold was extracted as the bromide complex into MIBK saturated with hydrobromic acid. A 10 μl sample was pipetted onto a graphite rod for measurement at 242.8 nm. A detection limit of 7 ng cm^{-2} of film was reported. The enzyme solution serizyme has been used to extract palladium from films [53]. Two dm^2 of film were extracted at 37°C, the film removed and the extract evaporated to dryness. The residue was treated with nitric acid/hydrogen peroxide, re-evaporated, treated with nitric/sulphuric/perchloric acid mixture and again evaporated before adding potassium bromide/hydrobromic acid. The palladium was extracted into toluene using 0.2 M dibutyl sulphide solution. The toluene was injected into a graphite furnace for ashing at 500°C and atomisation at 2700°C. This enabled very low levels of palladium, e.g. 1 ng cm^{-2}, to be determined in the presence of many other elements. Such enzyme digestion procedures to remove the film are likely to be generally applicable and more meaningful as the metal content of the paper or polymer backing may be included in procedures involving ashing.

References pp. 440—445

IV. CATALYSTS

The importance of catalysts in our energy and pollution conscious age is growing. Many catalysts depend for their activity on low levels of rather exotic metals, while even trace surface levels of elements such as lead may impair their activity. Thus there is plenty of scope for atomic absorption spectrometry. Many homogeneous catalysts are deposited on an alumina base, thus obtaining dissolution is not always easy and some interference in the air/acetylene flame may be encountered. A leaching procedure (e.g. with nitric acid) to dissolve adsorbed trace metals may be used to circumvent these problems.

Catalysts are widely utilised in the oil industry and various dissolution procedures, usually involving hydrofluoric acid, are used. Interestingly, few of these procedures have been published, presumably to avoid release of information about the catalysts. Labrecque [54] has published a method for determining cobalt, molybdenum and nickel in hydrodesulphurisation catalysts used in petroleum refining. Molybdenum/cobalt catalysts formed by impregnating γ-alumina, stabilised with 5% silica, have different activity to molybdenum/cobalt/nickel catalysts. A sample (0.5 g) was dissolved in concentrated sulphuric acid (20 ml) and hydrofluoric acid (at first 10 drops but later 5 ml) in a Teflon beaker. The beaker was heated for 45 min at 90°C and more hydrofluoric acid (10 ml) added if needed to clear a high silica content. The solution was heated at 90°C for 45 min to remove residual hydrofluoric acid. Molybdenum and cobalt were determined in a nitrous oxide/acetylene flame and because some enhancement was observed from the very high levels of aluminium present this was added to all the standards, or standard addition may be used. Nickel was determined in the air/acetylene flame and no enhancement was observed for this element.

Concern about air pollution has led to extensive investigations of catalysts for automobile exhausts. Elements such as lead (a potential poison), palladium and platinum can be determined after dissolution of the catalyst in acids such as aqua regia. Janouskova et al. [55] reported a detection limit of 0.8 ng for platinum in a catalyst using graphite furnace atomisation and 265.9 nm.

Many catalysts used in chemical synthesis can be treated in the same way, often the nitrous oxide/acetylene flame is used because of the refractory nature of the elements to be determined. Harrington and Bramstedt [56] have determined rhenium in electro-chemical surface catalysts by stripping the coating with molten potassium hydroxide/potassium nitrate. This melt was extracted with hydrochloric acid, the residue was fused with sodium peroxide for further rhenium determination. Titanium, being the substrate on which the catalyst was coated, was added to the standards, an air/acetylene flame and 343.3 nm were used for the finish.

V. SEMI-CONDUCTORS

These represent a class of generally easily dissolved chemicals in which even very low levels of trace metals may be deleterious. There is also some interest in determining major element levels, often to distinguish different semi-conductor types. The low levels are best determined using the graphite furnace and most interest centres around possible matrix interferences. It is very important to avoid contamination during digestion and for this reason enclosed PTFE lined pressure digestion vessels may be used. Various methods have also been proposed for layer by layer determination of trace metals and this has been achieved by etching, electrochemical oxidation followed by dissolution, ion-beam bombardment and the use of organic solvents. For example a 1:70 mixture of hydrofluoric and nitric acids has been used to etch epitaxial silicon layers for the determination of antimony [57]. This solution was evaporated almost to dryness, nitric acid (0.2 ml) added and the sample again evaporated. After repeating the nitric acid dissolution twice the residue was dissolved in water (20—50 µl). The analysis was completed in a graphite furnace with drying at 100°C, ashing at 420°C and atomisation at 2050°C.

Pelosi and Attolini [58] have developed a solvent extraction/graphite furnace method to determine gallium impurities in $ZnIn_2S_4$ semi-conductors. A few mg of sample were dissolved in 1 M nitric acid with heating; it was not necessary to separate the colloidal sulphur formed. Cupferron was used as the chelating agent gallium cupferrate being non-volatile, for extraction into butanol at pH 4.7 (ammonium acetate buffer). A 20 µl aliquot was dried at 80°C (25 s), ashed by a programmed temperature rise from 220°C to 630°C with atomisation at 2635°C (5 s) and measurement at 287.4 nm. Dittrich et al. have published an extensive series of reports on trace determinations in AIII BV type semi-conductors e.g. [59—62]. Mixtures of hydrochloric acid/nitric acid (1:1) were generally used as the solvent for materials such as GaAs, GaP, InAs with graphite furnace atomisation under carefully controlled conditions to avoid losses of analyte, e.g. as cadmium, zinc and manganese chloride species. Interestingly, they have found that whereas tellurium was best determined by graphite furnace atomisation [63] hydride generation with atomisation in a silica tube was preferred for the determination of selenium [64]. Nakahara and Musha have noted that indium [65] and gallium [66] may more sensitively be determined in argon (or nitrogen)/hydrogen entrained air flames than in conventional flames and the many interferences observed may be overcome by adding magnesium iodide and magnesium chloride respectively. Semi-conductors such as SbGa, GaP, $Al_xGa_{1-x}P$, $In_xGa_{1-x}Sb$ and $Ga_xIn_{1-x}P$ (10—30 mg) were dissolved in concentrated hydrofluoric acid (5 ml) and concentrated nitric acid (5 ml) in a polythene bottle on a water bath. Excellent detection limits were reported (for indium 0.025 ppm at 303.9 nm), but for major component analysis this hardly seems necessary

References pp. 440—445

and many workers will probably prefer to use the more conventional and trusted air/acetylene flame.

VI. ELECTROPLATING SOLUTIONS

There is a need to check the concentration of major metals, any metal containing additives (such as brighteners) and trace impurities, in electroplating baths to ensure good deposition. Impurities may build up in the baths following introduction of the items to be plated, commercial salts and tap waters, and these need to be monitored carefully. A number of ancillary applications such as determining plating thicknesses by leaching off the plate, or controlling the composition of alloys (e.g. the lead/tin ratio in a tinning bath by dissolving the alloy in hydrofluoric and nitric acids) are useful in the same industry. Silver-plated ware used for foodstuffs is subject to legislation in a number of countries to control toxic metals being leached into consumables. The legislation usually requires that acetic acid (e.g. 4% v/v in water) be stood in the food container for twenty four hours and then the arsenic, antimony, cadmium and lead contents determined. Levels around 1 ppm are typically quoted although higher levels of lead may be allowed. This analysis can be performed by flame atomic absorption. In our laboratory we have studied a number of teapots and found by controlling the surfaces exposed to the leaching solution that arsenic and cadmium originate from the brass base-metal, lead from the solder used in the spout (where it is difficult to "throw" the plate properly) and antimony from brighteners used.

When major components of plating solutions are determined, large dilutions may be required (e.g. a factor of 5000) to bring the sample into the normal working range for flame analysis. Such dilutions will, however, minimise any interference effects and viscosity effects from additives, and are thus to be preferred to the use of less sensitive lines or burner rotation. The above interference effects may be important in the determination of trace metal levels and attempts should be made to match the standards for major component levels, or to use the method of standard additions. Solvent extraction to remove the analyte from the matrix may be necessary.

Price [67] has reviewed the application of atomic absorption to a variety of plating solutions. Iron, lead and zinc are reported as the main impurities in cyanide copper-plating baths which may contain up to 200 g l^{-1} of copper sulphate; a twenty-fold dilution of the sample for trace determination is recommended. Nickel baths may contain 60 g l^{-1} of nickel and it may be necessary to monitor copper, zinc, iron, lead, chromium, calcium and magnesium at the ppm level. The standard addition method is probably best for such an application. Zinc has been extracted with trioctylamine-hydrochloride when present in the range 0.03—10 μg ml^{-1} in a nickel plating solution [68]. The zinc was re-extracted back into 1 M nitric acid for measurement in the air/acetylene flame at 213.9 nm. Lead was extracted from nickel plating solutions by Byr'ko et al. [69] using a 0.2% solution of hexahydro-azepinium hexahydro-azepine-1-carbodithioate into amyl acetate. Chromium plating solutions present interesting industrial hygiene problems because of the need to monitor hexavalent chromium in the atmosphere. The actual plating solution may contain 500 g of chromic oxide per litre and this acid solution must be monitored for the pick up of copper, iron, nickel and zinc. As chromium may interfere with iron determinations it is advisable to ensure the standards contain an equivalent amount of chromium. Zinc cyanide baths and acid zinc baths present few problems, as zinc (even at 500 fold excess) does not interfere with the determination of copper, iron, lead and tin which are tolerable only up to 0.1 g l^{-1}. There is conversely an apparent need to destroy the cyanide complex of nickel when analysing a cadmium cyanide bath, although copper does not apparently suffer from cyanide depression [70]. The complex can be degraded with a mixture of sulphuric and perchloric acids. Silver plating solutions are usually made from solutions of silver in potassium cyanide and these are amenable to flame analysis provided the advice concerning silver acetylide given in section III is followed.

Some unusual interference effects in the determination of gold in plating solutions have been reported. These appear to arise from cyanide complexes and can be overcome by fuming with a sulphuric/perchloric acid mixture or evaporating an acidified (with hydrochloric acid) sample to dryness and taking up the residue in aqua regia. Several workers have reported extracting gold from cyanide waste solutions for flame determination. The gold(I) may be oxidised to gold(III) and the chloro-complex extracted from hydrochloric acid into MIBK [71]. The organic solvent can then be aspirated into an air/acetylene flame. Butler et al. [72] have reported the automation of a similar chemical treatment. Cobalt, copper, iron, nickel and zinc may be determined directly in an air/acetylene flame in gold cyanide plating baths [67] but for chromium and lead, standard addition methods are probably best. Rhodium plating baths contain up to 5 g l^{-1} of rhodium sulphate and while trace elements, e.g. cobalt, iron and nickel, can be determined without difficulty in the air/ acetylene flame [67] the rhodium content is best determined in the nitrous oxide/acetylene flame, provided an ionisation buffer is used.

Anode coatings used in chlorine/caustic soda cells may consist of baked layers containing ruthenium and iridium on a titanium substrate. These metals have been determined by fusing the coating with potassium hydroxide/potassium nitrate [73]. The melt was dissolved in hydrochloric acid and titanium added as a buffer for ruthenium (measurement at 349.9 nm) and potassium for iridium (measurement at 285.0 nm) to overcome signal depressions in the air/acetylene flame.

References pp. 440—445

VII. COSMETICS, DETERGENTS AND HOUSEHOLD PRODUCTS

Soaps consist essentially of organic compounds (see section II.B.2), detergents contain high concentrations of phosphates and sulphates, while cosmetics may contain further inorganic species. Although problems may arise from the presence of large amounts of phosphate and sulphate in flame analysis, where the use of lanthanum as a releasing agent may be essential, the presence of oxyanions is favourable for graphite furnace analysis. In a furnace, lengthy ashing times may be essential to remove non-specific absorption, arising from the pyrolysis of orthophosphates. Unless care is taken, this may result in loss of volatile metals.

Pardhan and Ottaway [74] have determined cadmium, chromium, copper, iron, lead and nickel down to $1\,\mu g\,g^{-1}$ in detergents by simple dissolution in water at a concentration of $1\,mg\,ml^{-1}$ and graphite furnace atomic absorption. The use of low detergent concentrations was essential to overcome background absorption interferences. These workers determined the same elements in soaps. In this case $1\,mg\,ml^{-1}$ sample solution was again used to prevent excessive spreading of the sample in the tube. It was found necessary to add orthophosphoric acid to the sample to overcome a substantial depression on the lead signal in the soap matrix. The high level of phosphorus in detergents means that it may be determined directly by flame atomisation [75]. The adulteration of virgin olive oil, used in soap manufacture, with refined or other oils can be detected by monitoring the sodium level [76]; the level is in the range $0.2-75.5\,\mu g\,g^{-1}$. The oil can be solubilised with absolute alcohol $(1 + 1)$ and diluted with butan-2-one. The standards were prepared from olive oil and sodium oleate. A linear range from 3—1000 ppm of sodium oleate in oil was reported using the air/acetylene flame.

Cosmetics and toiletries pose a variety of different analytical problems. The potentially violent reaction between lanoline and perchloric/nitric acid should be noted [45]. Drying of the sample is necessary to achieve a dry-weight basis result. Mario and Gerner [77] thoroughly dried hand lotion in a round bottom glass flask containing glass beads prior to the determination of silicon for another reason. After drying, benzene (100 ml) was added and this was filtered, a further aliquot of benzene (50 ml) was used to complete the extraction. The silicon can be determined in a nitrous oxide/acetylene flame using organic standards also dissolved in benzene. Zinc oxide and carbonate are frequent components of cosmetics and these raw materials can simply be dissolved in hydrochloric or nitric acid for the determination of toxic contaminants, such as cadmium, and zinc. Bismuth oxychloride is used in a number of preparations and Okamoto et al. [78] have determined lead in this, hair, deodorant and shaving cream sprays and lipstick. Spray (5 g for the hair and deodorant, 1 g for the shaving cream) was dissolved in ethanol (50 ml). The bismuth oxychloride (1 g or 5 g for low levels) was dissolved in 6 M hydrochloric acid (15 ml) and diluted to 100 ml with 0.5 M hydrochloric acid. The lipsticks were ashed at 500°C before extraction with first 20 ml and then 10 ml of 2 M hydrochloric acid. The lead content can be determined in an air/acetylene flame using background correction. For the spray samples, the standards should be dissolved in ethanol also. Bismuth may be present in lipstick in the range 0.50—8.5% and in eye-shadow in the range 12—14%, according to Gladney [79], who developed a simple carbon furnace procedure using nickel to prevent loss of bismuth. The sample was dissolved in hydrochloric/nitric/sulphuric acids and diluted with a nickel solution. A 50 µl aliquot was dried at 100°C for 30 s, charred at 1000°C for 40 s and atomised at 2400°C for 7 s.

Trace metals should be monitored in toothpastes for quality control and a particular check must be kept on lead levels. According to British Standard 5136 [80] there should not be more than 5 mg lead kg^{-1} in toothpaste. The B.S. involves slurrying the toothpaste (2 g) with ethanol (10—15 ml) and then evaporating the solvent. The dried paste is placed in a muffle furnace at 100°C and the temperature raised to 450°C in 50°C steps to avoid ignition. A few drops of concentrated nitric acid are added to the dark ashes and these are reheated at 450°C for 0.5—1 h. Then water (5 ml) and 5 M nitric acid (10 ml) are added, the mixture boiled for 5 min and filtered. The extraction is repeated a second time. For a flame atomic absorption finish, the pH is adjusted to 3—4 with ammonia solution, APDC solution added (2 ml of 1% w/v in acetone/water) and after 5 min MIBK added (10 ml). After shaking for 1 min, the organic layer may be aspirated into an air/acetylene flame with measurement at 283.3 nm. The standards should be prepared in the same way. Presumably, a similar approach could be used for other trace metals, although some workers prefer a simple acid extraction (e.g. 100 ml toothpaste with 10 ml of 1 N nitric acid, diluted to 100 ml before aspiration [81]).

Many other household products can be analysed in similar ways to those described above for chemicals. Household bleach is essentially an inorganic chemical. There has been concern expressed about mercury levels in hypochlorite bleach because of the way it is manufactured. The cold vapour reduction/aeration method referred to above is a good way of determining low mercury levels with minimal matrix problems [82]. In the past organomercurial compounds have been used (e.g. as bactericides) in some household products; these may be selectively determined by extraction with an organic solvent (e.g. carbon tetrachloride or benzene), and then application of the cold-vapour method following the addition of cysteine acetate, or by coupled gas chromatography/atomic absorption [83].

VIII. PHARMACEUTICAL PRODUCTS AND DRUGS

Most modern therapeutic drugs are organic compounds. Such pharmaceuticals may contain metallic ions as impurities, arising from starting materials, reagents, catalysts or contamination, and these have to be monitored in the products and raw materials to ensure that control limits are not exceeded. American [84], British [85] and European [86] Pharmacopoeias identify metals such as: aluminium, arsenic, barium, calcium, chromium, copper, iron, lead, lithium, magnesium, manganese, mercury, potassium, silver, sodium, palladium, selenium, strontium, tin, zinc and heavy metals, as potential impurities. Additionally, trace metal impurities may be indicative of the origin of preparations, this is of particular significance with regard to drugs of abuse. Some mineral salts and organometallic derivatives are of pharmacological interest and several metallic compounds are in pharmaceutical use. Elements which may be of therapeutic use include: aluminium, antimony, arsenic, barium, bismuth, calcium, chromium, copper, gold, iron, lithium, magnesium, mercury, potassium, silicon, silver, sodium, titanium and zinc. A large number of reports have also been made of indirect determination of organic compounds used in drugs via complexation and atomic absorption spectrometry. Thus there is considerable scope for atomic absorption applications in this field and clearly use in the pharmaceutical industry is increasing. Several informative reviews have recently appeared [87—91].

Very often the drugs are indirectly soluble in water, simple organic solvents or dilute acids. If an organic solvent is used, the standards must be prepared in the same solvent. Hydrochloric acid at concentrations similar to that found in the stomach is a preferred acid solvent. On occasion direct determination using in situ treatment in a graphite furnace (e.g. for oily injectable solutions) is used.

A. Determination of metallic concentrations of significance

Sodium and potassium can usually be determined with greater precision by flame atomic absorption than by atomic emission. These elements must be accurately and precisely determined in products such as infusion fluids and dialysis solutions [92, 93]. Samples should be adjusted to 0.1 to 20 μg ml^{-1} for sodium and 0.1 to 6 μg ml^{-1} for potassium and an air/acetylene, air/propane or entrained air/hydrogen flame used. A large excess (e.g. 1000 μg ml^{-1}) of an ionisation buffer such as cesium should be added to all standards and samples, as otherwise large errors may be observed [88]. Lithium must be determined in tablets of lithium carbonate which can simply be dissolved in 10% hydrochloric acid. Again an ionisation suppressor and the air/acetylene flame are generally used. In the treatment of certain mental illnesses, drugs containing rubidium or cesium may be administered. This can present a greater challenge as ionisation problems are again severe, 1000 μg ml^{-1} of potassium should be added to all standards and samples, and instrumental sensitivity at the resonance lines (rubidium 780.0, cesium 852.1 nm) may be exceedingly poor. A 'red-sensitive' photomultiplier should be fitted if problems are encountered.

Magnesium salts may be used for a variety of therapeutic purposes (e.g. the treatment of hypermotivity, spasmophilia), as well as the more well known uses of antacid activity (as the oxide or carbonate), as a purgative (as the hydroxide or sulphate) or as an excipient (as the stearate). Samples may be dissolved in hydrochloric acid, concentrated if necessary. In the case of preparations with a high organic content gentle ashing prior to dissolution may be useful. An air/acetylene flame and standard conditions can be used. Rousselet and Thuillier [87] recommend the use of a spectral buffer of lanthanum chloride ($50 \, g \, l^{-1}$), sodium chloride ($10 \, g \, l^{-1}$) and potassium chloride ($2 \, g \, l^{-1}$) which will eliminate chemical and ionisation interferences. Calcium salts are also widely used in tablets, syrups, suspensions and injections. Generally, these are soluble in hydrochloric acid and the use of the above buffer, or 5% lanthanum nitrate [94], with standard air/acetylene conditions is recommended. Calcium pantothenate, a B vitamin, may be determined via its calcium content at a level of 12 mg per capsule by the method of standard additions [88], provided calcium is not otherwise present in the formulation.

Anti-acids, astringents and antiseptic agents may contain a variety of aluminium salts. Organic salts, alumina, the hydroxide and phosphates may be attacked with concentrated hydrochloric acid and diluted to bring the aluminium concentration into the range $10-50 \, \mu g \, ml^{-1}$. Alternative procedures for antacids using hydrochloric/nitric acid [67] and extraction with 4 M hydrochloric acid [95] have been proposed. For silicates, the sample is best taken up in perchloric/hydrofluoric acid, evaporated to dryness to remove silica, and then the residue dissolved in warm hydrochloric acid [87]. In each case the nitrous oxide/acetylene flame is the preferred atom cell, and the method of standard additions may be used to minimise any errors arising from lateral diffusion.

Often the chemical speciation of iron, and hence its availability, is significant in iron supplements. Thus atomic absorption may be preceded by separation procedures [96]. To measure total iron in multi-vitamin preparations the sample can be taken up in hydrochloric acid and the determination made in an air/acetylene flame. Care must be taken with high mineral content samples to avoid inter-element interferences and the method of standard additions may be employed. Similar comments may be made about the determination of manganese and copper [97, 98], except that interferences are here less likely, but there may be a requirement for lower levels to be determined.

Vitamin B_{12} (cyanocobalamin) can be determined by its cobalt content (4.35%) and given the low concentrations ($0.2 \, \mu g$ cyanocobalamin per tablet has been reported [88]) and coloured solutions sometimes encountered, atomic absorption is a useful method for this determination. Care must obviously be taken if cobalt was used to prepare the vitamin, otherwise standard additions and a conventional air/acetylene flame can be used. If the concentration permits, the tablets may be dissolved in hot water, ethanol or preferably hydrochloric acid [99, 100]. Lower concentrations can be determined by graphite furnace [101, 102]. Peck [101] determined cobalt in the range 15—20 ng ml^{-1} on a 20 µl sample with a 30 s dry stage (150°C), 30 s char stage (675°C) and 40 s atomise stage (2700°C). Whitlock et al. [103] determined vitamin B_{12} in dry feeds by dissolving the sample in 200 ml, adding EDTA (5 g) and adjusting the pH to 7 with ammonia solution. The cobalt was then concentrated on charcoal (5 g) which was filtered and ashed. The ash was taken up in 5 N nitric acid (3 ml) before aspiration into an air/acetylene flame. Organic components present may enhance the absorbance of cobalt in the flame and either the method of standard additions or standards prepared from cyanocobalamin should be used. If necessary, solvent extraction can be used and the extraction of the 8-hydroxyquinolate into chloroform has been recommended [87].

Zinc is used in ointments and eye-lotions and is a constituent of different forms of insulin. In the former type of applications zinc oxide, zinc stearate and zinc undecanoate may be encountered in a variety of creams, ointments and pastes. Moody and Taylor [104] dissolved the residue from such samples after ether extraction (1 g in 5 ml ether) in concentrated hydrochloric acid. After dilution, the determination can be completed at 213.9 nm in the air/acetylene flame where interferences are not normally encountered. Various analytical techniques for determining zinc in insulin injections have been critically compared [105]; atomic absorption was preferred as being accurate, fast and precise. Spielhotz and Toralballa [106] reported a method capable of determining low levels of zinc in insulin. The sample (5 mg) was suspended in water (10 ml), 1 drop of 6 M hydrochloric acid was added to effect dissolution. After making up to 50 ml the determination was completed using an air/acetylene flame. Alternatively protamine insulin solution (1 ml) may be diluted to 50 ml after the addition of 1 drop of acid.

Although their use has recently come under scrutiny, a number of mercury compounds, both organic and inorganic, are used as diuretics, disinfectants and bacteriocides. Acid digestion may be used to release the mercury and the analysis completed via the cold vapour reduction/aeration method (see section II.A above). Unless great care is exercised, mercury may be lost during the hot stages of the digestion and Miller [88] has proposed digesting ointments and creams (100 mg) in nitric acid (3 ml) using a PTFE lined bomb for 1 h at 120°C. Organo-mercurial preservatives used in eye drops and multi-dose injections to prevent microbial growth (e.g. phenyl-mercuric acetate and nitrate, mersalyl and thiomersal) can be acid digested apparently without loss. When four digestion procedures were compared [107] a mixture of sulphuric/perchloric acid (5:2) was preferred as most rapid and interference free. Thompson and Hoffman [108] demonstrated recoveries in the range 98—103% when samples were digested with concentrated hydrochloric acid. Different optimal heating conditions were reported for different compounds. For ointments, an etheral dispersion may be extracted with 10% hydrochloric acid, The use of the cold vapour method here appears to be interference free.

The antirheumatic activity of gold salts ensures their continued therapeutic use. Preparations such as injectable solutions can be analysed by flame atomic absorption using air/acetylene and the 242.8 nm line. Standards can be prepared from tetrachloroauric acid. It seems likely that care is necessary both in storing solutions and adjusting flame conditions in order to obtain good results, as often reported in gold determinations. Certainly, conflicting successes have been reported for the analysis of sodium aurothiopropanol sulphonate [87, 109, 110]. Increasing use is being made of complex platinum salts in chemotherapy, e.g. cis-diaminodichloroplatinum. The drugs are generally water soluble and it is probably advisable to use a similar platinum complex for the standards if an air/acetylene flame is used. Alternatively, less problems of speciation will be encountered with the nitrous oxide/acetylene flame. In either case 265.9 nm is the recommended line.

Silicon may be present in various forms in some preparations, e.g. as 'tri-silicate' in antacids, polydimethyl-siloxane or silicone oil (see also section VII, ref. 77). The sample may be evaporated to dryness and the residue fused with sodium bicarbonate or another fusion agent, or taken up in hydrofluoric acid provided strong heating and loss of silicon is avoided. A nitrous oxide/acetylene flame must be used. Chromium may be found in disinfectants and antiseptics. Unless dilution of the sample is possible the use of the injection-cup technique (see section II.A) may be preferable as otherwise large amounts of corrosive salts such as sodium hypochlorite will be aspirated. If iron is also present it may be necessary to use a nitrous oxide/acetylene flame. Arsenic in arsenamide and lead arsenate preparations can be determined by boiling the sample in 5% nitric acid and aspiration of the sample [111]. Better sensitivity would be obtained using hydride generation (see section II.A).

B. Determination of impurities

The quality control of pharmaceuticals is particularly important. Care must be taken to limit the levels of toxic metals in the final product. The acid dissolution procedures described above (e.g. 6 M hydrochloric acid) are often equally applicable for the determination of impurities. Complete destruction of the matrix by wet oxidation or dry ashing may be necessary to obtain a completely independent method. Raw materials, catalysts, preparative equipment and containers are all possible sources of contamination. Lead, arsenic, mercury, copper, iron, zinc and several other metals may be subject to prescribed limits. Greater sensitivity is often required for lead and arsenic determinations and this can be achieved by electrothermal atomisation. Kovar et al. [112] brought samples into solution using 65% nitric acid under pressure at 170—180°C and, after adding ammonium and lanthanum nitrate, determined arsenic in the range 10—200 ng in a graphite furnace at 193.7 nm. Titanium dioxide pigments used in pharmaceuticals can be brought into concentrated sulphuric acid (10 ml) solution following fusion of 0.1 g pigment with 5 g potassium hydrogen sulphate and 10 mg glucose [113]. Solvent extraction concentration in MIBK was necessary for antimony determination.

While in Europe most limit tests use the method of standard additions, the United States Pharmacopoeia [84] requires that an aliquot of the test element equal to the set limit be added to the sample. If the response of the sample solution is less than the difference between the sample solution and sample plus aliquot (control) solution, it passes the test. Such a test may be used to limit sodium in other alkali salts. Atomic absorption spectrometry using the air/acetylene flame has been shown to be sufficient to test lead contamination in bismuth subcarbonate [114] and in zinc oxide and carbonate [115]. Miller [88] has reviewed such applications.

Palladium is employed as a catalyst in the manufacture of carbenicillin sodium, a semisynthetic penicillin. There is a limit of 25 μg ml^{-1} palladium in the final antibiotic. Atomic absorption, using air/acetylene and the 247.6 nm line, is the recommended procedure to test for this limit [85]. The powdered product can be formulated as a 10% aqueous solution but at such a high salt concentration, automatic background correction is strongly advised. Copper is another element used as a catalyst, e.g. in the preparation of anaesthetics and local analgesics. Since these agents are oxidisable by copper, a limit of 5 μg ml^{-1} is set. The sample should be dissolved in the minimum of hydrochloric acid (e.g. 0.4 g in 10 ml). Flame atomic absorption (air/acetylene, 324.7 nm) offers sufficient sensitivity for this analysis. Streptomycine has been reported to be contaminated by silicon during preparation [116]. This determination may be carried out by the method of standard additions using a nitrous oxide/acetylene flame, following dilution of the streptomycine or dissolution by the fusion procedure described in section VIII.A. Lithium, aluminium and osmium are other possible impurities which may be introduced into drugs during synthesis and which may be monitored by atomic absorption. Pro and Brunelle [117] have been able to detect the origin of illicit heroin by trace metal analysis.

Remarkably little use has been made of electrothermal atomisation in pharmaceutical analysis despite its probably ready applicability. Barium is a favourable element for furnace atomisation as this offers the simplest way of overcoming calcium interferences. The determination of barium in products intended for parenteral administration has been reported [118]. The drug was diluted and injected directly into a pyrolytic carbon-coated tube. The programme was dry 40 s (125°C), char 15 s (1000°C), atomise 4 s (2900°C). Levels as low as 10 ng ml^{-1} of barium, contaminant from glass, plasticisers (e.g. barium stearate) or fillers (barium sulphate), could be determined. Drug stabilising agents such as sodium hydrogen sulphite and sodium chloride enhanced the signal, but errors were avoided by adding these to the standards as well.

TABLE 3

INDIRECT METHODS FOR THE DETERMINATION OF DRUGS

Drug or compounds	Analyte metal	Notes	Reference
Alkaloids (e.g. brucine, strychnine)	Mo	Extract phosphomolybdate complex into MIBK	119—122
Amino-acids	Cu	Extract ion association complex into MIBK	123—124
p-Aminobenzoic acid	Cu	Extract ion association complex into MIBK	125
Azepines	Cu	Extract ion association complex into MIBK	126
Barbiturates	Cu	Precipitate complex, redissolve	127
Benzylpenicillin	Cd	Extract ion association complex into nitrobenzene	128
Chinoform (clioquinol)	Zn	Complex formed in MIBK	129
Chlorpheniramine maleate	Cu	Extract complex into chloroform	130
Chorprothixene	Cr	Precipitate reineckate	131
Ethambutol	Cu	Extract complex with ketone	132
Flufenamic acid	Cu	Extract complex with isopropyl acetate	133
Folic acid (vitamin M)	Ni	Oxidise, extract ion association complex into MIBK	134
Halogen containing compounds	Ag	Acidify, precipitate Ag halide	135
Isonicotinyl hydrazine (isoniazid)	Cu	Extract complex into MIBK	136
Methamphetamine hydrochloride	Bi	Precipitate complex	137
Noscapine	Cr	Extract reineckate into chloroform	138
Quinoline	Zn	Extract chelate into MIBK	129
Sodium EDTA	Ni	Add Ni, precipitate excess	139
Vitamin B_1	Pb	Desulphurise with plumbite	140

C. Indirect methods

Many modern drugs form extractable complexes with metals. This has been exploited in a large number of published indirect methods for determining such drugs. While the final measurement of the metals so complexed provides a rapid and simple finish, many of the methods are likely to be non-specific. Often related drugs will react with the same metal, and some metals, such as copper, have indeed been advocated for many different species. Careful control of pH must be exercised during complexation to optimise recoveries and selectivity. Thus, such methods are often of academic rather than practical interest. If however the identity of the drug is known, either via production information or specific tests, and interferents are absent, such tests may have a role to play when speed and simplicity are of the essence. With these comments in mind several of the published procedures are presented in Table 3.

IX. FUNGICIDES

Organometallic-containing fungicides are used in a number of treatments. Given the associated toxicity problems close analytical control is advisable. Wood preservatives may contain copper, chromium and arsenic compounds as part of their formulation. Atomic absorption procedures are recommended in British Standard 5666 [141] for the analysis of wood preservatives and treated timber. Arsenic, chromium and copper can be leached from the wood or the preservative after digestion with sulphuric acid/hydrogen peroxide at 75°C. After filtration, sodium sulphate solution is added and chromium determined in a nitrous oxide/acetylene flame, arsenic in an argon/hydrogen entrained air flame and copper in an air/acetylene flame. Fungicides (e.g. zineb, ziram, ferbam and maneb) containing zinc, iron and manganese were analysed by Gudzinowicz and Luciano [142]. The fungicides were acid hydrolysed with hydrochloric acid, which proved to be as effective as extraction with pyridine in chloroform. Standards were prepared from the corresponding metal cyclohexanebutyrates. Some broad spectrum turf fungicides contain cadmium at the 0.4—12% level, so that when 1—4 g of the fungicide is digested with a few millilitres of hot concentrated nitric acid and made up to 100 ml, the level in the analysis solution is typically 100—500 $\mu g\,ml^{-1}$. In their method, Jung and Clarke [143] overcame the problem of such high concentrations being several hundred times above the normal working range by using the 326.1 nm line which is about one thousand times less sensitive.

X. PAINTS AND PIGMENTS

Metals may be introduced into paints as the principal constituents in pigments, in drying oils, in pesticides or antifouling compounds and as markers in security paints. The need to monitor metal levels arises from the effects trace metals may have upon the colour or quality of the paint, its toxicity or in forensic applications. Many countries have legislation specifying the maximum permissible levels for toxic elements in toys. The British Toy (Safety) Regulations 1974 [144] sets dry weight limits in paint or varnishes on toys for soluble barium of 500 ppm, soluble antimony and

and chromium of 250 ppm, soluble arsenic, cadmium and mercury of 100 ppm and the limit for total lead is 2500 ppm. The statutory procedure for determining soluble metal content is intended to approximate to the dissolving power of human gastric juices. A finely powdered sample is mixed with 50 times its weight of hydrochloric acid (0.07 M) at room temperature (20—22°C). The solution is shaken for an hour and then left to stand for the same time, before filtration and then analysis which may be by flame atomic absorption. Mercury and arsenic are probably best determined in the filtrate by cold-vapour reduction/aeration and hydride generation respectively. Very often total metal, rather than soluble metal information is required. Dissolution of certain paint types is difficult and the various possible approaches have been reviewed [67, 145, 146].

Oils and thinners may be diluted with white spirit or MIBK, usually by at least a factor of 10 to promote efficient nebulisation. Metal napthenates can be used as standards [67]. Paints as well as the oils above may be diluted with methyl isobutyl ketone (MIBK) for direct injection into a graphite furnace.

Many paints require vigorous pre-treatment before the matrix can be destroyed. Even after considerable oxidation it may be necessary to filter off titanium dioxide. Some paints are, however, more amenable to acid dissolution. Eider [147] reported that it was possible to determine barium, cadmium, calcium, cobalt, lead, magnesium, manganese, mercury, tin, zinc and zirconium in vinyl additives and paint (1 g) by dissolution in hot concentrated nitric acid (10 ml). The mixture was filtered and made up to 100 ml. Liquid samples were again dissolved in MIBK (1 g in 100 ml). Air/acetylene and nitrous oxide/acetylene flames were used to complete the analysis as appropriate. Porter [148] reported losses of lead during the dry ashing of alkyd and latex paints and therefore recommended dissolution in either nitric/perchloric acids or nitric acid alone.

Most popular procedures involve an initial ashing of the paint at moderate temperatures followed by acid digestion. Given the lengthy and sometimes vigorous wet digestion procedures often required for paints, my experience is that a combined dry and wet ashing procedure is less time consuming and more effective. The ashing temperature should not be allowed to rise above 450°C to avoid losses of volatile elements. Paint from objects may be scraped off, using a sharp scalpel, into a pre-weighed silica dish. After weighing, typically 100—500 mg, a similar amount of sodium carbonate as moderator may be added [149]. The dishes should be heated in 50° steps to 450°C in a muffle furnace. After cooling, attempt to take the ash up in concentrated hydrochloric (1 ml) and nitric (1 ml) acids, if necessary 1—3 ml of perchloric acid along with further nitric acid may be added. The resultant clear solution may be made up to 25 ml with dilute hydrochloric acid. Most metals of interest, the obvious exception being mercury, may be determined by flame atomic absorption on the final solution. Pyrolysis of the paint and collecting the vapours in a strongly oxidising solution of

References pp. 440—445

permanganate/sulphuric acid is probably the best approach for mercury [150], with cold-vapour reduction/aeration finish. In the development of National Bureau of Standards SRM 1579, lead was determined in paint scrapes at percentage levels by ashing at 450°C/550°C, with subsequent extraction of the ash with acids and ammonium acetate for a flame finish [151]. Nitric acid and ammonium acetate have been recommended for leaching lead and cadmium from such extracts [152]. The oxygen flask method has been proposed as an alternative approach for micro-samples [145].

Refractory oxides in pigments may still cause some problems. Titanium dioxide pigments can be fused with potassium bisulphate and glucose prior to dissolution in sulphuric acid [153]. Chromium in pigments from art paintings has been determined by flame atomic absorption following fusion with sodium peroxide/potassium carbonate/sodium carbonate (2:1:1) and extraction with hydrochloric acid [154]. It was found useful to treat the sample with hydrofluoric acid to remove silica. Noga [155] found that when chromium in coatings was determined by fusing chromium trioxide with sodium carbonate/sodium metaborate (2:1), samples with inconveniently high salt contents were obtained. Consequently they ashed the paint (2—3 g of latex or oil-base) in an aluminium dish at 450°C. The ash was digested in a pressure bomb with 5 ml of an oxidising solution (0.2 g potassium permanganate in 100 ml 50% sulphuric acid). The excess permanganate after digestion was removed with a 0.1% w/v sodium azide solution prior to analysis in a nitrous oxide/acetylene flame.

Price's review [145] gives an excellent account of the mode of occurrence of metals in major constituents in paints and some recommended operating conditions for flame atomic absorption. It may be increasingly necessary to add to the list more "exotic" metals because of their growing use as markers in security paints.

If only analysis of the pigments is desired the sample may be dissolved in an organic solvent. Centrifugation allows the collection of the insoluble pigments for subsequent acid dissolution or alkali fusion. If necessary solvent extraction may be used to remove the trace metals from salts introduced during fusion. The determination of pigments in coloured magazines is described in section XI. Lead has been determined in wax crayons by direct insertion of sample (0.15 mg) into a graphite furnace [156]. The programme used was dry 100°C (60 s), char 500°C (120 s to eliminate all smoke), atomise 2300°C (10 s). The results using the 283.3 nm line compared well with a more conventional procedure of ashing at 500°C and taking up in nitric acid. Pellerin and Goulle [157] have determined cadmium, copper, lead and zinc in dyes and antioxidants, primarily used for drugs and foodstuffs. The sample was wet ashed with sulphuric acid and dissolved in nitric acid or nitric acid/ethanol. Either a graphite furnace or an air/acetylene flame was used as appropriate to the levels required.

Very often a rapid answer is required as to whether a paint does or does not contain high levels of lead. The above methods lack rapidity and there is a place for a rapid screening method, to be followed if necessary by a more precise method. Use of the microsampling cup technique first described by Delves [158] has therefore been reported by several workers [159—162]. Henn [159] used a MIBK-paint or water-paint slurry for oil- and water-based paints respectively. The sample (20 μl of a 1% slurry or less to minimise background absorption) was pipetted into the nickel cups which were dried and inserted into the flame in the conventional microsampling procedure. Standards were prepared from an oil standard diluted with MIBK. The reported detection limit was 5 μg g^{-1}. Mitchell et al. [160] scraped the paint (5—7 mg) directly into the nickel cups which were weighed and then inserted into the flame. The insensitive 287.3 nm lead line was used to enable determinations in the region of legal significance without dilution. Holak [163] reported a spot test using a sample on a nichrome wire rather than in a cup. If the result proved positive (i.e. the level of lead gave rise to concern), the analysis was repeated using nitric acid digestion in a PTFE lined bomb and determination using the sampling boat procedure. A variation of this method was to extract the paint with nitric acid and place a 10 μl sample on a platinum/rhodium loop [164]. The solvent was first evaporated electrically before insertion into an air/acetylene flame.

XI. PAPER AND PULP

Paper and pulp, being based on plant materials represent particularly favourable matrices for atomic absorption analysis by flame or by furnace. Such applications have been reviewed in the specialist literature [165—168]. Differences of opinion exist as to whether wet or dry ashing is to be preferred for flame analysis, but increasingly, paper samples are being added to graphite furnaces without any sample pre-treatment.

A method has been reported for the determination of calcium, copper, iron, magnesium, potassium, sodium and zinc in cellulose [169]. The sample (10 g) was air-dried and then ashed at 575°C until all the carbon was removed. Hydrochloric acid (5 ml of 6 M) was added to the residue and evaporated to dryness twice before taking up the sample in a third aliquot, diluting to 100 ml and aspiration into an air/acetylene flame. It is likely that volatile elements such as cadmium may be lost at such an elevated ashing temperature and temperatures below 500°C may be preferable. Alternatively wet ashing with nitric acid has been proposed for the determination of aluminium, cadmium, potassium and zinc in pressed boards [170] or sodium in gypsum glass board [171]. For the determination of lead in confection wrappers, the sample may be treated with concentrated nitric acid at 70—80°C and diluted for flame analysis [172]. In the full method, the wrapper was wiped clean with a damp tissue, cut up to 0.5 × 0.5 mm pieces and dried at 110°C (for paper, for plastic 80°C) for 1 h. The sample (0.5 g) was heated with concentrated nitric acid (1 ml) at

References pp. 440—445

90—100°C for 10—20 min without drying. Ammonium nitrate (1 ml of 0.1 M) and aluminium nitrate (1 ml of 0.1 M) were added, the mixture heated at 110°C for 2 h until dry. The residue was ashed for 12 h at 450°C and the ash dissolved in hot nitric acid (1 ml). Lead levels up to 30000 $\mu g\,g^{-1}$ were found in wrappers with yellow or orange pigments. High toxic-metal levels may also be found in coloured magazines and even children's comics. A sample of the paper may be dried, ground and extracted with warm hydrochloric acid to simulate gastric juices. In this way barium has been determined in the extract using the nitrous oxide/acetylene flame with potassium chloride added as ionisation suppressor, and cadmium, chromium and lead using an air/acetylene flame [173].

The direct determination of trace metals in paper and paper products following solid sample insertion into the graphite furnace has been reported by several workers [174—177]. In the method of Kerber et al. [174] copper, iron, manganese and silicon were determined in milligram quantities of finished paper. A furnace programme of dry (0 s), char (80 s at 1100°C) and atomise (8 s at 2400°C/2650°C for silicon only) was used. The agreement with results obtained by flame atomic absorption was good. A high frequency induction furnace has been used for the oxygen free ashing and then atomisation of samples of pulp and paper [175]. Interest was centred on metals of concern to users such as the food and photographic industry (e.g. cadmium, copper, lead and manganese), but pulp for photographic, letterhead, chromatographic, filter and grease-proof paper were studied. Discs (5.5 mm diameter) were punched from the paper and the furnace programme was: dry (60 s at 80°C), ash (60 s at 300°C for cadmium and lead, 460°C for copper and manganese); atomise (30 s at 1500°C for cadmium, 1600°C for manganese, 1900°C for copper and lead). Pattern recognition techniques have been used to identify 16 different papers following acquisition of data on 10 elements by atomic absorption [177]. For antimony, cadmium, chromium, cobalt, copper, lead, magnesium and silver, discs (3 mm diameter, of 150 punched the first 10 were discarded to preclude contamination) were dried and ashed directly in a mini-furnace (at 550°C for cadmium, lead and silver, at 800°C for antimony, at 1100°C for the other elements). Iron and magnesium were determined by ashing larger pieces (5 cm^2) at 260°C overnight and injecting aliquots of this ash dissolved in concentrated nitric acid (5 ml).

XII. TEXTILES, FIBRES AND LEATHER

There is interest in trace metal levels in both natural and synthetic fibres and fabrics but perhaps most interest is in synthetic fibres as these may contain residues of catalysts, treatments or stabilising agents. Reviews have been published of trace-metal analysis of rayon, polyamide, polyester and polypropylene fibres [178] and of cotton fabrics, especially for flame retardants [179]. Some elements may be determined by simple acid extraction or by ashing the fibres and dissolving the ash in hydrochloric or nitric acid. The latter procedure is especially useful for synthetic fibres. Certain synthetic fibres can be dissolved in an organic solvent which may be sprayed directly into a flame provided it does not harm the spray chamber and the standards are made up in the same solvent, e.g. cellulose acetate in MIBK, polyacrylonitrile in dimethyl formamide, nylon (polyamide) in formic acid, wool in 5% sodium hydroxide, cotton and cellulose in 72% sulphuric acid.

Price [180] has reported a special wet digestor for synthetic fibres. The sample (5—10 g) was shredded, charred and digested in the apparatus with concentrated sulphuric acid (10 ml). Hydrogen peroxide or nitric acid was added to complete the decomposition. Antimony, chromium, copper, iron, lead, manganese, tin and zinc can then be determined using normal flame conditions. To determine gold the sample was ashed at 700°C, dissolved in hydrochloric acid (10 ml of 6 M) and concentrated nitric acid (0.5 ml). The latter acid was fumed off and the chlorocomplex extracted into MIBK for flame analysis. Slavin's procedure [181] involved ashing and dissolution in nitric acid (1 + 3 with water). In this way, 2 g of fibre were brought into 10 ml solution to determine not only catalyst residues but also blend ratios. Alder and Bucklow [182] were able to determine chromium, copper and zinc in carbon cloth without ashing. The sample was ground to less than 200 μm, shaken with 2% nitric acid and 0.1% sodium hexametaphosphate. The resultant suspension was sprayed into a flame. The present author has determined trace iron, a catalyst residue, in carbon fibre using electrothermal atomisation [183]. The sample was ashed in an r.f. excited oxygen plasma asher and the residue taken up in the minimum of acid for greatest sensitivity.

Trace metals can be leached out of some natural fibres. To determine copper in webbing, rope and cotton duck, Simonian [184] warmed, to near boiling, 2—3 g fabric with 80—90 ml of 0.5 M hydrochloric acid. The sample was filtered and made up to 200 ml before aspiration into an air/acetylene flame.

Metal salts may be used in the treatment of wool. Flame methods for the determination of aluminium [185], barium, chromium, copper, mercury, strontium, tin, zinc [186] and zirconium [187] in wool have been published. Standard additions to wool cleaned by soaking and washing it with disodium EDTA (800 ml of 0.5 M for 30 g wool with soaking for 3 days and double washing) was used as the calibration technique. This compensated for interferences from hydrochloric acid and amino-acids. The samples were equilibrated to a constant humidity for 24 h and then 0.3 g sealed with 5 ml of constant boiling point hydrochloric acid in a glass tube. The tubes were placed in an oven at 110°C for 20 h. The nitrous oxide/acetylene flame was used for the determination of aluminium and zirconium. Sulphate, phosphate, citrate and silicate have been found to interfere in the determination of titanium and zirconium in fire-proofed wool [188]. These flame

References pp. 440—445

interferences were overcome by the use of excess iron(III) as releasing agent. Alternatively solutions may be prepared from wool samples (0.5 g) using sodium hydroxide solution (15 ml of 5% w/v) [67] although some interference from sodium may be encountered.

The long established use of chromium salts in the tanning of leather has been reflected in interest in the determination of chromium in leather extracts. Generally the leather is leached with acids or complexing agents such as oxine to remove the metal salts. Della Monica and McDowell [189] recommended leather strips be air dried at 80°C for 24 h and ground to pass a 10 mesh screen. The powder was vacuum dried at 70°C for 16 h and then 100 mg refluxed with 25 ml 2 M hydrochloric acid. The filtered acid was aspirated into a flame. For chromium a fuel-rich air/acetylene flame or a nitrous oxide/acetylene flame is to be recommended.

XIII. POLYMERS AND RUBBER

Many of the remarks made in the previous section concerning fibres can be applied to the analysis of plastics. Some polymers are soluble in organic solvents and samples may be prepared for direct aspiration into a flame in this way, e.g. MIBK is a suitable solvent for polyesters, polystyrene, polysiloxanes, cellulose acetate and butyrate; dimethyl formamide for polyacrylonitrile; dimethyl acetamide for polycarbonates and polyvinyl chloride; cyclohexanone for polyvinyl chloride and polyvinyl acetate; formic acid for polyamides; and methanol for polyethers. These organic solutions may alternatively be injected into a graphite furnace. Otherwise, polymers may be wet or dry ashed and the resultant ash dissolved in acid. An approach which is attracting increasing interest is the direct insertion of solid samples into a graphite furnace.

Oliver [190] recommends the dissolution of the polymer if possible (see above) but in other cases a wet ashing procedure was used. The sample was heated with 2—3 ml of concentrated sulphuric acid and then hydrogen peroxide added drop-wise until the organic matter was destroyed. Twenty elements were determined in a 2% solution of polymer. Polymers may be dispersed in an organic solvent and trace metals removed by leaching with an appropriate aqueous solution, preferably the procedure should be repeated more than once to ensure complete extraction. To determine antimony in fire-retardant polypropylene, the sample was dispersed in xylene and extracted with 6 M hydrochloric acid under reflux [191]. The filtered acid layer was combined with two further extracts prior to aspiration into the air/acetylene flame and measurement at 217.6 nm. Martinie and Schilt [45] reported that nylon would dissolve completely in perchloric/nitric acid digestion but potentially explosive problems were encountered in the dissolution of Amberlite resins and rubber.

Price [67] reported a method for the dissolution of polyvinyl chloride

(1 g) in perchloric acid (20 ml). After gentle heating, the sample was evaporated to fumes, and 5% lanthanum chloride solution (20 ml) added. Calcium and magnesium were determined in the air/acetylene flame on the filtrate. Any residue was fused with potassium hydroxide (5 g) and dissolved in hydrochloric acid (5 ml).

Extractable metals in plastics (e.g. those used as kitchen utensils) can be determined following a suitable acid leach. Cadmium has been determined in this way using either an ethanol/acetic acid/water (10 + 3 + 87) or diethyl ether/4% acetic acid extractant [192]. The extract was evaporated to dryness and redissolved in 1 M nitric acid prior to aspiration into an air/acetylene flame and measured at 228.8 nm.

Dry ashing at elevated temperatures is a more rapid procedure for involatile element determinations. A variety of acids, but principally hydrochloric and nitric, have been used to take up the ash. Kometani [193] has studied the effect of temperature on volatilisation losses of alkali salts during the ashing of a tetrafluoroethylene fluorocarbon resin. PTFE cannot be wet digested but starts to decompose at 450°C in air. A temperature of 525°C is needed to remove all the carbon. Sodium was lost at even moderate temperatures, but it appeared possible to ash up to 500°C without losses of most group I or II elements. Aluminium, iron and titanium were measured in polypropylene following ashing of 10 g of sample in a platinum crucible [194]. The crucible was heated slowly at first and then at 800°C for 30 min, 0.9 g sodium carbonate was added and the residue fused. The residue was taken up in 7 ml of 3 N sulphuric acid and made up to 10 ml with the same acid. Similar concentrations of sodium carbonate and sulphuric acid were added to the standards. A quartz vessel has been used for the controlled temperature ashing of polyamide resins [195]. Matsuo [196] has reported the use of the oxygen flask combustion method to bring plastics into solution. Other ashing procedures which can be used include low temperature ashing with radio-frequency excited oxygen plasmas.

Solid sampling has many advantages including minimal sample pretreatment times, minimal errors or contamination and avoidance of the dilution associated with dissolution. These advantages are particularly useful in polymer analysis where low levels may be significant and the large molecules involved may be difficult, or slow, to dissolve. Hence the considerable interest in the direct insertion of solid samples of plastic into graphite furnace atomisers. Henn [197] inserted samples of polyacrylamides (up to 5 mg) directly into a tube furnace with a tantalum sampling spoon. Chromium, copper and iron were determined down to the 0.01 $\mu g\,g^{-1}$ level using a programme of dry 20 s (125°C); char 60 s (Cr 1350°C, Cu 700°C, Fe 1000°C); atomise 15 s (Cr 2700°C, Cu and Fe 2500°C). Background correction was used along with the chromium 357.9 nm, copper 324.7 nm and iron 248.3 nm lines. Good agreement was found with a dissolution method, in which the polymers were dissolved in 0.16 M nitric acid, to give a 0.25 to 1% solution, and a 50 μl aliquot pipetted into the furnace, except that solid sampling gave results about 10% lower than by dissolution. Since this difference was not due to contamination it must be concluded that slightly lower sensitivity was found with the solid samples. Kerber et al. [174] have used a similar approach to determine aluminium, copper and iron in plastics. Limits set on the level of lead in pharmaceutical containers can be checked in a similar way [198]. A piece of washed container was cut, weighed and inserted in a carbon cup atomiser. The sample was ashed at 800°C and atomised at 1600°C, prior to lead determination in the range 0.03--1 ppm. Aqueous calibration proved possible but variations in background absorption signals required modifications to the furnace programme for different plastics. Two or three stage ashing was found necessary and probably ramp modes would be even more useful. Particular problems were encountered with polyvinyl chloride and polypropylene where an atomisation temperature of 1800°C was used.

Rubbers and adhesives are amenable to most of the treatments discussed above. Zinc can be extracted from vulcanisates with ethylacetoacetate and determined at 213.9 nm using an air/acetylene flame [199]. By heating rubber mixtures and adhesives first at 105°C and then to constant weight at 800°C, samples have been analysed for cobalt, copper, iron, manganese, nickel and zinc [200]. The residue was fused with sodium carbonate/borax (2:1), dissolved in hydrochloric acid (1:1) and an air/acetylene flame used. Selenium and tellurium used as organic accelerators, in the vulcanisation process for natural and synthetic rubbers, can be extracted with hydrochloric acid followed by nitric acid [201]. After boiling, cooling and dilution, the leachates were sprayed into an air/acetylene flame. Since tin(II) nitrate is added to the tyre pre-dip of resorcinol-formaldehyde adhesive a novel method for the monitoring of tyre cord-dip pick-up has been reported [202]. The dip was analysed for tin using a graphite furnace.

XIV. NUCLEAR INDUSTRY

Atomic absorption is being increasingly applied in the energy industries including those associated with nuclear energy. While radiochemical monitoring is frequently used, several metals at the trace level may be best monitored by atomic absorption. Particular problems arise when the matrix is radioactive and stringent safety precautions may be required. Electrothermal atomisation in an enclosed system, such as a glove box, is preferable to flame atomisation. The determination of daughter nuclides in fuels, or fuel processing liquors, is one potential application. In view of the high radioactivity of the solutions and low levels of metals of interest either selective solvent extraction or ion exchange, to separate the analyte metals from the uranium or plutonium matrix, is advisable. An example of the latter approach is the determination of lead in uranium products [203] by adsorption on Dowex 1—X8 from 2 M hydrobromic acid. The lead was eluted with

6 M hydrochloric acid prior to flame atomic absorption. Walker and Vita [204] determined aluminium, cadmium, calcium, chromium, cobalt, copper, iron, lead, magnesium, manganese, sodium, potassium, nickel and zinc in uranium compounds after removing the uranium with n-tributylphosphate in carbon tetrachloride. Uranic oxide, uranium trioxide and uranium dioxide (10 g) were dissolved in 6 M nitric acid (100 ml). It was necessary to fume with perchloric acid to dissolve any uranium—chromium compounds. Uranium hexafluoride, oxyfluoride and tetrafluoride were hydrolysed in water before being dissolved as the oxides. The aqueous residue after the solvent extraction was evaporated to dryness and taken up in 0.2 M hydrochloric acid prior to flame analysis. As well as overcoming problems from radioisotopes such as uranium-235 and concentrating the samples, these methods overcome problems from background absorption, varying uptake rates and occlusions which might be encountered in the presence of matrix. Ruthenium has been determined in nuclear waste materials at sub-nanogram levels by solvent extraction with APDC in amyl acetate and using a graphite furnace [205]. Most matrices (e.g. condensates, scrubber solutions, waste feeds) were amenable to simple hydrochloric acid treatment but solid waste was fused with sodium peroxide. An atomisation temperatue of $2800°C$ and the 349.9 nm line were used.

The "burn-up" of boron containing neutron absorbers can be monitored by determining the increasing lithium concentration (actually ^{7}Li) by atomic absorption at the 670.7 nm line. Samples (1—10 mg) have been digested with nitric acid in a micro-Carius tube at $250°C$ [206] and the lithium thus extracted.

XV. WASTE MATERIALS

While suitable sites remain, landfill will be a popular method for the disposal of waste materials. Provided proper precautions are taken, a "dilute and disperse" policy can be justified, not only as inexpensive but as relatively harmless. Indeed, environmental objections can easily be raised against alternative procedures such as incineration and dumping at sea. Toxic metals form perhaps the largest group of potential hazards, e.g. as regards leaching or from stack gases. Peck [207] has compared various digestion procedures for refuse. In the proposed method, eleven elements were determined (aluminium, calcium, copper, iron, lead, magnesium, nickel, potassium, sodium, tin and zinc) by flame atomic absorption. Domestic refuse was ground and blended before ashing (6 g) at $480°C$. The ash was digested with hydrofluoric/nitric acid (3:2) followed by nitric/perchloric acid (5:2) with heating to fuming on each occasion. The residue was dissolved in hydrochloric acid and made up to 50 ml. Unfortunately, a number of volatile elements will be lost in the above procedure and we have found an alternative proposed by Haynes [208] to be an extremely useful ashing procedure for elements such as antimony, arsenic and selenium prior to electrothermal atomisation. In the determination of arsenic and antimony in combustible domestic refuse, magnesium nitrate was used as an ashing aid and nickel nitrate as a volatilisation suppressant. To the sample (2 g) was added 15 ml of 10% magnesium nitrate and 5 ml of 25% nickel nitrate with 3—5 ml of 95% ethanol as a wetting agent. After drying at 120°C for 2—3 h and 160°C for 4 h, the mixture was ashed at 450—500°C for 4 h. Water (5—10 ml), nitric acid (10 ml) and hydrofluoric acid (1—3 ml) were added and the mixture heated to dryness. The residue was carefully taken up in nitric acid (10 ml) and hydrogen peroxide (10 ml of 30%), filtered and subjected to electrothermal atomisation. Recoveries of 98 to 95% were reported. The first stage of this digestion can successfully be modified to the direct determination of arsenic in solid samples added to the furnace as slurries in magnesium and nickel nitrate [209].

XVI. ARCHAEOLOGY

One of the lesser known areas where atomic absorption is proving to be particularly useful is archaeology. It is used in survey work (e.g. to identify the sources of objects and trade routes, to identify the raw materials used, to study ancient technology) and in order to detect forgeries. The role of atomic absorption in these areas has been reviewed [210, 211]. While non-destructive techniques are clearly preferred in this area of work, the high sensitivity and low sample requirements of atomic absorption commend themselves, particularly when it is recalled that techniques such as X-ray fluorescence require clean surfaces. Drilling a 1 mm diameter hole may well cause less damage than removing corrosion layers, or a patina, of several mm^2.

Atomic absorption is suitable for the analysis of several types of ancient inorganic materials, e.g. metals and alloys, silicates and minerals. Only a few milligrams of sample are required; typically 10 mg may be dissolved in 25 ml for analysis. Electrothermal methods may require even less sample and are thus very attractive in this field. Often papers describing results obtained by atomic absorption give little or no analytical details. Table 4 lists some of these publications to illustrate the potential scope of atomic absorption spectrometry in archaeology, but the review of Hughes et al. [210] remains the best source of experimental detail.

The analysis of copper alloys provides a good example of some the problems encountered in archaeological analysis. Sampling presents the major difficulty, not only must the sample taken be small (10 mg or less), but problems of inhomogeneity arising from segregation during primitive casting, surface embellishments and corrosion, may be encountered. Workers at the British Museum used a portable hand-held drill which operates off 12 V batteries to collect only the shiny inner metal turnings [210]. Size 60

TABLE 4

ARCHAEOLOGICAL APPLICATIONS OF ATOMIC ABSORPTION SPECTROMETRY

Material examined	Reference
Ancient bronzes	211, 212
Ancient glasses	213, 214
Ceramics	210
Copper ores and artifacts	215
Flint mine products	216, 217
Glass blowing tubes	218, 219
Gold coinage	220
Gold torcs	221
Iron age axes	222
Iron age bronze collars	223
Iron swords	224
Roman horse trappings	225
Silver coinage	226
Stained glass	227, 228

drill bits of hardened steel were used. To obtain 10 mg, about 5 mm penetration was required. Samples can usually be dissolved readily in 1 ml aqua regia (1 nitric acid + 3 hydrochloric acid) with gentle heating to 60°C to dissolve tin. After cooling, 1 ml more of aqua regia may be added and then the solution made up to 25 ml with distilled water. Some enhancement by copper of lead, silver, tin and zinc flame absorbances has been reported [210] and, therefore, appropriate amounts of copper should be added to the standards. Normal air/acetylene operating conditions are suitable for the determination of antimony, bismuth, cobalt, nickel, silver, tin in brasses and bronzes and also for copper and lead, but the use of insensitive lines (e.g. copper 244.2 nm) may be necessary to avoid sample dilution. For zinc, the sensitivity usually requires reduction by burner rotation. The use of the injection-cup technique (see section II.A) will help to conserve the sample. It may be necessary to use a graphite furnace to obtain sufficient sensitivity for arsenic, especially considering that a copper matrix is favourable for the electrothermal atomisation of arsenic but likely to cause large signal suppressions in the hydride generation technique.

Lead alloys (50 mg for flame analysis, 10 mg for furnace atomisation) can be dissolved in 1 ml of distilled water with 0.5 ml nitric acid added. After warming, 5 ml of concentrated hydrochloric acid is added to dissolve the tin content, any precipitated lead chloride may be removed by centrifugation. Nickel, silver, tin and zinc may be determined in this solution using a conventional air/acetylene flame. Silver is usually the element of most interest in lead, but the proportion of tin in pewter is significant. The lead content can be determined in the original nitric acid solution but it is difficult to measure tin and lead in the same solution.

Silver alloys present similar problems and the silver content can be determined after dissolution of 5—10 mg in 1 ml of 50% nitric acid. The tin and gold present will appear as a black residue which may be taken up in 1.5 ml concentrated hydrochloric acid, which of course precipitates silver chloride. The precipitate is removed by centrifugation. Thus again two solvents are required to determine the normal range of elements, e.g. bismuth, copper, gold, lead, silver and tin.

In the case of gold alloys, samples are presented in a form best suited to sampling by scraping off 2—4 mg of shavings. These may be taken up by warming to 60°C with 1 ml of concentrated nitric acid and 0.5 ml distilled water, 3 ml of concentrated hydrochloric acid is added with warming until the gold starts to dissolve when a further 3 ml hydrochloric acid is added. After dissolution a further 6 ml of hydrochloric acid is added, to ensure retention of gold and silver as the chlorocomplexes, and then the sample made up to 25 ml. Copper, gold and silver can conveniently be determined in this solution, remembering that if dilution is necessary the hydrochloric acid level must be maintained.

Iron is best sampled by drillings (10 mg) which may be dissolved in aqua regia (1 ml). Standards should be prepared in a similar way and conventional flame techniques then usually suffice.

In the analysis of silicates a total sample weight of 10—15 mg has been recommended [210]. Samples can be taken as chips, which can be powdered in a mortar, or by abrasion using a manual or mechanical wheel. The choice of method for a particular application will probably depend upon a consideration of the different contamination each may introduce. Silicates can either be dissolved in hydrofluoric/perchloric acids or fused, e.g. with lithium metaborate. The former approach will result in the loss of silicon and must be performed only with the appropriate safety precautions. If the material contains calcium carbonate or organic matter a dried sample (10—15 mg) should be evaporated to dryness with concentrated hydrochloric acid (1 ml) and concentrated nitric acid (1 ml). To the sample, preferably in a platinum crucible, 40% hydrofluoric acid (2 ml) should be added and the crucible heated at 100°C until fumes appear. After allowing all the fluoride to evaporate, 60% perchloric acid (2 ml) is added. With gentle heating, the sample will dissolve and can then be made up to 50 ml. In the fusion procedure the sample (10—15 mg) in the crucible is mixed intimately with lithium metaborate (100 mg) and fused at 900°C to give a clear melt. After cooling, the melt may be taken up in 3% nitric acid (7 ml). Blanks and matched standards are essential for both procedures. The air/acetylene flame can be used to determine antimony, chromium, cobalt, copper, iron, lead, lithium, manganese, nickel, potassium, sodium and tin. The nitrous oxide/acetylene flame should be used for aluminium, calcium, magnesium, silicon and titanium. A phosphorus figure may be obtained using a graphite furnace with added lanthanum. An accurate determination of the potassium concentration allows data obtained in thermoluminescence dating to be corrected for the presence of potassium-40 [229].

Flint may be sampled with a hollow-cored trepanning drill [210]. A solid cylinder of about 1 g weight, necessary as only trace metals are usually measured, is obtained. The flint may be cleaned by brief immersion in aqua regia, washed, dried, and dissolved in 40% hydrofluoric acid (10 ml) at 70°C in a PTFE beaker. The excess fluoride is evaporated and the dry residue taken up in concentrated nitric acid (0.5 ml) and evaporated again to remove any carbon. This residue is then dissolved in 60% perchloric acid (2 ml) with gentle heating and made up to 50 ml. Most rocks will dissolve with this procedure, although preliminary enhancement of surface area is sometimes advisable, in which case the core may be heated to red heat in a covered crucible and dropped in 20 ml distilled water. The shattered fragments can be collected and ground more readily.

It seems likely that the range and scope of archaeological applications will continue to increase, particularly using graphite furnaces, and as the significance of more exotic elements, such as the lanthanoids and higher weight alkali metals, becomes apparent.

XVII. GASES

Most samples arrive in the laboratory as solids or liquids, although occasionally there is a requirement to determine the analyte in a gas. Trace metal particulates can be collected on appropriate filters by drawing a metered amount of air through them. This application is a common environmental problem and is discussed elsewhere in this book (Chapter 4c). If the metals are present as truly gaseous components then it will be necessary to collect them in a suitable solution via an impinger. There have been several studies of alkyllead compounds in air and here the speciation of the lead is important. Such applications therefore involve the collection of the air on a suitable absorbent (e.g. activated charcoal) and then desorption into a coupled gas chromatography-atomic absorption system. Such a system, capable of detecting 17 pg of lead, which consisted of a heated interface tube from the gas chromatograph which led to a T-piece supporting a miniature hydrogen diffusion flame has been described [230]. The atoms from this flame passed into a ceramic tube, itself heated by an air/acetylene flame, and down which the hollow cathode light beam shone. Carbonyls are another form of compound by which metals can be encountered in the gaseous form. Denholm et al. [231] have determined iron and nickel carbonyls in town gas.

Only mercury and the noble gases are commonly encountered as atoms in the gaseous state. This property of mercury is commonly exploited in the cold vapour, reduction-aeration method described elsewhere in this text. Monitors for mercury vapour are commercially available which consist

References pp. 440—445

often of a mercury vapour lamp, a sample cell, a filter and photocell. Any atomic absorption instrument can be used to monitor mercury in the gas phase using a long path-length tube. The mercury vapours can be pre-concentrated on activated charcoal or by amalgamation onto gold and released by heating, or into acidified potassium permanganate and released by reduction and aeration. Goleb [232] has reviewed the problems of determining the noble gases. Most of the noble gas lines occur in the vacuum ultra-violet region of the spectrum, and thus, either vacuum spectrometers or non-resonance lines from metastable excited states (rather than ground state lines) in the near ultra-violet and visible regions (e.g. for argon at 811.5 nm or neon at 640.2 nm) must be used. Geissler tubes can be used as the emission sources or neon or argon filled hollow cathode lamps (for neon and argon respectively). The sample can be conveniently held in a long path-length tube with quartz windows.

XVIII. ISOTOPES

Atomic absorption offers a more practical opportunity for determining isotopic composition than atomic emission. Useful reviews of the possibilities of the technique have appeared in two books [233, 234]. Isotopic analysis is in theory possible provided that highly enriched isotope sources are available, the absorption line width available is less than the isotopic displacement and for a given isotope the nuclear spin hyperfine components must be partially resolved from the other isotopic components of the absorption line. In the simplest possible case, for an element with two isotopes, the lamp is prepared from the first isotope and only this isotope in the atom cell will absorb the radiation. The procedure can then be repeated with a lamp prepared with the second isotope. Effectively this is an extension of the impressive selectivity of atomic absorption, because of the classic lock and key effect, treating the different isotopes as different analytes.

This very simple case is rare because of the small wavelength shifts involved but has been observed for mercury-202 [235]. A cooled low-pressure mercury-202 microwave source and a low-pressure mercury vapour atom cell were used, to ensure minimal line broadening (0.0002 nm at the 253.7 nm line). Only mercury-202 and mercury-200 could be determined in this way, as the other isotopic lines showed overlap, and even this was only possible as, for mercury, low-temperature, low-pressure atom cells can be used.

More frequently the absorption line profiles overlap partially and, additionally, to obtain isotopically pure sources is difficult. Thus, part of the absorption observed will be due to the other isotope. Using a natural lithium hollow cathode lamp and a lamp made from enriched (93%) lithium-6, Chapman and Dale [236] were able to determine lithium isotopic abundances using an air/acetylene flame absorption cell. At the 670.8 nm line the isotopic shift, between lithium-6 and lithium-7 is 0.015 nm and the fine structure makes the "line" appear as a triplet, the central feature being composed of contributions from both isotopes. The lamps were aligned on the optical axis of a conventional spectrometer, i.e. the second in the place of the background correction continuum lamp. Lithium carbonate standards of known isotopic composition were used and the instrument calibrated by plotting the isotopic proportion of each isotope against the difference in the two observed absorbances. Other workers have also reported the determination of lithium isotopic ratios using flame atom cells [237, 238]. By using a water-cooled, low-pressure, hollow-cathode discharge tube as the absorption cell, Goleb and Yokoyama [239] obtained much finer lines (ca. 0.001 nm wide) and hence more accurate measurements. Only microgram amounts of sample were needed and this was not lost or spread using the hollow-cathode cell. The sources were hollow-cathode lamps containing lithium-6 and lithium-7 in helium carrier gas, again at low-pressure with water cooling.

Goleb [240, 241] has determined uranium-235 and uranium-238 in a corresponding manner. Again a water-cooled sputtering cell was used as the atom reservoir and water-cooled hollow-cathode lamps as sources. Samples could be placed as chips, compounds or oxides in the emission source and standards were placed in the absorption source. The 415.3 and 502.7 nm uranium lines were used as uranium-235 has negligible fine structure at this line but appreciable isotopic shifts (0.007 and 0.010 nm respectively) are observed. A linear calibration curve of the isotopic abundance of uranium-238 versus the percentage transmission through the absorption cell was obtained. In a similar way Kirchhoff [242] used the lead 283.3 nm line to determine the lead-206 and lead-208 isotopic ratios in samples also excited in a hollow-cathode lamp, with pure lead-208 (or lead-206) in the atom cell. When pure isotope hollow cathode lamps are used and the samples sprayed into a conventional air/acetylene flame, greater problems from line overlaps are observed [243].

Cases have been observed where the isotopic line absorption profiles completely overlap, e.g. boron-10 and -11 in a krypton-filled lamp at 249.7 nm [244]. Hannaford and Lowe [245] later showed that this was caused by an unusually large Doppler half-width induced by the fill-gas, and, if neon is used, the 208.9 and 209.0 nm lines can allow the determination of boron-10 and boron-11 isotope ratios. The 208.89/208.96 nm doublet was found to be more useful than the 249.68/249.77 nm doublet. Enriched isotope hollow-cathode lamps were used as sources. A sputtering cell was preferred to a nitrous oxide/acetylene flame as the atom reservoir, as it could be water-cooled to reduce broadening and solid samples could be used, thus avoiding the slow dissolution in nitric acid of samples of boron-10 used as a neutron absorber in reactor technology.

It perhaps should be stressed that while with specialist equipment accurate

References pp. 440—445

measurements of isotopic abundances may be made, this is constrained to those elements where resonance lines show an isotopic shift comparable to, or greater than, the absorption line profile under experimental conditions. This, therefore, limits the application of this technique to light elements (e.g. lithium and boron, discussed above, and oxygen [246]), and very heavy elements with appreciable nuclear charge (e.g. mercury, lead and uranium, discussed above). Further complications arise when elements have more than two major isotopes and in the accurate determination of minor constituents. Therefore, while this technique can usefully be applied in certain situations of interest to the nuclear industry, it does not presently offer a realistic alternative to more complex mass-spectrometric procedures for most elements.

REFERENCES

1. J. C. van Loon and C. M. Parissis, Analyst, 94 (1969) 1057.
2. E. Sebastiani, K. Ohls and G. Riemer, Fresenius Z. Anal. Chem., 264 (1973) 105.
3. D. C. Manning and W. Slavin, Anal. Chem., 50 (1978) 1234.
4. L. Ebdon, A. T. Ellis and R. W. Ward, Talanta, 29 (1982) 297.
5. IUPAC Analytical Chemistry Division, Commission on Micro-chemical Techniques and Trace Analysis, Pure Appl. Chem., 49 (1977) 893.
6. T. Y. Kometani, Anal. Chem., 49 (1977) 2289.
7. F. J. Langmyhr and J. T. Hakedal, Anal. Chim. Acta, 83 (1976) 127.
8. G. E. Batley and J. P. Matousek, Paper presented to Joint Meeting American and Japanese Chemical Societies, Honolulu, Hawaii, April 2—6, 1979.
9. F. O. Jensen, J. Dolezal and F. J. Langmyhr, Anal. Chim. Acta, 72 (1974) 245.
10. Y. Thomassen, B. V. Larsen, F. J. Langmyhr and W. Lund, Anal. Chim. Acta, 83 (1976) 103.
11. W. Lund and B. V. Larsen, Anal. Chim. Acta, 70 (1974) 299.
12. W. Lund and B. V. Larsen, Anal. Chim. Acta, 72 (1974) 57.
13. W. Lund, B. V. Larsen and N. Gunderson, Anal. Chim. Acta, 81 (1976) 319.
14. E. Jackwerth, Fresenius Z. Anal. Chem., 271 (1974) 120.
15. E. Jackwerth and H. Berndt, Anal. Chim. Acta, 74 (1975) 299.
16. E. Wunderlich and W. Haedeler, Fresenius Z. Anal. Chem., 284 (1977) 19.
17. P. N. W. Young, Analyst, 99 (1974) 588.
18. E. Jackwerth and P. G. Willmer, Spectrochim. Acta, Part B, 33 (1978) 343.
19. C. A. Watson, Ammonium Pyrrolidine Dithiocarbamate, Monograph 74, Hopkin and Williams, Chadwell Heath, England, October, 1971.
20. M. Briska and D. Hoffmeister, Talanta, 20 (1973) 895.
21. M. Yanagisawa, M. Suzuki and T. Takeuchi, Talanta, 14 (1967) 933.
22. M. Suzuki, M. Yanagisawa and T. Takeuchi, Talanta, 12 (1965) 989.
23. J. W. Edwards, G. D. Lominac and R. P. Buck, Anal. Chim. Acta, 57 (1971) 257.
24. M. S. Cresser, Solvent Extraction in Flame Spectroscopic Analysis, Butterworth, London, 1978.
25. W. W. White and P. J. Murphy, Anal. Chem., 49 (1977) 255.
26. I. May and L. P. Greenland, Anal. Chem., 49 (1977) 2376.
27. M. H. Jones and J. T. Woodcock, Anal. Chim. Acta, 69 (1974) 275.
28. C. Manoliu, B. Tomi and F. Petruc, Rev. Chim., 24 (1973) 991.
29. K. Dittrich and G. Liesch, Talanta, 20 (1973) 691.
30. G. Duncan and R. J. Herridge, Talanta, 17 (1970) 766.

31 G. Braca, G. Sbrana, G. Scandiffio and R. Cioni, At. Absorpt. Newsl., 14 (1975) 39.
32 M. A. Leonard and W. J. Swindall, Analyst, 98 (1973) 133.
33 J. P. Macquet, J. Hubert and T. Theophanides, Anal. Chim. Acta, 72 (1974) 251.
34 J. P. Macquet and T. Theophanides, Anal. Chim. Acta, 72 (1974) 261.
35 P. Thavornyutikarn, J. Organomet. Chem., 51 (1973) 237.
36 D. T. Burns, F. Glockling, V. B. Mahale and W. J. Swindall, Analyst, 103 (1978) 985.
37 E. J. Wood, R. Gonzalez, J. A. Blanco and A. O. Rucci, Talanta, 23 (1976) 473.
38 M. Bigois and R. Levy, Talanta, 23 (1976) 119.
39 I. M. Citron and H. Martins, Am. Lab., 6 (1979) 58.
40 M. S. Vigler, A. Strecker and A. Varnes, Appl. Spectrosc., 32 (1978) 60.
41 Analytical Methods Committee, Analyst, 104 (1979) 1070.
42 Analytical Methods Committee, Analyst, 104 (1979) 778.
43 Analytical Methods Committee, Analyst, 92 (1967) 403.
44 Analytical Methods Committee, Analyst, 101 (1976) 62.
45 G. D. Martinie and A. A. Schilt, Anal. Chem., 48 (1976) 70.
46 A. Ford, B. Young and C. Meloan, J. Agric. Food Chem., 22 (1974) 1034.
47 J. Perkins, Analyst, 88 (1963) 324.
48 R. L. Scott, At. Absorpt. Newsl., 9 (1970) 46.
49 C. Ling, Phot. Sci. Eng., 11 (1967) 200.
50 N. Kunimine and H. Kawada, Bunseki Kagaku, 16 (1967) 185.
51 N. Kunimine and H. Kawada, Bunseki Kagaku, 16 (1967) 189.
52 K. Dittrich and W. Mothes, Talanta, 22 (1975) 318.
53 P. Ambrosetti and F. Librici, At. Absorpt. Newsl., 18 (1979) 38.
54 J. J. Labrecque, Appl. Spectrosc., 30 (1976) 625.
55 J. Janouskova, M. Nehasilova and V. Sychra, At. Absorpt. Newsl., 12 (1973) 161.
56 D. E. Harrington and W. R. Bramstedt, At. Absorpt. Newsl., 15 (1976) 125.
57 N. F. Zakharchuk, I. G. Yudelevich, E. V. Alimova, L. A. Terenteva and N. F. Beizel, Zh. Anal. Khim., 33 (1978) 1977.
58 C. Pelosi and G. Attolini, Anal. Chim. Acta, 84 (1976) 179.
59 K. Dittrich and W. Zeppan, Talanta, 22 (1975) 299.
60 K. Dittrich, Talanta, 24 (1977) 725.
61 K. Dittrich, Talanta, 24 (1977) 735.
62 K. Dittrich, W. Mothes and P. Weber, Spectrochim. Acta, Part B, 33 (1978) 325.
63 K. Dittrich and H. Vogel, Talanta, 26 (1979) 737.
64 K. Dittrich, B. Vorberg and H. Wolters, Talanta, 26 (1979) 747.
65 T. Nakahara and S. Musha, Anal. Chim. Acta, 80 (1975) 47.
66 T. Nakahara and S. Musha, Anal. Chim. Acta, 75 (1975) 305.
67 W. J. Price, Spectrochemical Analysis by Atomic Absorption, Heyden, London, 1979.
68 E. S. Ioffe, V. S. Prisenko and I. I. Romazonova, Zavod. Lab., 40 (1974) 358.
69 V. M. Byr'ko, L. F. Prishchepov and I. A. Shikheeva, Zavod. Lab., 41 (1975) 525.
70 C. M. Whittington and J. B. Willis, Plating, 51 (1964) 767.
71 F. W. E. Strelow, E. C. Feast, P. M. Mathews, C. J. C. Bothman and C. R. Van Zyl, Anal. Chem., 38 (1966) 115.
72 L. R. P. Butler, J. A. Brink and S. A. Engelbrecht, Inst. Min. Metall., Trans. Sect. C, 76 (1967) 188.
73 D. E. Harrington and W. R. Bramstedt, Talanta, 22 (1975) 411.
74 S. I. Pardhan and J. M. Ottaway, Proc. Anal. Div. Chem. Soc., 12 (1975) 291.
75 G. C. Tarabella, G. I. Spielholtz and R. J. Steinberg, Mikrochim. Acta, (1972) 484.
76 D. Gegiou, Analyst, 99 (1974) 745.
77 E. Mario and R. E. Gerner, J. Pharm. Sci., 57 (1968) 1243.
78 M. Okamoto, M. Konda, I. Matsumoto and Y. Miya, J. Soc. Cosmet. Chem., 22 (1971) 589.
79 E. S. Gladney, At. Absorpt. Newsl., 16 (1977) 114.

80 British Standard 5136, Specification for Toothpastes, British Standards Institution, 2 Park Street, London W1A 2BS, 1981.
81 H. Konig and E. Walldorf, Fresenius Z. Anal. Chem., 289 (1978) 177.
82 L. Ebdon, J. R. Wilkinson and K. W. Jackson, Anal. Chim. Acta, 128 (1981) 45.
83 L. Ebdon, R. W. Ward and D. A. Leathard, Anal. Proc., 19 (1982) 110.
84 United States Pharmacopoeia, XIX edn., Mack Printing Co., Easton, PA, 1975.
85 British Pharmacopoeia, 27th edn., HMSO, London, 1973.
86 European Pharmacopoeia, Vol. III, Masionneuve, France,1975.
87 F. Rousselet and F. Thuillier, Prog. Anal. At. Spectrosc., 1 (1979) 353.
88 J. H. M. Miller, Int. Lab., (July/Aug. 1978) 37.
89 P. B. Bondo, Guidel. Anal. Toxicol. Programs, 2 (1977) 171.
90 V. V. Tkachuk, Farm. Zh. (Kiev), 6 (1978) 35.
91 J. Pawlaczyk and M. Makowska, Pol. Pharm., 36 (1979) 59.
92 G. F. Hazebroucq, in 3rd International Conference on Atomic Absorption and Atomic Fluorescence Spectrometry, Hilger, London, 1971, pp. 577—580.
93 R. V. Smith and M. A. Nessen, J. Pharm. Sci., 60 (1971) 907.
94 B. A. Dalrymple and C. T. Kenner, J. Pharm. Sci., 58 (1969) 604.
95 P. P. Kharkhanis and J. R. Anfinsen, J. Assoc. Off. Anal. Chem., 56 (1973) 358.
96 S. L. Ali and D. Steinbach, Pharm. Ztg., 124 (1979) 1422.
97 E. Van Den Ereckout and P. de Moerloose, Pharm. Weekbl., 106 (1971) 749.
98 Y. S. Chae, J. P. Vacik and W. H. Shelver, J. Pharm. Sci., 62 (1973) 1838.
99 Y. Kidani, K. Takeda and H. Koike, Bunseki Kagaku, 22 (1973) 719.
100 F. J. Diaz, Anal. Chim. Acta, 58 (1972) 455.
101 E. Peck, Anal. Lett., B11 (1978) 103.
102 P. O. Kosonen, A. M. Salonen and A. L. Nieminen, Finn. Chem. Lett., (1978) 136.
103 L. L. Whitlock, J. R. Melton and T. J. Billings, J. Assoc. Off. Anal. Chem., 59 (1976) 580.
104 R. R. Moody and R. B. Taylor, J. Pharm. Pharmacol., 24 (1972) 848.
105 K. Szivos, L. Polos, L. Bezev and E. Pungor, Acta Pharm. Hung., 43 (1973) 90.
106 G. I. Spielholtz and G. C. Toralballa, Analyst, 94 (1969) 1072.
107 I. T. Calder and J. H. M. Miller, J. Pharm. Pharmacol., 28 (1976) Suppl. 25P.
108 R. D. Thompson and T. J. Hoffman, J. Pharm. Sci., 64 (1975) 1863.
109 H. T. Smart and D. J. Campbell, Can. J. Pharm. Sci., 4 (1974) 73.
110 F. Rousselet, V. Courtois and M. L. Girard, Analusis, 3 (1975) 132.
111 J. R. Leaton, J. Assoc. Off. Anal. Chem., 53 (1970) 237.
112 K. A. Kovar, W. Lautenschlager and R. Seidel, Dtsch. Apoth.-Ztg., 115 (1975) 1855.
113 J. C. Meranger and E. Somers, Analyst, 93 (1968) 799.
114 O. Joens and L. Toft, Acta Pharm. Chem. Sci. Ed., 1 (1973) 14.
115 A. Smit and K. Bache-Hansen, Acta Pharm. Suec., 10 (1973) 254.
116 R. J. Hurtubise, J. Pharm. Sci., 63 (1974) 1128.
117 M. J. Pro and R. L. Brunelle, J. Assoc. Off. Anal. Chem., 53 (1970) 1137.
118 F. J. Szydlowski and F. R. Vianzon, Anal. Lett., B11 (1978) 161.
119 S. J. Simon and D. F. Boltz, Microchem. J., 20 (1975) 468.
120 I. Mitsui and Y. Fujimura, J. Hyg. Chem., 21 (1975) 183.
121 T. Minamikawa, K. Matsumara, A. Kamei and M. Yamakawa, Bunseki Kagaku, 20 (1971) 1011.
122 T. Minamikawa and N. Yamagishi, Bunseki Kagaku, 22 (1973) 1058.
123 Y. Kidani, S. Uno and K. Inagaki, Bunseki Kagaku, 25 (1976) 514.
124 Y. Kidani, S. Uno and K. Inagaki, Bunseki Kagaku, 26 (1977) 158.
125 Y. Kidani, T. Saotome, K. Inagaki and H. Koike, Bunseki Kagaku, 24 (1975) 463.
126 J. Alary, A. Villet and A. Coeur, Ann. Pharm. Fr., 34 (1976) 419.
127 T. Mitsui and Y. Fujimura, Bunseki Kagaku, 24 (1975) 575.
128 Y. Kidani, K. Nakamura, K. Inagaki and H. Koike, Bunseki Kagaku, 24 (1975) 742.

129 Y. Kidani, K. Inagaki, N. Osugi and H. Koike, Bunseki Kagaku, 22 (1973) 892.
130 Y. Kidani, T. Saotome, M. Kato and H. Koike, Bunseki Kagaku, 23 (1974) 265.
131 S. Tommilehto, Acta Pharm. Fenn., 88 (1979) 25.
132 A. V. Kovatsis and M. A. Tsougas, Arzneim-Forsch., 28 (1978) 248.
133 T. Minamikawa, K. Sakai, N. Hashitani, E. Fukushima and N. Yamagishi, Chem. Pharm. Bull. (Tokyo), 21 (1973) 1632.
134 Y. Kidani, K. Nakamura and K. Inagaki, Bunseki Kagaku, 25 (1976) 509.
135 Y. Kidani, H. Takemura and H. Koike, Bunseki Kagaku, 22 (1973) 187.
136 Y. Kidani, K. Inagaki, T. Saotome and H. Koike, Bunseki Kagaku, 22 (1973) 896.
137 T. Mitsui, Y. Fujimura and T. Suzuki, Bunseki Kagaku, 24 (1975) 244.
138 T. Minamikawa and K. Matsumura, Yakugaku Zasshi, 96 (1976) 440.
139 R. Hurtubise, J. Pharm. Sci., 63 (1974) 1131.
140 S. S. M. Hassan, M. T. Zaki and M. H. Eldisouki, J. Assoc. Off. Anal. Chem., 62 (1979) 315.
141 British Standard 5666, Methods of Analysis of Wood Preservatives and Treated Timber, Part 3, Quantitative Analysis of Preservatives and Treated Timber Containing Copper, Chromium, Arsenic Formulations, British Standards Institution, 2 Park Street, London W1A 2BS, 1979.
142 B. J. Gudzinowicz and V. J. Luciano, J. Assoc. Off. Anal. Chem., 49 (1966) 1.
143 P. D. Jung and D. Clarke, J. Assoc. Off. Anal. Chem., 57 (1974) 379.
144 Statutory Instrument 1974 No. 1367, The Toys (Safety) Regulations 1974, HMSO, London.
145 W. J. Price, Paint, Oil Colour J., 21 (1970) 282.
146 J. K. Duffer, Am. Soc. Test. Mat. Spec. Tech. Publ., 500 (1972) 550.
147 N. G. Eider, Appl. Spectrosc., 25 (1971) 313.
148 W. K. Porter, J. Assoc. Off. Anal. Chem., 57 (1974) 614.
149 T. Hodson and D. W. Lord, J. Assoc. Pub. Anal., 9 (1971) 60.
150 L. Ebdon, J. R. Wilkinson and K. W. Jackson, Analyst, 107 (1982) 269.
151 B. Griefer, E. J. Maienthal, T. C. Rains and S. D. Rasberry, Nat. Bur. Stand. (U.S.), Spec. Publ., No. 260-45, March 1973.
152 K. Eng, Skand. Tidskr. Farg Lack, 21 (1975) 7.
153 J. C. Meranger and E. Somers, Analyst, 93 (1968) 799.
154 O. Beniot and G. Geiger, Bull. Liaison Lab. Ponts Chaussees, 79 (1975) 83.
155 R. J. Noga, Anal. Chem., 47 (1975) 332.
156 V. Bartocci, P. Cescon, F. Castellani, R. Riccioni and M. Gusteri, At. Absorpt. Newsl., 16 (1977) 57.
157 F. Pellerin and J. P. Goulle, Ann. Pharm. Fr., 35 (1977) 189.
158 H. T. Delves, Analyst, 95 (1970) 431.
159 E. L. Henn, At. Absorpt. Newsl., 12 (1973) 109.
160 D. G. Mitchell, K. M. Aldous and A. F. Ward, At. Absorpt. Newsl., 13 (1974) 121.
161 O. W. Lau and K. L. Li, Analyst, 100 (1975) 430.
162 E. L. Henn, Paint Varn. Prod., 63 (1973) 29.
163 W. Holak, Anal. Chim. Acta, 74 (1975) 216.
164 M. Kubota, D. W. Golightly and R. Mavrodineanu, Appl. Spectrosc., 30 (1976) 56.
165 P. O. Bethge and R. Radestroem, Sv. Papperstidn., 69 (1966) 772.
166 G. F. Wallace, The Application of Atomic Absorption Spectrophotometry to the Analysis of Pulp and Paper Products, Print. Reprogr./Test. Conf. (Pap.)., 1977, p. 169.
167 W. Griebenow, B. Werthmann and O. Toeppel, Papier (Darmstadt), 31 (1977) 503.
168 O. Toeppel, W. Griebenow and B. Werthmann, Papier (Darmstadt), 31 (1977) 508.
169 O. Ant-Wuorinen and A. Visapaa, Paperi Ja Puu, 48 (1966) 649.
170 R. L. Anderson and W. R. Scott, AIChE, Symp. Ser., 70 (1974) 106.
171 G. A. King, M. J. Ridge and G. S. Walker, J. Oil Colour Chem. Assoc., 57 (1974) 127.

172 D. Watkins, T. Corbyons, J. Bradshaw and J. D. Winefordner, Anal. Chim. Acta, 85 (1976) 403.
173 P. Minkkinen, Kem.-Kemi, 3 (1976) 282.
174 J. D. Kerber, A. Koch and G. E. Peterson, At. Absorpt. Newsl., 12 (1973) 104.
175 F. J. Langmyhr, Y. Thomassen and A. Massoumi, Anal. Chim. Acta, 68 (1974) 305.
176 V. Sychra, D. Kolihova and N. Dudova, Chem. Listy, 70 (1976) 737.
177 P. J. Sinon, B. C. Glessen and T. R. Copeland, Anal. Chem., 49 (1978) 2285.
178 C. Tonini, Tinctoria, 75 (1978) 160.
179 B. Piccolo and V. W. Tripp, U.S. Dep. Agric., Agric. Res. Serv., (Rep.), ARS 72—98 (1972) 45.
180 J. P. Price, At. Absorpt. Newsl., 11 (1972) 1.
181 W. Slavin, At. Absorpt. Newsl., 4 (1965) 192.
182 J. F. Alder and P. L. Bucklow, At. Absorpt. Newsl., 18 (1979) 123.
183 L. Ebdon, Ph.D. Thesis, University of London, 1971.
184 J. V. Simonian, At. Absorpt. Newsl., 7 (1968) 63.
185 F. R. Hartley and A. S. Inglis, Analyst, 92 (1967) 622.
186 F. R. Hartley and A. S. Inglis, Analyst, 93 (1968) 394.
187 A. S. Inglis and P. W. Nicholls, J. Text. Inst., 64 (1973) 445.
188 J. L. Hoare and P. E. Ingham, J. Text. Inst., 70 (1979) 76.
189 E. S. Della Monica and P. E. McDowell, J. Am. Leather Chem. Assoc., (1971) 21.
190 M. Oliver, Fresenius Z. Anal. Chem., 248 (1969) 145.
191 H. Oguro, Bunseki Kagaku, 24 (1975) 797.
192 I. Lewandowska, Rocz. Panstev. Zakl. Higi., 29 (1978) 295.
193 T. Y. Kometani, Anal. Chem., 38 (1966) 1596.
194 D. Druckman, At. Absorpt. Newsl., 6 (1967) 113.
195 K. Kuga, I. Sugaya and K. Tsujii, Bunseki Kagaku, 28 (1979) 201.
196 T. Matsuo, J. Shida and M. Motaki, Bunseki Kagaku, 81 (1969) 521.
197 E. L. Henn, Anal. Chim. Acta, 73 (1974) 273.
198 P. Girgis-Takla and I. Chroneos, Analyst, 103 (1978) 122.
199 T. Ishii and S. Musha, Bunseki Kagaku, 20 (1971) 489.
200 I. G. Putov, S. A. Popova and A. G. Brashnarova, Khim. Ind. (Sofia), 49 (1977) 16.
201 G. R. Supp and I. Gibbs, At. Absorpt. Newsl., 13 (1974) 71.
202 J. L. Firkins, Rubber Chem. Technol., 47 (1974) 448.
203 J. Korkisch and H. Gross, Mikrochim. Acta, II (1975) 413.
204 C. R. Walker and O. A. Vita, Anal. Chim. Acta, 43 (1968) 27.
205 C. M. Young and J. M. Baldwin, Microchem. J., 23 (1978) 265.
206 E. Glaeser, Kernenergie, 21 (1978) 235.
207 E. Peck, Anal. Lett., A12 (1979) 255.
208 B. W. Haynes, At. Absorpt. Newsl., 17 (1978) 49.
209 W. C. Pearce and L. Ebdon, Analyst, in press.
210 M. J. Hughes, M. R. Cowell and P. T. Craddock, Archaeometry, 18 (1976) 19.
211 W. A. Oddy, Eur. Spectrosc. News, 34 (1981) 31.
212 P. Clayton, J. Egypt. Archaeol., 58 (1972) 167.
213 R. H. Brill, A Chemical Analytical Round-Robin on Four Synthetic Ancient Glasses, Ninth International Congress of Glass: Artistic and Historic Communications, Paris, 1972, pp. 93—110.
214 M. R. Cowell and A. E. A. Werner, Analysis of Some Egyptian Glass, Annales de 6^e Congres de L'Association Internationale pour L'Histoire du Verre, Liege, 1974, pp. 295—298.
215 P. R. Fields, J. Milstead, E. Henrickson and R. Ramette, in Science and Archaeology, Cambridge, MA, 1971, pp. 131—143.
216 G. de G. Sieveking, P. T. Craddock, M. J. Hughes, P. Bush and J. Ferguson, Nature, 228 (1970) 251.

217 G. de G. Sieveking, P. Bush, J. Ferguson, P. T. Craddock, M. J. Hughes and M. R. Cowell, Archaeometry, 14 (1972) 151.
218 J. Lang and J. Price, J. Archaeol. Sci., 2 (1975) No. 4.
219 J. Price, Some Roman Glass From Spain, Annales de 6^e Congres de L'Association Internationale pour L'Histoire du Verre, Liege, 1974, pp. 65—84.
220 W. A. Oddy and F. Schweizer, in E. T. Hall and D. M. Metcalf (Eds.), Methods of Chemical and Metallurgical Investigation of Ancient Coinage, Royal Numismatic Society, London, 1972, pp. 171—182.
221 J. Brailsford and J. E. Stapley, Proc. Prehistoric Soc., 38 (1972) 219.
222 J. V. S. Megaw, Proc. Prehistoric Soc., 35 (1969) 358.
223 J. V. S. Megaw, P. T. Craddock and A. E. A. Werner, Br. Mus. Quart., 37 (1973) 70.
224 J. Lang and A. R. Williams, J. Archaeol. Sci., 2 (1975) 199.
225 P. T. Craddock, J. Lang and K. S. Painter, Br. Mus. Quart., 37 (1973) 9.
226 H. McKerrell and R. B. K. Stevenson, in E. T. Hall and D. M. Metcalf (Eds.), Methods of Chemical and Metallurgical Investigation of Ancient Coinage, Royal Numismatic Society, London, 1972, pp. 195—210.
227 R. H. Brill, J. Glass Stud., 12 (1970) 185.
228 C. D. Vassas, Chemical, Thermal Analysis and Physical Study of Glasses of Mediæval Stained Glass Windows, Ninth International Congress of Glass: Artistic and Historical Communications, Paris, 1972, pp. 241—266.
229 M. A. Seeley, J. Archaeol. Sci., 2 (1975) 17.
230 L. Ebdon, R. W. Ward and D. A. Leathard, Analyst, 107 (1982) 129.
231 A. B. Denholm, P. A. A. Beale and R. Palmer, J. Appl. Chem., 13 (1963) 576.
232 J. A. Goleb, in J. A. Dean and T. C. Rains (Eds.), Flame Emission and Atomic Absorption Spectrometry, Vol. 3, Elements and Matrices, Marcel Dekker, New York, 1975, Chapter 13.
233 J. Ramirez-Munoz, Atomic Absorption Spectroscopy, Elsevier, Amsterdam, 1968, p. 387.
234 K. C. Thompson and R. J. Reynolds, Atomic Absorption, Fluorescence and Flame Emission Spectroscopy, 2nd edn., Charles Griffin, London, 1978, pp. 306—311.
235 K. R. Osborn and H. E. Gunning, J. Opt. Soc. Am., 45 (1955) 552.
236 J. F. Chapman and L. S. Dale, Anal. Chim. Acta, 87 (1976) 91.
237 D. C. Manning and W. Slavin, At. Absorpt. Newsl., 1 (1962) 39.
238 A. N. Zaidel and E. P. Korennoi, Opt. Spectrosc., 10 (1961) 299.
239 J. A. Goleb and Y. Yokoyama, Anal. Chim. Acta, 30 (1964) 213.
240 J. A. Goleb, Anal. Chem., 35 (1963) 1978.
241 J. A. Goleb, Anal. Chim. Acta, 34 (1966) 135.
242 H. Kirchhoff, Spectrochim. Acta Part B, 24 (1969) 235.
243 W. H. Brimhall, Anal. Chem., 41 (1969) 1349.
244 J. A. Goleb, Anal. Chim. Acta, 36 (1966) 130.
245 P. Hannaford and R. M. Lowe, Anal. Chem., 49 (1977) 1852.
246 B. V. L'vov, V. I. Mosichev and S. A. Senyuta, Ind. Lab. (USSR), 11 (1963) 1404.